# Polynomial Paradigms

Trends and applications in science and engineering

**Online at: https://doi.org/10.1088/978-0-7503-5067-9**

# Polynomial Paradigms

Trends and applications in science and engineering

**Edited by**
**S Chakraverty**
*Department of Mathematics, National Institute of Technology Rourkela, Rourkela, India*

**IOP** Publishing, Bristol, UK

ISBN 978-0-7503-5067-9 (ebook)
ISBN 978-0-7503-5065-5 (print)
ISBN 978-0-7503-5068-6 (myPrint)
ISBN 978-0-7503-5066-2 (mobi)

DOI 10.1088/978-0-7503-5067-9

Version: 20221201

IOP ebooks

British Library Cataloguing-in-Publication Data: A catalogue record for this book is available from the British Library.

Published by IOP Publishing, wholly owned by The Institute of Physics, London

IOP Publishing, No.2 The Distillery, Glassfields, Avon Street, Bristol, BS2 0GR, UK

US Office: IOP Publishing, Inc., 190 North Independence Mall West, Suite 601, Philadelphia, PA 19106, USA

*To my School Friend, Late Sri Manmohan Dey!*

# Contents

# Preface

The importance of polynomials is well known for their ability to improve the properties and applicability in the interdisciplinary fields of various science and engineering problems. Different types of polynomials play important roles in developing numerical and analytical methods in physics, mathematics, engineering and industry.

Many problems arising in engineering and science are mathematically constructed by a variety of differential, integral and other algebraic equations. Moreover, special polynomials provide a new way to solve and investigate various equations often encountered in different practical problems.

Over the past decades, this area of research has received ever-increasing attention and has gained a growing momentum in modern topics, such as computational probability, numerical analysis, computational wave, fluid and structural dynamics, data assimilation, statistics, image and signal processing, artificial intelligence, etc, to name a few. Various types of polynomials are crucial to the stability of high-order numerical methods. These high-order numerical methods, originally formulated for ordinary and partial differential equations, have been extended to integral, integro-differential equations, etc, and to various application problems. The study of different powerful polynomials and corresponding analytical/semi-analytical/numerical methods help us deepen our understanding of more general application problems.

In view of the above scenarios and challenges, the purpose of this book is to report and review the recent developments and applications of different polynomials in numerical and analytical/semi-analytical methods for solving a variety of science and engineering problems. This book contains contributions from leading experts in areas ranging from basic theory and concepts of polynomials, mathematical modelling, mathematical physics, engineering, high-order numerical methods for differential, integral and integro-differential equations, artificial intelligence (AI), fuzzy and interval based models and beyond. This book contains fifteen chapters covering various aspects of polynomials to handle diversified problems.

**Chapter 1** deals with the theoretical concepts viz. sums of the series of reciprocals of the cubic polynomials with three, two and one zero roots and with other integer roots.

It is worth mentioning that the versatile numerical tool for solving engineering problems is the finite element method (FEM), which is either based on Lagrange or Hermite polynomials, in general. Among the several drawbacks of FEM one is clearly the domain decomposition that might lead to numerical instabilities for complex problems. Therefore, meshless methods have been introduced in **chapter 2** for their ability to consider arbitrary domain without a proper meshing and this brings many advantages in terms of computational cost and handling complex geometries in linear and nonlinear regimes. The chapter aims at investigating accuracy and reliability of meshless methods by considering several polynomial interpolations based on a class of well-known radial basis functions (RBFs).

Chapters 3–6 are related to application of polynomials in solving differential, fractional differential and fractional integro-differential equations. In particular, **chapter 3** addresses combining of fractal and fractional differentiation concepts to produce new differentiation operators. These newly generated operators are used in the Adams–Bashforth numerical method based on two-step and three-step Newton and Lagrange interpolation polynomials for solving two differential equations with variable orders. The exponential, trigonometric, and hyperbolic forms of variables are considered to be the order of the fractional derivative.

A polynomial-based numerical technique has been applied for solving singularly perturbed differential-difference equations in **chapter 4**. The original differential-difference problem is reduced to an analogous singularly perturbed boundary value problem via Taylor series expansion. The reduced problem is then solved using a polynomial-based numerical approximation on Shishkin type meshes. Here, the numerical approximation comprises a midpoint upwind approximation in the region of coarse mesh and a cubic spline approximation in the region of fine mesh.

It may be noted that, Leptospirosis is one of the most common and widespread zoonotic infections in the world. It is a worldwide public health problem and the greatest concern for humid tropical and subtropical regions. Accordingly, in **chapter 5**, an ordinary differential equation as deterministic model has been investigated via polynomial-based method to solve the related system of differential equations. It has been established that the model is mathematically well-posed to predict the impact of preventive and treatment based control interventions on the transmission dynamics of Leptospirosis. Qualitative analysis of the model is carried out by investigating the existence and positivity of the solutions.

Fractional order Volterra–Fredholm integro-differential equations are handled in **chapter 6** with the help of Chebyshev and Bernstein polynomials in the Adomian decomposition method (ADM). Corresponding convergence analysis and the error bounds are also incorporated. Several example problems are included for the validation of the method(s).

Chapters 7–9 include solutions of dynamical and wave propagation problems with the use of polynomials. As such, the Rayleigh–Ritz method is employed to investigate the dynamical characteristics of nanobeam embedded in the Winkler–Pasternak elastic foundation in **chapter 7**, where orthogonal polynomials such as Chebyshev, Legendre, and Hermite are used as shape functions. The main advantages of these polynomials are orthogonality, making the technique computationally efficient, and avoiding ill-conditioning for higher degree polynomials. The governing differential equations of the present model are obtained by incorporating Hamilton's principle, and the frequency parameters are computed for classical boundary conditions.

**Chapter 8** illustrates the propagation of anti-plane wave in a model consisting of three layers coated with a thin film. The first and third layers are composed of piezoelectric–viscoelastic (PV) and piezoelectric–poroelastic (PP) materials, respectively. The second layer lying between these piezo-composite layers is an integration of these two materials and is hence named as the buffer layer. In order to solve the governing equations of the buffer, the material parameters and the displacement and potential functions are considered as polynomials of infinite order, which are solved using the power series approach.

In **chapter 9**, an interesting problem of robot dynamics is considered. The goal of this chapter has been to solve the governing differential equation of the single-link robot using the polynomial approximation method. In order to achieve this, the Galerkin method has been used. Further, a novel fuzzy model has been developed to describe the uncertain environment and the associated first few (fuzzy) frequencies are obtained by solving the equations using the fuzzy based Galerkin method.

In order to deal with uncertainty, the probability/interval/fuzzy theory approach may be beneficial. However, the application of probability theory in uncertain cases may be ineffective because it may generate less reliable results for small data sets. Accordingly, interval and fuzzy theory approach may be used to handle the uncertainties. Moreover, because of the uncertain membership grade, the use of type-2 fuzzy numbers over type-1 fuzzy numbers in uncertain problems is more realistic. As such, a numerical approach viz. polynomials based collocation method is developed in **chapter 10** to solve type-2 fuzzy differential equations. In this regard, the polynomials in the assumed solution of the collocation method are considered as type-2 fuzzy polynomials. Few numerical examples and one application problem have been solved to demonstrate the validity of the proposed technique.

**Chapter 11** investigates the applicability of Shannon probabilistic entropy in uncertainty analysis for some linear elasticity problems. The finite difference method is implemented for the Euler–Bernoulli beam model resting on the Winkler foundation with random compliance. The random polynomials serve further in triple stochastic analyses, where Monte Carlo simulation, the generalized iterative stochastic perturbation technique, as well as the semi-analytical probabilistic approach are employed to calculate the first four probabilistic coefficients of the structural responses. The Monte Carlo simulation is engaged to calculate Shannon probabilistic entropy fluctuations for various input uncertainty levels and to contrast it with the resulting coefficient of variation for the extreme deflections.

Use of polynomials are demonstrated in the AI/machine intelligence field in chapters 12 to 14. As such, **chapter 12** addresses different aspects and reviews of AI, in particular polynomials in hybrid AI are discussed that help us to optimize, forecasting, predicting events such as ARMA, linear regression, multiple linear regression, polynomial linear regression, genetic algorithm, etc.

Different neural models are analysed in **chapter 13** for estimating the condition of multidimensional poverty in an Indian state viz. Odisha. In this regard, two functional link neural networks (FLNN) such as the Chebyshev neural network (ChNN) and Legendre neural network (LeNN) based on Chebyshev and Legendre polynomials are developed. Further, a single hidden layer artificial neural network (ANN) has also been developed. Open access data from NITI Aayog, India has been utilized. The numerical efficiency and accuracy of the FLNN models are investigated by multidimensional poverty data. It is concluded that FLNNs are comparatively better approximators and computationally efficient compared to traditional ANN.

The idea behind data smoothing is that it can identify simplified changes to help in predicting different trends and patterns. Smoothing techniques may be utilised extensively to solve complex practical problems. In view of the above, **chapter 14** includes smoothing approaches to solve support vector machine problems reformulated as unconstrained optimization problems with fourth-order polynomial function and smoothing parameters. This technique is known as the polynomial smooth support vector machine (PSSVM). PSSVM is solved using a rapid Broyden–Fletcher–Goldfarb–Shanno (BFGS) method that converges globally.

It is important to note that polynomial systems are at the heart of a wide range of scientific fields. Multiple forms of polynomials are often involved in attempts to reduce currently open or hard problems to problems already solved. Because of their ever-increasing importance in many practical applications, polynomials span the entire spectrum of various disciplines of mathematics, science and engineering. In practice, a recurring problem is to determine the roots of polynomials under parametric uncertainty or imprecision. Interval mathematics is a subtle body of principles and methods for manipulating systems involving quantifiable uncertainties. As such, systematic investigation of uncertain polynomials is done in **chapter 15**. Rigorous mathematical foundation for interval and set-valued polynomials and theories of interval algebra and interval polynomials are addressed here.

The readers of this challenging book are academics, research and industry. In academics, this book will be useful for graduates, post graduates and researchers of various sciences and engineering fields of most of the universities throughout the globe. The Editor believes the integrated, challenging and holistic approaches about the use of polynomials in all of the chapters will certainly benefit the readers for their future studies and research. The Editor would like to thank all the authors for their excellent efforts and support in contributing and submitting the chapters on time. Finally, I would also like to thank the IOP team for their help and support throughout this project.

S Chakraverty
Editor

# Acknowledgement

The editor greatly appreciates the support, encouragement and patience provided by his family members in particular, his wife Shewli, and daughters Shreyati and Susprihaa. Further, the book may not have been possible without the blessings of his beloved parents the late Sh. Birendra K Chakraborty and the late Parul Chakraborty. I would also like to acknowledge the support of NIT Rourkela administration.

My sincere acknowledgement goes to the reviewers for their fruitful suggestions and appreciations in the book proposal. I greatly appreciate the support and help of all the chapter contributors for submitting their contributions on time and I am very much thankful for their excellent efforts. The whole team of IOP is also greatly appreciated for their support, cooperation and help which has allowed this book to be published according to the schedule. Finally, I am greatly indebted to the authors/researchers mentioned in the bibliography/reference sections given at the end of each chapter.

# Editor biography

## S Chakraverty

Prof. S Chakraverty has 30 years of experience as a researcher and teacher. Presently he is working in the Department of Mathematics (Applied Mathematics Group), National Institute of Technology Rourkela, Odisha as a Senior (Higher Administrative Grade) Professor. Prior to this, he was with CSIR-Central Building Research Institute, Roorkee, India. After graduating from St. Columba's College (Ranchi University), his career started at the University of Roorkee (Now, Indian Institute of Technology Roorkee) and did an MSc (Mathematics) & MPhil (Computer Applications) from there securing First positions in the university. Dr Chakraverty received his PhD from IIT Roorkee in 1993. Thereafter he did his post-doctoral research at the Institute of Sound and Vibration Research (ISVR), University of Southampton, UK and at the Faculty of Engineering and Computer Science, Concordia University, Canada. He was also a visiting professor at the Concordia and McGill universities, Canada, during 1997–1999 and visiting professor of University of Johannesburg, South Africa during 2011–2014. He has authored/co-authored/edited 30 books, published 418 research papers (to date) in journals and conferences and two books are ongoing. He is on the Editorial Boards of various International Journals, Book Series and Conferences. Prof. Chakraverty is the Chief Editor of the "International Journal of Fuzzy Computation and Modelling" (IJFCM), Inderscience Publisher, Switzerland (http://www.inderscience.com/ijfcm), Associate Editor of "Computational Methods in Structural Engineering, Frontiers in Built Environment" and "Curved and Layered Structures (De Gruyter)" and happens to be an Editorial Board member of "Springer Nature Applied Sciences", "IGI Research Insights Books", "Springer Book Series of Modeling and Optimization in Science and Technologies", "Coupled Systems Mechanics (Techno Press)", "Journal of Composites Science (MDPI)", "Engineering Research Express (IOP)" and "Applications and Applied Mathematics: An International Journal". He is also a reviewer of around 50 national and international Journals of repute and he was the President of the Section of Mathematical Sciences (including Statistics) of "Indian Science Congress" (2015–2016) and was the Vice President—"Orissa Mathematical Society" (2011–2013). Prof. Chakraverty is the recipient of prestigious awards viz. Indian National Science Academy (INSA) nomination under International Collaboration/Bilateral Exchange Program (with the Czech Republic), Platinum Jubilee ISCA Lecture Award (2014), CSIR Young Scientist Award (1997), BOYSCAST Fellow. (DST), UCOST Young Scientist Award (2007, 2008), Golden Jubilee Director's (CBRI) Award (2001), INSA International Bilateral Exchange Award ([2010–2011 (selected but could not undertake), 2015 (selected)], Roorkee University Gold Medals (1987, 1988) for first positions in MSc and MPhil (Comp. Appl.) etc.

He is in the list of 2% world scientists recently (2020, 2021) in the Artificial Intelligence & Image Processing category based on an independent study done by Stanford University scientists.

Prof. Chakraverty received IOP Publishing Top Cited Paper Awards (2021 and 2022) for one of the most cited articles from India, published across the entire IOP Publishing journal portfolio within the past three years (2018 to 2020 and 2019 to 2021). It also features in the top 1% of most cited papers in the Materials subject category. This data is from the citations recorded in Web of Science.

He has already guided 21 PhD students and 13 are ongoing. Prof. Chakraverty has undertaken around 16 research projects as Principle Investigator funded by international and national agencies totaling about Rs.1.5 crores. He has hoisted around 8 international students with different international/national fellowships to work in his group as PDF, PhD, visiting researchers for different periods. A good number of international and national Conferences, Workshops, and Training programs have also been organised by him. His present research area includes Differential Equations (Ordinary, Partial and Fractional), Numerical Analysis and Computational Methods, Structural Dynamics (FGM, Nano) and Fluid Dynamics, Mathematical and Uncertainty Modeling, Soft Computing and Machine Intelligence (Artificial Neural Network, Fuzzy, Interval and Affine Computations).

# List of contributors

**Majid Amirfakhrian**
University of Calgary, Canada

**E A Bakare**
Department of Mathematics, Federal University Oye Ekiti, Ekiti State, Nigeria

**S Chakraverty**
Department of Mathematics, National Institute of Technology Rourkela, India

**Hend Dawood**
Department of Mathematics, Faculty of Science Cairo University, Egypt

**Yasser Dawood**
Cairo Lab for Cognitive Science, Linguistics and Pragmatics Novum Academy, Egypt

**Francesco Fabbrocino**
Pegaso Telematic University, Italy

**Nicholas Fantuzzi**
University of Bologna, Italy

**Dineshkumar Harursampath**
Department of Aerospace Engineering, Indian Institute of Science, India

**S Hoskova-Mayerova**
Department of Mathematics and Physics, University of Defence in Brno, Czech Republic

**Rajarama Mohan Jena**
Department of Mathematics, National Institute of Technology Rourkela, India

**Marcin Kamiński**
Department of Structural Mechanics Faculty of Civil Engineering, Architecture and Transportation Łódź University of Technology, Poland

**Sandeep Kumar**
Department of Mathematics, National Institute of Technology Rourkela, India

**Raimondo Luciano**
Parthenope University, Italy

**Vinyas Mahesh**
Department of Mechanical Engineering, National Institute of Technology, Silchar, India

**Bhubaneswari Mishra**
Department of Mathematics, National Institute of Technology Rourkela, India

**Dhabaleswar Mohapatra**
Department of Mathematics, National Institute of Technology Rourkela, India

**Jugal Mohapatra**
Department of Mathematics, National Institute of Technology Rourkela, India

**Van Thien Nguyen**
Department of Mathematics, FPT University, Ho Chi Minh City, Vietnam

**G N Nkem**
Department of Mathematics, University of Ibadan, Ibadan, Nigeria

**O S Obabiyi**
Department of Mathematics, University of Ibadan, Ibadan, Nigeria

**Abhilipsa Panda**
Department of Mathematics, National Institute of Technology Rourkela, India

**Sathiskumar A Ponnusami**
Department of Mechanical Engineering and Aeronautics, City, University of London, UK

**Radovan Potůček**
Department of Mathematics and Physics, Faculty of Military Technology, University of Defence, Brno, Czech Republic

**Priya Rao**
Department of Mathematics, National Institute of Technology Rourkela, India

**Debanik Roy**
Division of Remote Handling & Robotics, Bhabha Atomic Research Centre and Homi Bhabha National Institute, Department of Atomic Energy, India

**Arup Kumar Sahoo**
Department of Mathematics, National Institute of Technology Rourkela, India

**Subal Ranjan Sahu**
Department of Mathematics at C V Raman Global University, India

**Serena Saitta**
Pegaso Telematic University, Italy

**Saba Sajadi**
Islamic Azad University, Iran

**A K Singh**
Department of Mathematics and Computing, Indian Institute of Technology (Indian School of Mines), Dhanbad, Jharkhand, India

**Sonam Singh**
Department of Mathematics and Computing Indian Institute of Technology (Indian School of Mines), Dhanbad, Jharkhand, India

**Riccardo Vescovini**
Politecnico di Milano, Italy

**Shengda Zeng**
Guangxi Colleges and Universities, Key Laboratory of Complex System Optimization and Big Data Processing, Yulin Normal University, Yulin, Guangxi, China

# Chapter 1

# Formulas for the sums of the series of reciprocals of the cubic polynomials with integer roots, at least one zero

**Radovan Potůček**

There is no branch of mathematics, however abstract, which may not some day be applied to phenomena of the real world.

—Nikolai Ivanovich Lobachevsky

## 1.1 Introduction

Before we summarize and build on the author's previous papers [1–8], let us recall the basic terms from the topics infinite series and polynomials.

The series

$$\sum_{k=1}^{\infty} a_k = a_1 + a_2 + a_3 + \cdots$$

*converges* to a limit $s$ if and only if the sequence of partial sums $\{s_n\}$ converges to $s$, i.e. $\lim_{n\to\infty} s_n = s$. We say that the series $\sum_{k=1}^{\infty} a_k$ has a *sum* $s$ and write $\sum_{k=1}^{\infty} a_k = s$.

The *sum of the reciprocals* of some positive integers is generally the sum of unit fractions. For example the sum of the reciprocals of the square numbers (the *Basel problem*) is $\pi^2/6$:

$$\sum_{k=1}^{\infty} \frac{1}{k^2} = \frac{1}{1^2} + \frac{1}{2^2} + \frac{1}{3^2} + \frac{1}{4^2} + \cdots = \frac{\pi^2}{6} = \zeta(2) \doteq 1.644\,934,$$

where

$$\zeta(s) = \sum_{k=1}^{\infty} \frac{1}{k^s}$$

doi:10.1088/978-0-7503-5067-9ch1

is the *Riemann zeta function*, which is a function of a complex variable $s$. This series converges when the real part of $s$ is greater than 1.

In [1] for the sum $s(-bi, bi)$ of the series of reciprocals of the quadratic polynomials with purely imaginary conjugate roots $\pm bi$, i.e. for the sum of the series $\sum_{k=1}^{\infty}\frac{1}{k^2+b^2}$, is derived a formula

$$s(-bi, bi) = \frac{\pi}{2b}\cdot\frac{e^{2\pi b}+1}{e^{2\pi b}-1} - \frac{1}{2b^2},$$

so, for example,

$$\begin{aligned} s(-i, i) &= \sum_{k=1}^{\infty}\frac{1}{k^2+1} = \frac{1}{1^2+1} + \frac{1}{2^2+1} + \frac{1}{3^2+1} + \frac{1}{4^2+1} + \cdots \\ &= \frac{\pi}{2}\cdot\frac{e^{2\pi}+1}{e^{2\pi}-1} - \frac{1}{2} \doteq 1.076\,674. \end{aligned}$$

The *nth harmonic number* is the sum of the reciprocals of the first $n$ natural numbers:

$$H(n) = \sum_{k=1}^{n}\frac{1}{k} = 1 + \frac{1}{2} + \frac{1}{3} + \cdots + \frac{1}{n},$$

$H_0$ being defined as 0. Basic information about harmonic numbers can be found e.g. in the website [9] or in [10]. For $n = 1, 2, \dots, 10$ we get table 1.1.

The *generalized harmonic number* $H(n, r)$ *of order* $n$ *in power* $r$ is the sum

$$H(n, r) = \sum_{k=1}^{n}\frac{1}{k^r} = 1 + \frac{1}{2^r} + \frac{1}{3^r} + \cdots + \frac{1}{n^r}.$$

For $r = 2$ and $n = 1, 2, \dots, 10$ we get table 1.2.

Evidently, $H(n, 1) = H(n)$ are harmonic numbers. Generalized harmonic number of order $n$ in power 2 can be written as a function of harmonic numbers using formula (see [9])

**Table 1.1.** The values of harmonic numbers $H(n)$ for $n = 1, 2, \dots, 10$.

| $n$ | 1 | 2 | 3 | 4 | 5 | 6 | 7 | 8 | 9 | 10 |
|---|---|---|---|---|---|---|---|---|---|---|
| $H(n)$ | 1 | $\frac{3}{2}$ | $\frac{11}{6}$ | $\frac{25}{12}$ | $\frac{137}{60}$ | $\frac{49}{20}$ | $\frac{363}{140}$ | $\frac{761}{280}$ | $\frac{7129}{2520}$ | $\frac{7381}{2520}$ |

**Table 1.2.** The values of generalized harmonic numbers $H(n, 2)$ for $n = 1, 2, \dots, 10$.

| $n$ | 1 | 2 | 3 | 4 | 5 | 6 | 7 | 8 | 9 | 10 |
|---|---|---|---|---|---|---|---|---|---|---|
| $H(n, 2)$ | 1 | $\frac{5}{4}$ | $\frac{49}{36}$ | $\frac{205}{144}$ | $\frac{5269}{3600}$ | $\frac{5369}{3600}$ | $\frac{266\,681}{176\,400}$ | $\frac{107\,7749}{705\,600}$ | $\frac{771\,817}{352\,800}$ | $\frac{196\,8329}{127\,0080}$ |

$$H(n, 2) = \sum_{k=1}^{n-1} \frac{H(k)}{k(k+1)} + \frac{H(n)}{n}.$$

The *telescoping series* is a series whose partial sums eventually only have a fixed number of terms after cancellation. For example, the series $\sum_{k=1}^{\infty} \frac{1}{k(k+3)}$ simplifies, after partial fraction decomposition $\frac{1}{k(k+3)} = \frac{1}{3}\left(\frac{1}{k} - \frac{1}{k+3}\right)$, as

$$\begin{aligned}
\sum_{k=1}^{\infty} \frac{1}{k(k+3)} &= \frac{1}{3}\sum_{k=1}^{\infty}\left(\frac{1}{k} - \frac{1}{k+3}\right) = \frac{1}{3}\lim_{n\to\infty}\sum_{k=1}^{n}\left(\frac{1}{k} - \frac{1}{k+3}\right) \\
&= \frac{1}{3}\lim_{n\to\infty}\left[\left(1 - \frac{1}{4}\right) + \left(\frac{1}{2} - \frac{1}{5}\right) + \left(\frac{1}{3} - \frac{1}{6}\right) + \left(\frac{1}{4} - \frac{1}{7}\right) + \cdots\cdots\right. \\
&\quad \left. + \left(\frac{1}{n-3} - \frac{1}{n}\right) + \left(\frac{1}{n-2} - \frac{1}{n+1}\right) + \left(\frac{1}{n-1} - \frac{1}{n+2}\right) + \left(\frac{1}{n} - \frac{1}{n+3}\right)\right] \\
&= \frac{1}{3}\lim_{n\to\infty}\left(1 + \frac{1}{2} + \frac{1}{3} - \frac{1}{n+1} - \frac{1}{n+2} - \frac{1}{n+3}\right) = \frac{1}{3}H(3) = \frac{11}{18} \doteq 0.611111.
\end{aligned}$$

Let us recall that a *cubic polynomial* is a function of a generally complex variable $x$ of the form

$$P(x) = ax^3 + bx^2 + cx + d,$$

where constants $a, b, c, d, a \neq 0$, are generally complex numbers and are called the *coefficients* of the polynomial $P(x)$.
Solutions of the corresponding *cubic equation*

$$ax^3 + bx^2 + cx + d = 0$$

are called *roots* of the cubic polynomial $P(x)$. Note that if all of the coefficients $a, b, c, d$ are real numbers, then the cubic polynomial has at least one real root.

It is often convenient to reduce a cubic equation so that its leading coefficient $a$ is one. This is done by dividing both sides of the cubic equation by $a$, which is always possible since $a$ is non-zero. This produces so named *reduced cubic equation*

$$x^3 + px^2 + qx + r = 0,$$

where $p = b/a$, $q = c/a$ and $r = d/a$. This *monic cubic equation* has the same solutions as the original cubic equation $ax^3 + bx^2 + cx + d = 0$. Without loss of generality we will only deal with reduced cubic equations and corresponding reduced cubic polynomial

$$P(x) = x^3 + px^2 + qx + r.$$

If we know the roots $x_1, x_2, x_3$ of the reduced cubic polynomial then it can be expressed in the *factored form*

$$P(x) = (x - x_1)(x - x_2)(x - x_3).$$

## 1.2 Integer roots of the special reduced cubic polynomial

Now, we will deal with integer roots of the special reduced cubic polynomial, that is, in our case, with integer roots of the reduced cubic polynomial with at least one zero root.

It is clear that if the polynomial $P(x) = (x - x_1)(x - x_2)(x - x_3)$ has the zero root of degree 3, i.e. if $x_1 = x_2 = x_3 = 0$, we can write

$$P(x) = x^3.$$

If $P(x)$ has the zero root of degree 2, i.e. if $x_1 = x_2 = 0$, we can write

$$P(x) = x^2(x - x_3) = x^3 + px^2,$$

where $p = -x_3$. It is evident that $P(x)$ has integer roots if and only if the non-zero coefficient $p$ is an integer.

And if the polynomial $P(x)$ has only one zero root, i.e. if $x_1 = 0$, we can write

$$P(x) = x(x - x_2)(x - x_3) = x(x^2 + px + q),$$

where $p = -(x_2 + x_3)$ and $q = x_2x_3$. Next, suppose that $x_2 \leqslant x_3$.

Now, let us consider a reduced cubic polynomial with two non-zero integer roots and with one zero root and corresponding reduced cubic equation of the form

$$x^3 + px^2 + qx = x(x^2 + px + q) = 0 \tag{1.1}$$

with non-zero integer coefficients $p, q$. It is a well-known fact that reduced cubic equation (1.1) has, in addition to the root $x_1 = 0$, also the roots $x_{2,3}$ of the form

$$x_{2,3} = \frac{-p \pm \sqrt{p^2 - 4q}}{2},$$

so

$$x_2 = \frac{-p - \sqrt{p^2 - 4q}}{2} \quad \text{and} \quad x_3 = \frac{-p + \sqrt{p^2 - 4q}}{2}.$$

If the roots $x_2$ and $x_3$ are to be integers $m$ and $n$, then equalities

$$2\,m = -p - \sqrt{p^2 - 4q} \quad \text{and} \quad 2n = -p + \sqrt{p^2 - 4q}$$

must hold. When we raise these equalities to the square, we receive the equalities

$$(2\,m + p)^2 = p^2 - 4q \quad \text{and} \quad (2n + p)^2 = p^2 - 4q.$$

After simplification we get the relations

$$m^2 + pm + q = 0 \quad \text{and} \quad n^2 + pn + q = 0,$$

whence for appropriate integer $k$ we get the condition for coefficients $p, q$ of equation (1.1)

$$q = -k(k + p).$$

So the reduced cubic equation (1.1) with zero root $x_1$ has non-zero integer roots $x_{2,3}$ if and only if the quadratic equation

$$x^2 + px + q = 0 \tag{1.2}$$

has integer roots $x_{2,3}$. Equation (1.1) then has the form

$$x^3 + px^2 - k(k + p)x = 0$$

for appropriate integer $k$.

**Example 1.1.** Decide which of the two following reduced cubic equations has two non-zero integer roots in addition to the zero root:

$$\text{(a) } x^3 + 8x^2 + 14x = 0, \quad \text{(b) } x^3 - 9x^2 - 36x = 0.$$

**Solution 1.1.** (a) We have $p = 8$ and $q = 14$. The equation $x^3 + 8x^2 + 14x = 0$ has evidently the root $x_1 = 0$. Now we find out whether the quadratic equation (1.2), i.e. the equation $x^2 + 8x + 14 = 0$, has integer roots.

The equation $q = -k(k + p)$, i.e. the equation $14 = -k(k + 8)$, is not satisfied for any integer $k$, because none of the four integer factorizations $14 = 1 \cdot 14 = 2 \cdot 7 = (-1) \cdot (-14) = (-2) \cdot (-7)$ obviously satisfies this equation. So given the reduced cubic equation $x^3 + 8x^2 + 14x = 0$ does not have another two integer roots.

If we use the quadratic formula to check this result, we get its non-integer roots

$$x_{2,3} = \frac{-8 \pm \sqrt{64 - 56}}{2} = \frac{-8 \pm \sqrt{8}}{2} = \left\langle \begin{matrix} -4 - \sqrt{2} \\ -4 + \sqrt{2} \end{matrix} \right. .$$

(b) We have $p = -9$ and $q = -36$. The equation $x^3 - 9x^2 - 36x = 0$ has evidently the root $x_1 = 0$. Again, we find out whether quadratic equation (1.2), i.e. the equation $x^2 - 9x - 36 = 0$, has integer roots.

The equation $q = -k(k + p)$, i.e. the equation $-36 = -k(k - 9)$, whence $k^2 - 9k - 36 = 0$, i.e. $(k + 3)(k - 12) = 0$, is satisfied for $k = -3$ and for $k = 12$. So given reduced cubic equation $x^3 - 9x^2 - 36x = 0$ has another two integer roots.

If we use the quadratic formula to check this result, we get its integer roots

$$x_{2,3} = \frac{-9 \pm \sqrt{81 + 144}}{2} = \frac{-9 \pm \sqrt{225}}{2} = \left\langle \begin{matrix} \dfrac{-9 - 15}{2} = -12 \\ \dfrac{-9 + 15}{2} = 3 \end{matrix} \right. .$$

## 1.3 The case of triple zero root

Assume that the series of reciprocals of the cubic polynomials has a triple zero root. So it is the series of the form

$$\sum_{k=1}^{\infty} \frac{1}{k^3}. \tag{1.3}$$

It is a well-known fact that the sum of the series (1.3), denoted further by $s(0, 0, 0)$, is the so-called *Apéry's constant* which is given by the formula

$$s(0, 0, 0) = \zeta(3) = \sum_{k=1}^{\infty} \frac{1}{k^3} = \frac{1}{1^3} + \frac{1}{2^3} + \frac{1}{3^3} + \frac{1}{4^3} + \cdots \doteq 1.202\,057.$$

The solution of this case, which results in the Apéry's constant $\zeta(3)$, has been well known since the 18th century, when it was dealt with by *Leonhard Euler* (1707–1783), who in 1735 calculated this constant to 16 decimal places. In the 19th century, the greatest credit for the knowledge of Riemann's zeta function belongs to *Bernhard Riemann* (1826–1866), who in 1859 published a key treatise on this function.

Further development of knowledge of Riemann's zeta function occurred in the 20th century, when in 1978 *Roger Apéry* (1916–1994) proved that the constant named after him is an irrational number. Interestingly, there are a number of representations of the Apéry's constant $\zeta(3)$ using series, alternating series and integrals. The simplest ones are the following three representations (see e.g. [11]):

$$\zeta(3) = \frac{8}{7} \sum_{k=0}^{\infty} \frac{1}{(2k+1)^3}, \quad \zeta(3) = \frac{4}{3} \sum_{k=0}^{\infty} \frac{(-1)^k}{(k+1)^3}, \quad \zeta(3) = \frac{2}{3} \int_0^{\infty} \frac{x^2}{e^x + 1}\, dx.$$

## 1.4 The case of double zero root

Now, let us consider the series of reciprocals of reduced cubic polynomials of the form

$$k^3 - ak^2 = k^2(k - a)$$

with double zero root and one non-zero integer root $a$, i.e. the series

$$\sum_{\substack{k=1 \\ k \neq a}}^{\infty} \frac{1}{k^2(k-a)}, \tag{1.4}$$

and let us determine its sum. We differentiate two cases—a case of negative integer root $a$ and a case of positive integer root $a$.

### 1.4.1 The case of double zero root and one negative integer root

First, let us assume the case of a negative integer root $a$ and denote the sum of the series (1.4) by $s(a, 0, 0)$. Then, after partial fraction decomposition, we get the $k$th term

$$a_k = \frac{1}{k^2(k-a)} = \frac{K}{k^2} + \frac{L}{k} + \frac{M}{k-a},$$

where coefficients $K$, $L$, $M$ are real constants. In [5] it is derived that

$$a_k = \frac{1}{k^2(k-a)} = -\frac{1}{a} \cdot \frac{1}{k^2} + \frac{1}{a^2}\left(\frac{1}{k-a} - \frac{1}{k}\right). \tag{1.5}$$

**Example 1.2.** Using the $n$th partial sum calculate the sum $s(-4, 0, 0)$ of the series

$$\sum_{k=1}^{\infty} \frac{1}{k^2(k+4)}.$$

According to equation (1.5), where $a = -4$, we have

$$a_k = \frac{1}{k^2(k+4)} = \frac{1}{4} \cdot \frac{1}{k^2} + \frac{1}{16}\left(\frac{1}{k+4} - \frac{1}{k}\right),$$

so the $n$th partial sum $s_n(-4, 0, 0) = a_1 + a_2 + \cdots + a_n$ is

$$\begin{aligned} s_n(-4, 0, 0) &= \frac{1}{4}\left(\frac{1}{1^2} + \frac{1}{2^2} + \cdots + \frac{1}{n^2}\right) \\ &+ \frac{1}{16}\left[\left(\frac{1}{5} - \frac{1}{1}\right) + \left(\frac{1}{6} - \frac{1}{2}\right) + \left(\frac{1}{7} - \frac{1}{3}\right) + \left(\frac{1}{8} - \frac{1}{4}\right) + \left(\frac{1}{9} - \frac{1}{5}\right) + \cdots \right. \\ &+ \left(\frac{1}{n} - \frac{1}{n-4}\right) + \left(\frac{1}{n+1} - \frac{1}{n-3}\right) + \left(\frac{1}{n+2} - \frac{1}{n-2}\right) + \left(\frac{1}{n+3} - \frac{1}{n-1}\right) \\ &\left. + \left(\frac{1}{n+4} - \frac{1}{n}\right)\right] \\ &= \frac{1}{4}H(n, 2) + \frac{1}{16}\left(-\frac{1}{1} - \frac{1}{2} - \frac{1}{3} - \frac{1}{4} + \frac{1}{n+1} + \frac{1}{n+2} + \frac{1}{n+3} + \frac{1}{n+4}\right). \end{aligned}$$

Considering basic facts about limits

$$\lim_{n\to\infty} \frac{1}{c} = \frac{1}{c} \quad \text{and} \quad \lim_{n\to\infty} \frac{1}{n+c} = 0, \tag{1.6}$$

where $c$ is an arbitrary integer, and that

$$\lim_{n\to\infty} H(n, 2) = \lim_{n\to\infty} \sum_{k=1}^{n} \frac{1}{k^2} = \sum_{k=1}^{\infty} \frac{1}{k^2} = \frac{\pi^2}{6}$$

we get

$$\begin{aligned} s(-4, 0, 0) &= \lim_{n\to\infty} s_n(-4, 0, 0) \\ &= \frac{1}{4} \lim_{n\to\infty} H(n, 2) \\ &+ \frac{1}{16} \lim_{n\to\infty}\left(-\frac{1}{1} - \frac{1}{2} - \frac{1}{3} - \frac{1}{4} + \frac{1}{n+1} + \frac{1}{n+2} + \frac{1}{n+3} + \frac{1}{n+4}\right) \\ &= \frac{1}{4} \cdot \frac{\pi^2}{6} - \frac{H(4)}{16} = \frac{\pi^2}{24} - \frac{25}{12 \cdot 16} = \frac{8\pi^2 - 25}{192} \doteq 0.281\,025. \end{aligned}$$

Clearly, for arbitrary negative integer $a$ is $A = -a$ positive integer, so $k - a = k + A$ are also positive integers for all $k = 1, 2, \ldots, n$. After cancellation of all of the inner terms we get the $n$th partial sum $s_n = a_1 + a_2 + \cdots + a_n$ in a form

$$s_n(a, 0, 0) = s_n(-A, 0, 0) = \sum_{k=1}^{n}\left[\frac{1}{A}\cdot\frac{1}{k^2} + \frac{1}{A^2}\left(\frac{1}{k+A} - \frac{1}{k}\right)\right] = \frac{1}{A}\left(\frac{1}{1^2} + \frac{1}{2^2} + \cdots + \frac{1}{n^2}\right)$$
$$+ \frac{1}{a^2}\left[\left(\frac{1}{1+A} - \frac{1}{1}\right) + \left(\frac{1}{2+A} - \frac{1}{2}\right) + \left(\frac{1}{3+A} - \frac{1}{3}\right) + \cdots\right.$$
$$+ \left(\frac{1}{2A} - \frac{1}{A}\right) + \left(\frac{1}{2A+1} - \frac{1}{A+1}\right) + \cdots$$
$$+ \left(\frac{1}{n} - \frac{1}{n-A}\right) + \left(\frac{1}{n+1} - \frac{1}{n-A+1}\right) + \cdots\cdots$$
$$\left. + \left(\frac{1}{n+A-2} - \frac{1}{n-2}\right) + \left(\frac{1}{n+A-1} - \frac{1}{n-1}\right) + \left(\frac{1}{n+A} - \frac{1}{n}\right)\right]$$
$$= \frac{H(n, 2)}{A} + \frac{1}{A^2}\left(-\frac{1}{1} - \frac{1}{2} - \cdots - \frac{1}{A} + \frac{1}{n+1} + \frac{1}{n+2} + \cdots + \frac{1}{n+A}\right).$$

Considering (1.6), we generally have

$$s(a, 0, 0) = \lim_{n\to\infty} s_n(a, 0, 0)$$
$$= -\frac{1}{a}\lim_{n\to\infty} H(n, 2)$$
$$+ \frac{1}{a^2}\lim_{n\to\infty}\left(-\frac{1}{1} - \frac{1}{2} - \cdots - \frac{1}{-a} + \frac{1}{n+1} + \frac{1}{n+2} + \cdots + \frac{1}{n-a}\right)$$
$$= -\frac{1}{a}\cdot\frac{\pi^2}{6} - \frac{H(-a)}{a^2}.$$

Altogether, we derived this statement:

**Theorem 1.1.** The series $\sum_{k=1}^{\infty}\frac{1}{k^2(k-a)}$ has for a negative integer $a$ the sum in the form

$$s(a, 0, 0) = -\frac{\pi^2}{6a} - \frac{H(-a)}{a^2}, \tag{1.7}$$

where $H(n)$ is the $n$th harmonic number.

So that e.g.

$$s(-1, 0, 0) = \sum_{k=1}^{\infty}\frac{1}{k^2(k+1)} = \frac{\pi^2}{6} - 1, \quad s(-2, 0, 0) = \sum_{k=1}^{\infty}\frac{1}{k^2(k+2)} = \frac{\pi^2}{12} - \frac{3}{8},$$

and

$$s(-4, 0, 0) = \sum_{k=1}^{\infty}\frac{1}{k^2(k+4)} = \frac{\pi^2}{24} - \frac{25}{12\cdot 16} = \frac{8\pi^2 - 25}{192} \doteq 0.281\,025,$$

as was calculated in example 1.2. From the equality $\sum_{k=1}^{\infty}\frac{1}{k^2(k+1)} = \frac{\pi^2}{6} - 1$ we can derive that

$$\pi = \sqrt{6\left(1 + \sum_{k=1}^{\infty} \frac{1}{k^2(k+1)}\right)}.$$

Using the first hundred thousand summands in Maple 2020 calculation we get

$$\pi \doteq 3.141\,592\,653\,54.$$

This approximate value differs from the number $\pi = 3.141\,592\,653\,59\ldots$ by less than $10^{-10}$.

### 1.4.2 The case of double zero root and one positive integer root

Now, let us assume the case of a positive integer root $a$ and denote the sum of the series (1.4) by $s(0, 0, a)$. After partial fraction decomposition, we also get, as in the previous case,

$$a_k = \frac{1}{k^2(k-a)} = -\frac{1}{a} \cdot \frac{1}{k^2} + \frac{1}{a^2}\left(\frac{1}{k-a} - \frac{1}{k}\right).$$

Note that the summation index $k$ must now be different from $a$ to avoid division by zero.

**Example 1.3.** Using the $n$th partial sum calculate the sum $s(0, 0, 4)$ of the series

$$\sum_{\substack{k=1 \\ k \neq 4}}^{\infty} \frac{1}{k^2(k-4)}.$$

According to equation (1.5), where $a = 4$, we have

$$a_k = \frac{1}{k^2(k-4)} = \frac{1}{4} \cdot \frac{1}{k^2} + \frac{1}{16}\left(\frac{1}{k-4} - \frac{1}{k}\right), \qquad k \neq 4,$$

so the $n$th partial sum $s_n(0, 0, 4) = a_1 + a_2 + a_3 + a_5 + \cdots + a_n$ is

$$\begin{aligned}
s_n(0, 0, 4) = &-\frac{1}{4}\left(\frac{1}{1^2} + \frac{1}{2^2} + \frac{1}{3^2} + \frac{1}{5^2} + \cdots + \frac{1}{n^2}\right) \\
&+ \frac{1}{16}\left[\left(\frac{1}{-3} - \frac{1}{1}\right) + \left(\frac{1}{-2} - \frac{1}{2}\right) + \left(\frac{1}{-1} - \frac{1}{3}\right) + \left(\frac{1}{1} - \frac{1}{5}\right) + \left(\frac{1}{2} - \frac{1}{6}\right) + \left(\frac{1}{3} - \frac{1}{7}\right)\right. \\
&+ \left(\frac{1}{4} - \frac{1}{8}\right) + \left(\frac{1}{5} - \frac{1}{9}\right) + \left(\frac{1}{6} - \frac{1}{10}\right) + \cdots \\
&+ \left(\frac{1}{n-8} - \frac{1}{n-4}\right) + \left(\frac{1}{n-7} - \frac{1}{n-3}\right) + \left(\frac{1}{n-6} - \frac{1}{n-2}\right) + \left(\frac{1}{n-5} - \frac{1}{n-1}\right) \\
&\left. + \left(\frac{1}{n-4} - \frac{1}{n}\right)\right] \\
= &-\frac{H(n, 2)}{4} + \frac{1}{4^3} + \frac{1}{16}\left(-\frac{1}{1} - \frac{1}{2} - \frac{1}{3} + \frac{1}{4} - \frac{1}{n-3} - \frac{1}{n-2} - \frac{1}{n-1} - \frac{1}{n}\right).
\end{aligned}$$

As in the previous case, concerning one negative integer root, we get

$$s(0,0,4)=\lim_{n\to\infty}s_n(0,0,4)=-\frac{1}{4}\lim_{n\to\infty}H(n,2)+\frac{1}{64}$$
$$-\frac{1}{16}\lim_{n\to\infty}\left(\frac{1}{1}+\frac{1}{2}+\frac{1}{3}-\frac{1}{4}+\frac{1}{n-3}+\frac{1}{n-2}+\frac{1}{n-1}+\frac{1}{n}\right)-\frac{1}{4}\cdot\frac{\pi^2}{6}+\frac{2}{64}-\frac{H(3)}{16}$$
$$=-\frac{1}{4}\cdot\frac{\pi^2}{6}+\frac{2}{64}-\frac{H(3)}{16}=-\frac{\pi^2}{24}+\frac{1}{32}-\frac{11}{6\cdot 16}=\frac{-4\pi^2+3-11}{96}\doteq-0.494\,567.$$

For arbitrary positive integer $a$, after cancellation of all of the inner terms, we get the $n$th partial sum $s_n=a_1+a_2+\cdots+a_{a-1}+a_{a+1}+\cdots+a_n$ in a form

$$s_n(0,0,a)=\sum_{\substack{k=1\\k\neq a}}^{n}\left[-\frac{1}{a}\cdot\frac{1}{k^2}+\frac{1}{a^2}\left(\frac{1}{k-a}-\frac{1}{k}\right)\right]=-\frac{1}{a}\left(\frac{1}{1^2}+\frac{1}{2^2}+\cdots+\frac{1}{n^2}\right)+\frac{1}{a}\cdot\frac{1}{a^2}$$
$$+\frac{1}{a^2}\left[\left(\frac{1}{1-a}-\frac{1}{1}\right)+\left(\frac{1}{2-a}-\frac{1}{2}\right)+\left(\frac{1}{3-a}-\frac{1}{3}\right)+\cdots\right.$$
$$+\left(\frac{1}{-3}-\frac{1}{a-3}\right)+\left(\frac{1}{-2}-\frac{1}{a-2}\right)+\left(\frac{1}{-1}-\frac{1}{a-1}\right)+\left(\frac{1}{1}-\frac{1}{a+1}\right)$$
$$+\left(\frac{1}{2}-\frac{1}{a+2}\right)+\left(\frac{1}{3}-\frac{1}{a+3}\right)+\cdots\cdots$$
$$+\left(\frac{1}{a-2}-\frac{1}{2a-2}\right)+\left(\frac{1}{a-1}-\frac{1}{2a-1}\right)+\left(\frac{1}{a}-\frac{1}{2a}\right)+\left(\frac{1}{a+1}-\frac{1}{2a+1}\right)$$
$$+\cdots\cdots+\left(\frac{1}{n-2a}-\frac{1}{n-a}\right)+\left(\frac{1}{n-2a+1}-\frac{1}{n-a+1}\right)$$
$$+\left(\frac{1}{n-2a+2}-\frac{1}{n-a+2}\right)+\cdots\cdots$$
$$\left.+\left(\frac{1}{n-a-2}-\frac{1}{n-2}\right)+\left(\frac{1}{n-a-1}-\frac{1}{n-1}\right)+\left(\frac{1}{n-a}-\frac{1}{n}\right)\right]$$
$$=-\frac{H(n,2)}{a}+\frac{1}{a^3}+\frac{1}{a^2}\left(-\frac{1}{1}-\frac{1}{2}-\cdots-\frac{1}{a-1}+\frac{1}{a}-\frac{1}{n-a+1}-\frac{1}{n-a+2}-\cdots-\frac{1}{n}\right).$$

Therefore we generally have

$$s(0,0,a)=\lim_{n\to\infty}s_n(0,0,a)=-\frac{1}{a}\lim_{n\to\infty}H(n,2)+\frac{2}{a^3}$$
$$-\frac{1}{a^2}\lim_{n\to\infty}\left(\frac{1}{1}+\frac{1}{2}+\cdots+\frac{1}{a-1}+\frac{1}{n-a+1}+\frac{1}{n-a+2}+\cdots+\frac{1}{n}\right).$$

In total, we derived the following statement:

**Theorem 1.2.** The series $\sum_{\substack{k=1\\k\neq a}}^{\infty}\frac{1}{k^2(k-a)}$ has for a positive integer $a$ the sum in the form

$$s(0,0,a)=-\frac{\pi^2}{6a}-\frac{H(a-1)}{a^2}+\frac{2}{a^3},\tag{1.8}$$

where $H(n)$ is the $n$th harmonic number.

So that e.g.

$$s(0,0,1)=\sum_{k=2}^{\infty}\frac{1}{k^2(k-1)}=2-\frac{\pi^2}{6},\quad s(0,0,2)=\sum_{k=1 k\neq 2}^{\infty}\frac{1}{k^2(k-2)}=-\frac{\pi^2}{12}$$

(the sum $s(0, 0, 1)$ is the only positive sum $s(0, 0, a)$ for positive $a$), and

$$s(0,0,4)=\sum_{k=1 k\neq 4}^{\infty}\frac{1}{k^2(k-4)}=-\frac{\pi^2}{24}-\frac{11}{96}+\frac{1}{32}\doteq -0.494\,567,$$

as was calculated in example 1.3. From the equality $\sum_{k=2}^{\infty}\frac{1}{k^2(k-1)}=2-\frac{\pi^2}{6}$ we can derive that

$$\pi=\sqrt{6\left(2-\sum_{k=2}^{\infty}\frac{1}{k^2(k-1)}\right)}.$$

Using the first hundred thousand summands in Maple 2020 calculation we get

$$\pi\doteq 3.141\,592\,653\,63.$$

This approximate value differs from the number $\pi = 3.141\,592\,653\,59\ldots$ by less than $10^{-10}$.

### 1.4.3 The case of double zero root and one arbitrary integer root

Because formulas (1.7) and (1.8), i.e. the formulas

$$s(a,0,0)=-\frac{\pi^2}{6a}-\frac{H(-a)}{a^2}\quad\text{and}\quad s(0,0,a)=-\frac{\pi^2}{6a}-\frac{H(a-1)}{a^2}+\frac{2}{a^3},$$

are very similar, they can be obviously expressed using the function $\sigma$ of non-zero integer $x$

$$\sigma(x)=\frac{\operatorname{sgn}(x)+1}{2}=\begin{cases}0 & \text{for } x \text{ negative}\\ 1 & \text{for } x \text{ positive}\end{cases}\tag{1.9}$$

by the only summary general formula given in the following statement:

**Theorem 1.3.** The series $\sum_{\substack{k=1\\ k\neq\alpha}}^{\infty}\frac{1}{k^2(k-\alpha)}$ has for a non-zero integer $\alpha$ the sum in the form

$$s(0,0,\alpha)=-\frac{\pi^2}{6\alpha}-\frac{H[|\alpha|-\sigma(\alpha)]}{\alpha^2}+\frac{2\sigma(\alpha)}{\alpha^3},\tag{1.10}$$

where $H(n)$ is the $n$th harmonic number.

**Remark 1.1.** Let us note that derived summary general formula (1.10) has the form

$$s(0,0,\alpha)=\frac{k}{\alpha}+\frac{\ell}{\alpha^2}+\frac{m}{\alpha^3}.$$

Now, we determine the unknown coefficients $k$, $\ell$, $m$ by means of the least squares method in Maple 2020. For negative integer $\alpha$ we use the following statements and their outputs (figure 1.1):

```
> with(CurveFitting):  with(plots):
> s:=x->-Pi*Pi/(6*x) - harmonic(abs(x)-(signum(x)+1)/2)/(x*x)
  + (signum(x)+1)/(x*x*x):
> data:=[seq([x,s(x)], x=-50..-1)]:
> L1:=LeastSquares(data,x,curve=k/x+l/(x*x)+m/(x*x*x)):  evalf(%);
```

$$-\frac{0.7855228538}{x^3}-\frac{1.650917992}{x^2}-\frac{1.510830207}{x}$$

```
> plot([data,L1], x=-50..-1, color=[blue,green],style=[point,line]);
```

Similarly, for positive integer $\alpha$ we use these statements and their outputs (figure 1.2):

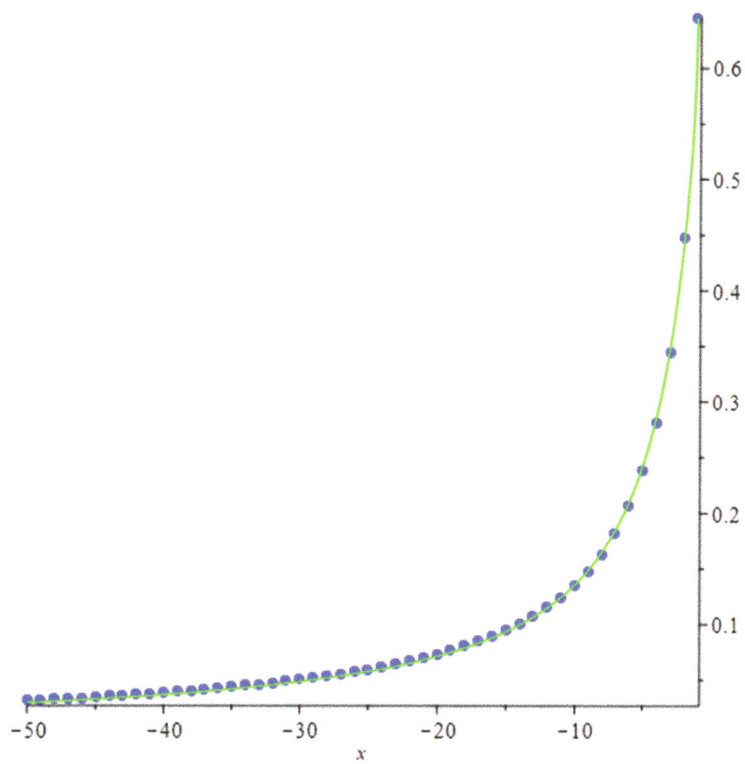


**Figure 1.1.** Dot chart of $s(0, 0, \alpha)$ values for integer $\alpha < 0$ and approximation curve graph.

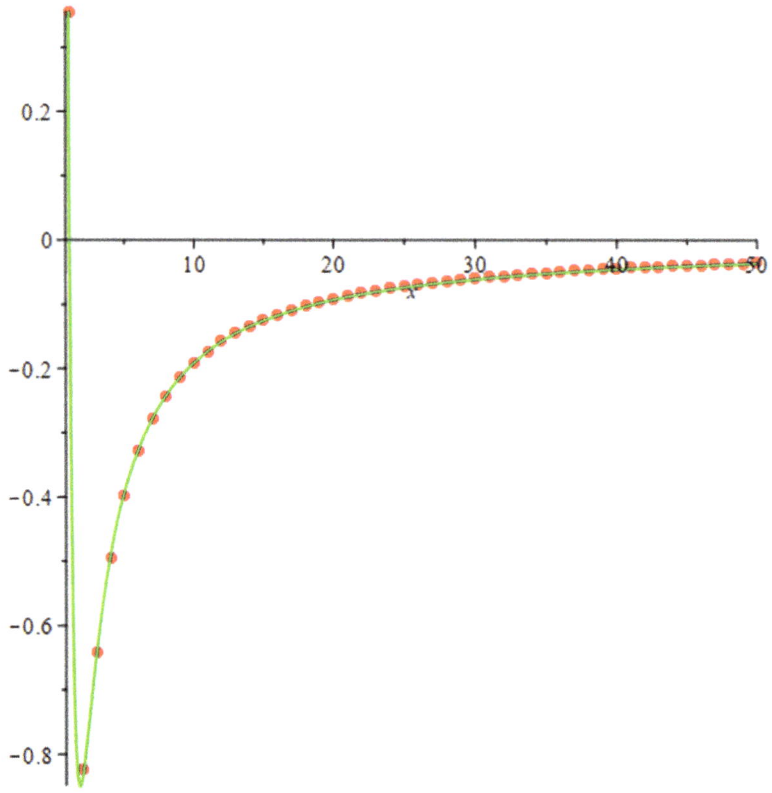


**Figure 1.2.** Dot chart of $s(0, 0, \alpha)$ values for integer $\alpha > 0$ and approximation curve graph.

```
> with(CurveFitting):  with(plots):
> s:=x->-Pi*Pi/(6*x) - harmonic(abs(x)-(signum(x)+1)/2)/(x*x)
  + (signum(x)+1)/(x*x*x):
> data:=[seq([x,s(x)], x=1..50)]:
> L2:=LeastSquares(data,x,curve=k/x+l/(x*x)+m/(x*x*x)):  evalf(%);
```

$$\frac{3.785522854}{x^3} - \frac{1.650917992}{x^2} - \frac{1.779037929}{x}$$

```
> plot([data,L2], x=1..50, color=[red,green],style=[point,line]);
```

For negative integer $a$ so we can write very roughly

$$s(a, 0, 0) \doteq -\frac{7}{9a} - \frac{5}{3a^2} - \frac{3}{2a^3} \tag{1.11}$$

and for positive integer $a$ we get the approximate formula

$$s(0, 0, a) \doteq -\frac{7}{4a} - \frac{5}{3a^2} + \frac{19}{5a^3}. \tag{1.12}$$

For these approximate formulas, we used the convert commands

```
> convert(-0.78552328538,rational,3);
> convert(-1.650917992,rational,3);
> convert(-1.510830207,rational,3);
```

in Maple 2020. It is interesting that by Maple the calculated second coefficient $\ell = -1.650\,917\,992$ is the same for both $a < 0$ and $a > 0$.

## 1.5 The case of one zero root

Assume that the series of reciprocals of the cubic polynomials has one zero root. So, let us consider the series of reciprocals of reduced cubic polynomials of the form

$$k^3 - (a+b)k^2 + abk = k(k-a)(k-b)$$

with one zero root and two non-zero integer roots $a$ and $b$, i.e. the series

$$\sum_{k=1 k\neq a,b}^{\infty} \frac{1}{k(k-a)(k-b)},$$

and let us determine its sum. We differentiate three basic cases with two different non-zero integer roots: $a < b < 0$, $a < 0 < b$ and $0 < a < b$. We will also analyze two more special cases with double non-zero integer roots: $a = b < 0$ and $0 < a = b$.

### 1.5.1 The case of two different negative integer roots and one zero root

Let us consider the series of reciprocals of the reduced cubic polynomials with one zero and two different negative integer roots $a < b < 0$, i.e. the series

$$\sum_{k=1}^{\infty} \frac{1}{k(k-a)(k-b)}, \tag{1.13}$$

and let us determine its sum $s(a, b, 0)$. After partial fraction decomposition, we get the $k$th term in the form

$$a_k = \frac{1}{k(k-a)(k-b)} = \frac{K}{k} + \frac{L}{k-a} + \frac{M}{k-b},$$

where coefficients $K$, $L$, $M$ are real constants. In [4] it is derived that

$$a_k = \frac{1}{k(k-a)(k-b)} = \frac{1}{ab} \cdot \frac{1}{k} + \frac{1}{a(a-b)} \cdot \frac{1}{k-a} - \frac{1}{b(a-b)} \cdot \frac{1}{k-b}. \tag{1.14}$$

**Example 1.4.** Using the $n$th partial sum calculate the sum $s(-4, -2, 0)$ of the series

$$\sum_{k=1}^{\infty} \frac{1}{k(k+4)(k+2)}.$$

According to equation (1.14), where $a = -4$ and $b = -2$, we have

$$a_k = \frac{1}{k(k+4)(k+2)} = \frac{1}{8}\cdot\frac{1}{k} + \frac{1}{8}\cdot\frac{1}{k+4} - \frac{1}{4}\cdot\frac{1}{k+2}$$
$$= \frac{1}{8}\left(\frac{1}{k} + \frac{1}{k+4} - \frac{2}{k+2}\right),$$

so the $n$th partial sum $s_n(-4, -2, 0) = a_1 + a_2 + \cdots + a_n$ has the form

$$s_n(-4,-2,0) = \frac{1}{8}\left[\left(\frac{1}{1}+\frac{1}{5}-\frac{2}{3}\right)+\left(\frac{1}{2}+\frac{1}{6}-\frac{2}{4}\right)+\left(\frac{1}{3}+\frac{1}{7}-\frac{2}{5}\right)+\left(\frac{1}{4}+\frac{1}{8}-\frac{2}{6}\right)\right.$$
$$+\left(\frac{1}{5}+\frac{1}{9}-\frac{2}{7}\right)+\left(\frac{1}{6}+\frac{1}{10}-\frac{2}{8}\right)+\cdots+\left(\frac{1}{n-5}+\frac{1}{n-1}-\frac{2}{n-3}\right)$$
$$+\left(\frac{1}{n-4}+\frac{1}{n}-\frac{2}{n-2}\right)+\left(\frac{1}{n-3}+\frac{1}{n+1}-\frac{2}{n-1}\right)+\left(\frac{1}{n-2}+\frac{1}{n+2}-\frac{2}{n}\right)$$
$$\left.+\left(\frac{1}{n-1}+\frac{1}{n+3}-\frac{2}{n+1}\right)+\left(\frac{1}{n}+\frac{1}{n+4}-\frac{2}{n+2}\right)\right]$$
$$= \frac{1}{8}\left(\frac{1}{1}+\frac{1}{2}-\frac{1}{3}-\frac{1}{4}-\frac{1}{n+1}-\frac{1}{n+2}+\frac{1}{n+3}+\frac{1}{n+4}\right).$$

Considering (1.6), we get

$$s(-4,-2,0) = \lim_{n\to\infty} s_n(-4,-2,0) = \frac{1}{8}\cdot\frac{12+6-4-3}{12} = \frac{1}{8}\cdot\frac{11}{12} = \frac{11}{96} = 0.114$$
$$58\bar{3}.$$

Clearly, for arbitrary negative integers $a < b < 0$, $A = -a$ and $B = -b$ are positive integers, so $k - a = k + A$ and $k - b = k + B$ are also positive integers for all $k = 1, 2, \ldots, n$. Equation (1.14), that has the form

$$a_k = \frac{1}{k(k-a)(k-b)} = \frac{1}{ab}\cdot\frac{1}{k} + \frac{1}{a(a-b)}\cdot\frac{1}{k-a} - \frac{1}{b(a-b)}\cdot\frac{1}{k-b},$$

then can be rewritten in the form

$$a_k = \frac{1}{-A(-B)}\cdot\frac{1}{k} + \frac{1}{-A(-A+B)}\cdot\frac{1}{k+A} - \frac{1}{-B(-A+B)}\cdot\frac{1}{k+B}$$
$$= \frac{1}{AB}\cdot\frac{1}{k} + \frac{1}{A(A-B)}\cdot\frac{1}{k+A} - \frac{1}{B(A-B)}\cdot\frac{1}{k+B}$$
$$= \frac{1}{AB(A-B)}\left(\frac{A-B}{k} + \frac{B}{k+A} - \frac{A}{k+B}\right).$$

Since $A = B + (A - B)$, we have $\frac{A}{k+B} = \frac{B+(A-B)}{k+B} = \frac{B}{k+B} + \frac{A-B}{k+B}$, so we get

$$a_k = \frac{1}{AB(A-B)}\left(\frac{A-B}{k} + \frac{B}{k+A} - \frac{B}{k+B} - \frac{A-B}{k+B}\right)$$
$$= \frac{1}{AB(A-B)}\left[B\left(\frac{1}{k+A} - \frac{1}{k+B}\right) - (A-B)\left(\frac{1}{k+B} - \frac{1}{k}\right)\right].$$

Because $a < b < 0$, then $-A < -B < 0$, i.e. $A > B > 0$. Let us denote $d = A - B$, so $d$ is positive integer and $B = A - d$. Then the $n$th partial sum has now a form

$$s_n(a, b, 0) = s_n(-A, -B, 0) = \sum_{k=1}^{n}\left[\frac{1}{A(A-B)}\left(\frac{1}{k+A} - \frac{1}{k+B}\right) - \frac{1}{AB}\left(\frac{1}{k+B} - \frac{1}{k}\right)\right]$$
$$= \frac{1}{A(A-B)}\sum_{k=1}^{n}\left(\frac{1}{k+A} - \frac{1}{k+B}\right) - \frac{1}{AB}\sum_{k=1}^{n}\left(\frac{1}{k+B} - \frac{1}{k}\right).$$

The first sum can be written in the form

$$\sum_{k=1}^{n}\left(\frac{1}{k+A} - \frac{1}{k+B}\right) = \sum_{k=1}^{n}\left(\frac{1}{A+k} - \frac{1}{A+k-d}\right) = \left(\frac{1}{A+1} - \frac{1}{A+1-d}\right)$$
$$+\left(\frac{1}{A+2} - \frac{1}{A+2-d}\right) + \cdots + \left(\frac{1}{A+d} - \frac{1}{A}\right) + \left(\frac{1}{A+d+1} - \frac{1}{A+1}\right) + \cdots$$
$$\cdots + \left(\frac{1}{A+n-d-1} - \frac{1}{A+n-2d-1}\right) + \left(\frac{1}{A+n-d} - \frac{1}{A+n-2d}\right)$$
$$+\left(\frac{1}{A+n-d+1} - \frac{1}{A+n-2d+1}\right) + \cdots + \left(\frac{1}{A+n-1} - \frac{1}{A+n-d-1}\right)$$
$$+\left(\frac{1}{A+n} - \frac{1}{A+n-d}\right).$$

After cancellation of all terms we get this sum in the form

$$\sum_{k=1}^{n}\left(\frac{1}{k+A} - \frac{1}{k+B}\right) = -\frac{1}{A+1-d} - \frac{1}{A+2-d} - \cdots - \frac{1}{A}$$
$$+ \underbrace{\frac{1}{A+n-d+1} + \cdots + \frac{1}{A+n-1} + \frac{1}{A+n}}_{S(A,n,d)}$$
$$= -\left(\frac{1}{A+1-d} + \frac{1}{A+2-d} + \cdots + \frac{1}{A}\right) + S(A, n, d)$$
$$= -\left[\left(\frac{1}{1} + \frac{1}{2} + \cdots + \frac{1}{A}\right) - \left(\frac{1}{1} + \frac{1}{2} + \cdots + \frac{1}{A-d}\right)\right] + S(A, n, d)$$
$$= H(A-d) - H(A) + S(A, n, d) = H(B) - H(A) + S(A, n, d).$$

The second sum can be analogously written in the form

$$\sum_{k=1}^{n}\left(\frac{1}{k+B} - \frac{1}{k}\right) = \sum_{k=1}^{n}\left(\frac{1}{B+k} - \frac{1}{k}\right) = \left(\frac{1}{B+1} - \frac{1}{1}\right) + \left(\frac{1}{B+2} - \frac{1}{2}\right) + \cdots$$
$$\cdots + \left(\frac{1}{2B} - \frac{1}{B}\right) + \left(\frac{1}{2B+1} - \frac{1}{B+1}\right) + \cdots + \left(\frac{1}{n} - \frac{1}{n-B}\right)$$
$$+\left(\frac{1}{n+1} - \frac{1}{n-B+1}\right) + \cdots + \left(\frac{1}{B+n-1} - \frac{1}{n-1}\right) + \left(\frac{1}{B+n} - \frac{1}{n}\right).$$

After cancellation of all terms we get this sum in the form

$$\sum_{k=1}^{n}\left(\frac{1}{k+B}-\frac{1}{k}\right)=-\frac{1}{1}-\frac{1}{2}-\cdots-\frac{1}{B}+\underbrace{\frac{1}{n+1}+\cdots+\frac{1}{B+n-1}+\frac{1}{B+n}}_{S(B,n)}$$

$$=-\left(\frac{1}{1}+\frac{1}{2}+\cdots+\frac{1}{B}\right)+S(B,n)=S(B,n)-H(B).$$

Altogether we have

$$s_n(-A,-B,0)=\frac{1}{A(A-B)}[H(B)-H(A)+S(A,n,d)]-\frac{1}{AB}[S(B,n)-H(B)].$$

Considering (1.6), we have

$$\lim_{n\to\infty}S(A,n,d)=0 \quad\text{and}\quad \lim_{n\to\infty}S(B,n)=0,$$

so we get

$$s(a,b,0)=\lim_{n\to\infty}s_n(-A,-B,0)$$
$$=\frac{H(B)-H(A)}{A(A-B)}+\frac{H(B)}{AB}=\frac{H(A)}{A(B-A)}-\frac{H(B)}{B(B-A)}.$$

In this way, we derived the following statement:

**Theorem 1.4.** The series $\sum_{k=1}^{\infty}\frac{1}{k(k-a)(k-b)}$, where $a<b<0$ are negative integers, has the sum

$$s(a,b,0)=\frac{H(-a)}{a(b-a)}-\frac{H(-b)}{b(b-a)}, \tag{1.15}$$

where $H(n)$ is the $n$th harmonic number.

Let us note, for example, that

$$s(-4,-2,0)=-\frac{H(4)}{8}+\frac{H(2)}{4}=-\frac{1}{8}\cdot\frac{25}{12}+\frac{1}{4}\cdot\frac{3}{2}=\frac{3}{8}-\frac{25}{96}=\frac{36-25}{96}=\frac{11}{96}$$
$$=0.114\,58\bar{3},$$

as was calculated in example 1.4.

### 1.5.2 The case of one negative, one zero and one positive integer root

Let us consider the series of reciprocals of the cubic polynomials with one negative integer root $a$, one zero root and one positive integer root $b$, i.e. the series

$$\sum_{\substack{k=1\\k\neq b}}^{\infty}\frac{1}{k(k-a)(k-b)} \tag{1.16}$$

and let us determine its sum $s(a,0,b)$.

**Example 1.5.** Using the $n$th partial sum calculate the sum $s(-2, 0, 4)$ of the series

$$\sum_{k=1 k\neq 4}^{\infty} \frac{1}{k(k+2)(k-4)}.$$

By means of the result of the Maple 2020 command

```
> convert(1/(k*(k+2)*(k-4)),parfrac);
```

$$\frac{1}{12(k+2)} - \frac{1}{8k} + \frac{1}{24(k-4)}$$

we get the partial fraction decomposition of the $k$th term

$$a_k = \frac{1}{k(k+2)(k-4)}, \qquad k \neq 4,$$

in the form

$$a_k = \frac{1}{12(k+2)} - \frac{1}{8k} + \frac{1}{24(k-4)} = \frac{1}{24}\left(\frac{2}{k+2} - \frac{3}{k} + \frac{1}{k-4}\right).$$

The sum in parentheses we express in the following form:

$$a_k = \frac{1}{24}\left(\frac{2}{k+2} - \frac{2+1}{k} + \frac{1}{k-4}\right) = \frac{2}{24}\left(\frac{1}{k+2} - \frac{1}{k}\right) - \frac{1}{24}\left(\frac{1}{k} - \frac{1}{k-4}\right),$$

i.e.

$$a_k = \frac{1}{12}\left(\frac{1}{k+2} - \frac{1}{k}\right) - \frac{1}{24}\left(\frac{1}{k} - \frac{1}{k-4}\right).$$

So the $n$th partial sum $s_n(-2, 0, 4) = a_1 + a_2 + a_3 + a_5 + \cdots + a_n$ has the form

$$s_n(-2, 0, 4) = \frac{1}{12}\sum_{\substack{k=1 \\ k\neq 4}}^{\infty}\left(\frac{1}{k+2} - \frac{1}{k}\right) - \frac{1}{24}\sum_{\substack{k=1 \\ k\neq 4}}^{\infty}\left(\frac{1}{k} - \frac{1}{k-4}\right) = \frac{1}{12}s_n(-2, 0)$$
$$- \frac{1}{24}s_n(0, 4),$$

where

$$s_n(-2, 0) = \left(\frac{1}{3} - \frac{1}{1}\right) + \left(\frac{1}{4} - \frac{1}{2}\right) + \left(\frac{1}{5} - \frac{1}{3}\right) + \left(\frac{1}{7} - \frac{1}{5}\right) + \left(\frac{1}{8} - \frac{1}{6}\right) + \left(\frac{1}{9} - \frac{1}{7}\right) + \cdots$$
$$\cdots + \left(\frac{1}{n-1} - \frac{1}{n-3}\right) + \left(\frac{1}{n} - \frac{1}{n-2}\right) + \left(\frac{1}{n+1} - \frac{1}{n-1}\right) + \left(\frac{1}{n+2} - \frac{1}{n}\right)$$
$$= -\left(\frac{1}{1} + \frac{1}{2}\right) + \frac{1}{4} - \frac{1}{6} + \left(\frac{1}{n+1} + \frac{1}{n+2}\right)$$

and

$$\begin{aligned}s_n(0,4) &= \left(\frac{1}{1}+\frac{1}{3}\right)+\left(\frac{1}{2}+\frac{1}{2}\right)+\left(\frac{1}{3}+\frac{1}{1}\right)+\left(\frac{1}{5}-\frac{1}{1}\right)+\left(\frac{1}{6}-\frac{1}{2}\right)+\left(\frac{1}{7}-\frac{1}{3}\right)\\ &\quad+\left(\frac{1}{8}-\frac{1}{4}\right)+\left(\frac{1}{9}-\frac{1}{5}\right)+\cdots+\left(\frac{1}{n-5}-\frac{1}{n-9}\right)+\left(\frac{1}{n-4}-\frac{1}{n-8}\right)\\ &\quad\cdots+\left(\frac{1}{n-3}-\frac{1}{n-7}\right)+\left(\frac{1}{n-2}-\frac{1}{n-6}\right)+\left(\frac{1}{n-1}-\frac{1}{n-5}\right)+\left(\frac{1}{n}-\frac{1}{n-4}\right)\\ &= 2\left(\frac{1}{1}+\frac{1}{2}+\frac{1}{3}\right)-\left(\frac{1}{1}+\frac{1}{2}+\frac{1}{3}+\frac{1}{4}\right)+\left(\frac{1}{n-3}+\frac{1}{n-2}+\frac{1}{n-1}+\frac{1}{n}\right).\end{aligned}$$

Therefore, we have

$$\begin{aligned}s_n(-2,0,4) &= \frac{1}{12}\left[-\left(\frac{1}{1}+\frac{1}{2}\right)+\frac{1}{4}-\frac{1}{6}+\left(\frac{1}{n+1}+\frac{1}{n+2}\right)\right]\\ &\quad-\frac{1}{24}\left[2\left(\frac{1}{1}+\frac{1}{2}+\frac{1}{3}\right)-\left(\frac{1}{1}+\frac{1}{2}+\frac{1}{3}+\frac{1}{4}\right)\right.\\ &\quad\left.+\left(\frac{1}{n-3}+\frac{1}{n-2}+\frac{1}{n-1}+\frac{1}{n}\right)\right].\end{aligned}$$

Considering (1.6), we get

$$\begin{aligned}s(-2,0,4) &= \frac{1}{12}\left[-\left(\frac{1}{1}+\frac{1}{2}\right)-\left(\frac{1}{6}-\frac{1}{4}\right)\right]-\frac{1}{24}\left[\left(\frac{1}{1}+\frac{1}{2}+\frac{1}{3}\right)-\frac{1}{4}\right]\\ &= \frac{1}{12}\left[-H(2)+\frac{1}{12}\right]-\frac{1}{24}\left[H(3)-\frac{1}{4}\right]=\frac{1}{12}\left(-\frac{3}{2}+\frac{1}{12}\right)-\frac{1}{24}\left(\frac{11}{6}-\frac{1}{4}\right)\\ &= \frac{1}{12}\left(-\frac{17}{12}\right)-\frac{1}{24}\cdot\frac{19}{12}=-\frac{53}{288}=-0.184\,02\bar{7}.\end{aligned}$$

The sum $s(-2, 0, 4)$ we can also compute by means of Maple 2020 this way:

```
> evalf(sum(1/(k*(k+2)*(k-4)),k=1..3)+sum(1/(k*(k+2)*(k-4)),
  k=5..infinity),9);
```

$$-0.184027778$$

Before we derive the formula for the sum $s(a, 0, b)$, where $a$ is negative integer and $b$ is positive integer, we present three following lemmas, that have been proved in [7].

**Lemma 1.1.** Let $a < 0 < b$ be integers. Then a fraction $\frac{1}{k(k-a)(k-b)}$ can be rewritten in the form

$$\frac{1}{k(k-a)(k-b)}=\frac{1}{ab(a-b)}\left(\frac{b}{k-a}-\frac{b-a}{k}-\frac{a}{k-b}\right)$$

i.e. in the form

$$\frac{1}{k(k-a)(k-b)} = \frac{1}{a(a-b)}\left(\frac{1}{k-a} - \frac{1}{k}\right) + \frac{1}{b(a-b)}\left(\frac{1}{k} - \frac{1}{k-b}\right). \tag{1.17}$$

**Lemma 1.2.** Let $a < 0 < b$ be integers. Then it holds

$$\sum_{\substack{k=1 \\ k\neq b}}^{\infty}\left(\frac{1}{k-a} - \frac{1}{k}\right) = \frac{1}{b} - \frac{1}{b-a} - H(-a), \tag{1.18}$$

where $H(n)$ is the $n$th harmonic number.

**Lemma 1.3.** Let $a < 0 < b$ be integers. Then it holds

$$\sum_{\substack{k=1 \\ k\neq b}}^{\infty}\left(\frac{1}{k} - \frac{1}{k-b}\right) = H(b-1) - \frac{1}{b}, \tag{1.19}$$

where $H(n)$ is the $n$th harmonic number.

Now, we determine for integers $a < 0 < b$ the formula for the sum $s(a, 0, b)$ of the series (1.16) i.e. the series

$$\sum_{\substack{k=1 \\ k\neq b}}^{\infty}\frac{1}{k(k-a)(k-b)}.$$

According to (1.17) we can take the $k$th term $a_k$ of this series and write in the form

$$a_k = \frac{1}{a(a-b)}\left(\frac{1}{k-a} - \frac{1}{k}\right) + \frac{1}{b(a-b)}\left(\frac{1}{k} - \frac{1}{k-b}\right),$$

so the sum $s(a, 0, b)$ is

$$s(a, 0, b) = \frac{1}{a(a-b)}\sum_{\substack{k=1 \\ k\neq b}}^{\infty}\left(\frac{1}{k-a} - \frac{1}{k}\right) + \frac{1}{b(a-b)}\sum_{\substack{k=1 \\ k\neq b}}^{\infty}\left(\frac{1}{k} - \frac{1}{k-b}\right).$$

According to (1.18) and (1.19) we have

$$\begin{aligned} s(a, 0, b) &= \frac{1}{a(a-b)}\left[\frac{1}{b} - \frac{1}{b-a} - H(-a)\right] + \frac{1}{b(a-b)}\left[H(b-1) - \frac{1}{b}\right] \\ &= \frac{1}{a(a-b)}\left[\frac{a}{b(a-b)} - H(-a)\right] + \frac{1}{b(a-b)}\left[H(b-1) - \frac{1}{b}\right] \\ &= \frac{H(-a)}{a(b-a)} + \frac{1}{b(b-a)^2} - \frac{H(b-1)}{b(b-a)} + \frac{1}{b^2(b-a)} \\ &= \frac{H(-a)}{a(b-a)} - \frac{H(b-1)}{b(b-a)} + \frac{2b-a}{b^2(b-a)^2}. \end{aligned}$$

We derived the following statement:

**Theorem 1.5.** The series $\sum_{\substack{k=1\\k\neq b}}^{\infty} \frac{1}{k(k-a)(k-b)}$, where $a < 0 < b$ are integers, has the sum

$$s(a, 0, b) = \frac{H(-a)}{a(b-a)} - \frac{H(b-1)}{b(b-a)} + \frac{2b-a}{b^2(b-a)^2}, \tag{1.20}$$

where $H(n)$ is the $n$th harmonic number.

Let us note, for example, that

$$s(-2, 0, 4) = \frac{H(2)}{-12} - \frac{H(3)}{24} + \frac{10}{16 \cdot 36},$$

so

$$\begin{aligned} s(-2, 0, 4) &= -\frac{1}{12} \cdot \frac{3}{2} - \frac{1}{24} \cdot \frac{11}{6} + \frac{5}{24 \cdot 12} = \frac{-36 - 22 + 5}{24 \cdot 12} = -\frac{53}{288} \\ &= -0.184\,02\bar{7}, \end{aligned}$$

as was calculated in example 1.5.

### 1.5.3 The case of two different positive integer roots and one zero root

Let us consider the series of reciprocals of the reduced cubic polynomials with one zero and two different positive integer roots $0 < a < b$, i.e. the series

$$\sum_{\substack{k=1\\k\neq a,b}}^{\infty} \frac{1}{k(k-a)(k-b)}, \tag{1.21}$$

and let us determine its sum $s(0, a, b)$.

**Example 1.6.** Using the $n$th partial sum calculate the sum $s(0, 2, 4)$ of the series

$$\sum_{k=1 k\neq 2,4}^{\infty} \frac{1}{k(k-2)(k-4)}.$$

By means of the result of the Maple 2020 command

```
> convert(1/(k*(k-2)*(k-4)),parfrac);
```

$$\frac{1}{8k} - \frac{1}{4(k-2)} + \frac{1}{8(k-4)}$$

we get the partial fraction decomposition of the $k$th term

$$a_k = \frac{1}{k(k-2)(k-4)}, \qquad k \neq 2, 4,$$

in the form

$$a_k = \frac{1}{8k} - \frac{1}{4(k-2)} + \frac{1}{8(k-4)} = \frac{1}{8}\left(\frac{1}{k} - \frac{2}{k-2} + \frac{1}{k-4}\right).$$

The sum in parentheses we express in the following form:

$$a_k = \frac{1}{8}\left(\frac{1}{k} - \frac{1+1}{k-2} + \frac{1}{k-4}\right) = \frac{1}{8}\left(\frac{1}{k} - \frac{1}{k-2}\right) - \frac{1}{8}\left(\frac{1}{k-2} - \frac{1}{k-4}\right).$$

So the $n$th partial sum $s_n(0, 2, 4) = a_1 + a_3 + a_5 + a_6 + \cdots + a_n$ has the form

$$s_n(0, 2, 4) = \frac{1}{8}\sum_{\substack{k=1\\k\neq 2,4}}^{\infty}\left(\frac{1}{k} - \frac{1}{k-2}\right) - \frac{1}{8}\sum_{\substack{k=1\\k\neq 2,4}}^{\infty}\left(\frac{1}{k-2} - \frac{1}{k-4}\right)$$

$$= \frac{1}{8}[s_n(0, 2) - s_n(2, 4)],$$

where

$$s_n(0, 2) = \left(\frac{1}{1} + \frac{1}{1}\right) + \left(\frac{1}{3} - \frac{1}{1}\right) + \left(\frac{1}{5} - \frac{1}{3}\right) + \left(\frac{1}{6} - \frac{1}{4}\right) + \left(\frac{1}{7} - \frac{1}{5}\right) + \left(\frac{1}{8} - \frac{1}{6}\right) + \cdots$$

$$\cdots + \left(\frac{1}{n-3} - \frac{1}{n-5}\right) + \left(\frac{1}{n-2} - \frac{1}{n-4}\right) + \left(\frac{1}{n-1} - \frac{1}{n-3}\right) + \left(\frac{1}{n} - \frac{1}{n-2}\right)$$

$$= 2\cdot\frac{1}{1} - \frac{1}{1} - \frac{1}{4} + \frac{1}{n-1} + \frac{1}{n} = \frac{1}{1} - \frac{1}{4} + \frac{1}{n-1} + \frac{1}{n}$$

and

$$s_n(2, 4) = \left(-\frac{1}{1} + \frac{1}{3}\right) + \left(\frac{1}{1} + \frac{1}{1}\right) + \left(\frac{1}{3} - \frac{1}{1}\right) + \left(\frac{1}{4} - \frac{1}{2}\right) + \left(\frac{1}{5} - \frac{1}{3}\right) + \left(\frac{1}{6} - \frac{1}{4}\right)$$

$$+ \left(\frac{1}{7} - \frac{1}{5}\right) + \left(\frac{1}{8} - \frac{1}{6}\right) + \cdots + \left(\frac{1}{n-7} - \frac{1}{n-9}\right) + \left(\frac{1}{n-6} - \frac{1}{n-8}\right)$$

$$\cdots + \left(\frac{1}{n-5} - \frac{1}{n-7}\right) + \left(\frac{1}{n-4} - \frac{1}{n-6}\right) + \left(\frac{1}{n-3} - \frac{1}{n-5}\right) + \left(\frac{1}{n-2} - \frac{1}{n-4}\right)$$

$$= 2\left(-\frac{1}{1} + \frac{1}{3}\right) + 2\cdot\frac{1}{1} - \left(\frac{1}{2} + \frac{1}{3}\right) + \frac{1}{n-3} + \frac{1}{n-2} = \frac{2}{3} - \frac{5}{6} + \frac{1}{n-3} + \frac{1}{n-2}.$$

Therefore, we have

$$s_n(0, 2, 4) = \frac{1}{8}\left[\frac{1}{1} - \frac{1}{4} + \frac{1}{n-1} + \frac{1}{n} - \left(\frac{2}{3} - \frac{5}{6} + \frac{1}{n-3} + \frac{1}{n-2}\right)\right].$$

Considering (1.6), we get

$$s(0, 2, 4) = \frac{1}{8}\left(\frac{1}{1} - \frac{1}{4} - \frac{2}{3} + \frac{5}{6}\right) = \frac{1}{8}\cdot\frac{12 - 3 - 8 + 10}{12} = \frac{11}{96} = 0.114\,58\bar{3}.$$

The sum $s(0, 2, 4)$ we can also compute by means of Maple 2020 this way:

```
> evalf(1/(1*(1-2)*(1-4))+1/(3*(3-2)*(3-4))+sum(1/(k*(k-2)*(k-4)),
  k=5..infinity),9);
```

$$0.114583333$$

Before we derive the formula for the sum $s(0, a, b)$, where $0 < a < b$ are positive integers, we present three following lemmas, that have been proved in [8].

**Lemma 1.4.** Let $0 < a < b$ be positive integers. Then a fraction $\frac{1}{k(k-a)(k-b)}$, $k \neq 0, a, b$, can be rewritten in the form

$$\frac{1}{ab(b-a)}\left(\frac{b-a}{k} - \frac{b}{k-a} - \frac{a}{k-b}\right) = \frac{1}{ab(b-a)}\left(\frac{b-a}{k} - \frac{(b-a)+a}{k-a} - \frac{a}{k-b}\right),$$

i.e. in the form

$$\frac{1}{k(k-a)(k-b)} = \frac{1}{ab}\left(\frac{1}{k} - \frac{1}{k-a}\right) - \frac{1}{b(b-a)}\left(\frac{1}{k-a} - \frac{1}{k-b}\right). \tag{1.22}$$

**Lemma 1.5.** Let $0 < a < b$ be positive integers. Then it holds

$$\sum_{\substack{k=1 \\ k\neq a,b}}^{\infty}\left(\frac{1}{k} - \frac{1}{k-a}\right) = H(a-1) - \frac{1}{a} - \frac{1}{b} + \frac{1}{b-a}, \tag{1.23}$$

where $H(n)$ is the $n$th harmonic number.

**Lemma 1.6.** Let $0 < a < b$ be positive integers. Then it holds

$$\sum_{\substack{k=1 \\ k\neq a,b}}^{\infty}\left(\frac{1}{k-a} - \frac{1}{k-b}\right) = -H(a-1) + H(b-1) - \frac{2}{b-a}, \tag{1.24}$$

where $H(n)$ is the $n$th harmonic number.

Now, we determine for integers $0 < a < b$ the formula for the sum $s(0, a, b)$ of the series (1.21) i.e. the series

$$\sum_{\substack{k=1 \\ k\neq a,b}}^{\infty}\frac{1}{k(k-a)(k-b)}.$$

According to (1.22) we can the $k$th term $a_k$ of this series write in the form

$$a_k = \frac{1}{ab}\left(\frac{1}{k} - \frac{1}{k-a}\right) - \frac{1}{b(b-a)}\left(\frac{1}{k-a} - \frac{1}{k-b}\right),$$

so the sum $s(0, a, b)$ is

$$s(0, a, b) = \frac{1}{ab}\sum_{\substack{k=1\\k\neq a,b}}^{\infty}\left(\frac{1}{k} - \frac{1}{k-a}\right) - \frac{1}{b(b-a)}\sum_{\substack{k=1\\k\neq a,b}}^{\infty}\left(\frac{1}{k-a} - \frac{1}{k-b}\right).$$

According to (1.23) and (1.24) we have

$$\begin{aligned}
s(0, a, b) &= \frac{1}{ab}\left[H(a-1) - \frac{1}{a} - \frac{1}{b} + \frac{1}{b-a}\right] - \frac{1}{b(b-a)}\left[H(b-1) - H(a-1) - \frac{2}{b-a}\right]\\
&= \frac{1}{ab}\left[H(a-1) + \frac{a^2+ab-b^2}{ab(b-a)}\right] - \frac{1}{b(b-a)}\left[H(b-1) - H(a-1) - \frac{2}{b-a}\right]\\
&= \frac{H(a-1)}{ab} + \frac{a^2+ab-b^2}{a^2b^2(b-a)} - \frac{H(b-1)-H(a-1)}{b(b-a)} + \frac{2}{b(b-a)^2}\\
&= \frac{H(a-1)}{ab} - \frac{H(b-1)-H(a-1)}{b(b-a)} - \frac{a^3-2a^2b-2ab^2+b^3}{a^2b^2(b-a)^2}\\
&= \frac{H(a-1)}{a(b-a)} - \frac{H(b-1)}{b(b-a)} - \frac{a^3-2a^2b-2ab^2+b^3}{a^2b^2(b-a)^2}.
\end{aligned}$$

We derived the following statement:

**Theorem 1.6.** The series $\sum_{\substack{k=1\\k\neq a,\,b}}^{\infty} \frac{1}{k(k-a)(k-b)}$, where $0 < a < b$ are positive integers, has the sum

$$s(0, a, b) = \frac{H(a-1)}{a(b-a)} - \frac{H(b-1)}{b(b-a)} - \frac{a^3-2a^2b-2ab^2+b^3}{a^2b^2(b-a)^2}, \tag{1.25}$$

where $H(n)$ is the $n$th harmonic number.

Let us note, for example, that

$$s(0, 2, 4) = \frac{H(1)}{4} - \frac{H(3)}{8} - \frac{8-32-64+64}{4\cdot 16\cdot 4},$$

so

$$s(0, 2, 4) = \frac{1}{4} - \frac{1}{8}\cdot\frac{11}{6} + \frac{24}{8\cdot 32} = \frac{1}{4} - \frac{11}{48} + \frac{3}{32} = \frac{24-22+9}{96} = \frac{11}{96} = 0.114\,58\bar{3},$$

as was calculated in example 1.6.

Note that the equality of the results $s(-4, -2, 0) = 0.114\,58\bar{3}$ and $s(0, 2, 4) = 0.114\,58\bar{3}$ from examples 1.4 and 1.6 is not random, and using theorems 1.4 and 1.6, let us derive a corollary which states for which integers $a$, $b$, $a < b$, the equation $s(-b, -a, 0) = s(0, a, b)$ holds.

**Corollary 1.** The equation

$$s(-b, -a, 0) = s(0, a, b),$$

where $0 < a < b$ are positive integers, holds if and only if $b = 2a$.

Proof. By theorem 1.4 we have

$$s(-b, -a, 0) = \frac{H(b)}{b(a-b)} - \frac{H(a)}{a(a-b)} = \frac{H(a)}{a(b-a)} - \frac{H(b)}{b(b-a)}$$

and by theorem 1.6 it holds

$$s(0, a, b) = \frac{H(a-1)}{a(b-a)} - \frac{H(b-1)}{b(b-a)} - \frac{a^3 - 2a^2b - 2ab^2 + b^3}{a^2b^2(b-a)^2}.$$

By comparing the two right sides, we obtain an equation

$$\frac{H(a)}{a(b-a)} - \frac{H(b)}{b(b-a)} = \frac{H(a-1)}{a(b-a)} - \frac{H(b-1)}{b(b-a)} - \frac{a^3 - 2a^2b - 2ab^2 + b^3}{a^2b^2(b-a)^2}.$$

Multiplied by $a^2b^2(b-a)^2$, we obtain the equation

$$\begin{aligned} ab^2(b-a)H(a) - a^2b(b-a)H(b) = ab^2(b-a)H(a-1) - a^2b(b-a)H(b-1) \\ - a^3 + 2a^2b + 2ab^2 - b^3. \end{aligned}$$

With further modifications, assuming that $a \neq -b$, we gradually obtain the quadratic equation:

$$\begin{aligned} ab^2(b-a)[H(a) - H(a-1)] - a^2b(b-a)[H(b) - H(b-1)] &= -a^3 + 2a^2b + 2ab^2 - b^3 \\ ab^2(b-a) \cdot \frac{1}{a} - a^2b(b-a) \cdot \frac{1}{b} &= -a^3 + 2a^2b + 2ab^2 - b^3 \\ b^2(b-a) - a^2(b-a) &= -a^3 + 2a^2b + 2ab^2 - b^3 \\ b^3 - ab^2 - a^2b + a^3 &= -a^3 + 2a^2b + 2ab^2 - b^3 \\ 2a^3 + 2b^3 &= 3a^2b + 3ab^2 \\ 2(a+b)(a^2 - ab + b^2) &= 3ab(a+b) \\ 2a^2 - 2ab + 2b^2 &= 3ab \\ 2b^2 - 5ab + 2a^2 &= 0. \end{aligned}$$

If we look at the last equation as a quadratic equation with unknown $b$ and parameter $a$, then its roots have the form

$$b_{1,2} = \frac{5a \pm \sqrt{25a^2 - 16a^2}}{4} = \frac{5a \pm 3a}{2} = \left\langle \begin{array}{l} \left(\frac{a}{2}, \quad \text{but } a \not< \frac{a}{2}\right) \\ 2a. \end{array} \right.$$

We proved that the equality $s(-b, -a, 0) = s(0, a, b)$ holds for $b = 2a$. Note that in the second table of section 1.6, in addition to the case $s(-4, -2, 0) = s(0, 2, 4) \doteq 0.1146$, the case $s(-2, -1, 0) = s(0, 1, 2) = 0.2500$ is also given.

### 1.5.4 The case of one zero root and two arbitrary integer roots

Because formulas (1.15), (1.20) and (1.25), i.e. the formulas

$$s(a, b, 0) = \frac{H(-a)}{a(b-a)} - \frac{H(-b)}{b(b-a)},$$
$$s(a, 0, b) = \frac{H(-a)}{a(b-a)} - \frac{H(b-1)}{b(b-a)} + \frac{2b-a}{b^2(b-a)^2},$$
$$s(0, a, b) = \frac{H(a-1)}{a(b-a)} - \frac{H(b-1)}{b(b-a)} - \frac{a^3 - 2a^2b - 2ab^2 + b^3}{a^2b^2(b-a)^2},$$

are similar, it is likely to express them by the only general formula using the function of the non-zero integer $x$, we used previously for the summary general formula (1.10) for the case of double zero root and one arbitrary integer root.

This summary general formula is given in the following theorem:

**Theorem 1.7.** The series $\sum_{\substack{k=1\\ k\neq\alpha,\,\beta}}^{\infty} \frac{1}{k(k-\alpha)(k-\beta)}$ has for non-zero integers $\alpha$, $\beta$, $\alpha < \beta$, the sum in the form

$$s(0, \alpha, \beta) = \frac{H[|\alpha| - \sigma(\alpha)]}{\alpha(\beta-\alpha)} - \frac{H[|\beta| - \sigma(\beta)]}{\beta(\beta-\alpha)} + \frac{\sigma(\alpha)(2\alpha-\beta)}{\alpha^2(\beta-\alpha)^2} + \frac{\sigma(\beta)(2\beta-\alpha)}{\beta^2(\beta-\alpha)^2}, \tag{1.26}$$

where $H(n)$ is the $n$th harmonic number.

Proof. The formulas (1.15), (1.20) and (1.25) can be rewritten in the forms

$$\begin{aligned} s(a, b, 0) &= \frac{H(-a+0)}{a(b-a)} - \frac{H(-b+0)}{b(b-a)}, && a < b < 0,\\ s(a, 0, b) &= \frac{H(-a+0)}{a(b-a)} - \frac{H(b-1)}{b(b-a)} + \frac{2b-a}{b^2(b-a)^2}, && a < 0 < b,\\ s(0, a, b) &= \frac{H(a-1)}{a(b-a)} - \frac{H(b-1)}{b(b-a)} - \frac{a^2(a-2b)}{a^2b^2(b-a)^2} - \frac{b^2(b-2a)}{a^2b^2(b-a)^2}, && 0 < a < b, \end{aligned} \tag{1.27}$$

i.e. in the forms

$$\begin{aligned} s(a, b, 0) &= \frac{H[|a| - \sigma(a)]}{a(b-a)} - \frac{H[|b| - \sigma(b)]}{b(b-a)} + 0 + 0, && a < b < 0,\\ s(a, 0, b) &= \frac{H[|a| - \sigma(a)]}{a(b-a)} - \frac{H[|b| - \sigma(b)]}{b(b-a)} + 0 + \frac{2b-a}{b^2(b-a)^2}, && a < 0 < b,\\ s(0, a, b) &= \frac{H[|a| - \sigma(a)]}{a(b-a)} - \frac{H[|b| - \sigma(b)]}{b(b-a)} + \frac{2a-b}{a^2(b-a)^2} + \frac{2b-a}{b^2(b-a)^2}, && 0 < a < b. \end{aligned} \tag{1.28}$$

We can also write the third and fourth summands of these three formulas using the sigma function, which gives the formula for arbitrary non-zero integers $\alpha$, $\beta$, $\alpha < \beta$:

$$s(0, \alpha, \beta) = \frac{H[|\alpha| - \sigma(\alpha)]}{\alpha(\beta-\alpha)} - \frac{H[|\beta| - \sigma(\beta)]}{\beta(\beta-\alpha)} + \frac{\sigma(\alpha)(2\alpha-\beta)}{\alpha^2(\beta-\alpha)^2} + \frac{\sigma(\beta)(2\beta-\alpha)}{\beta^2(\beta-\alpha)^2}.$$

### 1.5.5 Three special cases of one zero and two integer roots

Now, let us consider the series of reciprocals of the reduced cubic polynomials with one zero and double negative root, double positive root and roots with opposite signs, i.e. the series

$$\sum_{k=1}^{\infty}\frac{1}{k)k-a)^2}, \qquad a<0, \tag{1.29}$$

$$\sum_{\substack{k=1\\k\neq a}}^{\infty}\frac{1}{k(k-a)^2}, \qquad a>0, \tag{1.30}$$

$$\sum_{\substack{k=1\\k\neq a,b}}^{\infty}\frac{1}{k(k+a)(k-a)}, \qquad a>0, \tag{1.31}$$

and let us determine their sums $s(a, a, 0)$, $s(0, a, a)$ and $s(-a, 0, a)$.

**Example 1.7.** Using the $n$th partial sum calculate the sum $s(-4, -4, 0)$ of the series

$$\sum_{k=1}^{\infty}\frac{1}{k(k+4)^2}.$$

In [3] it was derived that the $k$th term $a_k$ of the series (1.29) has the form

$$a_k=\frac{1}{k(k-a)^2}=\frac{1}{a^2}\left(\frac{1}{k}-\frac{1}{k-a}\right)+\frac{1}{a}\cdot\frac{1}{(k-a)^2}. \tag{1.32}$$

According to equation (1.32), where $a=-4$, we have

$$a_k=\frac{1}{k(k+4)^2}=\frac{1}{16}\left(\frac{1}{k}-\frac{1}{k+4}\right)+\frac{1}{4}\cdot\frac{1}{(k+4)^2},$$

so the $n$th partial sum $s_n(-4, -4, 0)=a_1+a_2+\cdots+a_n$ has the form

$$\begin{aligned}s_n(-4,-4,0)=&-\frac{1}{4}\left[\frac{1}{5^2}+\frac{1}{6^2}+\cdots+\frac{1}{(n+4)^2}\right]+\frac{1}{16}\left[\left(\frac{1}{1}-\frac{1}{5}\right)+\left(\frac{1}{2}-\frac{1}{6}\right)+\left(\frac{1}{3}-\frac{1}{7}\right)\right.\\&+\left(\frac{1}{4}-\frac{1}{8}\right)+\left(\frac{1}{5}-\frac{1}{9}\right)+\cdots+\left(\frac{1}{n-4}-\frac{1}{n}\right)+\left(\frac{1}{n-3}-\frac{1}{n+1}\right)\\&\left.+\left(\frac{1}{n-2}-\frac{1}{n+2}\right)+\left(\frac{1}{n-1}-\frac{1}{n+3}\right)+\left(\frac{1}{n}-\frac{1}{n+4}\right)\right]\\=&-\frac{1}{4}\left[\frac{1}{5^2}+\frac{1}{6^2}+\cdots+\frac{1}{(n+4)^2}\right]\\&+\frac{1}{16}\left(\frac{1}{1}+\frac{1}{2}+\frac{1}{3}+\frac{1}{4}-\frac{1}{n+1}-\frac{1}{n+2}-\frac{1}{n+3}-\frac{1}{n+4}\right).\end{aligned}$$

Considering (1.6) and the fact that

$$\lim_{n\to\infty} H(n,\,2) = \lim_{n\to\infty}\sum_{k=1}^{n}\frac{1}{k^2} = \sum_{k=1}^{\infty}\frac{1}{k^2} = \frac{\pi^2}{6},$$

whence

$$\lim_{n\to\infty}\left[\frac{1}{5^2}+\frac{1}{6^2}+\cdots+\frac{1}{(n+4)^2}\right] = \lim_{n\to\infty}[H(n+4,\,2) - H(4,\,2)] = \frac{\pi^2}{6} - \frac{205}{144},$$

we get

$$\begin{aligned} s(-4,\,-4,\,0) &= \lim_{n\to\infty} s_n(-4,\,-4,\,0) = -\frac{1}{4}\left(\frac{\pi^2}{6} - \frac{205}{144}\right) + \frac{H(4)}{16} = \frac{205 - 24\pi^2}{576} + \frac{25}{192} \\ &= \frac{205 - 24\pi^2 + 75}{576} = \frac{280 - 24\pi^2}{576} = \frac{35 - 3\pi^2}{72} \doteq 0.074\,878. \end{aligned}$$

**Example 1.8.** Using the $n$th partial sum calculate the sum $s(0,\,4,\,4)$ of the series

$$\sum_{\substack{k=1\\k\neq 4}}^{\infty}\frac{1}{k(k-4)^2}.$$

According to equation (1.32), where $a = 4$, we have

$$a_k = \frac{1}{k(k-4)^2} = \frac{1}{16}\left(\frac{1}{k} - \frac{1}{k-4}\right) + \frac{1}{4}\cdot\frac{1}{(k-4)^2},$$

so the $n$th partial sum $s_n(0,\,4,\,4) = a_1 + a_2 + a_3 + a_5 + \cdots + a_n$ has the form

$$\begin{aligned} s_n(0,\,4,\,4) &= \frac{1}{4}\left[\frac{1}{3^2}+\frac{1}{2^2}+\frac{1}{1^2}+\frac{1}{1^2}+\frac{1}{2^2}+\frac{1}{3^2}+\frac{1}{4^2}+\cdots+\frac{1}{(n-4)^2}\right] \\ &\quad + \frac{1}{16}\left[\left(\frac{1}{1}+\frac{1}{3}\right)+\left(\frac{1}{2}+\frac{1}{2}\right)+\left(\frac{1}{3}+\frac{1}{1}\right)+\left(\frac{1}{5}-\frac{1}{1}\right)+\left(\frac{1}{6}-\frac{1}{2}\right)\right. \\ &\quad + \left(\frac{1}{7}-\frac{1}{3}\right)+\left(\frac{1}{8}-\frac{1}{4}\right)+\left(\frac{1}{9}-\frac{1}{5}\right)+\cdots+\left(\frac{1}{n-4}-\frac{1}{n-8}\right) \\ &\quad \left.+\left(\frac{1}{n-3}-\frac{1}{n-7}\right)+\left(\frac{1}{n-2}-\frac{1}{n-6}\right)+\left(\frac{1}{n-1}-\frac{1}{n-5}\right)+\left(\frac{1}{n}-\frac{1}{n-4}\right)\right] \\ &= \frac{1}{4}\left[\frac{49}{36}+\frac{1}{1^2}+\frac{1}{2^2}+\cdots+\frac{1}{(n-4)^2}\right] \\ &\quad + \frac{1}{16}\left(\frac{11}{3}-\frac{1}{1}-\frac{1}{2}-\frac{1}{3}-\frac{1}{4}+\frac{1}{n-3}+\frac{1}{n-2}+\frac{1}{n-1}+\frac{1}{n}\right). \end{aligned}$$

Considering (1.6) and the fact that

$$\lim_{n\to\infty} H(n, 2) = \lim_{n\to\infty} \sum_{k=1}^{n} \frac{1}{k^2} = \sum_{k=1}^{\infty} \frac{1}{k^2} = \frac{\pi^2}{6},$$

we get

$$s(0, 4, 4) = \lim_{n\to\infty} s_n(0, 4, 4) = \frac{1}{4}\left(\frac{49}{36} + \frac{\pi^2}{6}\right) + \frac{1}{16}\left(\frac{11}{3} - H(4)\right) = \frac{49 + 6\pi^2}{144}$$

$$+ \frac{1}{16}\left(\frac{11}{3} - \frac{25}{12}\right) = \frac{49 + 6\pi^2}{144} + \frac{44 - 25}{192} = \frac{196 + 24\pi^2 + 57}{576}$$

$$= \frac{253 + 24\pi^2}{576} \doteq 0.850\,470.$$

By generalizing these two examples, the formulas for the sum $s(a, a, 0)$ and for the sum $s(0, a, a)$ were derived in [3]. It was derived that for double negative integer root $a$ of the reduced cubic polynomial it holds

$$s(a, a, 0) = \frac{\pi^2}{6a} - \frac{H(-a, 2)}{a} + \frac{H(-a)}{a^2} \tag{1.33}$$

and for double positive integer root $a$ of the reduced cubic polynomial it holds

$$s(0, a, a) = \frac{\pi^2}{6a} + \frac{H(a-1, 2)}{a} + \frac{H(a-1)}{a^2} - \frac{1}{a^3}, \tag{1.34}$$

where $H(n)$ is the $n$th harmonic number and $H(n, 2)$ is the $n$th generalized harmonic number.

At the same time, a summary general formula was derived for any double non-zero integer root $\alpha$ of the reduced cubic polynomial. This general formula containing the sigma function has the form

$$s(0, \alpha, \alpha) = \frac{\pi^2}{6\alpha} + \frac{H[|\alpha| - \sigma(\alpha), 2]}{|\alpha|} + \frac{H[|\alpha| - \sigma(\alpha)]}{\alpha^2} - \frac{\sigma(\alpha)}{\alpha^3}. \tag{1.35}$$

As a special case of the formula (1.20) for the sum of the series of reciprocals of the reduced cubic polynomials with one negative integer root $a$, one zero and one positive integer root $b$ that has the form

$$s(a, 0, b) = \frac{H(-a)}{a(b-a)} - \frac{H(b-1)}{b(b-a)} + \frac{2b - a}{b^2(b-a)^2},$$

let us state the following formula for the sum of the series of reciprocals of the reduced cubic polynomials with one zero root and two integer roots with opposite signs. Substituting $a = -a$ and $b = a$ we obtain

$$s(-a, 0, a) = \frac{H(a)}{-a(a+a)} - \frac{H(a-1)}{a(a+a)} + \frac{2a + a}{a^2(a+a)^2},$$

whence we get the formula

$$s(-a, 0, a) = \frac{3}{4a^3} - \frac{H(a) + H(a-1)}{2a^2}. \tag{1.36}$$

Let us note, for example, that

$$s(-4, 0, 4) = \frac{3}{4 \cdot 4^3} - \frac{H(4) + H(3)}{2 \cdot 4^2} = \frac{3}{256} - \frac{1}{32}\left(\frac{25}{12} + \frac{11}{6}\right),$$

so

$$\begin{aligned} s(-4, 0, 4) &= \frac{3}{256} - \frac{1}{32} \cdot \frac{25+22}{12} = \frac{3}{16 \cdot 16} - \frac{47}{16 \cdot 24} = \frac{9-94}{768} \\ &= -\frac{85}{768} = -0.110\,677\,08\bar{3}. \end{aligned}$$

**Remark 1.2.** Let us note that derived formula (1.36) has the form

$$s(-a, 0, a) = \frac{\ell}{a^2} + \frac{m}{a^3}.$$

Now, we determine the unknown coefficients $\ell$, $m$ by means of the least squares method in Maple 2020. For positive integer $a$ we use the following statements and their outputs (Figure 1.3):

```
> with(CurveFitting):  with(plots):
> s:=x->3/(4*x*x*x) - (harmonic(x)+harmonic(x-1))/(2*x*x):
> data:=[seq([x,s(x)], x=1..50)]:
> L1:=LeastSquares(data,x,curve=l/(x*x)+m/(x*x*x)):  evalf(%);
```

$$\frac{2.423362314}{x^2} - \frac{2.171406694}{x^3}$$

```
> plot([data,L1], x=1..50, color=[blue,green],style=[point,line]);
```

We can write very roughly

$$s(-a, 0, a) \doteq \frac{12}{5a^2} - \frac{13}{6a^3}. \tag{1.37}$$

For this approximate formula, we used the convert commands

```
> convert(2.423362314,rational,3);
> convert(-2.171406694,rational,3);
```

in Maple 2020.

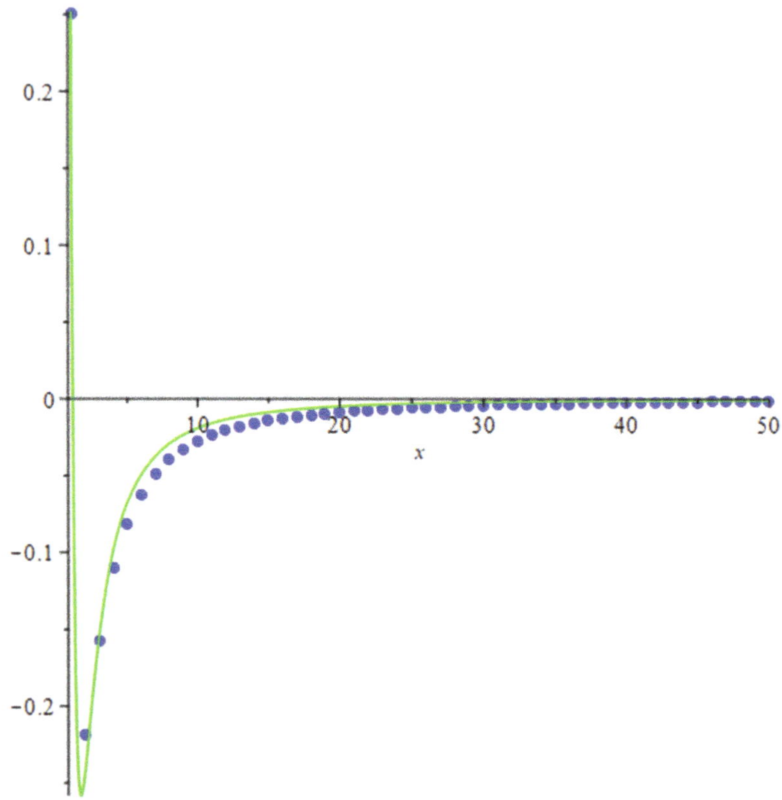


**Figure 1.3.** Dot chart of $s(-a, 0, a)$ values for integer $a > 0$ and approximation curve graph.

## 1.6 Some approximate values for seven types of the sums

In [3–5, 7, 8], the values of seven types of the sums $s(a, 0, 0)$ and $s(a, a, 0)$ for integer $a < 0$, $s(0, 0, a)$ and $s(0, a, a)$ for integer $a > 0$, and $s(a, b, 0)$, $s(a, 0, b)$ and $s(0, a, b)$ for non-zero integers $a < b$ were calculated using the procedures in computer algebra system Maple with an accuracy of $10^{-10}$.

The results and formulas derived in these articles were always numerically verified using approximate direct evaluation. For example, in [5] we used on the one hand an approximative direct evaluation of the sum

$$s(0, 0, \alpha, t) = \sum_{k=1 k\neq\alpha}^{t} \frac{1}{k^2(k - \alpha)}, \qquad \text{where } t = 500\,000,$$

using the basic programming language of Maple 2016, and on the other hand the formula (1.7) for evaluation of the sum $s(a, 0, 0)$ with negative integer $a$ and the formula (1.8) for evaluation of the sum $s(0, 0, a)$ with positive integer $a$.

We compared 20 pairs of these ways obtained sums $s(0, 0, \alpha, 500\,000)$ and the sums $s(a, 0, 0)$ and $s(0, 0, a)$ to verify the formulas (1.7) and (1.8). We used the simple procedure `rp300a` and the following two `for` statements:

```
> rp300a:=proc(a,t)
> local k,s02at,s02an,s02ap; s02at:=0;
> for k from 1 to t do
> if k<>a then s02at:=s02at+1/(k*k*(k-a))
> end if;
> end do;
> if a<0 then s02an:=-Pi*Pi/(6*a)-harmonic(-a)/(a*a);
> print("s(",a,")=",evalf[9](s02an));
> print("s(",a,t,")=",evalf[9](s02at),"d=",evalf[9](abs(s02at-s02an)))
> else s02ap:=-Pi*Pi/(6*a)+2/(a*a*a)-harmonic(a-1)/(a*a);
> print("s(",a,")=",evalf[9](s02ap));
> print("s(",a,t,")=",evalf[9](s02at),"d=",evalf[9](abs(s02at-s02ap)));
> end if;
> end proc:
```

```
> for i from -10 to -1 do
> rp300a(i,500000);
> end do;
```

```
> for i from 1 to 10 do
> rp300a(i,500000);
> end do;
```

Some of the values of the sums $s(a, 0, 0)$, $s(0, 0, a)$ and $s(a, b, 0)$, $s(a, a, 0)$, $s(a, 0, b)$, $s(0, a, b)$, $s(0, a, a)$ rounded to three decimals are stated in tables 1.3 and 1.4.

## 1.7 Conclusion

We dealt with the sums of the series of reciprocals of the cubic polynomials with three, two and one zero roots and with other integer roots.

In the introductory part of the chapter, the well-known case of triple zero root of the cubic polynomial was mentioned.

Further, the cases of a double zero root together with another simple negative and positive integer root were stated. Previously derived sub-formulas were

**Table 1.3.** Ten approximate values of the sums $s(a, 0, 0)$ and $s(0, 0, a)$ obtained by means of the formulas (1.7) and (1.8).

| $a$ | −5 | −4 | −3 | −2 | −1 |
|---|---|---|---|---|---|
| $s(a, 0, 0)$ | 0.238 | 0.281 | 0.345 | 0.448 | 0.645 |
| $a$ | 1 | 2 | 3 | 4 | 5 |
| $s(0, 0, a)$ | 0.355 | −0.823 | −0.641 | −0.495 | −0.396 |

**Table 1.4.** Fifty five approximate values of the sums $s(a, b, 0)$, $s(a, a, 0)$, $s(a, 0, b)$, $s(0, a, b)$ and $s(0, a, a)$ obtained by means of the formulas (1.15), (1.33), (1.20), (1.25) and (1.34).

| $a$–$b$ | $b = -5$ | $b = -4$ | $b = -3$ | $b = -2$ | $b = -1$ | $b = 1$ | $b = 2$ | $b = 3$ | $b = 4$ | $b = 5$ |
|---|---|---|---|---|---|---|---|---|---|---|
| $a = -5$ | 0.055 | 0.064 | 0.077 | 0.098 | 0.136 | 0.118 | −0.091 | −0.100 | −0.092 | −0.081 |
| $a = -4$ | × | 0.075 | 0.090 | 0.115 | 0.160 | 0.136 | −0.115 | −0.123 | −0.111 | −0.097 |
| $a = -3$ | × | × | 0.109 | 0.139 | 0.194 | 0.160 | −0.152 | −0.157 | −0.139 | −0.120 |
| $a = -2$ | × | × | × | 0.178 | 0.250 | 0.194 | −0.219 | −0.214 | −0.184 | −0.157 |
| $a = -1$ | × | × | × | × | 0.355 | 0.250 | −0.361 | −0.326 | −0.269 | −0.224 |
| $a = 1$ | × | × | × | × | × | 0.645 | 0.250 | −0.361 | −0.326 | −0.269 |
| $a = 2$ | × | × | × | × | × | × | 1.447 | 0.694 | 0.115 | 0.356 |
| $a = 3$ | × | × | × | × | × | × | × | 1.095 | 0.576 | 0.139 |
| $a = 4$ | × | × | × | × | × | × | × | × | 0.850 | 0.469 |
| $a = 5$ | × | × | × | × | × | × | × | × | × | 0.689 |

combined into one summary general formula for double zero root and another arbitrary integer root $\alpha$ using harmonic numbers $H(n)$ and the function $\sigma(n) = (\text{sgn}(n) + 1)/2$:

$$s(0, 0, \alpha) = -\frac{\pi^2}{6\alpha} - \frac{H[|\alpha| - \sigma(\alpha)]}{\alpha^2} + \frac{2\sigma(\alpha)}{\alpha^3}.$$

The next part of the chapter presented cases of a simple zero root and two different negative integer roots, two different positive integer roots and one negative and one positive integer root. Subsequently, the summary general formula was derived for one zero and two arbitrary integer roots $\alpha$, $\beta$, $\alpha < \beta$, using harmonic numbers $H(n)$ and the function $\sigma(n)$:

$$s(0, \alpha, \beta) = \frac{H[|\alpha| - \sigma(\alpha)]}{\alpha(\beta - \alpha)} - \frac{H[|\beta| - \sigma(\beta)]}{\beta(\beta - \alpha)} + \frac{\sigma(\alpha)(2\alpha - \beta)}{\alpha^2(\beta - \alpha)^2} + \frac{\sigma(\beta)(2\beta - \alpha)}{\beta^2(\beta - \alpha)^2}.$$

This was followed by a section on the case of one zero and one double integer root. The two sub-formulas for the double negative and double positive integer roots were combined into the summary general formula for one zero and one arbitrary double integer root $\alpha$, using harmonic numbers $H(n)$, generalized harmonic numbers $H(n, 2)$ and the function $\sigma(n)$:

$$s(0, \alpha, \alpha) = \frac{\pi^2}{6\alpha} + \frac{H[|\alpha| - \sigma(\alpha), 2]}{|\alpha|} + \frac{H[|\alpha| - \sigma(\alpha)]}{\alpha^2} - \frac{\sigma(\alpha)}{\alpha^3}.$$

In the final part of the chapter, some approximate values of the sums of individual series of reciprocals of the cubic polynomials were presented.

So, these series belong to special types of infinite series, such as geometric series, whose sums are given analytically by means of a simple formula.

## References


[1] Potuček R 2017 The sum of the series of reciprocals of the quadratic polynomials with purely imaginary conjugate roots *Math. Edu. Res. Appl.* **3** 17–23

[2] Potuček R 2018 The sum of the series of reciprocals of the cubic polynomials with three different positive integer roots *Math. Edu. Res. Appl.* **4** 1–8

[3] Potuček R 2019 The sum of the series of reciprocals of the cubic polynomials with one zero and double non-zero integer root *Math. Edu. Res. Appl.* **5** 9–15

[4] Potuček R 2019 The sum of the series of reciprocals of the cubic polynomials with one zero and two different negative integer roots *Proc. 6th Int. Conf. of Mathematics, Information Technologies and Applied Sciences (MITAV)* vol 1 *(University of Defence, Brno, Czech Republic, 2019)* pp 58–65

[5] Potuček R 2019 The sum of the series of reciprocals of the cubic polynomials with triple and double zero root *Proc. 18th Conf. on Applied Mathematics (APLIMAT)* vol 1 *(Slovak University Technology in Bratislava, Vydavatelstvo STU, Bratislava, Slovakia, 2019)* pp 959–67

[6] Potuček R 2020 The sum of the series of reciprocals of the cubic polynomials with three different negative integer roots *Proc. 19th Conf. on Applied Mathematics (APLIMAT* vol 1 *(Slovak University Technology in Bratislava, Publishing House SPEKTRUM STU, Bratislava, Slovakia, 2020)* pp 888–96

[7] Potuček R 2022 The sum of the series of reciprocals of the cubic polynomials with one negative, one zero and one positive integer roots *Proc. 9th Int. Conf. of Mathematics, Information Technologies and Applied Sciences (MITAV)* vol 1 *(University of Defence, Brno, Czech Republic, 2022)* p 10

[8] Potuček R 2022 The sum of the series of reciprocals of the cubic polynomials with one zero and two different positive integer roots *Math. Edu. Res. Appl.* **8** 10

[9] Wikipedia Contributors 2022 Harmonic number. Wikipedia, The Free Encyclopedia [cit. 2022-05-28] https://en.wikipedia.org/w/index.php?title=Harmonic_number&oldid=1072824510

[10] Benjamin A T, Preston G O and Quinn J J 2002 A Stirling encounter with harmonic numbers *Math. Mag.* **75** 95–103

[11] Wikipedia Contributors 2022 Apéry's constant. Wikipedia, The Free Encyclopedia [cit. 2022-05-28] https://en.wikipedia.org/wiki/Apery's_constant

# Chapter 2

## Polynomials for meshless methods in finding solutions in gradient elasticity problems

**N Fantuzzi, S Saitta, F Fabbrocino, R Vescovini and R Luciano**

The most common numerical tool for solving engineering problems is the finite element method (FEM), which is either based on Lagrange or Hermite polynomials in its native implementation. Among the several drawbacks of FEM one is clearly the domain decomposition which might lead to numerical instabilities for complex problems. Therefore, meshless methods have been introduced for their ability to consider arbitrary domain without a proper meshing and this brings many advantages in terms of computational cost and handling complex geometries in linear and nonlinear regimes. The present chapter aims at investigating accuracy and reliability of meshless methods by considering several polynomial interpolations based on a class of well-known radial basis functions (RBFs). Such novel numerical approaches have been tested within the framework of gradient elasticity where higher-order derivatives of the field variables are involved. Such problems in classical FEM are solved by using Hermite polynomials and many field variables, in contrast, the present collocation meshless methods can deal with the same problem with a reduced number of parameters leading to lower computational cost of the simulations.

### 2.1 Introduction

In the context of numerical analysis, the most used numerical method is the well-known finite element method (FEM). However, some competitors are quickly gaining a certain degree of success, in particular the mesh free methods. This family of numerical methods differs from FEM because the domain of interest is not actually discretized but represented by a randomly distributed set of nodes. The time consuming element of mesh generation is spared and a significant flexibility in adding or removing nodes for subsequent analysis is introduced. However, if the weak formulation of the problem is used, for the sole purpose of numerical integrals evaluation a mesh has to be introduced.

doi:10.1088/978-0-7503-5067-9ch2  

Mesh free methods have already been widely and successfully applied to classical structural mechanics problems [1–3]. As mentioned, they are a family of methods, meaning that several different methods are available, each one different from the others for the procedure used for the shape functions computation [1, 4–7]. Among them, the radial point interpolation method (RPIM) is one of the most affirmed since it possesses the Kronecker delta function property [8–10]. Moreover, it allows to introduce additional degrees of freedom needed to better represent the behaviour of a solid, thanks to its Hermite formulation [11].

As researchers' attention is moving to nanostructural components, this last property gains significance considering that micro and nanostructural elements [12–20] cannot be described by means of classical theories [21]. In fact, long range interactions between the material's particles strongly affect the mechanical behaviour of both MEMS (micro-electro-mechanical-systems) and NEMS (nano-electro-mechanical-systems) [22–33]. While these interactions are negligible at the macroscale, they cannot be ignored at the microscale.

The static [34–37] and dynamic [14, 38–46] behaviour of these structures has been studied by means of new nonlocal theories.

The static bending analyses of micro-sized beams modelled according to the Bernoulli–Euler beam theory, strain gradient elasticity and modified couple stress theories [26, 47, 48] show how the results obtained with higher-order elasticity theories are significantly different from the results calculated by means of classical elasticity theory [49]. Single-walled carbon nanotubes (SWCNTs) embedded in an elastic medium modelled according to higher-order shear deformation microbeam models in conjunction with modified strain gradient theory, when analysed in bending, showed a correlation between the bending behaviour and the small-size, stiffness of the elastic foundation with effects on shear deformation as well [50]. Among the most used non-classical theories, Eringen's nonlocal elasticity theory is one of the most used. The addition of parameters allows to improve its formulation to obtain an enhanced Eringen differential model which has been used to study the bending of nano/micro beams under concentrated and distributed loads and subjected to various types of boundary conditions [51, 52].

As for the dynamic analysis, symmetric porous functionally graded (FG) nano-beams carrying several fullerenes have been analysed to study their large-amplitude nonlinear vibrations [53] while vibrations of porous nanobeams have been studied using a meshless DQ method [54].

The size-dependent response of nanostructures is described using consistent variational theories of the higher-order nonlocal gradient elasticity which exploit nonlocal and two gradient characteristic lengths [55].

Mixed theories also make use of higher-order strain gradient theories which are combined with higher-order stress- and strain-driven mixture nonlocal elasticity theories [56–59].

Numerical analysis of laminated composite Kirchhoff nanoplates modelled according to the strain gradient theory have been thoroughly conducted using numerical FEM [60]. The membrane and bending degrees of freedom are approximated using Hermite interpolation functions using both the conforming and nonconforming approaches [61–64].

The Hermite–RPIM has been successfully applied to perform the static bending analysis of isotropic nanoplates modelled according to the second-order strain gradient theory [65, 66]. The same method has also been applied to the analysis of different kinds of composite nanoplates modelled according to the same theory [67].

## 2.2 Strain gradient plate model

The subject of the present investigation is represented by rectangular nanoplates represented through the in-plane size $a \times b$ and the out of plane thickness $z$ expressed in a Cartesian reference frame as shown in figure 2.1.

Both isotropic and composite plates are accounted for. In the case of a generic laminated nanoplate, the structure is considered to be made by $N_L$ thin layers. The generic $k$th ply has thickness $h_k = z_{k+1} - z_k$ where $z_k$ and $z_{k+1}$ are the ply's lower and upper bounds respectively. Hence, the overall thickness of the plate is computed as $h = \sum_{k=1}^{N_L} h_k$.

### 2.2.1 Kinematics

Classical laminated plate theory (CLPT) or Kirchhoff's theory states that the three-dimensional displacements $\mathbf{U} = \{U \; V \; W\}^{\mathsf{T}}$ can be written as a function of the degrees of freedom of the middle surface of the plate $\mathbf{u} = \{u \; v \; w\}^{\mathsf{T}}$:

$$\mathbf{U} = \mathbf{u} - z\mathbb{D}^{(s)}\mathbf{u} = \mathbf{u} - z\begin{bmatrix} 0 & 0 & \dfrac{\partial}{\partial x} \\ 0 & 0 & \dfrac{\partial}{\partial y} \\ 0 & 0 & 0 \end{bmatrix}\mathbf{u}. \tag{2.1}$$

Subsequently, the three-dimensional strains, $\boldsymbol{\varepsilon} = \{\varepsilon_{xx} \; \varepsilon_{yy} \; \gamma_{xy}\}^{\mathsf{T}}$, are developed as:

$$\boldsymbol{\varepsilon} = \boldsymbol{\varepsilon}^{(m)} - z\boldsymbol{\varepsilon}^{(b)} = \mathbb{D}^{(m)}\mathbf{u} - z\mathbb{D}^{(b)}\mathbf{u} = \begin{bmatrix} \dfrac{\partial}{\partial x} & 0 & 0 \\ 0 & \dfrac{\partial}{\partial y} & 0 \\ \dfrac{\partial}{\partial y} & \dfrac{\partial}{\partial x} & 0 \end{bmatrix}\mathbf{u} - z\begin{bmatrix} 0 & 0 & \dfrac{\partial^2}{\partial x^2} \\ 0 & 0 & \dfrac{\partial^2}{\partial y^2} \\ 0 & 0 & 2\dfrac{\partial^2}{\partial xy} \end{bmatrix} \tag{2.2}$$

where the membrane $\boldsymbol{\varepsilon}^{(m)}$ and bending $\boldsymbol{\varepsilon}^{(b)}$ are reported in their extended expressions.

### 2.2.2 Constitutive equation

The stress vector related to the $k$th ply, $\boldsymbol{\sigma}^{(k)} = \{\sigma_{xx} \; \sigma_{yy} \; \sigma_{xy}\}^{\mathsf{T}}$, in the plane stress case, can be evaluated by means of the constitutive law. However, the mechanical behaviour of nanoplates is affected by the material micro-structure and by the

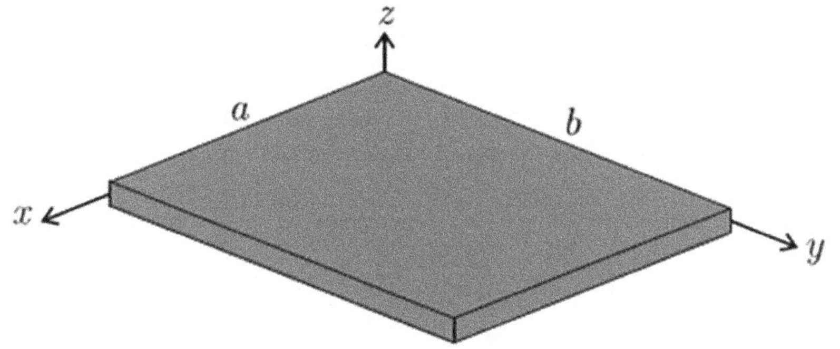


**Figure 2.1.** Generic plate geometry.

long range interaction forces between molecules which are far from each other. To account for the micro/macroscale interactions, a so-called internal length scale parameter $\ell$ is directly introduced in the constitutive equation.

$$\boldsymbol{\sigma}^{(k)} = (1 - \ell^2\Delta^2)\bar{\mathbf{Q}}^{(k)}\boldsymbol{\varepsilon} \tag{2.3}$$

where $\Delta^2(\cdot) = \partial^2(\cdot)/\partial y^2 + \partial^2(\cdot)/\partial x^2$ denotes the Laplacian operator and $\bar{\mathbf{Q}}$ is the classical plane stress reduced stiffness coefficient matrix for the $k$th layer characterised by a generic orientation $\theta^{(k)}$ [68].

### 2.2.3 Equilibrium equations

The system of governing equations is obtained by means of Hamilton's principle:

$$\delta U + \delta V - \delta K = 0 \tag{2.4}$$

where $\delta U$ is the variation of elastic energy, $\delta V$ is the potential of the external loads and $\delta K$ is the kinetic energy variation.

Recalling equations (2.2) and (2.3), the strain energy variation can be written as:

$$\begin{aligned}
\delta U = \int_\Omega \Big\{ & \left(\mathbb{D}^{(m)}\delta\mathbf{u}\right)^T (\mathbf{A}\mathbb{D}^{(m)} + \mathbf{B}\mathbb{D}^{(b)})\mathbf{u} + \left(\mathbb{D}^{(b)}\delta\mathbf{u}\right)^T \left(\mathbf{B}\mathbb{D}^{(m)} + \mathbf{D}\mathbb{D}^{(b)}\right)\mathbf{u} \\
& + \ell^2 \Big[ \left(\mathbb{D}_x^{(m)}\delta\mathbf{u}\right)^T \left(\mathbf{A}\mathbb{D}_x^{(m)} + \mathbf{B}\mathbb{D}_x^{(b)}\right)\mathbf{u} + \left(\mathbb{D}_y^{(m)}\delta\mathbf{u}\right)^T \left(\mathbf{A}\mathbb{D}_y^{(m)} + \mathbf{B}\mathbb{D}_y^{(b)}\right)\mathbf{u} \\
& + \left(\mathbb{D}_x^{(b)}\delta\mathbf{u}\right)^T \left(\mathbf{B}\mathbb{D}_x^{(m)} + \mathbf{D}\mathbb{D}_x^{(b)}\right)\mathbf{u} + \left(\mathbb{D}_y^{(b)}\delta\mathbf{u}\right)^T \left(\mathbf{B}\mathbb{D}_y^{(m)} + \mathbf{D}\mathbb{D}_y^{(b)}\right)\mathbf{u} \Big] \Big\} d\Omega
\end{aligned} \tag{2.5}$$

in which $\mathbb{D}_x^{(m)}$, $\mathbb{D}_y^{(m)}$, $\mathbb{D}_y^{(m)}$ and $\mathbb{D}_y^{(b)}$ are the derivatives with respect to the $x$ and $y$ coordinate of the differential operators introduced in equation (2.2) for the membrane and bending components respectively. The domain $\Omega$ is the reference domain which is taken as the plate middle surface and matrices $\mathbf{A}$, $\mathbf{B}$ and $\mathbf{D}$ are the equivalent stiffness matrices, which are obtaining by integration of matrix $\bar{\mathbf{Q}}$ across the thickness of the plate.

On the other hand, the potential of the external forces can be written as:

$$\delta V = -\int_{\Omega} \delta \mathbf{u}^{\mathrm{T}} \mathbf{q}_{\mathrm{ext}} d\Omega \tag{2.6}$$

where $\mathbf{q}_{\mathrm{ext}}$ represents the vector of external loads.

Putting together equations (2.5) and (2.6), the Hamilton's reads:

$$\begin{aligned} 0 = \int_{\Omega} \Big\{ & \left(\mathbb{D}^{(m)}\delta\mathbf{u}\right)^T \left(\mathbb{A}\mathbb{D}^{(m)} + \mathbb{B}\mathbb{D}^{(b)}\right)\mathbf{u} + \left(\mathbb{D}^{(b)}\delta\mathbf{u}\right)^T \left(\mathbb{B}\mathbb{D}^{(m)} + \mathbb{D}\mathbb{D}^{(b)}\right)\mathbf{u} \\ & + \ell^2 \Big[ \left(\mathbb{D}_x^{(m)}\delta\mathbf{u}\right)^T \left(\mathbb{A}\mathbb{D}_x^{(m)} + \mathbb{B}\mathbb{D}_x^{(b)}\right)\mathbf{u} + \left(\mathbb{D}_y^{(m)}\delta\mathbf{u}\right)^T \left(\mathbb{A}\mathbb{D}_y^{(m)} + \mathbb{B}\mathbb{D}_y^{(b)}\right)\mathbf{u} \\ & + \left(\mathbb{D}_x^{(b)}\delta\mathbf{u}\right)^T \left(\mathbb{B}\mathbb{D}_x^{(m)} + \mathbb{D}\mathbb{D}_x^{(b)}\right)\mathbf{u} + \left(\mathbb{D}_y^{(b)}\delta\mathbf{u}\right)^T \left(\mathbb{B}\mathbb{D}_y^{(m)} + \mathbb{D}\mathbb{D}_y^{(b)}\right)\mathbf{u} \Big] \\ & - \delta\mathbf{u}^T \mathbf{q}_{\mathrm{ext}} \Big\} d\Omega. \end{aligned} \tag{2.7}$$

While equation (2.7) is sufficient for the static analysis of nanoplates, further terms have to be added if dynamic analyses have to be performed.

For the generic composite nanoplate considered, recalling equation (2.1), the kinetic energy variation is written as:

$$\begin{aligned} \delta K &= \sum_{k=1}^{N_L} \int_{\Omega} \int_{z_k}^{z_{k+1}} \rho^{(k)} \delta \dot{\mathbf{U}}^{\mathrm{T}} \dot{\mathbf{U}} dz d\Omega \\ &= -\int_{\Omega} \Big\{ \delta\mathbf{u}^{\mathrm{T}} \mathbf{I}_0 \ddot{\mathbf{u}} + \delta\mathbf{u}^{\mathrm{T}} \mathbf{I}_1 (\mathbb{D}^{(s)} \ddot{\mathbf{u}}) + (\mathbb{D}^{(s)} \delta\mathbf{u})^{\mathrm{T}} \mathbf{I}_1 \ddot{\mathbf{u}} \\ &\quad + (\mathbb{D}^{(s)} \delta\mathbf{u})^{\mathrm{T}} \mathbf{I}_2 (\mathbb{D}^{(s)} \ddot{\mathbf{u}}) \Big\} d\Omega \end{aligned} \tag{2.8}$$

where

$$\mathbf{I}_i = \left( \sum_{k=1}^{N_L} \int_{\Omega} \int_{z_k}^{z_{k+1}} (-1)^i z^i \rho^{(k)} dz \right) \mathbf{I} = I_i \mathbf{I} \tag{2.9}$$

for $i = 0, 1, 2$, with $\mathbf{I}$ being the $3 \times 3$ identity matrix and $\rho^{(k)}$ the mass density of the $k$th ply.

On the other hand, to perform the stability analysis of a generic composite plate, the geometric stiffness matrix has to be introduced. To this aim, the von Kármán nonlinear strains are introduced:

$$\boldsymbol{\varepsilon}^{(vk)} = (\mathbb{D}^{(1)}\mathbf{u}) \circ (\mathbb{D}^{(1)}\mathbf{u}) = \begin{bmatrix} 0 & 0 & \dfrac{1}{2}\dfrac{\partial}{\partial x} \\ 0 & 0 & \dfrac{1}{2}\dfrac{\partial}{\partial y} \\ 0 & 0 & \dfrac{\partial}{\partial x} \end{bmatrix} \mathbf{u} \circ \begin{bmatrix} 0 & 0 & \dfrac{\partial}{\partial x} \\ 0 & 0 & \dfrac{\partial}{\partial y} \\ 0 & 0 & \dfrac{\partial}{\partial y} \end{bmatrix} \mathbf{u}. \tag{2.10}$$

Introducing these strains in the Hamilton's principle yields:

$$
\begin{aligned}
\delta U_{vk} &= \sum_{k=1}^{N_L} \int_{\Omega} \int_{z_k}^{z_{k+1}} \delta \boldsymbol{\varepsilon}^{(vk)\mathrm{T}} \boldsymbol{\sigma}^{(k)} dz d\Omega \\
&= \int_{\Omega} \{(\mathbb{D}^{(1)}\delta \mathbf{u})^{\mathrm{T}}[\mathbf{N}_d(\mathbb{D}^{(2)}\delta \mathbf{u})] + (\mathbb{D}^{(2)}\delta \mathbf{u})^{\mathrm{T}}[\mathbf{N}_d(\mathbb{D}^{(1)}\delta \mathbf{u})]\} d\Omega
\end{aligned}
\tag{2.11}
$$

where $\mathbf{N}_d$ is a diagonal matrix collecting the pre-stress condition:

$$
\mathbf{N}_d = \begin{bmatrix} \sigma_x^0 & 0 & 0 \\ 0 & \sigma_y^0 & 0 \\ 0 & 0 & \tau_{xy}^0 \end{bmatrix}.
\tag{2.12}
$$

Putting together the different contributions given in equations (2.5) to (2.11), the variational form of the governing equation, as a function of the displacement field, is written as:

$$
\begin{aligned}
0 = \int_{\Omega} \Big\{ & (\mathbb{D}^{(m)}\delta \mathbf{u})^T(\mathbf{A}\mathbb{D}^{(m)} + \mathbf{B}\mathbb{D}^{(b)})\mathbf{u} + (\mathbb{D}^{(b)}\delta \mathbf{u})^T(\mathbf{B}\mathbb{D}^{(m)} + \mathbf{D}\mathbb{D}^{(b)})\mathbf{u} \\
& + \ell^2 \Big[ \left(\mathbb{D}_x^{(m)}\delta \mathbf{u}\right)^T \left(\mathbf{A}\mathbb{D}_x^{(m)} + \mathbf{B}\mathbb{D}_x^{(b)}\right)\mathbf{u} + \left(\mathbb{D}_y^{(m)}\delta \mathbf{u}\right)^T \left(\mathbf{A}\mathbb{D}_y^{(m)} + \mathbf{B}\mathbb{D}_y^{(b)}\right)\mathbf{u} \\
& + \left(\mathbb{D}_x^{(b)}\delta \mathbf{u}\right)^T \left(\mathbf{B}\mathbb{D}_x^{(m)} + \mathbf{D}\mathbb{D}_x^{(b)}\right)\mathbf{u} + \left(\mathbb{D}_y^{(b)}\delta \mathbf{u}\right)^T \left(\mathbf{B}\mathbb{D}_y^{(m)} + \mathbf{D}\mathbb{D}_y^{(b)}\right)\mathbf{u} \Big] \\
& + (\mathbb{D}^{(1)}\delta \mathbf{u})^{\mathrm{T}}[\mathbf{N}_d(\mathbb{D}^{(2)}\delta \mathbf{u})] + (\mathbb{D}^{(2)}\delta \mathbf{u})^{\mathrm{T}}[\mathbf{N}_d(\mathbb{D}^{(1)}\delta \mathbf{u})] - \delta \mathbf{u}^T \mathbf{q}_{ext} \\
& + \delta \mathbf{u}^{\mathrm{T}} \mathbf{I}_0 \ddot{\mathbf{u}} + \delta \mathbf{u}^{\mathrm{T}} \mathbf{I}_1(\mathbb{D}^{(s)}\ddot{\mathbf{u}}) + (\mathbb{D}^{(s)}\delta \mathbf{u})^{\mathrm{T}} \mathbf{I}_1 \ddot{\mathbf{u}} + (\mathbb{D}^{(s)}\delta \mathbf{u})^{\mathrm{T}} \mathbf{I}_2(\mathbb{D}^{(s)}\ddot{\mathbf{u}}) \Big\} d\Omega.
\end{aligned}
\tag{2.13}
$$

## 2.3 Mesh free approach

The numerical analysis is performed by means of a mesh free radial point interpolation method (RPIM). The main feature of the mesh free method consists in the absence of a mesh for the purpose of domain discretization. In fact, the domain and its boundaries are represented, not discretized, by a set of nodes which can be either regularly or randomly distributed. Differently from other mesh free methods, the RPIM has the advantage of possessing the Kronecker delta function property which allows for an easy imposition of the essential boundary conditions.

However, the RPIM shape functions do not already embed all the necessary degrees of freedom at the node which, in turn, poses an issue for the imposition of the boundary conditions. In the case of a traditional Kirchhoff composite plate, the degrees of freedom considered are the in-plane displacements $u$ and $v$ and the transverse displacement $w$. But as a consequence of the strain gradient theory introducing higher-order derivatives in the formulation, the first derivatives of the transverse deflection, $w_{,x}$ and $w_{,y}$, are also taken as variables for the problem. Any higher-order derivative can be obtained by direct derivation. In the case of an isotropic plate, all of the previous degrees of freedom are accounted for except for

**Table 2.1.** Degrees of freedom considered for different kind of materials.

| Material | Degrees of freedom |
|---|---|
| Isotropic | $w\ w_{,x}\ w_{,y}$ |
| Composite | $u\ v\ w\ w_{,x}\ w_{,y}$ |

the in-plane coordinates. Table 2.1 summarises the degrees of freedom considered in the two different cases.

Due to the presence of the aforementioned higher-order derivatives in the formulation of the problem, rather than the classic RPIM the Hermite–RPIM proves to be more suitable for these kind of applications.

A generic field function $f(x, y)$ can be approximated by a linear combination of radial basis functions (RBF) and their derivatives at all the nodes falling within a local domain support domain:

$$\begin{aligned} f(x, y) &= \sum_{i=1}^{n} R_i(x, y)a_i + \sum_{i=1}^{n} R_{i,x}(x, y)a_i^x + \sum_{i=1}^{n} R_{i,y}(x, y)a_i^y \\ &= \mathbf{R}^{\mathrm{T}}(\mathbf{x})\mathbf{a} + \mathbf{R}_{,x}^{\mathrm{T}}(\mathbf{x})\mathbf{a}^x + \mathbf{R}_{,y}^{\mathrm{T}}(\mathbf{x})\mathbf{a}^y \end{aligned} \tag{2.14}$$

where $R_i(x, y)$, $R_{i,x}(x, y)$ and $R_{i,y}$ are the RBF and its derivatives and $a_i$, $a_i^x$ and $a_i^y$ the corresponding unknown vectors of coefficients.

To obtain the derivative of the field function $f(x, y)$, it is assumed that it is equal to the derivative of its approximation function. Imposing that the field function and its derivative are satisfied at the $n$ nodes falling within the support domain, allows to obtain the vector of coefficients in equation (2.14). The resultant system of $3n$ linear equations can be written in matrix form as:

$$\mathcal{F} = \begin{Bmatrix} \mathbf{f} \\ \mathbf{f}_{,x} \\ \mathbf{f}_{,y} \end{Bmatrix} = \begin{bmatrix} \mathbf{R} & \mathbf{R}_{,x} & \mathbf{R}_{,y} \\ \mathbf{R}_{,x} & \mathbf{R}_{,xx} & \mathbf{R}_{,xy} \\ \mathbf{R}_{,y} & \mathbf{R}_{,xy} & \mathbf{R}_{,yy} \end{bmatrix} \begin{bmatrix} \mathbf{a} \\ \mathbf{a}^x \\ \mathbf{a}^y \end{bmatrix} = \mathbf{Ga} \tag{2.15}$$

where the vectors of function values $\mathbf{f}$, $\mathbf{f}_{,x}$ and $\mathbf{f}_{,y}$ are given as:

$$\begin{aligned} \mathbf{f} &= \left[f_1 \quad f_2 \cdots f_n\right]^{\mathrm{T}} \\ \mathbf{f}_{,x} &= \left[f_{1,x} \quad f_{2,x} \cdots f_{n,x}\right]^{\mathrm{T}} \\ \mathbf{f}_{,y} &= \left[f_{1,y} \quad f_{2,y} \cdots f_{n,y}\right]^{\mathrm{T}}. \end{aligned} \tag{2.16}$$

Solving equation (2.15) and substituting into equation (2.14) yields:

$$\begin{aligned} f(x, y) &= \left[\mathbf{R}^{\mathrm{T}} \quad \mathbf{R}_{,x}^{\mathrm{T}} \quad \mathbf{R}_{,y}^{\mathrm{T}}\right] \begin{bmatrix} \mathbf{a} \\ \mathbf{a}^x \\ \mathbf{a}^y \end{bmatrix} \\ &= \left[\mathbf{R}^{\mathrm{T}} \quad \mathbf{R}_{,x}^{\mathrm{T}} \quad \mathbf{R}_{,y}^{\mathrm{T}}\right]\mathbf{G}^{-1}\mathcal{F}. \end{aligned} \tag{2.17}$$

It follows that the Hermite–RPIM shape functions are written as:

$$\begin{aligned}\mathbf{\Phi}^{\mathrm{T}}(\mathbf{x}) &= \left[\mathbf{R}^{\mathrm{T}}\ \mathbf{R}_{,x}^{\mathrm{T}}\ \mathbf{R}_{,y}^{\mathrm{T}}\right]\mathbf{G}^{-1} = [\boldsymbol{\phi}\ \ \boldsymbol{\phi}^{x}\ \ \boldsymbol{\phi}^{y}] \\ &= \left[\varphi_1\cdots\varphi_i\cdots\varphi_n\ \ \varphi_1^{x}\cdots\varphi_i^{x}\cdots\varphi_n^{x}\ \ \varphi_1^{y}\cdots\varphi_i^{y}\cdots\varphi_n^{y}\right].\end{aligned} \tag{2.18}$$

### 2.3.1 RBF

The mathematics community provided numerous forms of RBFs among which, the four shown in table 2.2 are the most used [69].

The RBF are functions of the distance $r$ between two points $\mathbf{x}$ and $\mathbf{x}_i$. For a 2D problem, $r$ is expressed as:

$$r = \left[(x - x_i)^2 + (y - y_i)^2\right]^{1/2}. \tag{2.19}$$

Among the numerous RBFs available, in this work the multiquadrics (MQ) RBF is used for a number of reasons [69].

First, for MQ-RBF, the nodal distribution doesn't significantly affect the surface fitting accuracy meaning that the RPIM will not have stability issues for random and irregular nodal distributions. Moreover, the condition number of matrix $\mathbf{G}$ is much lower for the MQ than the one obtained, for instance, with the EXP basis function, which is another reason why the MQ-RBF is preferred.

As suggested in [69], instead of the traditional form shown in table 2.2 the MQ-RBF is here used with dimensionless parameters. The modification affects the parameter $C$ which is now written as $C = \alpha_C d_c$, where $d_c$ is the average nodal spacing. It follows that the MQ-RBF depends on $q$ and $\alpha_C$, which are nondimensional coefficients that have to be tuned to achieve proper accuracy of the results.

An example of what the shape functions look like as computed with this method is given in figure 2.2. To compute them, a generic squared domain of size $L = 1$ is represented using nine regularly distributed nodes: one node is placed on each corner, one is in the middle of each side and one is in the centre of the domain. All of the nodes are used to construct the shape functions. Figure 2.2 shows the shape functions and their first derivatives with respect to $x$ and $y$ relative to the nodes in the bottom left corner, in the middle left side, in the top right corner and in the centre respectively. For this particular case, the dimensionless parameters of the MQ-RBF are taken as $C = 1$ and $q = 0.05$.

**Table 2.2.** Most common forms of radial basis functions.

| Name | Expression |
|---|---|
| Multiquadrics (MQ) | $R_i(x, y) = (r_i^2 + C^2)^q = [(x - x_i)^2 + (y - y_i)^2 + C^2]^q$ |
| Gaussian (EXP) | $R_i(x, y) = \exp(-cr_i^2) = \exp\{-c[(x - x_i)^2 + (y - y_i)^2]\}$ |
| Thin plate spline (TPS) | $R_i(x, y) = r_i^\eta = [(x - x_i)^2 + (y - y_i)^2]^\eta$ |
| Logarithmic RBF | $R_i(x, y) = r_i^\eta \log r_i$ |

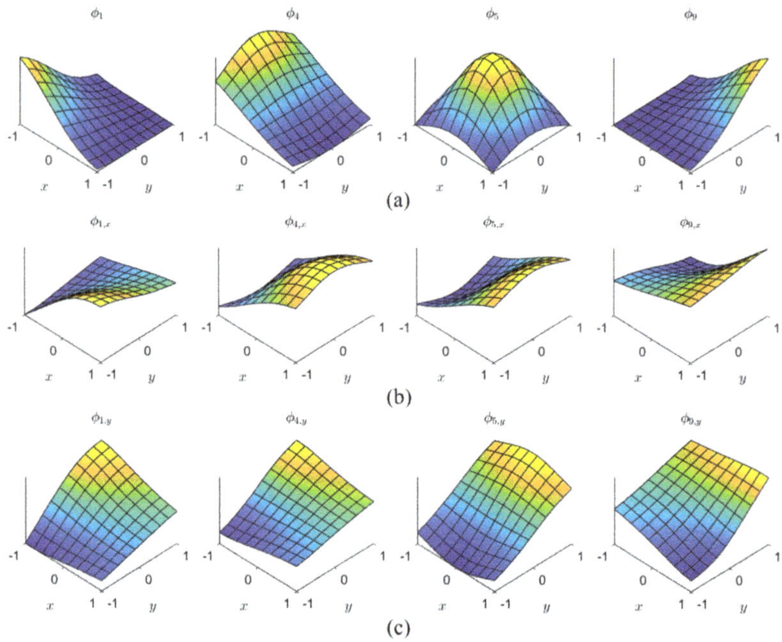


**Figure 2.2.** (a) Shape functions. (b) Shape functions derivative with respect to $x$. (c) Shape functions derivative with respect to $y$.

### 2.3.2 Discretized equations

The steps following the computation of the shape functions are the same as in traditional FEM. The approximation is directly introduced in the weak form of the governing equation of motion.

Since the latter, given in equation (2.13), is written as a function of the displacement, it follows that the field variable to be approximated by means of the RPIM is the displacement field.

This means that the vector of field function $\mathcal{F}$ will contain, at each node, the degrees of freedom shown in table 2.1 according to the different type of material considered.

As the approximation is directly introduced in equation (2.13), the following notation is introduced for further clarity in the writing of the final linear system of equations:

$$\mathbf{u} = \begin{bmatrix} \mathbf{\Phi}_u & 0 & 0 \\ 0 & \mathbf{\Phi}_v & 0 \\ 0 & 0 & \mathbf{\Phi}_w \end{bmatrix} \begin{Bmatrix} \tilde{\mathbf{u}} \\ \tilde{\mathbf{v}} \\ \tilde{\mathbf{w}} \end{Bmatrix} = \mathbf{\Phi}\mathbf{d} \tag{2.20}$$

where $\mathbf{\Phi}$ is the matrix collecting the Hermite–RPIM shape functions and $\mathbf{d}$ is the vector collecting the degrees of freedom at each node.

The final form of the variational statement is given as:

$$
\begin{aligned}
&\int_{\Omega_s} \Big\{(\mathbb{D}^{(m)}\boldsymbol{\Phi})^\top \mathbf{A}(\mathbb{D}^{(m)}\boldsymbol{\Phi}) + (\mathbb{D}^{(m)}\boldsymbol{\Phi})^\top \mathbf{B}(\mathbb{D}^{(b)}\boldsymbol{\Phi}) + (\mathbb{D}^{(b)}\boldsymbol{\Phi})^\top \mathbf{D}(\mathbb{D}^{(b)}\boldsymbol{\Phi}) \\
&\quad + \ell^2\Big[(\mathbb{D}_x^{(m)}\boldsymbol{\Phi})^\top \mathbf{A}(\mathbb{D}_x^{(m)}\boldsymbol{\Phi}) + (\mathbb{D}_x^{(m)}\boldsymbol{\Phi})^\top \mathbf{B}(\mathbb{D}_x^{(b)}\boldsymbol{\Phi}) + (\mathbb{D}_y^{(m)}\boldsymbol{\Phi})^\top \mathbf{A}(\mathbb{D}_y^{(m)}\boldsymbol{\Phi}) \\
&\quad + (\mathbb{D}_y^{(m)}\boldsymbol{\Phi})^\top \mathbf{B}(\mathbb{D}_y^{(b)}\boldsymbol{\Phi}) + (\mathbb{D}_x^{(b)}\boldsymbol{\Phi})^\top \mathbf{B}(\mathbb{D}_x^{(m)}\boldsymbol{\Phi}) + (\mathbb{D}_x^{(b)}\boldsymbol{\Phi})^\top \mathbf{D}(\mathbb{D}_x^{(b)}\boldsymbol{\Phi}) \\
&\quad + (\mathbb{D}_y^{(b)}\boldsymbol{\Phi})^\top \mathbf{B}(\mathbb{D}_y^{(m)}\boldsymbol{\Phi}) + (\mathbb{D}_y^{(b)}\boldsymbol{\Phi})^\top \mathbf{D}(\mathbb{D}_y^{(b)}\boldsymbol{\Phi})\Big]\Big\} d\Omega_s \mathbf{d} \\
&\quad + \int_{\Omega_s} \Big\{(\mathbb{D}^{(1)}\boldsymbol{\Phi})^\top \mathbf{N}_d(\mathbb{D}^{(2)}\boldsymbol{\Phi}) + (\mathbb{D}^{(2)}\boldsymbol{\Phi})^\top \mathbf{N}_d(\mathbb{D}^{(1)}\boldsymbol{\Phi})\Big\} d\Omega_s \mathbf{d} \\
&\quad + \int_{\Omega_s} \Big\{\boldsymbol{\Phi}^\top \mathbf{I}_0 \boldsymbol{\Phi} + \boldsymbol{\Phi}^\top \mathbf{I}_1(\mathbb{D}^{(s)}\boldsymbol{\Phi}) + (\mathbb{D}^{(s)}\boldsymbol{\Phi})^\top \mathbf{I}_1(\mathbb{D}^{(s)}\boldsymbol{\Phi})\Big\} d\Omega_s \ddot{\mathbf{d}} \\
&\quad = \int_{\Omega_s} \boldsymbol{\Phi}^\top \mathbf{q}_{\text{ext}} d\Omega_s
\end{aligned}
\tag{2.21}
$$

where $\Omega_s$ is the support domain centred in the current node. The previous equation is written for every node in the domain and the contributions are assembled as in conventional FEM:

$$
(\mathbf{K} + \mathbf{K}_G)\mathbf{d} + \mathbf{M}\ddot{\mathbf{d}} = \mathbf{F} \tag{2.22}
$$

where $\mathbf{K}$ is the stiffness matrix, $\mathbf{K}_G$ is the geometric stiffness matrix, $\mathbf{M}$ is the mass matrix and $\mathbf{F}$ is the load vector.

Note that even if a mesh is not required to construct the shape functions, the approximation is directly introduced in the weak form of the governing equations, which means that there are integrals to be evaluated. Hence, for the sole purpose of integrals numerical evaluation, a mesh called background mesh is introduced. The background mesh and the nodal distribution are two separate objects, they are in no way linked to one another.

## 2.4 Numerical analyses and verification

This section aims to show the results of the numerical analysis conducted on different kinds of isotropic and composite nanoplates. Bending, free vibrations and buckling analyses are performed and numerical accuracy and convergence of results is shown.

As mentioned, the mesh free approximation procedure requires the definition of a support domain. At each step, the shape functions are built using only the nodes falling within such moving domain. While it can be centred in the nodes, it is common practice to centre the support domain in the Gauss points used for the numerical integration. The analyses shown in the following sections will make use of this last technique.

### 2.4.1 Isotropic plates

First, isotropic Kirchhoff's nanoplates modelled according to the second-order strain gradient theory are analysed in bending. Since isotropic materials don't introduce coupling between the in-plane and out of plane behaviour, the in-plane degrees of freedom are not considered, in agreement with table 2.1.

The nanoplate is subjected to a transverse external load $q_z$, has a squared geometry and its thickness is taken as $h = 0.34$ nm. Young's modulus and Poisson's ratio are taken as 1100 GPa and 0.3 respectively. The plate is analysed considering five different kinds of boundary conditions: SSSS, CCCC, SCSC, SFSF, SCSF. Table 2.3 provides a description of the fundamental kinds of constraints which are used as a combination on the nanoplate. Note that the four edges are identified by the values of the physical coordinates $x$ and $y$.

The support domain, introduced by the mesh free RPIM, is centred in the Gauss points and its size $d_s$ is given by the traditional definition [69] $d_s = \alpha_s d_c$ where $d_c$ is the average nodal spacing $d_c = \sqrt{\Delta x^2 + \Delta y^2}$ and $\alpha_s$ is a dimensionless parameter to be tuned.

The nanoplate is represented by a set of regularly distributed nodes. Convergence of the results is assessed accounting for several different nodal densities, namely $3 \times 3$, $5 \times 5$, $7 \times 7$, $11 \times 11$. Different values of the internal length scale parameter $\ell$ are also accounted for: 0, 0.2, 0.5, 1.

The integration is performed using a background mesh whose nodes are considered to be coincident with the nodes used to represent the problem domain. Moreover, a $2 \times 2$ Gauss integration scheme is used in each cell of the background mesh.

The results of the analysis, the static deflection of the plate, is computed in its nondimensional form as:

$$\bar{w} = \frac{1000wD}{q_z a^4} \tag{2.23}$$

where $w$ is the maximum (central) plate deflection, $q_z$ is the magnitude of the external load and $D$ is the bending rigidity $D = Eh^3/12(1 - \nu^2)$.

The results are compared in terms of percentage error as shown in tables 2.4–2.7:

$$\text{err}_{\%} = 100\frac{|w_e - \bar{w}|}{w_e} \tag{2.24}$$

where $\bar{w}$ is the result of the analysis and $w_e$ is the exact solution taken from [35].

The results shown in tables 2.4–2.7 are affected by a number of phenomenon.

**Table 2.3.** Essential boundary conditions considered.

| BCs | $x = 0, a$ | $y = 0, b$ |
|---|---|---|
| Supported | $v = w = \frac{\partial w}{\partial y} = 0$ | $u = w = \frac{\partial w}{\partial x} = 0$ |
| Clamped | $u = v = w = \frac{\partial w}{\partial x} = \frac{\partial w}{\partial y} = 0$ | $u = v = w = \frac{\partial w}{\partial x} = \frac{\partial w}{\partial y} = 0$ |
| Free | No variables involved | No variables involved |

**Table 2.4.** Values of $\bar{w}$ for 3 × 3 nodal distribution, obtained for $\alpha_C = 3$, $q = 1.3$, $\alpha_s = 2$.

| BC | $\ell$ (nm) | Exact | Result | Error (%) |
|---|---|---|---|---|
| SSSS | 0 | 4.062 4 | 2.862 2 | 29.544 1 |
| | 0.2 | 4.033 0 | 2.833 8 | 29.734 7 |
| | 0.5 | 3.884 4 | 2.695 6 | 30.604 5 |
| | 1 | 3.423 1 | 2.314 2 | 32.394 6 |
| CCCC | 0 | 1.265 3 | 1.071 6 | 15.308 6 |
| | 0.2 | 1.233 3 | 1.055 5 | 14.416 6 |
| | 0.5 | 1.097 9 | 0.978 5 | 10.875 3 |
| | 1 | 0.794 6 | 0.776 2 | 2.315 6 |
| SCSC | 0 | 1.917 1 | 1.548 1 | 19.247 8 |
| | 0.2 | 1.878 3 | 1.526 9 | 18.708 4 |
| | 0.5 | 1.709 3 | 1.424 7 | 16.650 1 |
| | 1 | 1.304 0 | 1.151 9 | 11.664 1 |
| SFSF | 0 | 15.011 3 | 13.508 9 | 10.008 5 |
| | 0.2 | 14.947 0 | 13.451 1 | 10.008 0 |
| | 0.5 | 14.616 5 | 13.171 1 | 9.888 8 |
| | 1 | 13.545 1 | 12.395 7 | 8.485 7 |
| SCSF | 0 | 11.235 9 | 10.826 2 | 3.646 3 |
| | 0.2 | 11.170 3 | 10.763 5 | 3.641 8 |
| | 0.5 | 10.845 4 | 10.456 8 | 3.583 1 |
| | 1 | 9.841 6 | 9.573 1 | 2.728 2 |

First of all, the choice to consider the collocation nodes coincident with the nodes of the background mesh influences the results in terms of mesh refinement. In fact, a smaller nodal density implies a coarser mesh is used for the integration and vice versa.

There are three nondimensional parameters, introduced by the presented mesh free formulation, that require tuning: $\alpha_C$, $q$ and $\alpha_s$. While these parameters are kept constant for different kinds of boundary conditions, they vary as the nodal density changes. This choice, although not very convenient, is made for the purpose of showing the influence on the results of the nodal density and refinement of the background mesh only. The parameters are chosen to be the most suitable combination for the specific nodal distribution, so that the analysis is performed under some of the best achievable conditions in each case. Of course, an optimal combination of parameters may be identified and kept constant as the nodal density varies. However, this would introduce new sources of numerical errors. Therefore, for this particular analysis, the focus is centred on the influence that the number of collocation nodes and refinement of the background mesh have on the accuracy of the results.

The convergence trend is graphically shown in figure 2.3.

**Table 2.5.** Values of $\bar{w}$ for 5 × 5 nodal distribution, obtained for $\alpha_C = 2.38$, $q = 0.01$, $\alpha_s = 2.4$.

| BC | $\ell$ (nm) | Exact | Result | Error (%) |
|---|---|---|---|---|
| SSSS | 0 | 4.062 4 | 3.854 9 | 5.107 8 |
| | 0.2 | 4.033 0 | 3.826 2 | 5.127 7 |
| | 0.5 | 3.884 4 | 3.684 4 | 5.148 8 |
| | 1 | 3.423 1 | 3.273 6 | 4.367 4 |
| CCCC | 0 | 1.265 3 | 1.305 8 | 3.200 8 |
| | 0.2 | 1.233 3 | 1.277 4 | 3.575 8 |
| | 0.5 | 1.097 9 | 1.146 6 | 4.435 7 |
| | 1 | 0.794 6 | 0.838 4 | 5.512 2 |
| SCSC | 0 | 1.917 1 | 1.885 1 | 1.669 2 |
| | 0.2 | 1.878 3 | 1.850 9 | 1.458 8 |
| | 0.5 | 1.709 3 | 1.690 3 | 1.111 6 |
| | 1 | 1.304 0 | 1.291 4 | 0.966 3 |
| SFSF | 0 | 15.011 3 | 14.872 1 | 0.927 3 |
| | 0.2 | 14.947 0 | 14.808 7 | 0.925 3 |
| | 0.5 | 14.616 5 | 14.514 4 | 0.698 5 |
| | 1 | 13.545 1 | 13.729 0 | 1.357 7 |
| SCSF | 0 | 11.235 9 | 11.226 5 | 0.083 7 |
| | 0.2 | 11.170 3 | 11.162 3 | 0.071 6 |
| | 0.5 | 10.845 4 | 10.852 9 | 0.069 2 |
| | 1 | 9.841 6 | 9.952 7 | 1.128 9 |

It can be observed how the convergence trend is not strictly monotonic as in classical theories. In fact, while in the case $\ell = 0$ the convergence is still monotonic, a different behaviour can be observed when $\ell$ assumes any value. The explanation of this phenomenon has to be searched in the relative influence the nodal coordinates and the nodal density have on the numerical results.

It comes naturally to understand how representing the nodal density used to represent the domain influences the accuracy of the results. In the RPIM, the nodal spacing directly enters the computation of derivatives of any order. Moreover, it has been explained how the strain gradient theory introduces higher-order derivatives in the formulation of the problem, which translates in the necessity of computing higher-order derivatives of the RBF. As a consequence, the spatial coordinates of the nodes are multiplied several times, once per each derivative order, for each other. Small nodal densities, meaning less nodes, imply larger coordinates value meaning that larger numbers appear in the shape functions. Increasing the nodal density causes the nodal coordinates to get smaller and smaller, yielding larger errors when introduced in the shape functions.

The balancing of these two aspects, nodal density and magnitude of the nodal coordinates, may cause the alteration in the trend exhibited by the convergence.

**Table 2.6.** Values of $\bar{w}$ for 7 × 7 nodal distribution, obtained for $\alpha_C = 1.06$, $q = 1.3$, $\alpha_s = 3$.

| BC | $\ell$ (nm) | Exact | Result | Error (%) |
|---|---|---|---|---|
| SSSS | 0 | 4.062 4 | 3.961 7 | 2.478 8 |
| | 0.2 | 4.033 0 | 3.931 0 | 2.529 1 |
| | 0.5 | 3.884 4 | 3.781 5 | 2.649 1 |
| | 1 | 3.423 1 | 3.346 1 | 2.249 4 |
| CCCC | 0 | 1.265 3 | 1.279 6 | 1.130 2 |
| | 0.2 | 1.233 3 | 1.247 0 | 1.110 8 |
| | 0.5 | 1.097 9 | 1.117 7 | 1.803 4 |
| | 1 | 0.794 6 | 0.830 5 | 4.518 0 |
| SCSC | 0 | 1.917 1 | 1.907 2 | 0.516 4 |
| | 0.2 | 1.878 3 | 1.866 8 | 0.612 3 |
| | 0.5 | 1.709 3 | 1.703 5 | 0.339 3 |
| | 1 | 1.304 0 | 1.322 0 | 1.380 4 |
| SFSF | 0 | 15.011 3 | 14.904 9 | 0.708 8 |
| | 0.2 | 14.947 0 | 14.822 8 | 0.830 9 |
| | 0.5 | 14.616 5 | 14.506 3 | 0.753 9 |
| | 1 | 13.545 1 | 13.732 7 | 1.385 0 |
| SCSF | 0 | 11.235 9 | 11.201 9 | 0.302 6 |
| | 0.2 | 11.170 3 | 11.125 5 | 0.401 1 |
| | 0.5 | 10.845 4 | 10.805 8 | 0.365 1 |
| | 1 | 9.841 6 | 9.924 8 | 0.845 4 |

**Table 2.7.** Values of $\bar{w}$ for 11 × 11 nodal distribution, obtained for $\alpha_C = 1.06$, $q = 1.3$, $\alpha_s = 2.3$.

| BC | $\ell$ (nm) | Exact | Result | Error (%) |
|---|---|---|---|---|
| SSSS | 0 | 4.062 4 | 4.047 2 | 0.374 2 |
| | 0.2 | 4.033 0 | 4.017 4 | 0.386 8 |
| | 0.5 | 3.884 4 | 3.871 1 | 0.342 4 |
| | 1 | 3.423 1 | 3.442 4 | 0.563 8 |
| CCCC | 0 | 1.265 3 | 1.279 4 | 1.114 4 |
| | 0.2 | 1.233 3 | 1.246 4 | 1.062 2 |
| | 0.5 | 1.097 9 | 1.108 4 | 0.956 4 |
| | 1 | 0.794 6 | 0.801 8 | 0.906 1 |
| SCSC | 0 | 1.917 1 | 1.927 2 | 0.526 8 |
| | 0.2 | 1.878 3 | 1.887 2 | 0.473 8 |
| | 0.5 | 1.709 3 | 1.716 3 | 0.409 5 |
| | 1 | 1.304 0 | 1.311 1 | 0.544 5 |
| SFSF | 0 | 15.011 3 | 15.019 8 | 0.056 6 |
| | 0.2 | 14.947 0 | 14.943 0 | 0.021 4 |
| | 0.5 | 14.616 5 | 14.617 8 | 0.008 9 |
| | 1 | 13.545 1 | 13.731 5 | 1.376 1 |
| SCSF | 0 | 11.235 9 | 11.226 5 | 0.131 7 |
| | 0.2 | 11.170 3 | 11.162 3 | 0.051 0 |
| | 0.5 | 10.845 4 | 10.852 9 | 0.012 0 |
| | 1 | 9.841 6 | 9.952 7 | 0.689 9 |

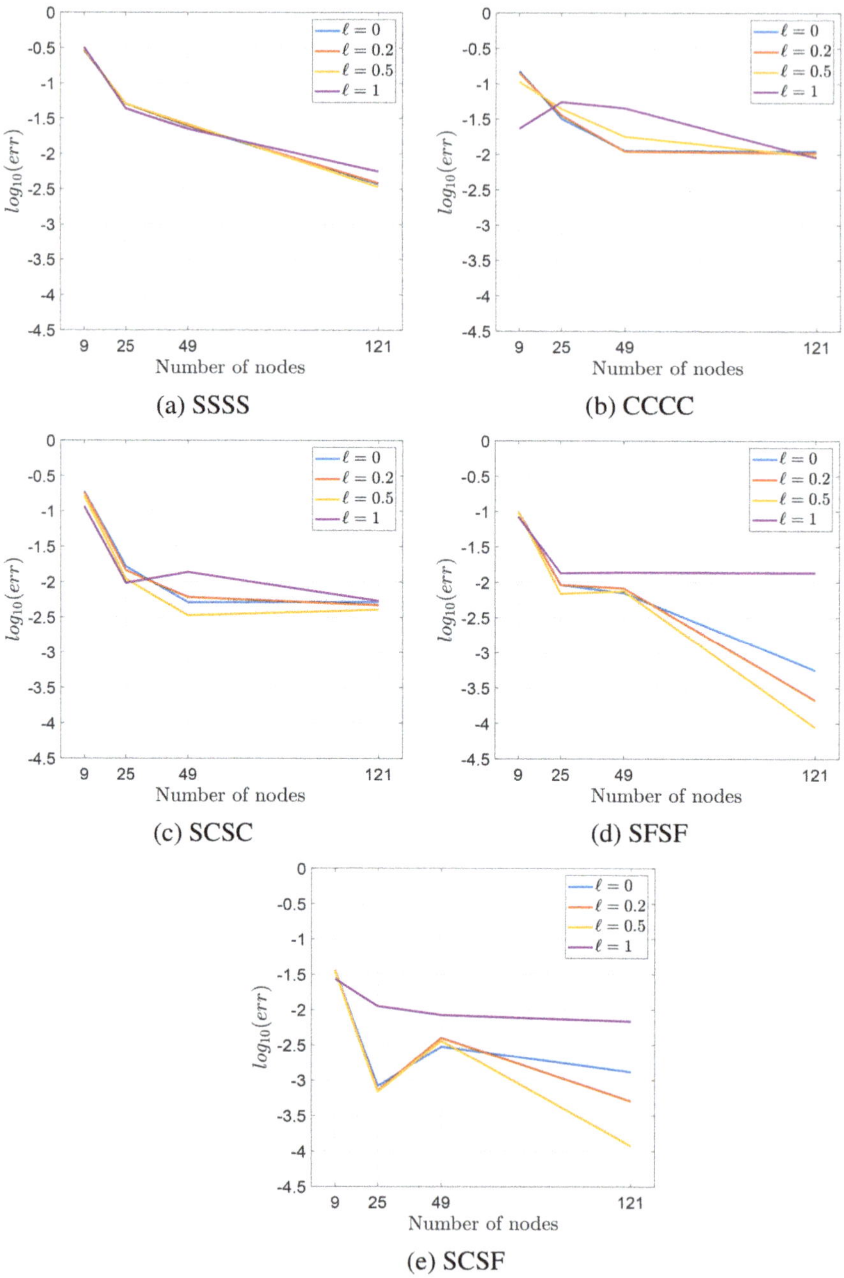


**Figure 2.3.** Convergence of static analysis of isotropic nanoplates with different boundary conditions and regular nodal distribution.

Observing the results obtained for the largest nodal density, it can be observed how the value of the error never even reaches 2%. Moreover, as expected, the error value tends to increase as the value of the $\ell$ parameter increases. As the value of $\ell$

increases, the weight of the higher-order derivatives in the computations increases as well, causing the degradation of accuracy of the results of the analysis.

The same analysis is now repeated on nanoplates characterised by different geometries. Four different values of the aspect ratio $b/a$ are accounted for, 0.5, 1, 2, 3. The analysis is performed considering $15 \times 15$ regularly distributed nodes to represent the domain and a $3 \times 3$ Gauss integration scheme.

The results, shown in tables 2.8–2.11, are in agreement with the reference solution [35] and present a behaviour coherent with the one described above.

Additionally, a graphical representation of the variation of the behaviour with the number of nodes is also presented in figures 2.4–2.7.

*2.4.1.1 Random nodal distribution*

It was explained how the nodal distribution and the background mesh are two separate objects, independent from each other. This section will show this concept through an example in which the nodes of the background mesh are not coincident with the nodes used to represent the problem domain.

A squared isotropic nanoplate is represented by a randomly distributed set of nodes. An initially evenly spaced population of nodes is randomly perturbed in the $x$

**Table 2.8.** Nondimensional values of $\bar{w}$ for $b/a = 0.5$ obtained for $\alpha_C = 1$, $q = 1.01$ and $\alpha_s = 2.3$.

| BC | $\ell$ (nm) | Analytical | Present | Error (%) |
|---|---|---|---|---|
| SSSS | 0 | 0.633 0 | 0.632 1 | 0.142 2 |
| | 0.2 | 0.621 8 | 0.621 4 | 0.064 3 |
| | 0.5 | 0.568 0 | 0.577 3 | 1.637 3 |
| | 1 | 0.431 6 | 0.478 7 | 10.912 9 |
| CCCC | 0 | 0.158 3 | 0.159 7 | 0.884 4 |
| | 0.2 | 0.147 7 | 0.148 2 | 0.338 5 |
| | 0.5 | 0.110 9 | 0.110 8 | 0.090 2 |
| | 1 | 0.059 0 | 0.058 6 | 0.678 0 |
| SCSC | 0 | 0.163 2 | 0.164 5 | 0.796 6 |
| | 0.2 | 0.152 3 | 0.152 8 | 0.328 3 |
| | 0.5 | 0.114 6 | 0.114 5 | 0.087 3 |
| | 1 | 0.061 2 | 0.060 8 | 0.653 6 |
| SFSF | 0 | 14.644 6 | 14.426 9 | 1.486 6 |
| | 0.2 | 14.581 4 | 14.348 6 | 1.596 6 |
| | 0.5 | 14.256 1 | 13.989 3 | 1.871 5 |
| | 1 | 13.200 5 | 13.003 3 | 1.493 9 |
| SCSF | 0 | 3.639 2 | 3.633 2 | 0.164 9 |
| | 0.2 | 3.597 6 | 3.583 0 | 0.405 8 |
| | 0.5 | 3.410 2 | 3.376 4 | 0.991 1 |
| | 1 | 2.911 1 | 2.833 0 | 2.682 8 |

**Table 2.9.** Nondimensional values of $\bar{w}$ for $b/a = 1$ obtained for $\alpha_C = 2.25$, $q = 0.1$ and $\alpha_s = 2.2$.

| BC | $\ell$ (nm) | Analytical | Present | Error (%) |
|---|---|---|---|---|
| SSSS | 0 | 4.062 4 | 4.048 1 | 0.352 0 |
| | 0.2 | 4.033 0 | 4.019 7 | 0.329 8 |
| | 0.5 | 3.884 4 | 3.882 4 | 0.051 5 |
| | 1 | 3.423 1 | 3.479 4 | 1.513 2 |
| CCCC | 0 | 1.265 3 | 1.268 8 | 0.276 6 |
| | 0.2 | 1.233 3 | 1.235 7 | 0.340 5 |
| | 0.5 | 1.097 9 | 1.106 1 | 0.746 9 |
| | 1 | 0.794 6 | 0.810 1 | 1.950 7 |
| SCSC | 0 | 1.917 1 | 1.917 4 | 0.015 6 |
| | 0.2 | 1.878 3 | 1.879 7 | 0.074 5 |
| | 0.5 | 1.709 3 | 1.716 9 | 0.444 6 |
| | 1 | 1.304 0 | 1.325 5 | 1.648 8 |
| SFSF | 0 | 15.011 3 | 14.983 7 | 0.183 9 |
| | 0.2 | 14.947 0 | 14.918 1 | 0.193 3 |
| | 0.5 | 14.616 5 | 14.624 4 | 0.054 0 |
| | 1 | 13.541 5 | 13.813 0 | 1.977 8 |
| SCSF | 0 | 11.235 9 | 11.226 5 | 0.083 7 |
| | 0.2 | 11.170 3 | 11.158 9 | 0.102 1 |
| | 0.5 | 10.845 4 | 10.843 1 | 0.021 2 |
| | 1 | 9.841 6 | 9.923 4 | 0.831 2 |

**Table 2.10.** Nondimensional values of $\bar{w}$ for $b/a = 2$ obtained for $\alpha_C = 1.45$, $q = 0.4$ and $\alpha_s = 2.2$.

| BC | $\ell$ (nm) | Analytical | Present | Error (%) |
|---|---|---|---|---|
| SSSS | 0 | 10.128 7 | 10.065 8 | 0.621 0 |
| | 0.2 | 10.083 3 | 10.020 1 | 0.626 8 |
| | 0.5 | 9.850 2 | 9.810 1 | 0.407 1 |
| | 1 | 9.088 6 | 9.192 5 | 1.143 2 |
| CCCC | 0 | 2.533 0 | 2.539 3 | 0.248 7 |
| | 0.2 | 2.486 1 | 2.493 3 | 0.289 6 |
| | 0.5 | 2.279 9 | 2.298 3 | 0.807 1 |
| | 1 | 1.774 1 | 1.816 8 | 2.406 9 |
| SCSC | 0 | 8.445 0 | 8.416 4 | 0.338 7 |
| | 0.2 | 8.388 7 | 8.357 3 | 0.374 3 |
| | 0.5 | 8.114 5 | 8.085 8 | 0.353 7 |
| | 1 | 7.282 0 | 7.315 5 | 0.460 0 |
| SFSF | 0 | 15.202 2 | 15.261 4 | 0.389 4 |
| | 0.2 | 15.137 3 | 15.164 1 | 0.177 0 |
| | 0.5 | 14.804 0 | 14.738 7 | 0.441 1 |
| | 1 | 13.721 1 | 13.662 4 | 0.427 8 |
| SCSF | 0 | 14.949 1 | 15.016 9 | 0.453 5 |
| | 0.2 | 14.883 8 | 14.918 8 | 0.235 2 |
| | 0.5 | 14.549 3 | 14.488 9 | 0.415 1 |
| | 1 | 13.465 8 | 13.387 9 | 0.578 5 |

**Table 2.11.** Nondimensional values of $\bar{w}$ for $b/a = 3$ obtained for $\alpha_C = 1.5$, $q = -0.25$ and $\alpha_s = 2.5$.

| BC | $\ell$ (nm) | Analytical | Present | Error (%) |
|---|---|---|---|---|
| SSSS | 0 | 12.232 8 | 12.321 1 | 0.721 8 |
| | 0.2 | 12.183 9 | 12.270 6 | 0.711 6 |
| | 0.5 | 11.932 4 | 12.054 1 | 1.019 9 |
| | 1 | 11.104 6 | 11.479 3 | 3.374 3 |
| CCCC | 0 | 2.617 2 | 2.666 0 | 1.864 6 |
| | 0.2 | 2.570 6 | 2.904 1 | 1.303 2 |
| | 0.5 | 2.363 9 | 2.389 3 | 1.074 5 |
| | 1 | 1.851 8 | 1.868 7 | 0.912 6 |
| SCSC | 0 | 11.681 3 | 11.771 8 | 0.774 7 |
| | 0.2 | 11.628 6 | 11.717 2 | 0.761 9 |
| | 0.5 | 11.362 3 | 11.479 7 | 1.033 2 |
| | 1 | 10.503 9 | 10.838 2 | 3.182 6 |
| SFSF | 0 | 15.218 1 | 15.510 1 | 1.918 8 |
| | 0.2 | 15.153 2 | 15.322 3 | 1.115 9 |
| | 0.5 | 14.819 6 | 14.615 2 | 1.379 3 |
| | 1 | 13.736 1 | 13.067 7 | 4.866 0 |
| SCSF | 0 | 15.203 5 | 15.498 4 | 1.939 7 |
| | 0.2 | 15.138 5 | 15.310 9 | 1.138 8 |
| | 0.5 | 14.804 9 | 14.603 7 | 1.359 0 |
| | 1 | 13.721 2 | 13.054 8 | 4.856 7 |

and $y$ directions by the introduction of a small perturbation having a maximum magnitude of 0.3. Nodal densities of $5 \times 5$, $9 \times 9$, $11 \times 11$ and $15 \times 15$ nodes are accounted for and perturbed as described. An example of what the background mesh and nodal distribution look like is provided in figure 2.8 for the $15 \times 15$ nodal density case.

Simply-supported and clamped boundary conditions are considered and the results are computed in their nondimensional form as expressed in equation (2.23) and compared in terms of percentage error as given in (2.24).

The convergence results are obtained by keeping the nondimensional parameters $C$ and $q$ introduced by the RPIM constant as the nodal density varies. The size $\alpha_s$ of the support domain, however, is adapted to the changing number of nodes and is considered to vary between 1.8 and 2.4.

The background mesh is considered to have the same number of nodes as the collocation points but it is considered regular. Moreover, a $3 \times 3$ Gauss integration rule is employed for numerical integrals evaluation.

Results are presented in tables 2.12–2.15.

The trend of the error remains coherent with the previous analyses and convergence is ensured as well. It appears that using a regular or random nodal

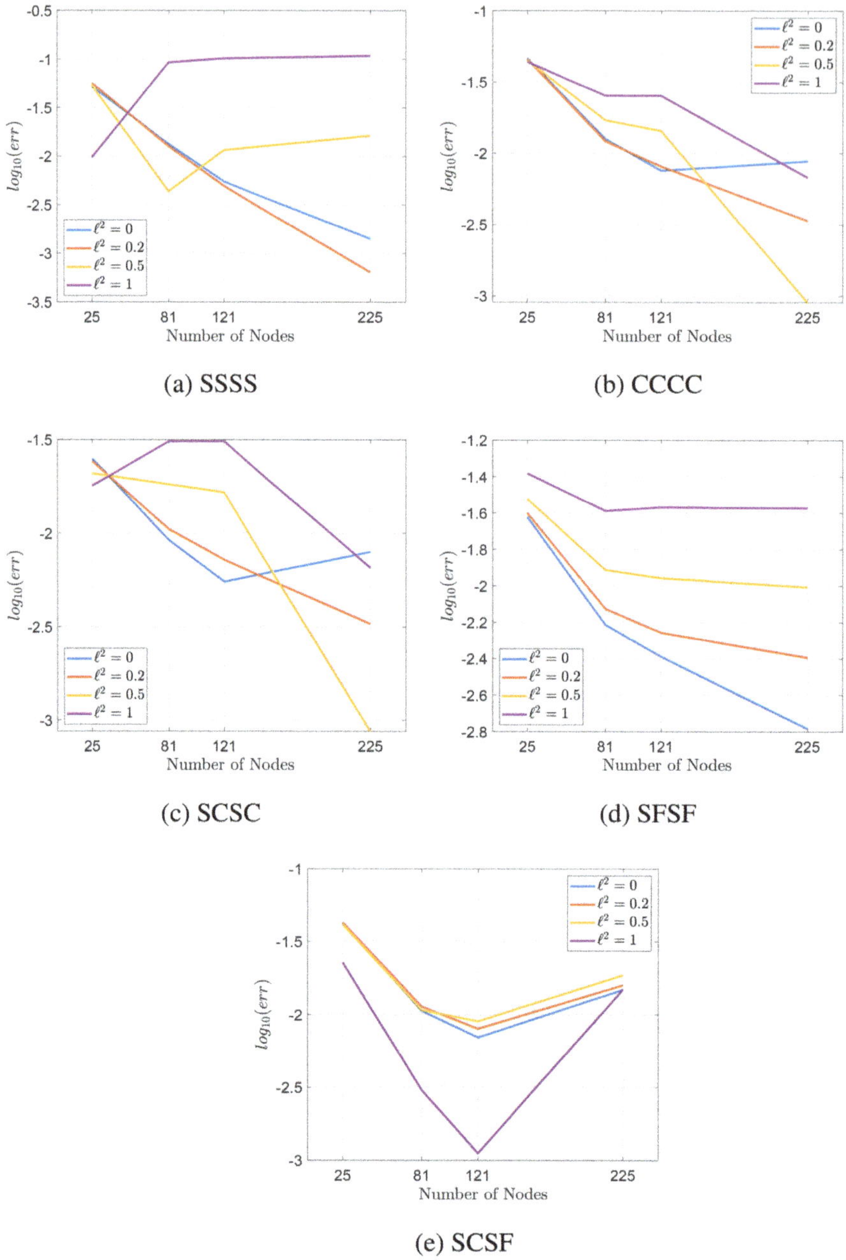


**Figure 2.4.** Convergence analysis for isotropic nanoplates with ratio $b/a = 0.5$.

distribution does not significantly affects the accuracy of the final results, which remains high. Note that the observations on the influence of the nonlocal parameter $\ell$ as well as that of the high order derivatives involved in the calculation still stand.

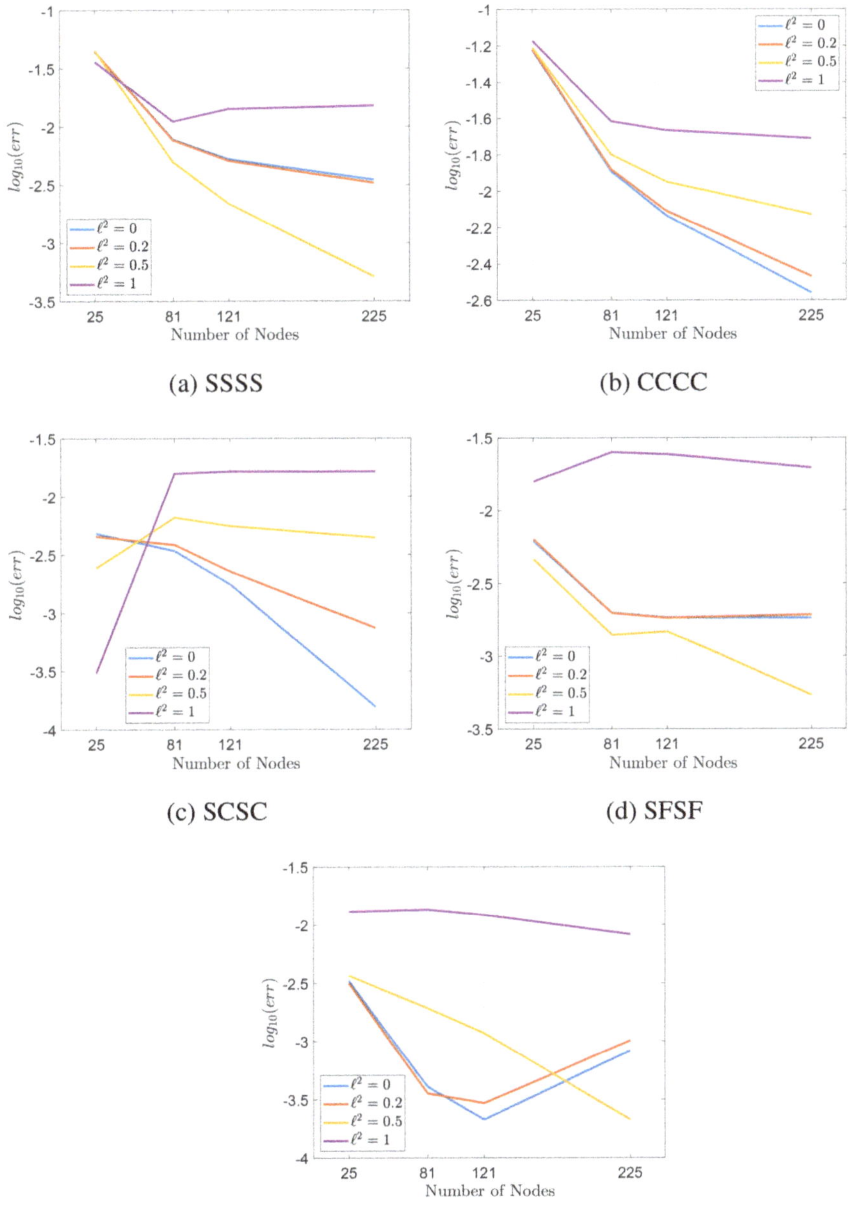


(a) SSSS (b) CCCC

(c) SCSC (d) SFSF

(e) SCSF

**Figure 2.5.** Convergence analysis for isotropic nanoplates with ratio $b/a = 1$.

## 2.4.2 Composite plates

Static and dynamic analysis of composite nanoplates are now performed. Different kinds of cross-ply nanoplates are analysed in bending, free vibrations and buckling while different material characteristic and lamination sequences are accounted for.

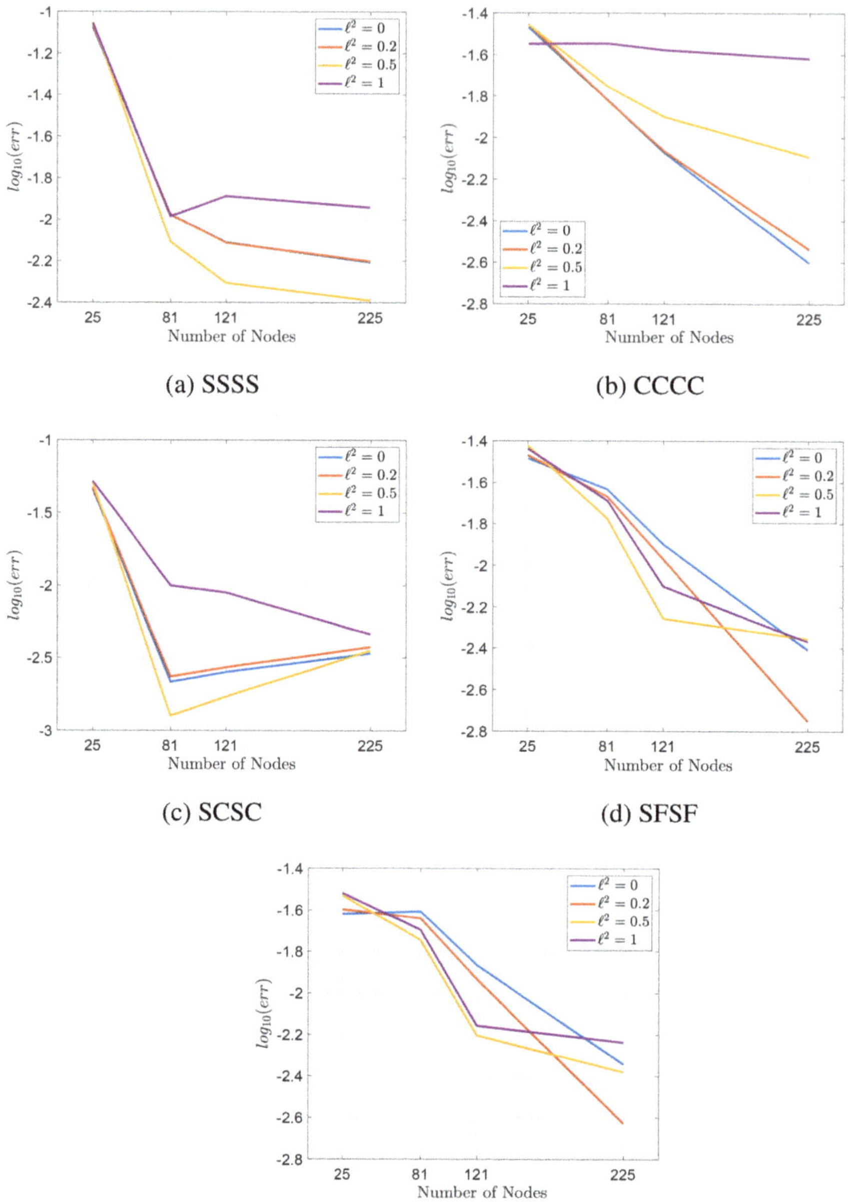


(a) SSSS (b) CCCC

(c) SCSC (d) SFSF

(e) SCSF

**Figure 2.6.** Convergence analysis for isotropic nanoplates with ratio $b/a = 2$.

*2.4.2.1 Static analysis*

The first analysis is performed on a squared simply-supported nanoplate subjected to a sinusoidal load. Different lamination schemes are studied: 0, (0/90), $(0/90)_2$ and $(0/90)_4$. The material property used are taken as $E_1/E_2 = 25$, $\nu_{12} = 0.25$, $G_{12} = G_{13} = 0.5E_2$, $G_{23} = 0.2E_2$ while the thickness is given by $h = a/100$.

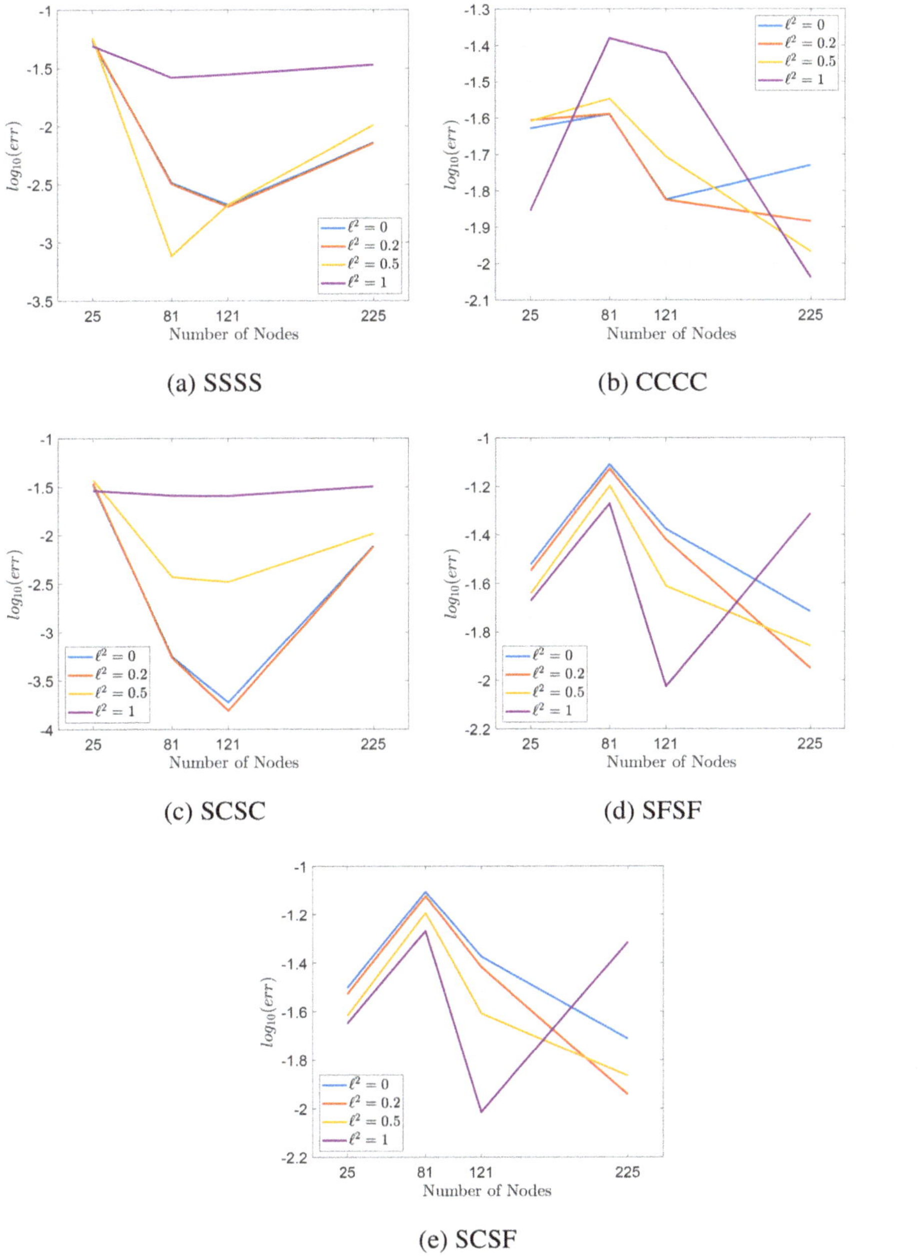


**Figure 2.7.** Convergence analysis for isotropic nanoplates with ratio $b/a = 3$.

Results are obtained in terms of nondimensional mid-deflection:

$$\bar{w} = w_0 \frac{E_2 h^3}{q_s a^4} \tag{2.25}$$

where $q_s$ is the magnitude of the sinusoidal load.

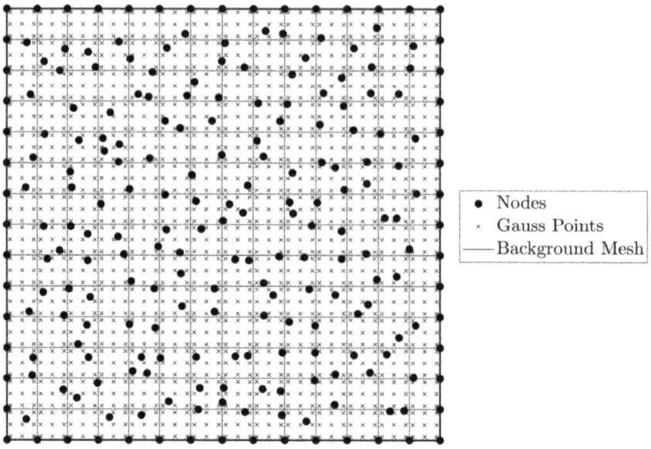


**Figure 2.8.** Square plate discretised by a random nodal distribution. A regular mesh is employed for integration.

**Table 2.12.** Nondimensional values of $\bar{w}$ for SSSS and CCCC square nanoplate obtained for $C = 1.42, q = 0.9$ and $5 \times 5$ randomly distributed nodes.

| Boundary conditions | $\ell$ (nm) | Reference solution | Result | Error (%) |
|---|---|---|---|---|
| SSSS | 0 | 4.062 4 | 3.660 9 | 9.883 3 |
| | 0.2 | 4.033 0 | 3.602 8 | 10.667 0 |
| | 0.5 | 3.884 4 | 3.340 8 | 13.994 4 |
| | 1 | 3.423 1 | 2.665 5 | 22.132 0 |
| CCCC | 0 | 1.265 3 | 1.280 1 | 1.169 7 |
| | 0.2 | 1.233 3 | 1.234 1 | 0.064 9 |
| | 0.5 | 1.097 9 | 1.059 0 | 3.543 1 |
| | 1 | 0.794 6 | 0.722 0 | 9.136 7 |

**Table 2.13.** Nondimensional values of $\bar{w}$ for SSSS and CCCC square nanoplate obtained for $C = 1.42, q = 0.9$ and $9 \times 9$ randomly distributed nodes.

| Boundary conditions | $\ell$ (*nm*) | Reference solution | Result | Error (%) |
|---|---|---|---|---|
| SSSS | 0 | 4.062 4 | 4.010 5 | 1.277 6 |
| | 0.2 | 4.033 0 | 3.949 8 | 2.063 0 |
| | 0.5 | 3.884 4 | 3.754 1 | 3.354 4 |
| | 1 | 3.423 1 | 3.220 0 | 5.933 2 |
| CCCC | 0 | 1.265 3 | 1.265 0 | 0.023 7 |
| | 0.2 | 1.233 3 | 1.224 0 | 0.754 1 |
| | 0.5 | 1.097 9 | 1.070 2 | 2.523 0 |
| | 1 | 0.794 6 | 0.751 4 | 5.436 7 |

**Table 2.14.** Nondimensional values of $\bar{w}$ for SSSS and CCCC square nanoplate obtained for $C = 1.42, q = 0.9$ and $11 \times 11$ randomly distributed nodes.

| Boundary conditions | $\ell$ (*nm*) | Reference solution | Result | Error (%) |
|---|---|---|---|---|
| SSSS | 0 | 4.062 4 | 4.065 1 | 0.066 5 |
| | 0.2 | 4.033 0 | 4.008 9 | 0.597 6 |
| | 0.5 | 3.884 4 | 3.825 5 | 1.516 3 |
| | 1 | 3.423 1 | 3.314 1 | 3.184 2 |
| CCCC | 0 | 1.265 3 | 1.266 5 | 0.094 8 |
| | 0.2 | 1.233 3 | 1.230 0 | 0.267 6 |
| | 0.5 | 1.097 9 | 1.083 4 | 1.320 7 |
| | 1 | 0.794 6 | 0.768 7 | 3.259 5 |

**Table 2.15.** Nondimensional values of $\bar{w}$ for SSSS and CCCC square nanoplate obtained for $C = 1.42, q = 0.9$ and $15 \times 15$ randomly distributed nodes.

| Boundary conditions | $\ell$ (*nm*) | Reference solution | Result | Error (%) |
|---|---|---|---|---|
| SSSS | 0 | 4.062 4 | 4.062 7 | 0.007 4 |
| | 0.2 | 4.033 0 | 4.021 3 | 0.290 1 |
| | 0.5 | 3.884 4 | 3.875 6 | 0.226 5 |
| | 1 | 3.423 1 | 3.454 9 | 0.929 0 |
| CCCC | 0 | 1.265 3 | 1.272 3 | 0.553 2 |
| | 0.2 | 1.233 3 | 1.231 0 | 0.186 5 |
| | 0.5 | 1.097 9 | 1.093 0 | 0.446 3 |
| | 1 | 0.794 6 | 0.790 6 | 0.503 4 |

A nodal density of $11 \times 11$ evenly distributed nodes is used for domain representation and the numerical integration is performed using $2 \times 2$ Gauss points. The results, shown in table 2.16, are compared in terms of percentage error as given in equation (2.24).

Similarly to what is observed in the isotropic case, the results are affected by the relative influence of the nodal density and the higher-order derivatives as well as the value of the parameter $\ell$. The RPIM seems to perform well even when applied to composite plates. In fact, it can be observed how the value of the percentage error never even reaches 4%.

#### *2.4.2.2 Buckling analysis*

Results of the dynamic analysis of simply-supported cross-ply plates are now presented.

First, the buckling analysis is performed considering plates subjected to both uniaxial and biaxial in-plane compressive loads. Each ply has mechanical properties

**Table 2.16.** Nondimensional values of $\bar{w}$ mid-deflection for composite nanoplates, obtained for $\alpha_C = 1.85$, $q = -1.6$, $\alpha_s = 2.4$.

| $\ell$ (nm) | Laminate | Reference [27] | Result | Error (%) |
|---|---|---|---|---|
| 0 | 0 | 0.004 312 | 0.004 349 | 0.858 071 |
| | (0/90) | 0.010 636 | 0.010 693 | 0.535 916 |
| | $(0/90)_2$ | 0.005 065 | 0.005 109 | 0.868 707 |
| | $(0/90)_4$ | 0.004 479 | 0.004 520 | 0.915 383 |
| 0.05 | 0 | 0.002 170 | 0.002 137 | 1.520 737 |
| | (0/90) | 0.003 931 | 0.004 033 | 2.594 760 |
| | $(0/90)_2$ | 0.002 444 | 0.002 351 | 3.805 237 |
| | $(0/90)_4$ | 0.002 233 | 0.002 167 | 2.955 665 |
| 0.1 | 0 | 0.001 450 | 0.001 490 | 2.758 621 |
| | (0/90) | 0.002 522 | 0.002 584 | 2.458 366 |
| | $(0/90)_2$ | 0.001 623 | 0.001 605 | 1.109 057 |
| | $(0/90)_4$ | 0.001 490 | 0.001 497 | 0.469 799 |

**Table 2.17.** Buckling of composite $(0/90)_2$ plate, $a/b = 0.5$, under uniaxial compressive load, represented by 15 × 15 nodes, with different $E_1/E_2$ and $\alpha_C = 1.52$, $q = -0.34$, $\alpha_s = 1.8$.

| $E_1/E_2$ | $\ell$ (nm) | Exact [62] | Results | Error (%) |
|---|---|---|---|---|
| 5 | 0 | 4.705 | 4.673 | 0.680 |
| | 0.05 | 7.608 | 7.674 | 0.868 |
| | 0.1 | 10.510 | 10.538 | 0.266 |
| 10 | 0 | 4.157 | 4.142 | 0.361 |
| | 0.05 | 6.720 | 6.783 | 0.938 |
| | 0.1 | 9.284 | 9.280 | 0.043 |
| 20 | 0 | 3.828 | 3.823 | 0.131 |
| | 0.05 | 6.189 | 6.252 | 1.018 |
| | 0.1 | 8.550 | 8.531 | 0.222 |
| 25 | 0 | 3.757 | 3.754 | 0.080 |
| | 0.05 | 6.074 | 6.138 | 1.054 |
| | 0.1 | 8.391 | 8.370 | 0.250 |
| 40 | 0 | 3.647 | 3.647 | 0.000 |
| | 0.05 | 5.896 | 5.961 | 1.102 |
| | 0.1 | 8.145 | 8.121 | 0.295 |

$G_{12} = 0.5E_2$, $\nu = 0.25$ and variable $E_1/E_2$, the thickness is considered equal for each ply and aspect ratio $a/b$ and $E_1/E_2$ ratio are variable. Different values of the nonlocal ratio $(\ell/a)^2$ are also considered, namely 0, 0.05 and 0.1. The plate is represented by 15 × 15 regularly distributed nodes and is characterised by a $(0/90)_2$ lamination scheme.

**Table 2.18.** Buckling of composite $(0/90)_2$ plate, $a/b = 0.5$, under biaxial compressive load, represented by $15 \times 15$ nodes, with different $E_1/E_2$ and $\alpha_C = 1.52$, $q = -0.34$, $\alpha_s = 1.8$.

| $E_1/E_2$ | $\ell$ (nm) | Exact [62] | Results | Error (%) |
|---|---|---|---|---|
| 5 | 0 | 3.764 | 3.749 | 0.399 |
| | 0.05 | 6.086 | 6.140 | 0.887 |
| | 0.1 | 8.408 | 8.430 | 0.262 |
| 10 | 0 | 3.325 | 3.322 | 0.090 |
| | 0.05 | 5.376 | 5.426 | 0.930 |
| | 0.1 | 7.428 | 7.423 | 0.067 |
| 20 | 0 | 3.062 | 3.066 | 0.131 |
| | 0.05 | 4.951 | 5.001 | 1.010 |
| | 0.1 | 6.840 | 6.823 | 0.249 |
| 25 | 0 | 3.005 | 3.011 | 0.200 |
| | 0.05 | 4.859 | 4.910 | 1.050 |
| | 0.1 | 6.713 | 6.694 | 0.283 |
| 40 | 0 | 2.917 | 2.925 | 0.274 |
| | 0.05 | 4.717 | 4.768 | 1.081 |
| | 0.1 | 6.516 | 6.495 | 0.322 |

**Table 2.19.** Buckling of composite $(0/90)_2$ plate, $a/b = 1$, under uniaxial compressive load, represented by $15 \times 15$ nodes, with different $E_1/E_2$ and $\alpha_C = 1.7$, $q = -0.5$, $\alpha_s = 2.8$.

| $E_1/E_2$ | $\ell$ (nm) | Exact [62] | Results | Error (%) |
|---|---|---|---|---|
| 5 | 0 | 2.643 | 2.630 | 0.492 |
| | 0.05 | 5.252 | 5.221 | 0.590 |
| | 0.1 | 7.861 | 7.708 | 1.946 |
| 10 | 0 | 2.189 | 2.181 | 0.366 |
| | 0.05 | 4.350 | 4.364 | 0.322 |
| | 0.1 | 6.551 | 6.447 | 0.983 |
| 20 | 0 | 1.923 | 1.917 | 0.312 |
| | 0.05 | 3.821 | 3.857 | 0.942 |
| | 0.1 | 5.719 | 5.700 | 0.332 |
| 25 | 0 | 1.866 | 1.861 | 0.268 |
| | 0.05 | 3.708 | 3.749 | 1.106 |
| | 0.1 | 5.550 | 5.540 | 0.180 |
| 40 | 0 | 1.778 | 1.774 | 0.225 |
| | 0.05 | 3.533 | 3.580 | 1.330 |
| | 0.1 | 5.289 | 5.291 | 0.038 |

**Table 2.20.** Buckling of composite $(0/90)_2$ plate, $a/b = 1$, under biaxial compressive load, represented by $15 \times 15$ nodes, with different $E_1/E_2$ and $\alpha_C = 1.7$, $q = -0.5$, $\alpha_s = 2.8$.

| $E_1/E_2$ | $\ell$ (nm) | Exact [62] | Results | Error (%) |
|---|---|---|---|---|
| 5 | 0 | 1.322 | 1.315 | 0.529 |
| | 0.05 | 2.626 | 2.611 | 0.571 |
| | 0.1 | 3.930 | 3.854 | 1.934 |
| 10 | 0 | 1.095 | 1.091 | 0.365 |
| | 0.05 | 2.175 | 2.182 | 0.322 |
| | 0.1 | 3.255 | 3.224 | 0.952 |
| 20 | 0 | 0.962 | 0.959 | 0.312 |
| | 0.05 | 1.911 | 1.929 | 0.942 |
| | 0.1 | 2.859 | 2.850 | 0.315 |
| 25 | 0 | 0.933 | 0.931 | 0.214 |
| | 0.05 | 1.854 | 1.874 | 1.079 |
| | 0.1 | 2.775 | 2.770 | 0.180 |
| 40 | 0 | 0.889 | 0.887 | 0.225 |
| | 0.05 | 1.767 | 1.790 | 1.302 |
| | 0.1 | 2.644 | 2.646 | 0.076 |

**Table 2.21.** Buckling of composite $(0/90)_2$ plate, $a/b = 1.5$, under uniaxial compressive load, represented by $15 \times 15$ nodes, with different $E_1/E_2$ and $\alpha_C = 1.42$, $q = 0.01$, $\alpha_s = 2.35$.

| $E_1/E_2$ | $\ell$ (nm) | Exact [62] | Results | Error (%) |
|---|---|---|---|---|
| 5 | 0 | 2.955 | 2.926 | 0.981 |
| | 0.05 | 8.539 | 8.524 | 0.176 |
| | 0.1 | 13.799 | 13.476 | 2.341 |
| 10 | 0 | 2.487 | 2.461 | 1.045 |
| | 0.05 | 7.281 | 7.296 | 0.206 |
| | 0.1 | 11.766 | 11.511 | 2.167 |
| 20 | 0 | 2.211 | 2.186 | 1.131 |
| | 0.05 | 6.538 | 6.573 | 0.535 |
| | 0.1 | 10.565 | 10.354 | 1.997 |
| 25 | 0 | 2.152 | 2.128 | 1.115 |
| | 0.05 | 6.378 | 6.418 | 0.627 |
| | 0.1 | 10.307 | 10.107 | 1.940 |
| 40 | 0 | 2.061 | 2.037 | 1.165 |
| | 0.05 | 6.132 | 6.178 | 0.750 |
| | 0.1 | 9.909 | 9.724 | 1.867 |

**Table 2.22.** Buckling of composite $(0/90)_2$ plate, $a/b = 1.5$, under biaxial compressive load, represented by $15 \times 15$ nodes, with different $E_1/E_2$ and $\alpha_C = 1.42$, $q = 0.01$, $\alpha_s = 2.35$.

| $\mathbf{E_1/E_2}$ | $\ell$ (nm) | Exact [62] | Results | Error (%) |
|---|---|---|---|---|
| 5 | 0 | 1.009 | 1.013 | 0.396 |
| | 0.05 | 2.628 | 2.621 | 0.266 |
| | 0.1 | 4.246 | 4.142 | 2.449 |
| 10 | 0 | 0.860 | 0.864 | 0.465 |
| | 0.05 | 2.240 | 2.243 | 0.134 |
| | 0.1 | 3.620 | 3.537 | 2.293 |
| 20 | 0 | 0.773 | 0.775 | 0.259 |
| | 0.05 | 2.012 | 2.021 | 0.447 |
| | 0.1 | 3.251 | 3.182 | 2.122 |
| 25 | 0 | 0.754 | 0.757 | 0.398 |
| | 0.05 | 1.963 | 1.973 | 0.509 |
| | 0.1 | 3.171 | 3.105 | 2.081 |
| 40 | 0 | 0.725 | 0.727 | 0.276 |
| | 0.05 | 1.887 | 1.899 | 0.636 |
| | 0.1 | 3.049 | 2.988 | 2.001 |

The critical load is computed in its nondimensional form as:

$$\bar{N} = \frac{\hat{N}_{xx} b^2}{\pi^2 D_{22}} \tag{2.26}$$

where $D_{22}$ is the bending stiffness appearing in matrix $\mathbf{D}$ as defined in [68].

Numerical results, shown in tables 2.17–2.22, are compared in terms of percentage error as:

$$\text{err}_{\%} = 100 \frac{|N_e - \bar{N}|}{N_e} \tag{2.27}$$

where $N_e$ is the reference solution as given in [62].

It can be noted how the results can deviate from the classical trend of having error magnitude increasing with an increase of the internal length scale parameter $\ell$. The phenomenon can be due to different factors.

One reason can be found in the parameters introduced by the RPIM, $q$ and $\alpha_C$ which appear due to the MQ radial basis function chosen and $\alpha_s$ which regulates the size of the support domain. The tuning of these parameters strongly affects the results and is identified as a first cause that may justify the aforementioned behaviour.

Another cause may be identified in the higher-order derivatives introduced by the strain gradient theory. Numerical errors are amplified by the use of a mesh

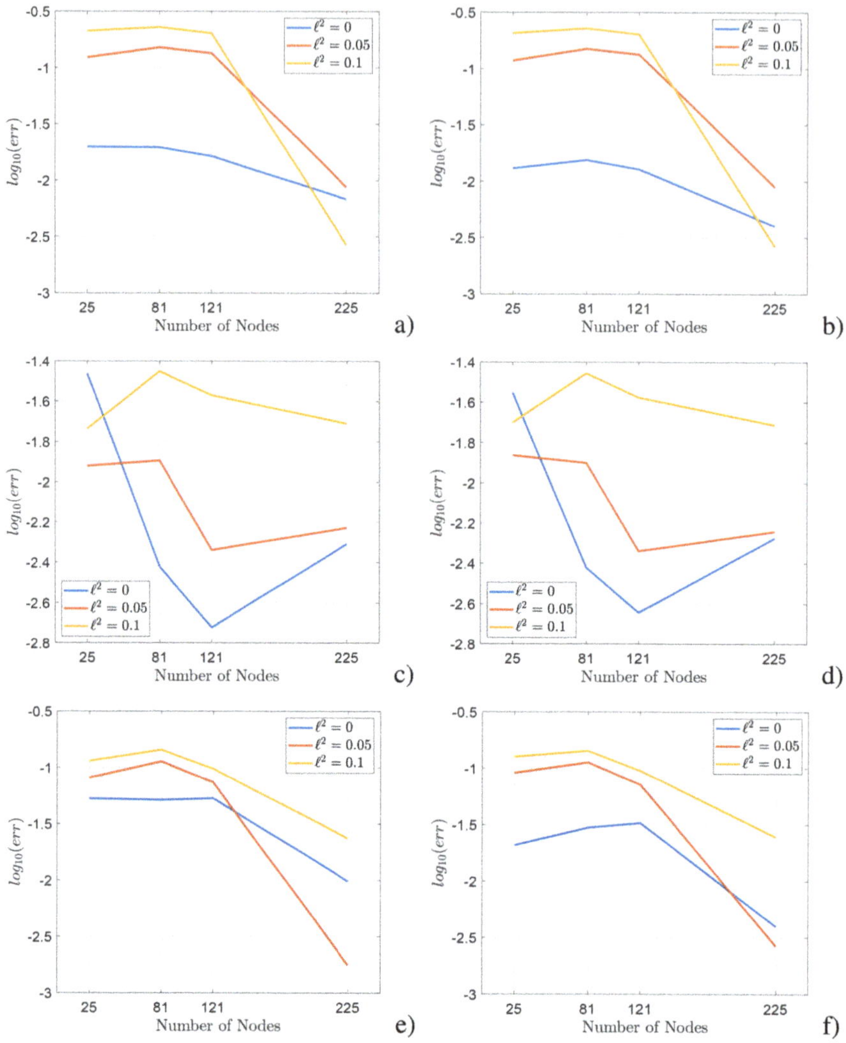


**Figure 2.9.** Convergence analysis for nanoplates with ratio $E_1/E_2 = 5$ analysed in buckling: (a) $a/b = 0.5$ uniaxial buckling, (b) $a/b = 0.5$ biaxial buckling, (c) $a/b = 1$ uniaxial buckling, (d) $a/b = 1$ biaxial buckling, (e) $a/b = 1.5$ uniaxial buckling, and (f) $a/b = 1.5$ biaxial buckling.

free method as the spatial coordinates of the nodes, as explained, appear to be multiplied several times, once per each derivative order. Variation of the aspect ratio $a/b$ while keeping the nodal density constant, implies that coordinates, hence numbers, of different magnitude enter the shape functions thus altering the traditional error trend. The convergence analysis shown in figures 2.9–2.13 provides a graphical representation of this phenomenon. However, it can be seen how the numerical results are overall in good agreement with the reference solution.

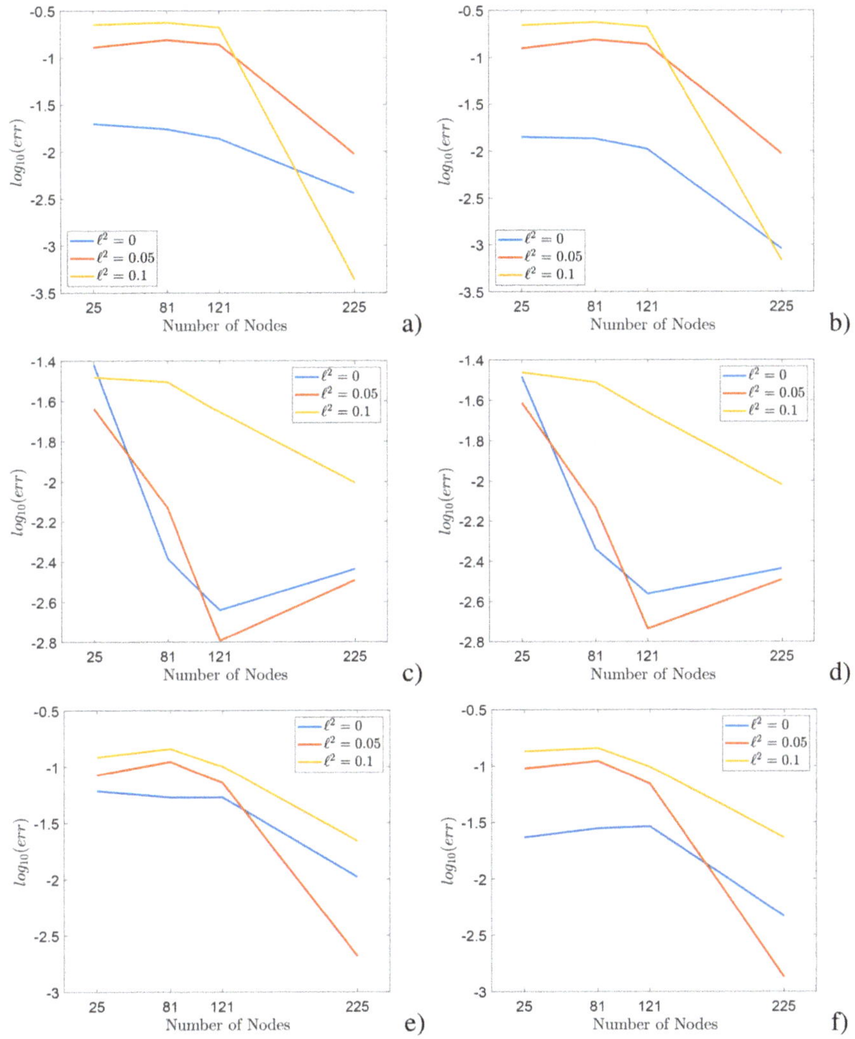


**Figure 2.10.** Convergence analysis for nanoplates with ratio $E_1/E_2 = 10$ analysed in buckling: (a) $a/b = 0.5$ uniaxial buckling, (b) $a/b = 0.5$ biaxial buckling, (c) $a/b = 1$ uniaxial buckling, (d) $a/b = 1$ biaxial buckling, (e) $a/b = 1.5$ uniaxial buckling, and (f) $a/b = 1.5$ biaxial buckling.

## 2.5 Dynamic analysis

The next analysis focuses on the free vibration of a simply-supported squared plate characterised by different lamination sequences. The plate has density $\rho = 2300$ kg m$^{-3}$ and mechanical properties $E_1/E_2 = 10$, $G_{12} = G_{13} = 0.5E_2$, $\nu_{12} = 0.25$. $11 \times 11$ regular nodes are used to represent the plate domain and a $2 \times 2$ Gauss integration scheme is used for integrals evaluation.

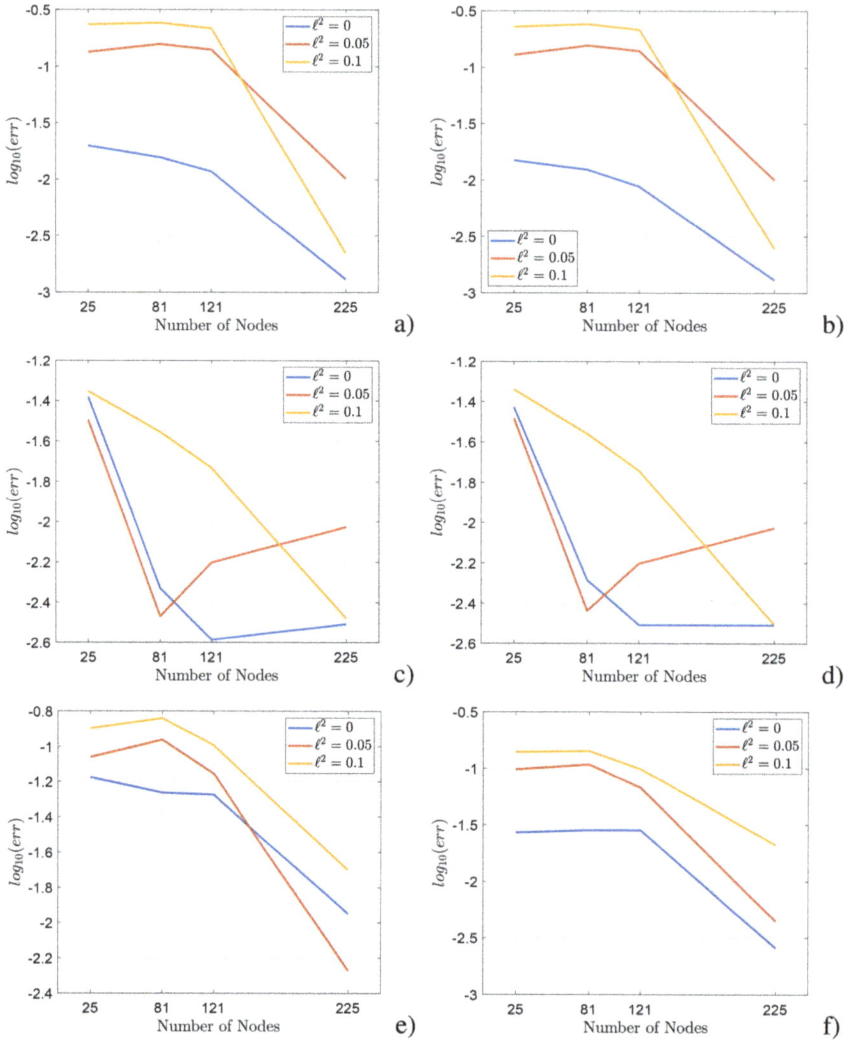


**Figure 2.11.** Convergence analysis for nanoplates with ratio $E_1/E_2 = 20$ analysed in buckling: (a) $a/b = 0.5$ uniaxial buckling, (b) $a/b = 0.5$ biaxial buckling, (c) $a/b = 1$ uniaxial buckling, (d) $a/b = 1$ biaxial buckling, (e) $a/b = 1.5$ uniaxial buckling, (f) $a/b = 1.5$ biaxial buckling.

The fundamental frequency is given in nondimensional form as:

$$\bar{\omega} = \frac{\omega b^2}{\pi^2}\sqrt{\frac{\rho h}{D_{22}}} \tag{2.28}$$

and results are compared in terms on percentage error as given in (2.27).

Results are shown in table 2.23 for different values of $(\ell/a)^2$ and lamination sequences 0, (0/90), $(0/90)_2$ and $(0/90)_4$.

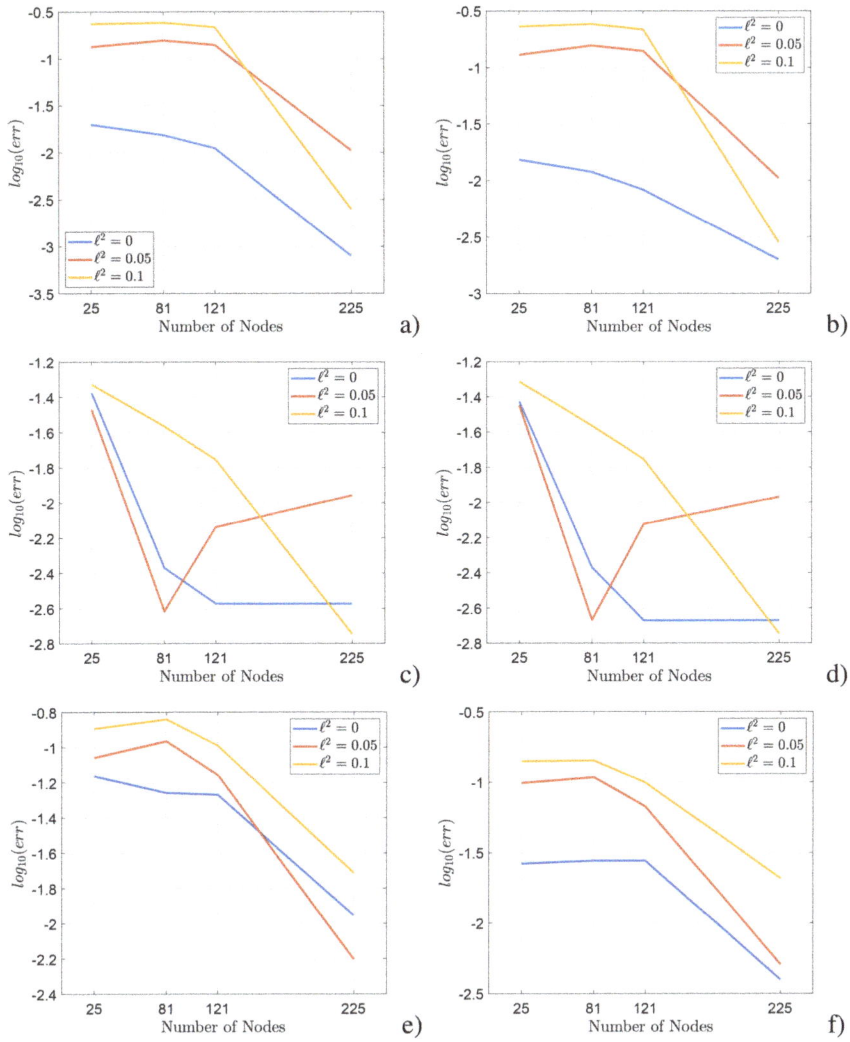


**Figure 2.12.** Convergence analysis for nanoplates with ratio $E_1/E_2 = 25$ analysed in buckling: (a) $a/b = 0.5$ uniaxial buckling, (b) $a/b = 0.5$ biaxial buckling, (c) $a/b = 1$ uniaxial buckling, (d) $a/b = 1$ biaxial buckling, (e) $a/b = 1.5$ uniaxial buckling, and (f) $a/b = 1.5$ biaxial buckling.

## 2.6 Conclusions

This work shows the procedure and results of the analysis of thin nanoplates modelled according to the second-order strain gradient theory performed using a Hermite radial point interpolation method. A thorough theoretical explanation of the RPIM is provided as well as a discussion on the topic of the radial basis functions used. Both isotropic and composite plates are studied. First, the static bending analysis of isotropic nanoplates of different geometries represented by a regular nodal distribution and subjected to different kinds of boundary conditions are

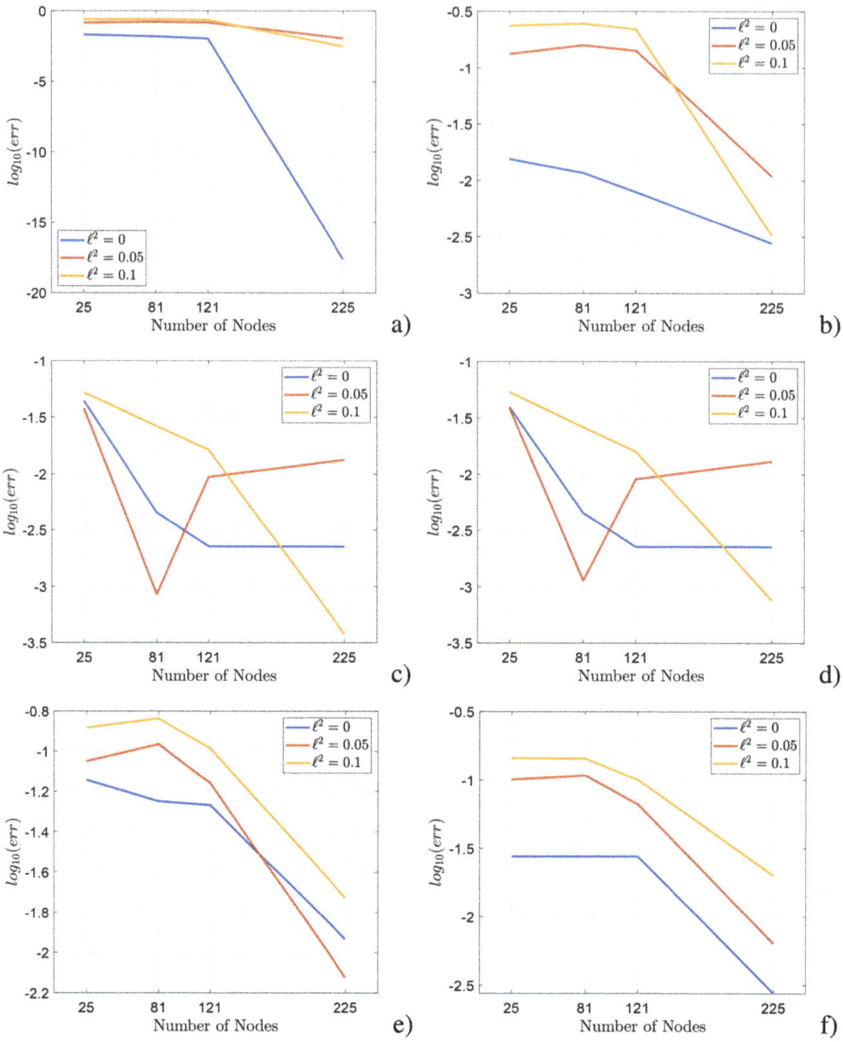


**Figure 2.13.** Convergence analysis for nanoplates with ratio $E_1/E_2 = 40$ analysed in buckling: (a) $a/b = 0.5$ uniaxial buckling, (b) $a/b = 0.5$ biaxial buckling, (c) $a/b = 1$ uniaxial buckling, (d) $a/b = 1$ biaxial buckling, (e) $a/b = 1.5$ uniaxial buckling, and (f) $a/b = 1.5$ biaxial buckling.

studied. The same analysis is then performed on fully clamped and simply-supported nanoplates, represented by nodes randomly scattered in the domain. In both cases convergence is studied by accounting for different nodal densities.

Next, the static bending analysis of composite simply-supported nanoplates is performed. A regular nodal distribution is used to represent the domain and different lamination sequences are accounted for. The free vibration analysis is then performed considering the same conditions. Finally, the buckling analysis is performed. Nanoplates with different geometries and different $E_1/E_2$ ratio are studied while subjected to both uniaxial and biaxial compressive loads. In all the

**Table 2.23.** Nondimensional values of $\bar{\omega}$ natural frequency for square composite nanoplates, 11 × 11 nodal distribution, obtained for $\alpha_C = 1.85$, $q = -1.7$, $\alpha_s = 2$.

| $\ell$ (nm) | Laminate | Reference [40] | Result | Error (%) |
|---|---|---|---|---|
| 0 | (0/90) | 1.183 | 1.182 | 0.085 |
| | $(0/90)_2$ | 1.479 | 1.476 | 0.203 |
| | $(0/90)_4$ | 1.545 | 1.540 | 0.324 |
| 0.05 | (0/90) | 1.888 | 1.866 | 1.165 |
| | $(0/90)_2$ | 2.132 | 2.164 | 1.501 |
| | $(0/90)_4$ | 2.189 | 2.221 | 1.462 |
| 0.1 | (0/90) | 2.333 | 2.307 | 1.114 |
| | $(0/90)_2$ | 2.614 | 2.619 | 0.191 |
| | $(0/90)_4$ | 2.679 | 2.677 | 0.075 |

aforementioned analyses, the RPIM provides reliable accurate results, no matter the geometry, the boundary conditions, the material or the loads applied to the structure.

## References


[1] Liu G R 2009 *Mesh Free Methods: Moving beyond the Finite Element Method* 2nd edn (Boca Raton, FL: CRC Press)

[2] Gu Y T and Liu G R 2001 A local point interpolation method for static and dynamic analysis of thin beams *Comput. Methods Appl. Mech. Eng.* **190** 5515–28

[3] Liu G R and Gu Y T 2001 A point interpolation method for two-dimensional solids *Int. J. Numer. Methods Eng.* **50** 937–51

[4] Liu G R and Gu Y T 2005 *An Introduction to Meshfree Methods and Their Programming* (Berlin: Springer)

[5] Liu G R *et al* 2009 An extended Galerkin weak form and a point interpolation method with continuous strain field and superconvergence using triangular mesh *Comput. Mech.* **43** 651–73

[6] Xu X *et al* 2010 A point interpolation method with locally smoothed strain field (PIM-LS2) for mechanics problems using triangular mesh *Finite Elem. Anal. Des.* **46** 862–74

[7] Liu G R and Gu Y T 2003 A matrix triangularization algorithm for the polynomial point interpolation method *Comput. Methods Appl. Mech. Eng* **192** 2269–95

[8] Wang J G and Liu G 2002 A point interpolation meshless method based on radial basis functions *Int. J. Numer. Methods Eng* **54** 1623–48

[9] Liu G R *et al* 2005 A meshfree radial point interpolation method (RPIM) for three-dimensional solids *Comput. Mech.* **36** 421–30

[10] Li Y *et al* 2020 A node-based smoothed radial point interpolation method with linear strain fields for vibration analysis of solids *Eng. Anal. Bound. Elem.* **114** 8–22

[11] Cui X, Liu G and Li G 2011 A smoothed Hermite radial point interpolation method for thin plate analysis *Arch. Appl. Mech.* **81** 1–18

[12] Akgöz B and Civalek O 2015 Bending analysis of FG microbeams resting on Winkler elastic foundation via strain gradient elasticity *Compos. Struct.* **134** 294–301

[13] Barretta R *et al* 2015 A gradient Eringen model for functionally graded nanorods *Compos. Struct.* **131** 1124–31

[14] Apuzzo A *et al* 2017 Free vibrations of Bernoulli–Euler nano-beams by the stress-driven nonlocal integral model *Composites* B **123** 105–11

[15] Akgöz B and Civalek O 2015 A microstructure-dependent sinusoidal plate model based on the strain gradient elasticity theory *Acta Mech.* **226** 2277–94

[16] Jankowski P *et al* 2021 On the piezoelectric effect on stability of symmetric FGM porous nanobeams *Compos. Struct.* **267** 113880

[17] Ali Faghidian S *et al* 2022 On the analytical and meshless numerical approaches to mixture stress gradient functionally graded nano-bar in tension *Eng. Anal. Bound. Elem.* **134** 571–80

[18] Čanaija M, Barretta R and Sciarra F M D 2016 A gradient elasticity model of Bernoulli–Euler nanobeams in non-isothermal environments *Eur. J. Mech.* A **55** 243–55

[19] Barretta R *et al* 2016 Functionally graded Timoshenko nanobeams: a novel nonlocal gradient formulation *Composites* B **100** 208–19

[20] Barretta R and Marotti de Sciarra F 2013 A nonlocal model for carbon nanotubes under axial loads *Adv. Mater. Sci. Eng.* **2013** 360935

[21] Woinowsky-Krieger S and Timoshenko S P 1959 *Theory of Plates and Shells* (McGraw-Hill: New York)

[22] Babu B and Patel B P 2019 A new computationally efficient finite element formulation for nanoplates using second-order strain gradient Kirchhoff's plate theory *Composites* B **168** 302–11

[23] Barretta R and Luciano R 2014 Exact solutions of isotropic viscoelastic functionally graded Kirchhoff plates *Compos. Struct.* **118** 448–54

[24] Farrahi G H, Faghidian S A and Smith D J 2009 Reconstruction of residual stresses in autofrettaged thick-walled tubes from limited measurements *Int. J. Press. Vessels Pip.* **86** 777–84

[25] Aifantis E C 1992 On the role of gradients in the localization of deformation and fracture *Int. J. Eng. Sci.* **30** 1279–99

[26] Akgöz B and Civalek O 2012 Analysis of micro-sized beams for various boundary conditions based on the strain gradient elasticity theory *Arch. Appl. Mech.* **82** 423–43

[27] Cornacchia F *et al* 2019 Solution for cross- and angle-ply laminated Kirchhoff nano plates in bending using strain gradient theory *Composites* B **173** 107006

[28] Xu B-B, Gao X-W and Cui M 2021 High precision simulation of thermal-mechanical problems in functionally graded materials by spectral element differential method *Compos. Struct.* **270** 114084

[29] Mariano P M *et al* 1999 Constitutive relations for elastic microcracked bodies: from a lattice model to a multifield continuum description *Int. J. Damage Mech.* **8** 153–73

[30] Leonetti L *et al* 2019 Scale effects in orthotropic composite assemblies as micropolar continua: a comparison between weak- and strong-form finite element solutions *Materials* **12** 758

[31] Trovalusci P and Augusti G 1998 A continuum model with microstructure for materials with flaws and inclusions *J. Phys. IV France* **08** 383–90

[32] Fantuzzi N, Trovalusci P and Dharasura S 2019 Mechanical behavior of anisotropic composite materials as micropolar continua *Front. Mater* **6** 59

[33] Mancusi G *et al* 2017 Size effect and dynamic properties of 2D lattice materials *Composites* B **112** 235–42
[34] Barretta R, Luciano R and Marotti de Sciarra F 2015 A fully gradient model for Euler-Bernoulli nanobeams *Math. Probl. Eng.* **2015** 495095
[35] Babu B and Patel B P 2019 Analytical solution for strain gradient elastic Kirchhoff rectangular plates under transverse static loading *Eur. J. Mech.* A **73** 101–11
[36] Luciano R and Willis J R 2001 Non-local constitutive response of a random laminate subjected to configuration-dependent body force *J. Mech. Phys. Solids* **49** 431–44
[37] Cornacchia F *et al* 2019 Tensile strength of the unbonded flexible pipes *Compos. Struct.* **218** 142–51
[38] Ferreira A J M *et al* 2005 Static deformations and vibration analysis of composite and sandwich plates using a layerwise theory and multiquadrics discretizations *Eng. Anal. Bound. Elem.* **29** 1104–14
[39] Apuzzo A *et al* 2017 A closed-form model for torsion of nanobeams with an enhanced nonlocal formulation *Composites* B **108** 315–24
[40] Cornacchia F *et al* 2021 Analytical solution of cross- and angle-ply nano plates with strain gradient theory for linear vibrations and buckling *Mech. Adv. Mater. Struct.* **28** 1201–15
[41] Alizadeh A *et al* 2021 A modified couple stress-based model for the nonlinear vibrational analysis of nano-disks using multiple scales method *J. Appl. Comput. Mech.* **2** 580–96
[42] Mahmure A *et al* 2021 Primary resonance of double-curved nanocomposite shells using nonlinear theory and multi-scales method: modeling and analytical solution *Int. J. Non Linear Mech* **137** 103–816
[43] Avey M *et al* 2021 Nonlinear vibration of multilayer shell-type structural elements with double curvature consisting of CNT patterned layers within different theories *Compos. Struct.* **275** 114401
[44] Deniz A *et al* 2021 Modeling and solution of large amplitude vibration problem of construction elements made of nanocomposites using shear deformation theory *Materials* **14** 3843
[45] Barretta R, Ali Faghidian S and Luciano R 2019 Longitudinal vibrations of nano-rods by stress-driven integral elasticity *Mech. Adv. Mater. Struct.* **26** 1307–15
[46] Numanoğlu H M, Akgöz B and Civalek C 2018 On dynamic analysis of nanorods *Int. J. Eng. Sci.* **130** 33–50
[47] Lazar M, Maugin G A and Aifantis E C 2006 Dislocations in second strain gradient elasticity *Int. J. Solids Struct.* **43** 1787–817
[48] Darban H *et al* 2020 Higher modes of buckling in shear deformable nanobeams *Int. J. Eng. Sci.* **154** 103338
[49] Barretta R and Luciano R 2015 Analogies between Kirchhoff plates and functionally graded Saint-Venant beams under torsion *Contin. Mech. Thermodyn* **27** 499–505
[50] Akgöz B and Civalek O 2016 Bending analysis of embedded carbon nanotubes resting on an elastic foundation using strain gradient theoryc *Acta Astronaut.* **119** 1–12
[51] Demir C and Civalek O 2017 On the analysis of microbeams *Int. J. Eng. Sci.* **121** 14–33
[52] Tuna M, Kirca M and Trovalusci P 2019 Deformation of atomic models and their equivalent continuum counterparts using Eringen's two-phase local/nonlocal model *Mech. Res. Commun.* **97** 26–32
[53] Kamil Żur K *et al* 2021 On the nonlinear dynamics of porous composite nanobeams connected with fullerenes *Compos. Struct.* **274** 114356

[54] Jankowski P, Żur K K and Farajpour A 2022 Analytical and meshless DQM approaches to free vibration analysis of symmetric FGM porous nanobeams with piezoelectric effect *Eng. Anal. Bound. Elem.* **136** 266–89
[55] Ali Faghidian S 2020 Higher-order nonlocal gradient elasticity: a consistent variational theory *Int. J. Eng. Sci.* **154** 103337
[56] Ali Faghidian S 2020 Higher order mixture nonlocal gradient theory of wave propagation *Math. Methods Appl. Sci.* https://doi.org/10.1002/mma.6885
[57] Ali Faghidian S, Kamil Żur K and Reddy J N 2022 A mixed variational framework for higher-order unified gradient elasticity *Int. J. Eng. Sci.* **170** 103603
[58] Ali Faghidian S 2015 Inverse determination of the regularized residual stress and eigenstrain fields due to surface peening *J. Strain Anal. Eng. Des.* **50** 84–91
[59] Faghidian S A *et al* 2012 Measurement, analysis and reconstruction of residual stresses *J. Strain Anal. Eng. Des.* **47** 254–64
[60] Ferreira A J M 2003 A formulation of the multiquadric radial basis function method for the analysis of laminated composite plates *Compos. Struct.* **59** 385–92
[61] Bacciocchi M, Fantuzzi N and Ferreira A J M 2020 Conforming and nonconforming laminated finite element Kirchhoff nanoplates in bending using strain gradient theory *Comput. Struct.* **239** 106322
[62] Bacciocchi M *et al* 2021 Linear eigenvalue analysis of laminated thin plates including the strain gradient effect by means of conforming and nonconforming rectangular finite elements *Comput. Struct.* **257** 106676
[63] Tocci Monaco G *et al* 2021 Trigonometric solution for the bending analysis of magneto-electro-elastic strain gradient nonlocal nanoplates in hygro-thermal environment *Mathematics* **9** 567
[64] Tocci Monaco G *et al* 2021 Critical temperatures for vibrations and buckling of magneto-electro-elastic nonlocal strain gradient plates *Nanomaterials* **11** 87
[65] Saitta S *et al* 2022 Radial Point Interpolation Method for isotropic nanoplates in bending using strain gradient theory *Int. J. Comput. Methods* https://doi.org/10.1142/S0219876222500232
[66] Saitta S *et al* 2022 Optimization of a Radial Point Interpolation Meshless strategy for strain gradient nanoplates *Eng. Anal. Bound. Elem.* **140** 70–8
[67] Fabbrocino F *et al* 2022 Meshless computational strategy for higher order strain gradient plate models *Math. Comput. Appl.* **27** 19
[68] Reddy J N 2004 *Mechanics of Laminated Composite Plates and Shells: Theory and Analysis* 2nd edn (Boca Raton, FL: CRC Press)
[69] Liu G R 2003 *Mesh Free Methods Moving Beyond the Finite Element Method* (Boca Raton, FL: CRC Press LLC)

# Chapter 3

# Numerical solution of fractal-fractional variable orders differential equations using two-step and three-step Newton and Lagrange interpolation polynomials

**Rajarama Mohan Jena, Shengda Zeng and Van Thien Nguyen**

Recently, fractal and fractional differentiation concepts have been combined to produce new differentiation operators. These newly developed operators can be generated using various kernels, namely power law, exponential, Mittag-Leffler, etc. These operators contain two parameters: one is for fractional-order, and the other is for fractal dimension. Further, the Adams–Bashforth method has proven to be an efficient numerical technique for solving linear and nonlinear differential equations, especially for non-integer orders. In the present analysis, we have applied these newly generated operators with the Adams–Bashforth numerical method based on two-step and three-step Newton and Lagrange interpolation polynomials for solving two differential equations with variable orders. The exponential, trigonometric, and hyperbolic forms of variables are considered to be the order of the fractional derivative. We have verified the presented techniques by comparing the present solutions with the exact solution for integer-order cases. The convergence study for variable-order and integer-order differential equations has also been carried out in terms of tables by reducing the step length values $h$.

## 3.1 Introduction

Recent studies have shown that fractional calculus is useful in many engineering and science fields. Several types of fractional derivative and integral concepts have recently been used in many areas [1–8]. Using mathematical models to simulate real-world problems is a new development in science and engineering. These models play an essential role since they are often used for prediction, control, and analysis. Mathematical models that describe real-world problems are equally complex. The

doi:10.1088/978-0-7503-5067-9ch3  

exact solutions of these models provide a significant contribution to explaining complex phenomena of physical problems. Nonlinear problems and the mathematical models that describe them cannot always be solved analytically, necessitating computational methods.

Nowadays, numerical methods have proven beneficial in science and engineering, owing to their ability to provide the approximate solution of nonlinear ordinary/partial/fractional differential equations. A numerical method can easily handle this when an analytical method fails to obtain the result. In the twenty-first century, their uses have been expanded in life sciences, medicine, industry, and natural sciences. Even researchers in the arts field have incorporated these numerical techniques as components of scientific computations. As humans strive to influence their environment by prediction, the advancement of computing power has been revolutionized the utility of realistic mechanical and mathematical models in all scientific and engineering areas, necessitating the necessary numerical methods to implement these dynamic mathematical models representing real-world problems [9–12]. It is worth mentioning that the idea of interpolation was introduced 2000 years ago. At that time, many scientists developed various numerical techniques such as Newton's method, the Lagrange interpolation polynomial, Gaussian elimination, etc [13–17]. One may also refer to [18–22] for some interesting numerical methods.

The non-local properties of fractional derivatives are the most important characteristics of fractional derivatives. This property allows the behavior of a phenomenon to be predicted by looking at its history from the beginning to the present. In this context, Reimann–Liouville fractional derivatives can be considered as primary fractional derivatives [23]. This derivative is primarily concerned with a singularity in the kernel. To address these shortcomings, Caputo and Fabrizio [24] proposed a new fractional definition known as the Caputo–Fabrizio operator with a non-singular kernel in 2015. In the form of Riemann–Liouville and Caputo, Atangana and Baleanu [25] introduced two new definitions of the Mittag-Leffler function as a non-singular and non-local kernel. The following are some of the defining characteristics of Atangana–Baleanu fractional derivatives: this derivative possesses both Markovian and non-Markovian properties. On the other hand, the Reimann–Liouville derivative is Markovian, whereas the Caputo–Fabrizio derivative is non-Markovian. The derivatives waiting time is at the same time power law, stretched exponential, and Brownian motion. On the other side, the Reimann–Liouville derivative is only power law, and the Caputo–Fabrizio derivative is exponential decay.

In this paper, we have applied the Adams–Bashforth process, which is based on two and three steps of Newton polynomial interpolation (NPI) and Lagrange polynomial interpolation (LPI), to solve the variable-order fractal-fractional differential equations described in the Caputo and Atangan–Baleanu sense. The manuscript is divided into the following sections: the definitions and notation of fractal-fractional differentiation and integrals are covered in section 3.2. In sections 3.3 and 3.4, the computational schemes for the variable-order fractal-fractional models have been described using NPI with power-law and Mittag-Leffler kernels, respectively. Similarly, using LPI with power-law and Mittag-Leffler kernels, the algorithm for variable-order fractal-fractional differential equations has been elaborated in sections 3.5 and 3.6. In section 3.7, two numerical examples are given, along with their solutions. Finally, section 3.8 contains concluding remarks.

## 3.2 Definitions and notations

This segment introduces a new variable-order fractal-fractional derivative that combines Caputo and Atangana–Baleanu derivatives [26–30].

**Definition 3.1.** Let $\theta(t)$ be a continuous and fractal differential function on an open interval $(a, b)$ with an order $\gamma$, then the fractal-fractional (FF) derivatives of $\theta(t)$ with order $\alpha(t)$ in the Caputo sense is defined as

$$^{FF-C}D_{0,t}^{\alpha(t),\gamma}\{\theta(t)\} = \frac{1}{\Gamma(m-\alpha(t))}\int_0^t (t-\tau)^{m-\alpha(t)-1}\left(\frac{d}{d\tau^{\gamma}}\theta(\tau)\right)d\tau, \tag{3.1}$$

where $m - 1 < \alpha(t), \gamma \leqslant m$, $\dfrac{d\theta(\tau)}{d\tau^{\gamma}} = \lim\limits_{t\to\tau}\dfrac{\theta(t)-\theta(\tau)}{t^{\gamma}-\tau^{\gamma}}$, and $m \in N$.

**Definition 3.2.** The Atangana–Baleanu fractal-fractional derivatives of $\theta(t)$ with order $\alpha(t)$ in Caputo senses is defined as

$$^{FF-ABC}D_{0,t}^{\alpha(t),\gamma}\{\theta(t)\} = \frac{AB(\alpha(t))}{1-\alpha(t)}\int_0^t E_{\alpha(t)}\left(-\frac{\alpha(t)}{1-\alpha(t)}(t-\tau)^{\alpha(t)}\right)\left(\frac{d}{d\tau^{\gamma}}\theta(\tau)\right)d\tau, \tag{3.2}$$

where $0 < \alpha(t), \gamma \leqslant 1$, and $AB(\alpha(t)) = 1 - \alpha(t) + \dfrac{\alpha(t)}{\Gamma(\alpha(t))}$.

**Definition 3.3.** The fractal-fractional Caputo integral of $\theta(t)$ with order $\alpha(t)$ is defined as

$$^{FF-c}I_{0,t}^{\alpha(t),\gamma}\{\theta(t)\} = \frac{\gamma}{\Gamma(\alpha(t))}\int_0^t (t-\tau)^{\alpha(t)-1}\,\tau^{\gamma-1}\,\theta(\tau)d\tau. \tag{3.3}$$

**Definition 3.4.** The fractal-fractional integral of $\theta(t)$ with order $\alpha(t)$ in Atangana–Baleanu Caputo sense is written as

$$^{FF-ABC}I_{0,t}^{\alpha(t),\gamma}\{\theta(t)\} = \frac{\gamma\,\alpha(t)}{AB(\alpha(t))\Gamma(\alpha(t))}\int_0^t \tau^{\gamma-1}(t-\tau)^{\alpha(t)-1}\,\theta(\tau)d\tau + \frac{\gamma(1-\alpha(t))t^{\gamma-1}}{AB(\alpha(t))}\theta(t). \tag{3.4}$$

## 3.3 Algorithm for fractal-fractional variable-order differential equation based on Newton interpolation polynomial via power-law type kernel

Let us consider the following variable-order fractal-fractional Cauchy problem in the Caputo sense

$$^{FF-C}D_{0,t}^{\alpha(t),\gamma}\theta(t) = \phi(t, \theta(t)). \tag{3.5}$$

Integrating equation (3.5), we have

$$\theta(t) - \theta(0) = \frac{\gamma}{\Gamma(\alpha(t))}\int_0^t \tau^{\gamma-1}\phi(\tau,\ \theta(\tau))(t-\tau)^{\alpha(t)-1}d\tau. \tag{3.6}$$

Let us consider $\gamma\ \tau^{\gamma-1}\phi(t,\ \theta(t)) = H(t,\ \theta(t))$. Thus, equation (3.6) reduces to

$$\chi(t) - \chi(0) = \frac{1}{\Gamma(\alpha(t))}\int_0^t F(\tau,\ \chi(\tau))(t-\tau)^{\alpha(t)-1}d\tau \tag{3.7}$$

where the function $H$ is nonlinear. At the point $t_{m+1}(=(m+1)h)$, we obtain

$$\theta(t_{m+1}) - \theta(0) = \frac{1}{\Gamma(\alpha(t_m))}\int_0^{t_{m+1}} H(\tau,\ \theta(\tau))(t_{m+1}-\tau)^{\alpha(t_m)-1}d\tau. \tag{3.8}$$

### 3.3.1 Implementation of two-step Newton interpolation polynomial

Equation (3.8) can be rewritten in summation form as

$$\theta(t_{m+1}) = \theta(0) + \frac{1}{\Gamma(\alpha(t_m))}\sum_{k=0}^{m}\int_{t_k}^{t_{k+1}} H(\tau,\ \theta(\tau))(t_{m+1}-\tau)^{\alpha(t_m)-1}d\tau. \tag{3.9}$$

Now, we replace the original equation $H(\tau,\ \theta(\tau))$ with the Newton polynomial, and we end up with the following expression

$$\theta^{m+1} = \theta^0 + \frac{1}{\Gamma(\alpha(t_m))}\sum_{k=2}^{m}\int_{t_k}^{t_{k+1}}\left\{\begin{array}{l} H\left(t_{k-2},\theta^{k-2}\right) + \dfrac{H\left(t_{k-1},\theta^{k-1}\right) - H\left(t_{k-2},\theta^{k-2}\right)}{h}(\tau - t_{k-2}) + \\ \dfrac{H\left(t_k,\theta^k\right) - 2H\left(t_{k-1},\theta^{k-1}\right) + H\left(t_{k-2},\theta^{k-2}\right)}{2!h^2}(\tau - t_{k-2})(\tau - t_{k-1}) \end{array}\right\}(t_{m+1}-\tau)^{\alpha(t_m)-1}d\tau \tag{3.10}$$

or

$$\theta^{m+1} = \theta^0 + \frac{1}{\Gamma(\alpha(t_m))}\sum_{k=2}^{m}\left\{\begin{array}{l} \displaystyle\int_{t_k}^{t_{k+1}} H\left(t_{k-2},\theta^{k-2}\right)(t_{m+1}-\tau)^{\alpha(t_m)-1}d\tau + \\ \displaystyle\int_{t_k}^{t_{k+1}} \frac{H\left(t_{k-1},\theta^{k-1}\right) - H\left(t_{k-2},\theta^{k-2}\right)}{h}(\tau - t_{k-2})(t_{m+1}-\tau)^{\alpha(t_{m1})-1}d\tau \\ \displaystyle +\int_{t_k}^{t_{k+1}} \frac{H\left(t_k,\theta^k\right) - 2H\left(t_{k-1},\theta^{k-1}\right) + H\left(t_{k-2},\theta^{k-2}\right)}{2!h^2}(\tau - t_{k-2})(\tau - t_{k-1}) \\ (t_{m+1}-\tau)^{\alpha(t_m)-1}d\tau \end{array}\right\}. \tag{3.11}$$

This leads to the following equality

$$\theta^{m+1} = \theta^0 + \left\{ \begin{array}{l} \dfrac{1}{\Gamma(\alpha(t_m))} \displaystyle\sum_{k=2}^{m} H\left(t_{k-2}, \theta^{k-2}\right) \int_{t_k}^{t_{k+1}} (t_{m+1} - \tau)^{\alpha(t_m)-1} d\tau + \\ \dfrac{1}{\Gamma(\alpha(t_m))} \displaystyle\sum_{k=2}^{m} \dfrac{H(t_{k-1}, \theta^{k-1}) - H(t_{k-2}, \theta^{k-2})}{h} \\ \displaystyle\int_{t_k}^{t_{k+1}} (\tau - t_{k-2})(t_{m+1} - \tau)^{\alpha(t_m)-1} d\tau \\ + \dfrac{1}{\Gamma(\alpha(t_m))} \displaystyle\sum_{k=2}^{m} \dfrac{H(t_k, \theta^k) - 2H(t_{k-1}, \theta^{k-1}) + H(t_{k-2}, \theta^{k-2})}{2!h^2} \\ \displaystyle\int_{t_k}^{t_{k+1}} \begin{array}{l} (\tau - t_{k-2})(\tau - t_{k-1}) \\ (t_{m+1} - \tau)^{\alpha(t_m)-1} d\tau \end{array} \end{array} \right\}. \tag{3.12}$$

The above integrals can be computed as follows

$$\int_{t_k}^{t_{k+1}} (t_{m+1} - \tau)^{\alpha(t_m)-1} d\tau = \frac{h^{\alpha(t_{m+1})}}{\alpha(t_{m+1})} \left\{ (m - k + 1)^{\alpha(t_m)} - (m - k)^{\alpha(t_m)} \right\} \tag{3.13}$$

$$\begin{array}{l} \displaystyle\int_{t_k}^{t_{k+1}} (\tau - t_{k-2})(t_{m+1} - \tau)^{\alpha(t_m)-1} d\tau = \dfrac{h^{\alpha(t_m)+1}}{\alpha(t_m)(\alpha(t_m) + 1)} \\ \qquad \left\{ \begin{array}{l} (m - k + 1)^{\alpha(t_m)}(m - k + 3+2\alpha(t_m)) \\ - (m - k)^{\alpha(t_m)}(m - k + 3+3\alpha(t_m)) \end{array} \right\} \end{array} \tag{3.14}$$

$$\begin{array}{l} \displaystyle\int_{t_k}^{t_{k+1}} (\tau - t_{k-2})(\tau - t_{k-1})(t_{m+1} - \tau)^{\alpha(t_m)-1} d\tau = \dfrac{h^{\alpha(t_m)+2}}{\alpha(t_m)(\alpha(t_m) + 1)(\alpha(t_m) + 2)} \\ \left[ \begin{array}{l} (m - k + 1)^{\alpha(t_m)} \left\{ 2(m - k)^2 + (3\alpha(t_m) + 10)(m - k) + 2\alpha(t_m)^2 + 9\alpha(t_m) + 12 \right\} \\ -(m - k)^{\alpha(t_m)} \left\{ 2(m - k)^2 + (5\alpha(t_m) + 10)(m - k) + 6\alpha(t_m)^2 + 18\alpha(t_m) + 12 \right\} \end{array} \right]. \end{array} \tag{3.15}$$

By the use of equations (3.13)–(3.15), equation (3.12) reduces to the following result

$$\begin{array}{l} \theta^{m+1} = \theta^0 + \dfrac{h^{\alpha(t_m)}}{\Gamma(\alpha(t_m) + 1)} \displaystyle\sum_{k=2}^{m} H(t_{k-2}, \theta^{k-2}) \left\{ (m - k + 1)^{\alpha(t_m)} - (m - k)^{\alpha(t_m)} \right\} \\ + \dfrac{h^{\alpha(t_m)}}{\Gamma(\alpha(t_m) + 2)} \displaystyle\sum_{k=2}^{m} H\left(t_{k-1}, \theta^{k-1}\right) - H\left(t_{k-2}, \theta^{k-2}\right) \\ \left\{ \begin{array}{l} (m - k + 1)^{\alpha(t_m)}(m - k + 3+2\alpha(t_m)) \\ -(m - k)^{\alpha(t_m)}(m - k + 3+3\alpha(t_m)) \end{array} \right\} \\ + \dfrac{h^{\alpha(t_m)}}{2\Gamma(\alpha(t_m) + 3)} \displaystyle\sum_{k=2}^{m} H(t_k, \theta^k) - 2H(t_{k-1}, \theta^{k-1}) + H(t_{k-2}, \theta^{k-2}) \\ \left[ \begin{array}{l} (m - k + 1)^{\alpha(t_m)} \left\{ 2(m - k)^2 + (3\alpha(t_m) + 10)(m - k) + 2\alpha(t_m)^2 + 9\alpha(t_m) + 12 \right\} \\ -(m - k)^{\alpha(t_m)} \left\{ 2(m - k)^2 + (5\alpha(t_m) + 10)(m - k) + 6\alpha(t_m)^2 + 18\alpha(t_m) + 12 \right\} \end{array} \right], \end{array} \tag{3.16}$$

which gives

$$\begin{aligned}
\theta^{m+1} &= \theta^0 + \frac{\gamma\, h^{\alpha(t_m)}}{\Gamma(\alpha(t_m)+1)} \sum_{k=2}^{m} t_{k-2}^{\gamma-1}\phi\left(t_{k-2}, \theta^{k-2}\right)\left\{(m-k+1)^{\alpha(t_m)} - (m-k)^{\alpha(t_m)}\right\} \\
&+ \frac{\gamma\, h^{\alpha(t_m)}}{\Gamma(\alpha(t_m)+2)} \sum_{k=2}^{m} t_{k-1}^{\gamma-1}\phi\left(t_{k-1}, \theta^{k-1}\right) - t_{k-2}^{\gamma-1}\phi\left(t_{k-2}, \theta^{k-2}\right) \\
&\left\{\begin{array}{l}(m-k+1)^{\alpha(t_m)}(m-k+3+2\alpha(t_m)) \\ -(m-k)^{\alpha(t_m)}(m-k+3+3\alpha(t_m))\end{array}\right\} + \\
&\frac{\gamma\, h^{\alpha(t_m)}}{2\Gamma(\alpha(t_m)+3)} \sum_{k=2}^{m}\left(t_k^{\gamma-1}\phi\left(t_k, \theta^k\right) - 2t_{k-1}^{\gamma-1}\phi\left(t_{k-1}, \theta^{k-1}\right) + t_{k-2}^{\gamma-1}\phi\left(t_{k-2}, \theta^{k-2}\right)\right) \\
&\left[\begin{array}{l}(m-k+1)^{\alpha(t_m)}\left\{2(m-k)^2 + (3\alpha(t_m)+10)(m-k) + 2\alpha(t_m)^2 + 9\alpha(t_m) + 12\right\} \\ -(m-k)^{\alpha(t_m)}\left\{2(m-k)^2 + (5\alpha(t_m)+10)(m-k) + 6\alpha(t_m)^2 + 18\alpha(t_m) + 12\right\}\end{array}\right].
\end{aligned} \tag{3.17}$$

### 3.3.2 Implementation of three-step Newton interpolation polynomial

Approximating the function $H(\tau, \theta(\tau))$ in equation (3.8) using three-step Newton interpolation polynomials, we have

$$\begin{aligned}
\theta^{m+1} &= \theta^0 + \frac{1}{\Gamma(\alpha(t_m))} \\
&\sum_{k=3}^{m}\int_{t_k}^{t_{k+1}}\left\{\begin{array}{l}
H(t_{k-3}, \theta^{k-3}) + \dfrac{H(t_{k-2}, \theta^{k-2}) - H(t_{k-3}, \theta^{k-3})}{h}(\tau - t_{k-3}) \\
+\dfrac{H(t_{k-1}, \theta^{k-1}) - 2H(t_{k-2}, \theta^{k-2}) + H(t_{k-3}, \theta^{k-3})}{2!h^2}(\tau - t_{k-3}) \\
(\tau - t_{k-2}) + \\
\dfrac{H(t_k, \theta^k) - 3H(t_{k-1}, \theta^{k-1}) + 3H(t_{k-2}, \theta^{k-2}) - H(t_{k-3}, \theta^{k-3})}{3!h^3}(\tau - t_{k-3}) \\
(\tau - t_{k-2})(\tau - t_{k-1})
\end{array}\right\} \\
&(t_{m+1} - \tau)^{\alpha(t_m)-1}d\tau
\end{aligned} \tag{3.18}$$

or

$$\theta^{m+1} = \theta^0 + \frac{1}{\Gamma(\alpha(t_m))}\sum_{k=3}^{m}\left\{\begin{array}{l}
\int_{t_k}^{t_{k+1}} H\left(t_{k-3}, \theta^{k-3}\right)(t_{m+1}-\tau)^{\alpha(t_m)-1}d\tau + \\
\int_{t_k}^{t_{k+1}} \dfrac{H\left(t_{k-2}, \theta^{k-2}\right) - H\left(t_{k-3}, \theta^{k-3}\right)}{h}(\tau - t_{k-3})(t_{m+1}-\tau)^{\alpha(t_m)-1}d\tau \\
+\int_{t_k}^{t_{k+1}} \dfrac{H\left(t_{k-1}, \theta^{k-1}\right) - 2H\left(t_{k-2}, \theta^{k-2}\right) + H\left(t_{k-3}, \theta^{k-3}\right)}{2!h^2}(\tau - t_{k-3})(\tau - t_{k-2}) \\
(t_{m+1}-\tau)^{\alpha(t_m)-1}d\tau \\
+\int_{t_k}^{t_{k+1}} \dfrac{H\left(t_k, \theta^k\right) - 3H\left(t_{k-1}, \theta^{k-1}\right) + 3H\left(t_{k-2}, \theta^{k-2}\right) - H\left(t_{k-3}, \theta^{k-3}\right)}{3!h^3} \\
(\tau - t_{k-3})(\tau - t_{k-2})(\tau - t_{k-1})(t_{m+1}-\tau)^{\alpha(t_m)-1}d\tau
\end{array}\right\}. \tag{3.19}$$

From equation (3.19), we obtain the following equality

$$\theta^{m+1} = \theta^0 + \left\{ \begin{array}{l} \frac{1}{\Gamma(\alpha(t_m))}\sum_{k=3}^{m} H(t_{k-3}, \theta^{k-3}) \int_{t_k}^{t_{k+1}} (t_{m+1} - \tau)^{\alpha(t_m)-1} d\tau + \\ \frac{1}{\Gamma(\alpha(t_m))}\sum_{k=3}^{m} \frac{H(t_{k-2}, \theta^{k-2}) - H(t_{k-3}, \theta^{k-3})}{h} \int_{t_k}^{t_{k+1}} (\tau - t_{k-3})(t_{m+1} - \tau)^{\alpha(t_m)-1} d\tau \\ + \frac{1}{\Gamma(\alpha(t_m))}\sum_{k=3}^{m} \frac{H(t_{k-1}, \theta^{k-1}) - 2H(t_{k-2}, \theta^{k-2}) + H(t_{k-3}, \theta^{k-3})}{2!h^2} \\ \int_{t_k}^{t_{k+1}} (\tau - t_{k-3})(\tau - t_{k-2})(t_{m+1} - \tau)^{\alpha(t_m)-1} d\tau \\ + \frac{1}{\Gamma(\alpha(t_m))}\sum_{k=3}^{m} \frac{H(t_k, \theta^k) - 3H(t_{k-1}, \theta^{k-1}) + 3H(t_{k-2}, \theta^{k-2}) - H(t_{k-3}, \theta^{k-3})}{3!h^3} \\ \int_{t_k}^{t_{k+1}} (\tau - t_{k-3})(\tau - t_{k-2})(\tau - t_{k-1})(t_{m+1} - \tau)^{\alpha(t_m)-1} d\tau \end{array} \right\}. \tag{3.20}$$

Further, the below integrals are to be evaluated

$$\int_{t_k}^{t_{k+1}} (t_{m+1} - \tau)^{\alpha(t_m)-1} d\tau = \frac{h^{\alpha(t_m)}}{\alpha(t_m)} \{(m - k + 1)^{\alpha(t_m)} - (m - k)^{\alpha(t_m)}\} \tag{3.21}$$

$$\begin{aligned} \int_{t_k}^{t_{k+1}} (\tau - t_{k-3})(t_{m+1} - \tau)^{\alpha(t_m)-1} d\tau = & \frac{h^{\alpha(t_m)+1}}{\alpha(t_m)(\alpha(t_m) + 1)} \\ & \left\{ \begin{array}{l} (m - k + 1)^{\alpha(t_m)}(m - k + 4+3\alpha(t_m)) \\ - (m - k)^{\alpha(t_m)}(m - k + 4+4\alpha(t_m)) \end{array} \right\} \end{aligned} \tag{3.22}$$

$$\begin{aligned} & \int_{t_k}^{t_{k+1}} (\tau - t_{k-3})(\tau - t_{k-2})(t_{m+1} - \tau)^{\alpha(t_m)-1} d\tau = \frac{h^{\alpha(t_m)+2}}{\alpha(t_m)(\alpha(t_m) + 1)(\alpha(t_m) + 2)} \\ & \left[ \begin{array}{l} (m - k + 1)^{\alpha(t_m)}\{2(m - k)^2 + (5\alpha(t_m) + 14)(m - k) + 6\alpha(t_m)^2 + 23\alpha(t_m) + 24\} \\ -(m - k)^{\alpha(t_m)}\{2(m - k)^2 + (7\alpha(t_m) + 14)(m - k) + 12\alpha(t_m)^2 + 36\alpha(t_m) + 24\} \end{array} \right] \end{aligned} \tag{3.23}$$

$$\begin{aligned} & \int_{t_k}^{t_{k+1}} (\tau - t_{k-3})(\tau - t_{k-2})(\tau - t_{k-1})(t_{m+1} - \tau)^{\alpha(t_m)-1} d\tau \\ & = -\frac{6h^{\alpha(t_m)+3}}{\alpha(t_m)(\alpha(t_m) + 1)(\alpha(t_m) + 2)(\alpha(t_m) + 3)} \\ & \left[ \begin{array}{l} \frac{(m - k + 1)^{\alpha(t_m)}}{6} \left\{ \begin{array}{l} -6\alpha(t_m)^3 + (11k - 11m - 47)\alpha(t_m)^2 + \left( \begin{array}{l} -12k^2 + (24m + 79)k- \\ 12m^2 - 79m - 133 \end{array} \right) \alpha(t_m) \\ -6(m - k + 2)(m - k + 3)(m - k + 4) \end{array} \right\} \\ +(m - k)^{\alpha(t_m)} \left\{ \begin{array}{l} 4\alpha(t_m)^3 + \left( \frac{-13}{3}k + \frac{13}{3}m + 24 \right) \alpha(t_m)^2 + \left( \begin{array}{l} 3k^2 + \left( -6m - \frac{65}{3} \right) k+ \\ 3m^2 + \frac{65}{3}m + 44 \end{array} \right) \alpha(t_m) \\ +(m - k + 2)(m - k + 3)(m - k + 4) \end{array} \right\} \end{array} \right]. \end{aligned} \tag{3.24}$$

Plugging equations (3.21)–(3.24) into equation (3.20), we get

$$
\begin{aligned}
\theta^{m+1} &= \theta^0 + \frac{h^{\alpha(t_m)}}{\Gamma(\alpha(t_m)+1)}\sum_{k=3}^{m} H\left(t_{k-3},\theta^{k-3}\right)\left\{(m-k+1)^{\alpha(t_m)} - (m-k)^{\alpha(t_m)}\right\}\\
&+\frac{h^{\alpha(t_m)}}{\Gamma(\alpha(t_m)+2)}\sum_{k=3}^{m}\left\{H\left(t_{k-2},\theta^{k-2}\right) - H\left(t_{k-3},\theta^{k-3}\right)\right\}\begin{Bmatrix}(m-k+1)^{\alpha(t_m)}(m-k+4+3\alpha(t_m))\\ -(m-k)^{\alpha(t_m)}(m-k+4+4\alpha(t_m))\end{Bmatrix}\\
&+\frac{h^{\alpha(t_m)}}{2\Gamma(\alpha(t_m)+3)}\sum_{k=3}^{m}\left\{H\left(t_{k-1},\theta^{k-1}\right) - 2H\left(t_{k-2},\theta^{k-2}\right) + H\left(t_{k-3},\theta^{k-3}\right)\right\}\\
&\begin{bmatrix}(m-k+1)^{\alpha(t_m)}\left\{2(m-k)^2 + (5\alpha(t_m)+14)(m-k) + 6\alpha(t_m)^2 + 23\alpha(t_m) + 24\right\}\\ -(m-k)^{\alpha(t_m)}\left\{2(m-k)^2 + (7\alpha(t_m)+14)(m-k) + 12\alpha(t_m)^2 + 36\alpha(t_m) + 24\right\}\end{bmatrix}\\
&-\frac{h^{\alpha(t_m)}}{\Gamma(\alpha(t_m)+4)}\sum_{k=3}^{m}\left\{H\left(t_k,\theta^k\right) - 3H\left(t_{k-1},\theta^{k-1}\right) + 3H\left(t_{k-2},\theta^{k-2}\right) - H\left(t_{k-3},\theta^{k-3}\right)\right\}\\
&\begin{bmatrix}\frac{(m-k+1)^{\alpha(t_m)}}{6}\begin{Bmatrix}-6\alpha(t_m)^3 + (11k-11m-47)\alpha(t_m)^2 + \begin{pmatrix}-12k^2 + (24m+79)k-\\ 12m^2 - 79m - 133\end{pmatrix}\alpha(t_m)\\ -6(m-k+2)(m-k+3)(m-k+4)\end{Bmatrix}\\ +(m-k)^{\alpha(t_m)}\begin{Bmatrix}4\alpha(t_m)^3 + \left(\frac{-13}{3}k + \frac{13}{3}m + 24\right)\alpha(t_m)^2 + \begin{pmatrix}3k^2 + \left(-6m - \frac{65}{3}\right)k+\\ 3m^2 + \frac{65}{3}m + 44\end{pmatrix}\alpha(t_m)\\ +(m-k+2)(m-k+3)(m-k+4)\end{Bmatrix}\end{bmatrix},
\end{aligned}
\tag{3.25}
$$

which becomes

$$
\begin{aligned}
\theta^{m+1} &= \theta^0 + \frac{\gamma\, h^{\alpha(t_m)}}{\Gamma(\alpha(t_m)+1)}\sum_{k=3}^{m} t_{k-3}^{\gamma-1}\phi\left(t_{k-3},\theta^{k-3}\right)\left\{(m-k+1)^{\alpha(t_m)} - (m-k)^{\alpha(t_m)}\right\}\\
&+\frac{\gamma\, h^{\alpha(t_m)}}{\Gamma(\alpha(t_m)+2)}\sum_{k=3}^{m}\left\{t_{k-2}^{\gamma-1}\phi\left(t_{k-2},\theta^{k-2}\right) - t_{k-3}^{\gamma-1}\phi\left(t_{k-3},\theta^{k-3}\right)\right\}\\
&\begin{Bmatrix}(m-k+1)^{\alpha(t_m)}(m-k+4+3\alpha(t_m))\\ -(m-k)^{\alpha(t_m)}(m-k+4+4\alpha(t_m))\end{Bmatrix}+\\
&\frac{\gamma\, h^{\alpha(t_m)}}{2\Gamma(\alpha(t_m)+3)}\sum_{k=3}^{m}\left\{t_{k-1}^{\gamma-1}\phi\left(t_{k-1},\theta^{k-1}\right) - 2t_{k-2}^{\gamma-1}\phi\left(t_{k-2},\theta^{k-2}\right) + t_{k-3}^{\gamma-1}\phi\left(t_{k-3},\theta^{k-3}\right)\right\}\\
&\begin{bmatrix}(m-k+1)^{\alpha(t_m)}\left\{2(m-k)^2 + (5\alpha(t_m)+14)(m-k) + 6\alpha(t_m)^2 + 23\alpha(t_m) + 24\right\}\\ -(m-k)^{\alpha(t_m)}\left\{2(m-k)^2 + (7\alpha(t_m)+14)(m-k) + 12\alpha(t_m)^2 + 36\alpha(t_m) + 24\right\}\end{bmatrix}\\
&-\frac{\gamma\, h^{\alpha(t_m)}}{\Gamma(\alpha(t_m)+4)}\sum_{k=3}^{m}\left\{t_k^{\gamma-1}\phi\left(t_k,\theta^k\right) - 3t_{k-1}^{\gamma-1}\phi\left(t_{k-1},\theta^{k-1}\right) + 3t_{k-2}^{\gamma-1}\phi\left(t_{k-2},\theta^{k-2}\right)\right.\\
&\quad\left. - t_{k-3}^{\gamma-1}\phi\left(t_{k-3},\theta^{k-3}\right)\right\}\\
&\begin{bmatrix}\frac{(m-k+1)^{\alpha(t_m)}}{6}\begin{Bmatrix}-6\alpha(t_m)^3 + (11k-11m-47)\alpha(t_m)^2 + \begin{pmatrix}-12k^2 + (24m+79)k-\\ 12m^2 - 79m - 133\end{pmatrix}\alpha(t_m)\\ -6(m-k+2)(m-k+3)(m-k+4)\end{Bmatrix}\\ +(m-k)^{\alpha(t_m)}\begin{Bmatrix}4\alpha(t_m)^3 + \left(\frac{-13}{3}k + \frac{13}{3}m + 24\right)\alpha(t_m)^2 + \begin{pmatrix}3k^2 + \left(-6m - \frac{65}{3}\right)k+\\ 3m^2 + \frac{65}{3}m + 44\end{pmatrix}\alpha(t_m)\\ +(m-k+2)(m-k+3)(m-k+4)\end{Bmatrix}\end{bmatrix}.
\end{aligned}
\tag{3.26}
$$

## 3.4 Algorithm for fractal-fractional variable-order differential equation based on Newton interpolation polynomials via Mittag-Leffler type kernel

We consider the following Cauchy problem in the ABC sense

$$ {}^{FF-ABC}D_{0,\,t}^{\alpha(t),\,\gamma}\theta(t) = \phi(t,\,\theta(t)). \tag{3.27} $$

Integrating the above equation (3.27), one may get

$$ \begin{aligned} \theta(t) - \theta(0) = \frac{1-\alpha(t)}{AB(\alpha(t))}\gamma\, t^{\gamma-1}\phi(t,\,\theta(t)) + \frac{\gamma\,\alpha(t)}{AB(\alpha(t))\Gamma(\alpha(t))} \\ \int_0^t \tau^{\gamma-1}\phi(\tau,\,\theta(\tau))(t-\tau)^{\alpha(t)-1}d\tau. \end{aligned} \tag{3.28} $$

For simplicity, let us assume $\gamma\, t^{\gamma-1}\phi(t,\,\theta(t)) = H(t,\ \theta(t))$. Thus, equation (3.28) reduces to

$$ \theta(t) - \theta(0) = \frac{1-\alpha(t)}{AB(\alpha(t))}H(t,\,\theta(t)) + \frac{\alpha(t)}{AB(\alpha(t))\,\Gamma(\alpha(t))}\int_0^t H(\tau,\ \theta(\tau))(t-\tau)^{\alpha(t)-1}d\tau. \tag{3.29} $$

At the point $t_{m+1}(=(m+1)h)$, we get

$$ \begin{aligned} \theta(t_{m+1}) - \theta(0) = \frac{1-\alpha(t_m)}{AB(\alpha(t_m))}H(t_m,\,\theta(t_m)) + \frac{\alpha(t_m)}{AB(\alpha(t_m))\Gamma(\alpha(t_m))} \\ \int_0^{t_{m+1}} H(\tau,\ \theta(\tau))(t_{m+1}-\tau)^{\alpha(t_m)-1}d\tau. \end{aligned} \tag{3.30} $$

### 3.4.1 Implementation of two-step Newton interpolation polynomial

Equation (3.30) is rewritten in summation form as

$$ \begin{aligned} \theta(t_{m+1}) = \theta(0) + \frac{1-\alpha(t_m)}{AB(\alpha(t_m))}H(t_m,\,\theta(t_m)) + \frac{\alpha(t_m)}{AB(\alpha(t_m))\Gamma(\alpha(t_m))} \\ \sum_{k=0}^{m}\int_{t_k}^{t_{k+1}} H(\tau,\ \theta(\tau))(t_{m+1}-\tau)^{\alpha(t_m)-1}d\tau. \end{aligned} \tag{3.31} $$

Let us approximate the function $H(t,\,\theta(t))$ with the Newton polynomial, and then we obtain

$$ \begin{aligned} \theta^{m+1} = \theta^0 + \frac{1-\alpha(t_m)}{AB(\alpha(t_m))}H(t_m,\,\theta(t_m)) + \frac{\alpha(t_m)}{AB(\alpha(t_m))\Gamma(\alpha(t_m))} \\ \sum_{k=2}^{m}\int_{t_k}^{t_{k+1}}\left\{\begin{matrix} H(t_{k-2},\,\theta^{k-2}) + \dfrac{H(t_{k-1},\,\theta^{k-1}) - H(t_{k-2},\,\theta^{k-2})}{h}(\tau - t_{k-2}) \\ + \dfrac{H(t_k,\,\theta^k) - 2H(t_{k-1},\,\theta^{k-1}) + H(t_{k-2},\,\theta^{k-2})}{2!h^2}(\tau - t_{k-2})(\tau - t_{k-1}) \end{matrix}\right\} \\ (t_{m+1}-\tau)^{\alpha(t_m)-1}d\tau. \end{aligned} \tag{3.32} $$

The above equation (3.32) leads to the following expression

$$\theta^{m+1} = \theta^0 + \frac{1-\alpha(t_m)}{AB(\alpha(t_m))}H(t_m, \theta(t_m)) + \frac{\alpha(t_m)}{AB(\alpha(t_m))\Gamma(\alpha(t_m))}$$

$$\sum_{k=2}^{m}\left\{\begin{array}{l} \int_{t_k}^{t_{k+1}} H(t_{k-2}, \theta^{k-2})(t_{m+1}-\tau)^{\alpha(t_m)-1}d\tau + \\ \int_{t_k}^{t_{k+1}} \frac{H(t_{k-1}, \theta^{k-1}) - H(t_{k-2}, \theta^{k-2})}{h}(\tau - t_{k-2})(t_{m+1}-\tau)^{\alpha(t_m)-1}d\tau \\ + \int_{t_k}^{t_{k+1}} \frac{H(t_k, \theta^k) - 2H(t_{k-1}, \theta^{k-1}) + H(t_{k-2}, \theta^{k-2})}{2!h^2}(\tau - t_{k-2})(\tau - t_{k-1}) \\ (t_{m+1}-\tau)^{\alpha(t_m)-1}d\tau \end{array}\right\}. \tag{3.33}$$

Using equations (3.13)–(3.15) in equation (3.33), one may obtain

$$\begin{aligned} \theta^{m+1} &= \theta^0 + \frac{1-\alpha(t_m)}{AB(\alpha(t_m))}H(t_m, \theta(t_m)) + \frac{\alpha(t_m)h^{\alpha(t_m)}}{AB(\alpha(t_m))\Gamma(\alpha(t_m)+1)}\sum_{k=2}^{m} H\left(t_{k-2}, \theta^{k-2}\right) \\ &\left\{(m-k+1)^{\alpha(t_m)} - (m-k)^{\alpha(t_m)}\right\} + \frac{\alpha(t_m)h^{\alpha(t_m)}}{AB(\alpha(t_m))\,\Gamma(\alpha(t_m)+2)} \\ &\sum_{k=2}^{m} H\left(t_{k-1}, \theta^{k-1}\right) - H\left(t_{k-2}, \theta^{k-2}\right) \\ &\left\{\begin{array}{l}(m-k+1)^{\alpha(t_m)}(m-k+3+2\alpha(t_m)) \\ -(m-k)^{\alpha(t_m)}(m-k+3+3\alpha(t_m))\end{array}\right\} + \frac{\alpha(t_m)h^{\alpha(t_m)}}{2AB(\alpha(t_m))\,\Gamma(\alpha(t_m)+3)} \\ &\sum_{k=2}^{m}\left(\begin{array}{l} H\left(t_k, \theta^k\right) - 2H\left(t_{k-1}, \theta^{k-1}\right) \\ +H\left(t_{k-2}, \theta^{k-2}\right)\end{array}\right) \\ &\left[\begin{array}{l} (m-k+1)^{\alpha(t_m)}\left\{\begin{array}{l}2(m-k)^2 + (3\alpha(t_m)+10)(m-k) \\ +2\alpha(t_m)^2 + 9\alpha(t_m) + 12\end{array}\right\} \\ -(m-k)^{\alpha(t_m)}\left\{\begin{array}{l}2(m-k)^2 + (5\alpha(t_m)+10)(m-k) \\ +6\alpha(t_m)^2 + 18\alpha(t_m) + 12\end{array}\right\} \end{array}\right]. \end{aligned} \tag{3.34}$$

Thus, we obtain

$$
\begin{aligned}
\theta^{m+1} = \theta^0 &+ \frac{1-\alpha(t_m)}{AB(\alpha(t_m))}\gamma\, t_m^{\gamma-1}\phi(t_m, \theta^m) + \frac{\gamma\, \alpha(t_m)h^{\alpha(t_m)}}{AB(\alpha(t_m))\,\Gamma(\alpha(t_m)+1)} \\
&\sum_{k=2}^{m} t_{k-2}^{\gamma-1}\phi\left(t_{k-2}, \theta^{k-2}\right) \\
&\left\{(m-k+1)^{\alpha(t_m)} - (m-k)^{\alpha(t_m)}\right\} + \frac{\gamma\, \alpha(t_m)h^{\alpha(t_m)}}{AB(\alpha(t_m))\Gamma(\alpha(t_m)+2)} \\
&\sum_{k=2}^{m} t_{k-1}^{\gamma-1}\phi\left(t_{k-1}, \theta^{k-1}\right) - t_{k-2}^{\gamma-1}\phi\left(t_{k-2}, \theta^{k-2}\right) \\
&\left\{\begin{matrix}(m-k+1)^{\alpha(t_m)}(m-k+3+2\alpha(t_m)) \\ -(m-k)^{\alpha(t_m)}(m-k+3+3\alpha(t_m))\end{matrix}\right\} + \frac{\gamma\, \alpha(t_m)h^{\alpha(t_m)}}{2AB(\alpha(t_m))\Gamma(\alpha(t_m)+3)} \\
&\sum_{k=2}^{m}\left(t_k^{\gamma-1}\phi\left(t_k, \theta^k\right) - 2t_{k-1}^{\gamma-1}\phi\left(t_{k-1}, \theta^{k-1}\right) + t_{k-2}^{\gamma-1}\phi\left(t_{k-2}, \theta^{k-2}\right)\right) \\
&\left[\begin{matrix}(m-k+1)^{\alpha(t_m)}\left\{2(m-k)^2 + (3\alpha(t_m)+10)(m-k) + 2\alpha(t_m)^2 + 9\alpha(t_m) + 12\right\} \\ -(m-k)^{\alpha(t_m)}\left\{2(m-k)^2 + (5\alpha(t_m)+10)(m-k) + 6\alpha(t_m)^2 + 18\alpha(t_m) + 12\right\}\end{matrix}\right].
\end{aligned}
\tag{3.35}
$$

### 3.4.2 Implementation of three-step Newton interpolation polynomial

For the approximation of function $H(t, \theta(t))$, we use the Newton interpolation polynomials as follows

$$
\begin{aligned}
P_k(\tau) &= H(t_{k-3}, \theta^{k-3}) + \frac{H(t_{k-2}, \theta^{k-2}) - H(t_{k-3}, \theta^{k-3})}{h}(\tau - t_{k-3}) \\
&+ \frac{H(t_{k-1}, \theta^{k-1}) - 2H(t_{k-2}, \theta^{k-2}) + H(t_{k-3}, \theta^{k-3})}{2!h^2}(\tau - t_{k-3})(\tau - t_{k-2}) \\
&+ \frac{H(t_k, \theta^k) - 3H(t_{k-1}, \theta^{k-1}) + 3H(t_{k-2}, \theta^{k-2}) - H(t_{k-3}, \theta^{k-3})}{3!h^3}(\tau - t_{k-3}) \\
&(\tau - t_{k-2})(\tau - t_{k-1}).
\end{aligned}
\tag{3.36}
$$

Putting equation (3.36) into equation (3.31), we get

$$
\begin{aligned}
\theta(t_{m+1}) = \theta^0 &+ \frac{1-\alpha(t_m)}{AB(\alpha(t_m))}H(t_m, \theta(t_m)) + \frac{\alpha(t_m)}{AB(\alpha(t_m))\,\Gamma(\alpha(t_m))} \\
&\sum_{k=3}^{m}\int_{t_k}^{t_{k+1}}\left\{\begin{matrix}H(t_{k-3}, \theta^{k-3}) + \dfrac{H(t_{k-2}, \theta^{k-2}) - H(t_{k-3}, \theta^{k-3})}{h}(\tau - t_{k-3}) \\ + \dfrac{H(t_{k-1}, \theta^{k-1}) - 2H(t_{k-2}, \theta^{k-2}) + H(t_{k-3}, \theta^{k-3})}{2!h^2}(\tau - t_{k-3}) \\ (\tau - t_{k-2}) + \\ \dfrac{H(t_k, \theta^k) - 3H(t_{k-1}, \theta^{k-1}) + 3H(t_{k-2}, \theta^{k-2}) - H(t_{k-3}, \theta^{k-3})}{3!h^3} \\ (\tau - t_{k-3}) \\ (\tau - t_{k-2})(\tau - t_{k-1})\end{matrix}\right\} \\
&(t_{m+1} - \tau)^{\alpha(t_m)-1}d\tau.
\end{aligned}
\tag{3.37}
$$

Simplifying equation (3.37) leads to

$$\theta^{m+1} = \theta^0 + \frac{1-\alpha(t_m)}{AB(\alpha(t_m))}H(t_m, \theta(t_m)) + \frac{\alpha(t_m)}{AB(\alpha(t_m))\,\Gamma(\alpha(t_m))}\sum_{k=3}^{m} H(t_{k-3}, \theta^{k-3})$$
$$\int_{t_k}^{t_{k+1}} (t_{m+1}-\tau)^{\alpha(t_m)-1}d\tau +$$
$$\frac{\alpha(t_m)}{AB(\alpha(t_m))\,\Gamma(\alpha(t_m))}\sum_{k=3}^{m}\frac{H(t_{k-2}, \theta^{k-2}) - H(t_{k-3}, \theta^{k-3})}{h}$$
$$\int_{t_k}^{t_{k+1}} (\tau - t_{k-3})(t_{m+1}-\tau)^{\alpha(t_m)-1}d\tau + \frac{\alpha(t_m)}{AB(\alpha(t_m))\,\Gamma(\alpha(t_m))}$$
$$\sum_{k=3}^{m}\left(\frac{H(t_{k-1}, \theta^{k-1}) - 2H(t_{k-2}, \theta^{k-2}) + H(t_{k-3}, \theta^{k-3})}{2!h^2}\right) \tag{3.38}$$
$$\int_{t_k}^{t_{k+1}} (\tau - t_{k-3})(\tau - t_{k-2})(t_{m+1}-\tau)^{\alpha(t_m)-1}d\tau$$
$$+\frac{\alpha(t_m)}{AB(\alpha(t_m))\,\Gamma(\alpha(t_m))}$$
$$\sum_{k=3}^{m}\frac{H(t_k, \theta^k) - 3H(t_{k-1}, \theta^{k-1}) + 3H(t_{k-2}, \theta^{k-2}) - H(t_{k-3}, \theta^{k-3})}{3!h^3}$$
$$\int_{t_k}^{t_{k+1}} (\tau - t_{k-3})(\tau - t_{k-2})(\tau - t_{k-1})(t_{m+1}-\tau)^{\alpha(t_m)-1}d\tau.$$

Substituting the evaluated integrals equations (3.21)–(3.24) into equation (3.38), one may have

$$\theta^{m+1} = \theta^0 + \frac{1-\alpha(t_m)}{AB(\alpha(t_m))}H(t_m, \theta(t_m)) + \frac{\alpha(t_m)h^{\alpha(t_m)}}{AB(\alpha(t_m))\,\Gamma(\alpha(t_m)+1)}\sum_{k=3}^{m} H\left(t_{k-3}, \theta^{k-3}\right)$$
$$\left\{(m-k+1)^{\alpha(t_m)} - (m-k)^{\alpha(t_m)}\right\} + \frac{\alpha(t_m)h^{\alpha(t_m)}}{AB(\alpha(t_m))\,\Gamma(\alpha(t_m)+2)}$$
$$\sum_{k=3}^{m}\left\{H\left(t_{k-2}, \theta^{k-2}\right) - H\left(t_{k-3}, \theta^{k-3}\right)\right\}$$
$$\left\{\begin{matrix}(m-k+1)^{\alpha(t_m)}(m-k+4+3\alpha(t_m))\\ -(m-k)^{\alpha(t_m)}(m-k+4+4\alpha(t_m))\end{matrix}\right\} + \frac{\alpha(t_m)h^{\alpha(t_m)}}{2AB(\alpha(t_m))\,\Gamma(\alpha(t_m)+3)}$$
$$\sum_{k=3}^{m}\left\{H\left(t_{k-1}, \theta^{k-1}\right) - 2H\left(t_{k-2}, \theta^{k-2}\right) + H\left(t_{k-3}, \theta^{k-3}\right)\right\}$$
$$\left[\begin{matrix}(m-k+1)^{\alpha(t_m)}\left\{2(m-k)^2 + (5\alpha(t_m)+14)(m-k) + 6\alpha(t_m)^2 + 23\alpha(t_m) + 24\right\}\\ -(m-k)^{\alpha(t_m)}\left\{2(m-k)^2 + (7\alpha(t_m)+14)(m-k) + 12\alpha(t_m)^2 + 36\alpha(t_m) + 24\right\}\end{matrix}\right] \tag{3.39}$$
$$-\frac{\alpha(t_m)h^{\alpha(t_m)}}{AB(\alpha(t_m))\,\Gamma(\alpha(t_m)+4)}\sum_{k=3}^{m}\left\{H\left(t_k, \theta^k\right) - 3H\left(t_{k-1}, \theta^{k-1}\right) + 3H\left(t_{k-2}, \theta^{k-2}\right)\right.$$
$$\left.-H\left(t_{k-3}, \theta^{k-3}\right)\right\}$$
$$\left[\begin{matrix}\frac{(m-k+1)^{\alpha(t_m)}}{6}\left\{\begin{matrix}-6\alpha(t_m)^3 + (11k - 11m - 47)\alpha(t_m)^2 + \begin{pmatrix}-12k^2 + (24m+79)k-\\ 12m^2 - 79m - 133\end{pmatrix}\alpha(t_m)\\ -6(m-k+2)(m-k+3)(m-k+4)\end{matrix}\right\}\\ +(m-k)^{\alpha(t_m)}\left\{\begin{matrix}4\alpha(t_m)^3 + \left(\frac{-13}{3}k + \frac{13}{3}m + 24\right)\alpha(t_m)^2 + \begin{pmatrix}3k^2 + \left(-6m - \frac{65}{3}\right)k+\\ 3m^2 + \frac{65}{3}m + 44\end{pmatrix}\alpha(t_m)\\ +(m-k+2)(m-k+3)(m-k+4)\end{matrix}\right\}\end{matrix}\right].$$

Thus, we have

$$\theta^{m+1}=\theta^0+\frac{1-\alpha(t_m)}{AB(\alpha(t_m))}\gamma\, t_m^{\gamma-1}\phi(t_m,\theta^m)+\frac{\gamma\,\alpha(t_m)h^{\alpha(t_m)}}{AB(\alpha(t_m))\,\Gamma(\alpha(t_m)+1)}\sum_{k=3}^{m}t_{k-3}^{\gamma-1}\phi\left(t_{k-3},\theta^{k-3}\right)$$
$$\left\{(m-k+1)^{\alpha(t_m)}-(m-k)^{\alpha(t_m)}\right\}+\frac{\gamma\,\alpha(t_m)h^{\alpha(t_m)}}{AB(\alpha(t_m))\,\Gamma(\alpha(t_m)+2)}$$
$$\sum_{k=3}^{m}\left\{t_{k-2}^{\gamma-1}\phi\left(t_{k-2},\theta^{k-2}\right)-t_{k-3}^{\gamma-1}\phi\left(t_{k-3},\theta^{k-3}\right)\right\}\left\{\begin{matrix}(m-k+1)^{\alpha(t_m)}(m-k+4+3\alpha(t_m))\\-(m-k)^{\alpha(t_m)}(m-k+4+4\alpha(t_m))\end{matrix}\right\}$$
$$+\frac{\gamma\,\alpha(t_m)h^{\alpha(t_m)}}{2AB(\alpha(t_m))\,\Gamma(\alpha(t_m)+3)}\sum_{k=3}^{m}\left\{t_{k-1}^{\gamma-1}\phi\left(t_{k-1},\theta^{k-1}\right)-2t_{k-2}^{\gamma-1}\phi\left(t_{k-2},\theta^{k-2}\right)\right.$$
$$\left.+t_{k-3}^{\gamma-1}\phi\left(t_{k-3},\theta^{k-3}\right)\right\}$$
$$\left[\begin{matrix}(m-k+1)^{\alpha(t_m)}\left\{2(m-k)^2+(5\alpha(t_m)+14)(m-k)+6\alpha(t_m)^2+23\alpha(t_m)+24\right\}\\-(m-k)^{\alpha(t_m)}\left\{2(m-k)^2+(7\alpha(t_m)+14)(m-k)+12\alpha(t_m)^2+36\alpha(t_m)+24\right\}\end{matrix}\right] \tag{3.40}$$
$$-\frac{\gamma\,\alpha(t_m)h^{\alpha(t_m)}}{AB(\alpha(t_m))\,\Gamma(\alpha(t_m)+4)}\sum_{k=3}^{m}\left\{\begin{matrix}t_k^{\gamma-1}\phi\left(t_k,\theta^k\right)-3t_{k-1}^{\gamma-1}\phi\left(t_{k-1},\theta^{k-1}\right)+3t_{k-2}^{\gamma-1}\phi\left(t_{k-2},\theta^{k-2}\right)\\-t_{k-3}^{\gamma-1}\phi\left(t_{k-3},\theta^{k-3}\right)\end{matrix}\right\}$$
$$\left[\begin{matrix}\frac{(m-k+1)^{\alpha(t_m)}}{6}\left\{\begin{matrix}-6\alpha(t_m)^3+(11k-11m-47)\alpha(t_m)^2+\left(\begin{matrix}-12k^2+(24m+79)k-\\12m^2-79m-133\end{matrix}\right)\alpha(t_m)\\-6(m-k+2)(m-k+3)(m-k+4)\end{matrix}\right\}\\+(m-k)^{\alpha(t_m)}\left\{\begin{matrix}4\alpha(t_m)^3+\left(\frac{-13}{3}k+\frac{13}{3}m+24\right)\alpha(t_m)^2+\left(\begin{matrix}3k^2+\left(-6m-\frac{65}{3}\right)k+\\3m^2+\frac{65}{3}m+44\end{matrix}\right)\alpha(t_m)\\+(m-k+2)(m-k+3)(m-k+4)\end{matrix}\right\}\end{matrix}\right].$$

## 3.5 Algorithm for fractal-fractional variable-order differential equation based on Lagrange polynomial interpolation via power-law type kernel

### 3.5.1 Implementation of two-step Lagrange interpolation polynomial

By replacing the original equation $H(\tau,\ \theta(\tau))$ in equation (3.9) with the Lagrange polynomial interpolation, the following expressions may be obtained

$$\theta^{m+1}=\theta^0+\frac{1}{\Gamma(\alpha(t_m))}\sum_{k=1}^{m}\int_{t_k}^{t_{k+1}}\left\{\frac{H(t_k,\theta^k)}{h}(\tau-t_{k-1})-\frac{H(t_{k-1},\theta^{k-1})}{h}(\tau-t_k)\right\}(t_{m+1}-\tau)^{\alpha(t_m)-1}d\tau \tag{3.41}$$

or

$$\theta^{m+1}=\theta^0+\frac{1}{\Gamma(\alpha(t_m))}\sum_{k=1}^{m}\left\{\begin{matrix}\int_{t_k}^{t_{k+1}}\frac{H(t_k,\theta^k)}{h}(\tau-t_{k-1})(t_{m+1}-\tau)^{\alpha(t_m)-1}d\tau\\-\int_{t_k}^{t_{k+1}}\frac{H(t_{k-1},\theta^{k-1})}{h}(\tau-t_k)(t_{m+1}-\tau)^{\alpha(t_m)-1}d\tau\end{matrix}\right\}. \tag{3.42}$$

For simplicity, equation (3.42) may again be written as

$$\theta^{m+1}=\theta^0+\left\{\begin{array}{l}\dfrac{1}{\Gamma(\alpha(t_m))}\displaystyle\sum_{k=1}^{m}\frac{H(t_k,\theta^k)}{h}\int_{t_k}^{t_{k+1}}(\tau-t_{k-1})(t_{m+1}-\tau)^{\alpha(t_m)-1}d\tau\\ -\dfrac{1}{\Gamma(\alpha(t_m))}\displaystyle\sum_{k=1}^{m}\frac{H(t_{k-1},\theta^{k-1})}{h}\int_{t_k}^{t_{k+1}}(\tau-t_k)(t_{m+1}-\tau)^{\alpha(t_m)-1}d\tau\end{array}\right\}. \quad (3.43)$$

The above integrals can be computed as follows

$$\int_{t_k}^{t_{k+1}}(\tau-t_{k-1})(t_{m+1}-\tau)^{\alpha(t_m)-1}d\tau=\frac{h^{\alpha(t_m)+1}}{\alpha(t_m)(\alpha(t_m)+1)}\left\{\begin{array}{l}(m-k+1)^{\alpha(t_m)}(m-k+2+\alpha(t_m))\\-(m-k)^{\alpha(t_m)}(m-k+2+2\alpha(t_m))\end{array}\right\} \quad (3.44)$$

$$\int_{t_k}^{t_{k+1}}(\tau-t_k)(t_{m+1}-\tau)^{\alpha(t_m)-1}d\tau=\frac{h^{\alpha(t_m)+1}}{\alpha(t_m)(\alpha(t_m)+1)}\left\{\begin{array}{l}(m-k+1)^{\alpha(t_m)+1}-(m-k)^{\alpha(t_m)}\\(m-k+1+\alpha(t_m))\end{array}\right\}. \quad (3.45)$$

By the use of equations (3.44) and (3.45), equation (3.43) reduces to the following result

$$\theta^{m+1}=\theta^0+\frac{h^{\alpha(t_m)}}{\Gamma(\alpha(t_m)+2)}\left[\sum_{k=1}^{m}\left\{\begin{array}{l}H(t_k,\theta^k)\left(\begin{array}{l}(m-k+1)^{\alpha(t_m)}(m-k+2+\alpha(t_m))\\-(m-k)^{\alpha(t_m)}(m-k+2+2\alpha(t_m))\end{array}\right)\\-\left\{\begin{array}{l}(m-k+1)^{\alpha(t_m)+1}-(m-k)^{\alpha(t_m)}\\(m-k+1+\alpha(t_m))\end{array}\right\}H(t_{k-1},\theta^{k-1})\end{array}\right\}\right]. \quad (3.46)$$

Thus, we obtain

$$\theta^{m+1}=\theta^0+\frac{\gamma\, h^{\alpha(t_m)}}{\Gamma(\alpha(t_m)+2)}\sum_{k=1}^{m}\left\{\begin{array}{l}t_k^{\gamma-1}\phi(t_k,\theta^k)\left(\begin{array}{l}(m-k+1)^{\alpha(t_m)}(m-k+2+\alpha(t_m))\\-(m-k)^{\alpha(t_m)}(m-k+2+2\alpha(t_m))\end{array}\right)\\-t_{k-1}^{\gamma-1}\phi(t_{k-1},\theta^{k-1})\left\{\begin{array}{l}(m-k+1)^{\alpha(t_m)+1}-(m-k)^{\alpha(t_m)}\\(m-k+1+\alpha(t_m))\end{array}\right\}\end{array}\right\}. \quad (3.47)$$

### 3.5.2 Implementation of three-step Lagrange interpolation polynomial

Approximating the function $H(\tau,\theta(\tau))$ in equation (3.9) using three-step Lagrange polynomial interpolation leads to

$$\theta^{m+1} = \theta^0 + \frac{1}{\Gamma(\alpha(t_m))} \sum_{k=2}^{m} \int_{t_k}^{t_{k+1}} \left\{ \begin{array}{l} \frac{H(t_k, \theta^k)}{2h^2}(\tau - t_{k-2})(\tau - t_{k-1}) \\ -\frac{H(t_{k-1}, \theta^{k-1})}{h^2}(\tau - t_{k-2})(\tau - t_k) \\ +\frac{H(t_{k-2}, \theta^{k-2})}{2h^2}(\tau - t_{k-1})(\tau - t_k) \end{array} \right\} \tag{3.48}$$

$$(t_{m+1} - \tau)^{\alpha(t_m)-1} d\tau.$$

From equation (3.48), we obtain the following equality

$$\theta^{m+1} = \theta^0 + \left\{ \begin{array}{l} \frac{1}{\Gamma(\alpha(t_m))} \sum_{k=2}^{m} \frac{H(t_k, \theta^k)}{2h^2} \int_{t_k}^{t_{k+1}} (\tau - t_{k-2})(\tau - t_{k-1})(t_{m+1} - \tau)^{\alpha(t_m)-1} d\tau \\ -\frac{1}{\Gamma(\alpha(t_m))} \sum_{k=2}^{m} \frac{H(t_{k-1}, \theta^{k-1})}{h^2} \int_{t_k}^{t_{k+1}} (\tau - t_{k-2})(\tau - t_k)(t_{m+1} - \tau)^{\alpha(t_m)-1} d\tau \\ + \frac{1}{\Gamma(\alpha(t_m))} \sum_{k=2}^{m} \frac{H(t_{k-2}, \theta^{k-2})}{2h^2} \int_{t_k}^{t_{k+1}} (\tau - t_{k-1})(\tau - t_k)(t_{m+1} - \tau)^{\alpha(t_m)-1} d\tau \end{array} \right\}. \tag{3.49}$$

Now, we need to evaluate the following integrals

$$\int_{t_k}^{t_{k+1}} (\tau - t_{k-2})(\tau - t_{k-1})(t_{m+1} - \tau)^{\alpha(t_m)-1} d\tau = \frac{h^{\alpha(t_m)+2}}{\alpha(t_m)(\alpha(t_m) + 1)(\alpha(t_m) + 2)}$$
$$\left[ \begin{array}{l} (m - k + 1)^{\alpha(t_m)} \left\{ 2(m - k)^2 + (3\alpha(t_m) + 10)(m - k) + 2\alpha(t_m)^2 + 9\alpha(t_m) + 12 \right\} \\ -(m - k)^{\alpha(t_m)} \left\{ 2(m - k)^2 + (5\alpha(t_m) + 10)(m - k) + 6\alpha(t_m)^2 + 18\alpha(t_m) + 12 \right\} \end{array} \right]. \tag{3.50}$$

$$\int_{t_k}^{t_{k+1}} (\tau - t_{k-2})(\tau - t_k)(t_{m+1} - \tau)^{\alpha(t_m)-1} d\tau = \frac{2h^{\alpha(t_m)+2}}{\alpha(t_m)(\alpha(t_m) + 1)(\alpha(t_m) + 2)}$$
$$\left[ \begin{array}{l} (m - k + 1)^{\alpha(t_m)} \left\{ (m - k + 1)(m - k + 3 + \alpha(t_m)) \right\} \\ -(m - k)^{\alpha(t_m)} \left\{ (m - k)^2 + (2\alpha(t_m) + 4)(m - k) + 1.5\alpha(t_m)^2 + 4.5\alpha(t_m) + 3 \right\} \end{array} \right]. \tag{3.51}$$

$$\int_{t_k}^{t_{k+1}} (\tau - t_{k-1})(\tau - t_k)(t_{m+1} - \tau)^{\alpha(t_m)-1} d\tau = \frac{2h^{\alpha(t_m)+2}}{\alpha(t_m)(\alpha(t_m) + 1)(\alpha(t_m) + 2)}$$
$$\left[ \begin{array}{l} (m - k + 1)^{\alpha(t_m)} \left\{ (m - k + 1)(m - k + 2 + 0.5\alpha(t_m)) \right\} \\ -(m - k)^{\alpha(t_m)} \left\{ (m - k)^2 + (1.5\alpha(t_m) + 3)(m - k) + \alpha(t_m)^2 + 3\alpha(t_m) + 2 \right\} \end{array} \right]. \tag{3.52}$$

Plugging equations (3.50)–(3.52) into equation (3.49), we get

$$\theta^{m+1} = \theta^0 + \frac{h^{\alpha(t_m)}}{2\Gamma(\alpha(t_m)+3)}\sum_{k=2}^{m} H(t_k, \theta^k)$$
$$\begin{bmatrix} (m-k+1)^{\alpha(t_m)}\left\{2(m-k)^2 + (3\alpha(t_m)+10)(m-k) + 2\alpha(t_m)^2 + 9\alpha(t_m) + 12\right\} \\ -(m-k)^{\alpha(t_m)}\left\{2(m-k)^2 + (5\alpha(t_m)+10)(m-k) + 6\alpha(t_m)^2 + 18\alpha(t_m) + 12\right\} \end{bmatrix} -$$
$$\frac{2h^{\alpha(t_m)}}{\Gamma(\alpha(t_m)+3)}\sum_{k=2}^{m} H(t_{k-1}, \theta^{k-1})$$
$$\begin{bmatrix} (m-k+1)^{\alpha(t_m)}\left\{(m-k+1)(m-k+3+\alpha(t_m))\right\} \\ -(m-k)^{\alpha(t_m)}\left\{(m-k)^2 + (2\alpha(t_m)+4)(m-k) + 1.5\alpha(t_m)^2 + 4.5\alpha(t_m) + 3\right\} \end{bmatrix} + \tag{3.53}$$
$$\frac{h^{\alpha(t_m)}}{\Gamma(\alpha(t_m)+3)}\sum_{k=2}^{m} H(t_{k-2}, \theta^{k-2})$$
$$\begin{bmatrix} (m-k+1)^{\alpha(t_m)}\left\{(m-k+1)(m-k+2+0.5\alpha(t_m))\right\} \\ -(m-k)^{\alpha(t_m)}\left\{(m-k)^2 + (1.5\alpha(t_m)+3)(m-k) + \alpha(t_m)^2 + 3\alpha(t_m) + 2\right\} \end{bmatrix}.$$

Thus, we have

$$\theta^{m+1} = \theta^0 + \frac{\gamma\, h^{\alpha(t_m)}}{2\Gamma(\alpha(t_m)+3)}\sum_{k=2}^{m} t_k^{\gamma-1}\phi(t_k, \theta^k)$$
$$\begin{bmatrix} (m-k+1)^{\alpha(t_m)}\left\{2(m-k)^2 + (3\alpha(t_m)+10)(m-k) + 2\alpha(t_m)^2 + 9\alpha(t_m) + 12\right\} \\ -(m-k)^{\alpha(t_m)}\left\{2(m-k)^2 + (5\alpha(t_m)+10)(m-k) + 6\alpha(t_m)^2 + 18\alpha(t_m) + 12\right\} \end{bmatrix} -$$
$$\frac{2\gamma\, h^{\alpha(t_m)}}{\Gamma(\alpha(t_m)+3)}\sum_{k=2}^{m} t_{k-1}^{\gamma-1}\phi(t_{k-1}, \theta^{k-1})$$
$$\begin{bmatrix} (m-k+1)^{\alpha(t_m)}\left\{(m-k+1)(m-k+3+\alpha(t_m))\right\} \\ -(m-k)^{\alpha(t_m)}\left\{(m-k)^2 + (2\alpha(t_m)+4)(m-k) + 1.5\alpha(t_m)^2 + 4.5\alpha(t_m) + 3\right\} \end{bmatrix} + \tag{3.54}$$
$$\frac{\gamma\, h^{\alpha(t_m)}}{\Gamma(\alpha(t_m)+3)}\sum_{k=2}^{m} t_{k-2}^{\gamma-1}\phi(t_{k-2}, \theta^{k-2})$$
$$\begin{bmatrix} (m-k+1)^{\alpha(t_m)}\left\{(m-k+1)(m-k+2+0.5\alpha(t_m))\right\} \\ -(m-k)^{\alpha(t_m)}\left\{(m-k)^2 + (1.5\alpha(t_m)+3)(m-k) + \alpha(t_m)^2 + 3\alpha(t_m) + 2\right\} \end{bmatrix}.$$

## 3.6 Algorithm for fractal-fractional variable-order differential equation based on Lagrange polynomial interpolation via Mittag-Leffler type kernel

### 3.6.1 Implementation of two-step Lagrange polynomial interpolation

Approximating the function $H(t, \theta(t))$ with the Lagrange polynomial interpolation, equation (3.31) reduces to

$$\theta^{m+1} = \theta^0 + \frac{1-\alpha(t_m)}{AB(\alpha(t_m))}H(t_m, \theta(t_m)) + \frac{\alpha(t_m)}{AB(\alpha(t_m))\Gamma(\alpha(t_m))}$$
$$\sum_{k=1}^{m}\int_{t_k}^{t_{k+1}}\left\{\frac{H(t_k, \theta^k)}{h}(\tau - t_{k-1}) - \frac{H(t_{k-1}, \theta^{k-1})}{h}(\tau - t_k)\right\}(t_{m+1}-\tau)^{\alpha(t_m)-1}d\tau \tag{3.55}$$

or

$$\theta^{m+1} = \theta^0 + \frac{1-\alpha(t_m)}{AB(\alpha(t_m))}H(t_m, \theta(t_m)) + \frac{\alpha(t_m)}{AB(\alpha(t_m))\Gamma(\alpha(t_m))}$$
$$\left\{\begin{aligned} &\sum_{k=1}^{m}\frac{H(t_k, \theta^k)}{h}\int_{t_k}^{t_{k+1}}(\tau - t_{k-1})(t_{m+1}-\tau)^{\alpha(t_m)-1}d\tau \\ &-\sum_{k=1}^{m}\frac{H(t_{k-1}, \theta^{k-1})}{h}\int_{t_k}^{t_{k+1}}(\tau - t_k)(t_{m+1}-\tau)^{\alpha(t_m)-1}d\tau \end{aligned}\right\}. \tag{3.56}$$

Using equations (3.44) and (3.45), equation (3.56) reduces to the following

$$\theta^{m+1} = \theta^0 + \frac{1-\alpha(t_m)}{AB(\alpha(t_m))}H(t_m, \theta(t_m))$$
$$+\frac{\alpha(t_m)h^{\alpha(t_m)}}{AB(\alpha(t_m))\Gamma(\alpha(t_m)+2)}\left[\sum_{k=1}^{m}\left\{\begin{aligned} &H\left(t_k, \theta^k\right)\begin{pmatrix}(m-k+1)^{\alpha(t_m)}(m-k+2+\alpha(t_m)) \\ -(m-k)^{\alpha(t_m)}(m-k+2+2\alpha(t_m))\end{pmatrix} \\ &-H\left(t_{k-1}, \theta^{k-1}\right)\begin{Bmatrix}(m-k+1)^{\alpha(t_m)+1}-(m-k)^{\alpha(t_m)} \\ (m-k+1+\alpha(t_m))\end{Bmatrix}\end{aligned}\right\}\right]. \tag{3.57}$$

$$\theta^{m+1} = \theta^0 + \frac{1-\alpha(t_m)}{AB(\alpha(t_m))}\gamma\, t_m^{\gamma-1}\phi(t_m, \theta^m)$$
$$+\frac{\gamma\,\alpha(t_m)h^{\alpha(t_m)}}{AB(\alpha(t_m))\Gamma(\alpha(t_m)+2)}\sum_{k=1}^{m}\left\{\begin{aligned} &t_k^{\gamma-1}\phi\left(t_k, \theta^k\right)\begin{pmatrix}(m-k+1)^{\alpha(t_m)}(m-k+2+\alpha(t_m)) \\ -(m-k)^{\alpha(t_m)}(m-k+2+2\alpha(t_m))\end{pmatrix} \\ &-\begin{Bmatrix}(m-k+1)^{\alpha(t_m)+1}-(m-k)^{\alpha(t_m)} \\ (m-k+1+\alpha(t_m))\end{Bmatrix}t_{k-1}^{\gamma-1}\phi\left(t_{k-1}, \theta^{k-1}\right)\end{aligned}\right\}. \tag{3.58}$$

### 3.6.2 Implementation of three-step Lagrange polynomial interpolation

Approximating the function $H(\tau, \theta(\tau))$ in equation (3.9) using three-step Lagrange polynomial interpolation in the Atangana–Baleanu sense, we have

$$\theta^{m+1} = \theta^0 + \frac{1-\alpha(t_m)}{AB(\alpha(t_m))}H(t_m, \theta(t_m))$$
$$+\frac{\alpha(t_m)}{AB(\alpha(t_m))\Gamma(\alpha(t_m))}\sum_{k=2}^{m}\int_{t_k}^{t_{k+1}}\left\{\begin{aligned} &\frac{H\left(t_k, \theta^k\right)}{2h^2}(\tau - t_{k-2})(\tau - t_{k-1}) \\ &-\frac{H\left(t_{k-1}, \theta^{k-1}\right)}{h^2}(\tau - t_{k-2})(\tau - t_k) \\ &+\frac{H\left(t_{k-2}, \theta^{k-2}\right)}{2h^2}(\tau - t_{k-1})(\tau - t_k)\end{aligned}\right\}(t_{m+1}-\tau)^{\alpha(t_m)-1}d\tau. \tag{3.59}$$

Rearranging equation (3.59), we get

$$\theta^{m+1} = \theta^0 + \frac{1-\alpha(t_m)}{AB(\alpha(t_m))}H(t_m, \theta(t_m)) + \frac{\alpha(t_m)}{AB(\alpha(t_m))\Gamma(\alpha(t_m))}$$
$$\left\{\begin{aligned} &\left(\sum_{k=2}^{m}\frac{H\left(t_k, \theta^k\right)}{2h^2}\int_{t_k}^{t_{k+1}}(\tau - t_{k-2})(\tau - t_{k-1})(t_{m+1}-\tau)^{\alpha(t_m)-1}d\tau\right) \\ &-\left(\sum_{k=2}^{m}\frac{H\left(t_{k-1}, \theta^{k-1}\right)}{h^2}\int_{t_k}^{t_{k+1}}(\tau - t_{k-2})(\tau - t_k)(t_{m+1}-\tau)^{\alpha(t_m)-1}d\tau\right) \\ &+\left(\sum_{k=2}^{m}\frac{H\left(t_{k-2}, \theta^{k-2}\right)}{2h^2}\int_{t_k}^{t_{k+1}}(\tau - t_{k-1})(\tau - t_k)(t_{m+1}-\tau)^{\alpha(t_m)-1}d\tau\right)\end{aligned}\right\}. \tag{3.60}$$

Substituting equations (3.50)–(3.52) in equation (3.60) and converting to the original function, we have

$$\begin{aligned}
\theta^{m+1} &= \theta^0 + \frac{1-\alpha(t_m)}{AB(\alpha(t_m))}\gamma\, t_m^{\gamma-1}\phi(t_m, \theta^m)\frac{\alpha(t_m)\,\gamma\, h^{\alpha(t_m)}}{2AB(\alpha(t_m))\,\Gamma(\alpha(t_m)+3)} \\
&\sum_{k=2}^{m} t_k^{\gamma-1}\phi(t_k, \theta^k) \\
&\left[\begin{array}{l}(m-k+1)^{\alpha(t_m)}\left\{2(m-k)^2+(3\alpha(t_m)+10)(m-k)+2\alpha(t_m)^2+9\alpha(t_m)+12\right\} \\ -(m-k)^{\alpha(t_m)}\left\{2(m-k)^2+(5\alpha(t_m)+10)(m-k)+6\alpha(t_m)^2+18\alpha(t_m)+12\right\}\end{array}\right] \\
&-\frac{\alpha(t_m)2\gamma\, h^{\alpha(t_m)}}{AB(\alpha(t_m))\Gamma(\alpha(t_m)+3)}\sum_{k=2}^{m} t_{k-1}^{\gamma-1}\phi(t_{k-1}, \theta^{k-1}) \\
&\left[\begin{array}{l}(m-k+1)^{\alpha(t_m)}\left\{(m-k+1)(m-k+3+\alpha(t_m))\right\} \\ -(m-k)^{\alpha(t_m)}\left\{(m-k)^2+(2\alpha(t_m)+4)(m-k)+1.5\alpha(t_m)^2+4.5\alpha(t_m)+3\right\}\end{array}\right] \\
&+\frac{\alpha(t_m)\,\gamma\, h^{\alpha(t_m)}}{AB(\alpha(t_m))\Gamma(\alpha(t_m)+3)}\sum_{k=2}^{m} t_{k-2}^{\gamma-1}\phi(t_{k-2}, \theta^{k-2}) \\
&\left[\begin{array}{l}(m-k+1)^{\alpha(t_m)}\left\{(m-k+1)(m-k+2+0.5\alpha(t_m))\right\} \\ -(m-k)^{\alpha(t_m)}\left\{(m-k)^2+(1.5\alpha(t_m)+3)(m-k)+\alpha(t_m)^2+3\alpha(t_m)+2\right\}\end{array}\right].
\end{aligned} \tag{3.61}$$

## 3.7 Numerical examples

Example 3.1. Let us consider the following fractal-fractional variable-order differential equation

$$\begin{cases} {}^{FF}D_{0,\,t}^{\alpha(t),\,\gamma}\theta(t) = \sin(t)\,\theta(t), \\ \theta(0) = e^{-1}. \end{cases} \tag{3.62}$$

The exact solution of equation (3.62) for the classical derivative is given as

$$\theta(t) = e^{-\cos(t)}. \tag{3.63}$$

Applying equations (3.17), (3.26), (3.35), and (3.40) for NPI, equations (3.47), (3.54), (3.58), and (3.61) for LPI and taking $\phi(t, \theta(t)) = \sin(t)\theta(t)$, we may find the solutions of equation (3.62). Figure 3.1 depicts the numerical solution comparison of example 3.1 described in the Caputo and Atangana–Baleanu sense. The convergence study for variable-order and absolute error between the exact and obtained solution for integer-order derivative are illustrated in tables 3.1–3.4 and 3.8–3.10 considering step length $h = 0.1, 0.01, 0.001$, and $t = 0$ to 1000. Figure 3.2 illustrates the solution plots of equation (3.62) at different variable-order fractional derivatives and fractal dimensions.

Example 3.2. We consider the following example

$$\begin{cases} {}^{FF}D_{0,\,t}^{\alpha(t),\,\gamma}\theta(t) = -\cos(2t)\,\theta^2(t), \\ \theta(0) = 1. \end{cases} \tag{3.64}$$

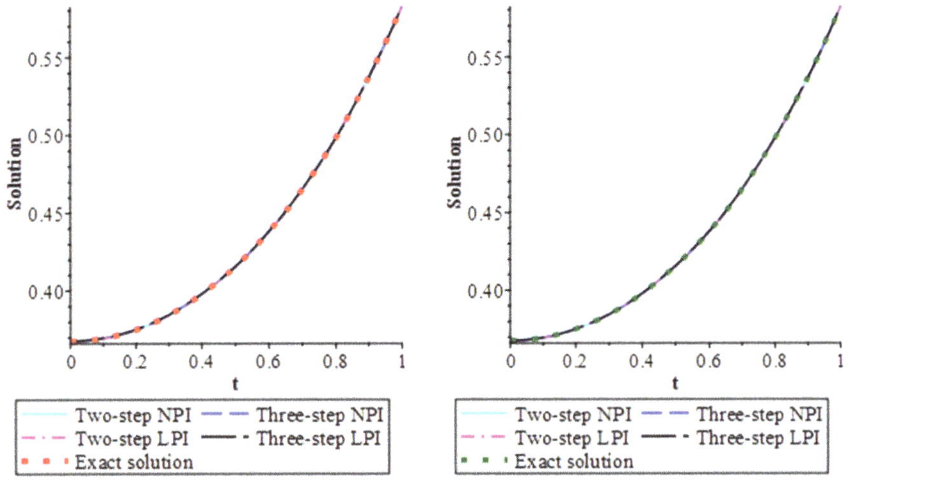


(a) (b)

**Figure 3.1.** Comparison solutions of example 3.1 defined in the (a) Caputo and (b) Atangana–Baleanu sense for integer-order derivatives.

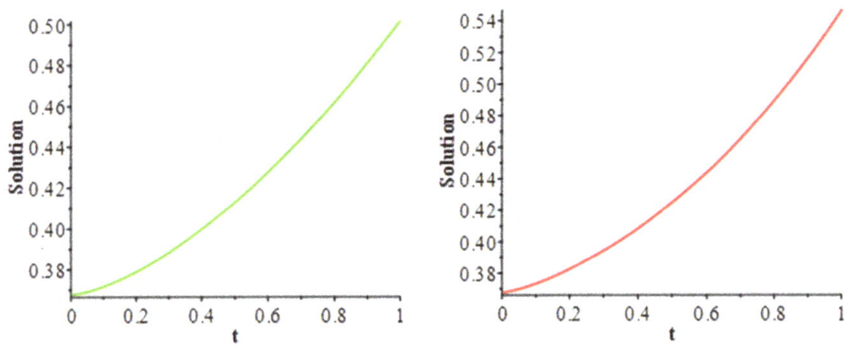


(a) (b)

**Figure 3.2.** Two-term LPI solutions plots of example 3.1 defined in the Caputo sense at (a) $\alpha(t) = 0.97{+}0.03\cos\left(\frac{t}{10}\right)$, $\gamma = 0.5$, (b) $\alpha(t) = \tanh(t+1)$, $\gamma = 0.7$.

The exact solution of equation (3.64) for the integer-order derivative is given as

$$\theta(t) = \frac{2}{2 + \sin(2t)}. \tag{3.65}$$

Implementing equations (3.17), (3.26), (3.35), and (3.40) for NPI, equations (3.47), (3.54), (3.58), and (3.61) for LPI and taking $\phi(t, \theta(t)) = -\cos(2t)\,\theta^2(t)$, we obtain the solutions of equation (3.64). Figure 3.3 demonstrates the numerical solution comparison of example 3.2 defined in the Caputo and Atangana–Baleanu

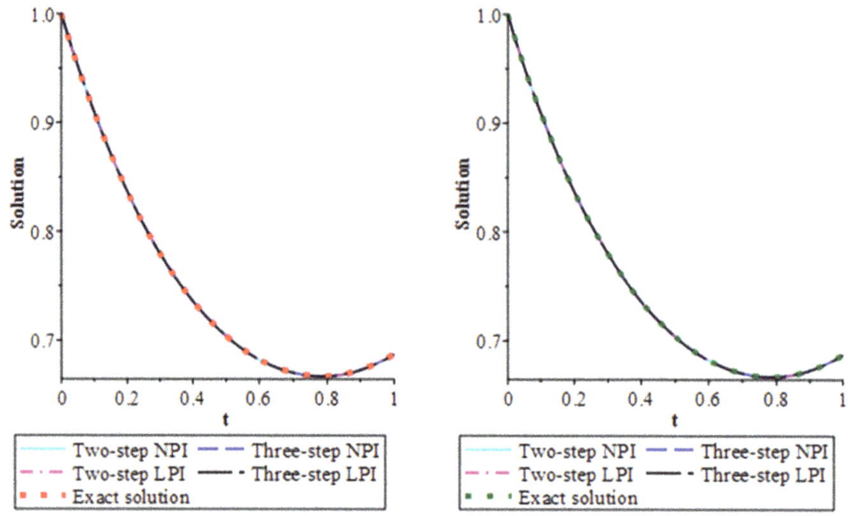


(a) (b)

**Figure 3.3.** Comparison solutions of example 3.2 defined in the (a) Caputo and (b) Atangana–Baleanu sense for integer-order derivatives.

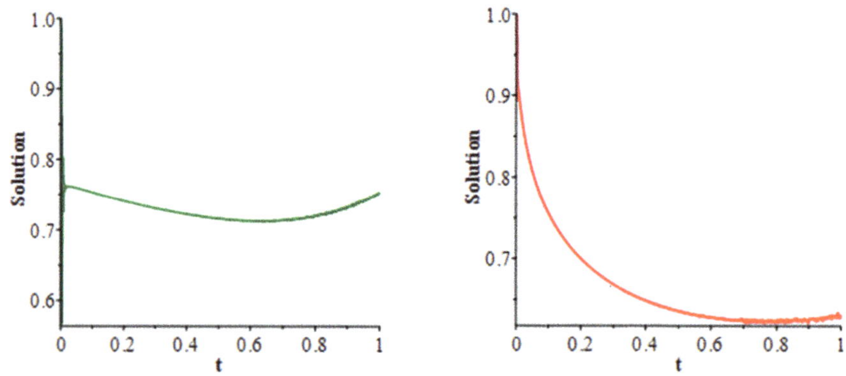


(a) (b)

**Figure 3.4.** Two-term NPI solutions plots of example 3.2 defined in the Atangana–Baleanu derivative sense (a) $\alpha(t) = \tanh(t + 1)$, $\gamma = 0.9$ (b)$\alpha(t) = 0.99 + \dfrac{0.01}{1 + \exp\left(-\frac{t}{4}\right)}$, $\gamma = 0.3$.

sense. Similarly, the convergence study for variable-order and absolute error between the exact and obtained solution for integer-order derivative are given in tables 3.5–3.7 and 3.11–3.13, considering step length $h = 0.1, 0.01, 0.001$, and $t = 0$ to 1000. Figure 3.4 presents the solution plots of equation (3.64) at various variable-order fractional derivatives and fractal dimensions.

**Table 3.1.** Two-step LPI and NPI solution and convergence study of example 3.1 at $\alpha(t) = 1$ and $\gamma = 1$.

| | | Power-law kernel type two-step LPI | | Mittag-Leffler kernel type two-step LPI | | Power-law kernel type two-step NPI | | Mittag-Leffler kernel type two-step NPI | | Exact solution |
|---|---|---|---|---|---|---|---|---|---|---|
| $t$ | Step Size $h$ | Solution $n = 1000$ | Absolute error | Solution $n = 1000$ | Absolute error | Solution $n = 1000$ | Absolute error | Solution $n = 1000$ | Absolute error | |
| $t = 0.2$ | 0.1 | 0.3678794412 | 0.74066711e-2 | 0.3678794412 | 0.7406671e-2 | 0.3678794412 | 0.7406671e-2 | 0.3678794412 | 0.7406671e-2 | **0.3752861123** |
| | 0.01 | 0.3745403892 | 0.7457231e-3 | 0.3745403892 | 0.7457231e-3 | 0.3744846933 | 0.801419e-3 | 0.3744846933 | 0.801419e-3 | |
| | 0.001 | **0.3752115520** | **0.745603e-4** | **0.3752115520** | **0.745603e-4** | **0.3752109953** | **0.75117e-4** | **0.3752109953** | **0.75117e-4** | |
| $t = 0.4$ | 0.1 | 0.3826792321 | 0.154172069e-1 | 0.3826792321 | 0.154172069e-1 | 0.3771306213 | 0.2096582e-1 | 0.3771306213 | 0.2096582e-1 | **0.3980964390** |
| | 0.01 | 0.3965452971 | 0.15511419e-2 | 0.3965452971 | 0.15511419e-2 | 0.3964880729 | 0.1608366e-2 | 0.3964880729 | 0.1608366e-2 | |
| | 0.001 | **0.3979414020** | **0.1550370e-3** | **0.3979414020** | **0.1550370e-3** | **0.3979408305** | **0.1556085e-3** | **0.3979408305** | **0.1556085e-3** | |
| $t = 0.6$ | 0.1 | 0.4134067064 | 0.246812325e-1 | 0.4134067064 | 0.246812325e-1 | 0.4075181181 | 0.3056982e-1 | 0.4075181181 | 0.3056982e-1 | **0.4380879389** |
| | 0.01 | 0.4356117898 | 0.24761491e-2 | 0.4356117898 | 0.24761491e-2 | 0.4355520063 | 0.2535933e-2 | 0.4355520063 | 0.2535933e-2 | |
| | 0.001 | **0.4378405472** | **0.2473917e-3** | **0.4378405472** | **0.2473917e-3** | **0.4378399506** | **0.2479883e-3** | **0.4378399506** | **0.2479883e-3** | |
| $t = 0.8$ | 0.1 | 0.4623963909 | 0.3582701e-1 | 0.4623963909 | 0.3582701e-1 | 0.4560961353 | 0.4212726e-1 | 0.4560961353 | 0.4212726e-1 | **0.4982233994** |
| | 0.01 | 0.4946437597 | 0.357964e-2 | 0.4946437597 | 0.357964e-2 | 0.4945802695 | 0.364313e-2 | 0.4945802695 | 0.364313e-2 | |
| | 0.001 | **0.4978659346** | **0.3574648e-3** | **0.4978659346** | **0.3574648e-3** | **0.4978653016** | **0.3580978e-3** | **0.4978653016** | **0.3580978e-3** | |
| $t = 1.0$ | 0.1 | 0.5331324073 | 0.494397e-1 | 0.5331324073 | 0.494397e-1 | 0.5262821324 | 0.5628998e-1 | 0.5262821324 | 0.5628998e-1 | **0.5825721108** |
| | 0.01 | 0.5776589970 | 0.4913114e-2 | 0.5776589970 | 0.4913114e-2 | 0.5775903362 | 0.4981775e-2 | 0.5775903362 | 0.4981775e-2 | |
| | 0.001 | **0.5820817768** | **0.490334e-3** | **0.5820817768** | **0.490334e-3** | **0.5820810923** | **0.4910185e-3** | **0.5820810923** | **0.4910185e-3** | |

**Table 3.2.** Two-step LPI and NPI solution and convergence study of example 3.1 at $\alpha(t) = 0.97 + 0.03\cos\left(\frac{t}{10}\right)$ and $\gamma = 1$.

| $t$ | Step Size $h$ | Power-law kernel type two-step LPI Solution $n = 1000$ | Mittag-Leffler kernel type two-step LPI Solution $n = 1000$ | Power-law kernel type two-step NPI Solution $n = 1000$ | Mittag-Leffler kernel type two-step NPI Solution $n = 1000$ |
|---|---|---|---|---|---|
| $t = 0.2$ | 0.1 | 0.3678794412 | 0.3678794412 | 0.3678794412 | 0.3678794412 |
| | 0.01 | 0.3745404736 | 0.3745407874 | 0.3744847774 | 0.3744850913 |
| | 0.001 | **0.3752116188** | **0.3752120363** | **0.3752111052** | **0.3752115242** |
| $t = 0.4$ | 0.1 | 0.3826794359 | 0.3826798450 | 0.3771307787 | 0.3771311982 |
| | 0.01 | 0.3965464919 | 0.3965494893 | 0.3964892708 | 0.3964922691 |
| | 0.001 | **0.3979436152** | **0.3979470540** | **0.3979421951** | **0.3979456631** |
| $t = 0.6$ | 0.1 | 0.4134086049 | 0.4134119371 | 0.4075199401 | 0.4075232958 |
| | 0.01 | 0.4356171758 | 0.4356284278 | 0.4355573922 | 0.4355686459 |
| | 0.001 | **0.4378510757** | **0.4378635294** | **0.4378458226** | **0.4378582781** |
| $t = 0.8$ | 0.1 | 0.4624037400 | 0.4624156474 | 0.4561034644 | 0.4561154250 |
| | 0.01 | 0.4946594737 | 0.4946893288 | 0.4945959044 | 0.4946257645 |
| | 0.001 | **0.4978848292** | **0.4979170042** | **0.4978819897** | **0.4979143629** |
| $t = 1.0$ | 0.1 | 0.5331522113 | 0.5331830567 | 0.5263021327 | 0.5263330899 |
| | 0.01 | 0.5776954970 | 0.5777616970 | 0.5776264710 | 0.5776926768 |
| | 0.001 | **0.5821835329** | **0.5822541809** | **0.5821191625** | **0.5821900750** |

**Table 3.3.** Two-step LPI and NPI solution and convergence study of example 3.1 at $\alpha(t) = \tanh(t + 1)$ and $\gamma = 1$.

| $t$ | Step Size $h$ | Power-law kernel type two-step LPI Solution $n = 1000$ | Mittag-Leffler kernel type two-step LPI Solution $n = 1000$ | Power-law kernel type two-step NPI Solution $n = 1000$ | Mittag-Leffler kernel type two-step NPI Solution $n = 1000$ |
|---|---|---|---|---|---|
| $t = 0.2$ | 0.1 | 0.3678794412 | 0.3678794412 | 0.3678794412 | 0.3678794412 |
| | 0.01 | 0.3783021623 | 0.3910322992 | 0.3782355761 | 0.3909700921 |
| | 0.001 | **0.3791011751** | **0.3927498597** | **0.3791005697** | **0.3927492957** |
| $t = 0.4$ | 0.1 | 0.3895912489 | 0.4022531447 | 0.3827278786 | 0.3952160909 |
| | 0.01 | 0.4041346707 | 0.4235227599 | 0.4040747883 | 0.4234645883 |
| | 0.001 | **0.4055247797** | **0.4253368673** | **0.4055241907** | **0.4253362845** |
| $t = 0.6$ | 0.1 | 0.4240285206 | 0.4430213708 | 0.4177603436 | 0.4367597681 |
| | 0.01 | 0.4456179190 | 0.4680277412 | 0.4455577756 | 0.4679681526 |
| | 0.001 | **0.4477415949** | **0.4703510599** | **0.4477459979** | **0.4703551989** |
| $t = 0.8$ | 0.1 | 0.4752441131 | 0.4973972765 | 0.4688099232 | 0.4908528407 |
| | 0.01 | 0.5061840755 | 0.5303956471 | 0.5061207948 | 0.5303324766 |
| | 0.001 | **0.5092718282** | **0.5336066352** | **0.5092535831** | **0.5335885986** |
| $t = 1.0$ | 0.1 | 0.5475128640 | 0.5717735496 | 0.5406031608 | 0.5646843045 |
| | 0.01 | 0.5904078308 | 0.6164468465 | 0.5903396617 | 0.6163787531 |
| | 0.001 | **0.5946215015** | **0.6207757554** | **0.5946525323** | **0.6208114336** |

**Table 3.4.** Two-step LPI and NPI solution and convergence study of example 3.1 at $\alpha(t) = \frac{0.01}{1 + e^{-\frac{t}{4}}} + 0.99$ and $\gamma = 1$.

| $t$ | Step Size $h$ | Power-law kernel type two-step LPI Solution $n = 1000$ | Mittag-Leffler kernel type two-step LPI Solution $n = 1000$ | Power-law kernel type two-step NPI Solution $n = 1000$ | Mittag-Leffler kernel type two-step NPI Solution $n = 1000$ |
|---|---|---|---|---|---|
| $t = 0.2$ | 0.1 | 0.3678794412 | 0.3678794412 | 0.3678794412 | 0.3678794412 |
| | 0.01 | 0.3746260937 | 0.3749442621 | 0.3745700763 | 0.3748883346 |
| | 0.001 | **0.3753039170** | **0.3756545370** | **0.3753033905** | **0.3756540118** |
| $t = 0.4$ | 0.1 | 0.3828474530 | 0.3831845716 | 0.3772608437 | 0.3776023894 |
| | 0.01 | 0.3968142906 | 0.3974907154 | 0.3967569547 | 0.3974334501 |
| | 0.001 | **0.3982197583** | **0.3989322545** | **0.3982193879** | **0.3989318836** |
| $t = 0.6$ | 0.1 | 0.4137965545 | 0.4144833532 | 0.4078913740 | 0.4085805050 |
| | 0.01 | 0.4361277598 | 0.4372191914 | 0.4360679574 | 0.4371594510 |
| | 0.001 | **0.4383740136** | **0.4395096186** | **0.4383680057** | **0.4395036354** |
| $t = 0.8$ | 0.1 | 0.4630686150 | 0.4641676968 | 0.4567633386 | 0.4578630574 |
| | 0.01 | 0.4954747626 | 0.4970918269 | 0.4954113443 | 0.4970284808 |
| | 0.001 | **0.4987279100** | **0.5004022611** | **0.4987114494** | **0.5003856898** |
| $t = 1.0$ | 0.1 | 0.5341618511 | 0.5357867221 | 0.5273156064 | 0.5289401904 |
| | 0.01 | 0.5788907795 | 0.5811996764 | 0.5788218232 | 0.5811308188 |
| | 0.001 | **0.5833789747** | **0.5857637999** | **0.5833313966** | **0.5857155476** |

**Table 3.5.** Two-step LPI and NPI solution and convergence study of example 3.2 at $\alpha(t) = 1$ and $\gamma = 1$.

| | | Power-law kernel type two-step LPI | | Mittag-Leffler kernel type two-step LPI | | Power-law kernel type two-step NPI | | Mittag-Leffler kernel type two-step NPI | | Exact solution |
|---|---|---|---|---|---|---|---|---|---|---|
| $t$ | Step Size $h$ | Solution $n = 1000$ | Absolute error | Solution $n = 1000$ | Absolute error | Solution $n = 1000$ | Absolute error | Solution $n = 1000$ | Absolute error | |
| $t = 0.2$ | 0.1 | 0.9 | 0.6297621e-1 | 0.9 | 0.6297621e-1 | 0.9 | 0.6297621e-1 | 0.9 | 0.6297621e-1 | **0.8370237914** |
| | 0.01 | 0.8508156492 | 0.1379186e-1 | 0.8508156492 | 0.1379186e-1 | 0.8581709178 | 0.2114713e-1 | 0.8581709178 | 0.2114713e-1 | |
| | 0.001 | **0.8383730067** | **0.1349215e-2** | **0.8383730067** | **0.1349215e-2** | **0.8390770041** | **0.2053213e-2** | **0.8390770041** | **0.2053213e-2** | |
| $t = 0.4$ | 0.1 | 0.8508837328 | 0.1148742 | 0.8508837328 | 0.1148742 | 0.9452992452 | 0.2092897 | 0.9452992452 | 0.2092897 | **0.7360095376** |
| | 0.01 | 0.7454366486 | 0.9427111e-2 | 0.7454366486 | 0.9427111e-2 | 0.7510614299 | 0.1505189e-1 | 0.7510614299 | 0.1505189e-1 | |
| | 0.001 | **0.7369310046** | **0.921467e-3** | **0.7369310046** | **0.921467e-3** | **0.7374747486** | **0.1465211e-2** | **0.7374747486** | **0.1465211e-2** | |
| $t = 0.6$ | 0.1 | 0.7639611532 | 0.8184201e-1 | 0.7639611532 | 0.8184201e-1 | 0.8287662466 | 0.1466471 | 0.8287662466 | 0.1466471 | **0.6821191468** |
| | 0.01 | 0.6886388592 | 0.6519712e-2 | 0.6886388592 | 0.6519712e-2 | 0.6934269623 | 0.1130782e-1 | 0.6934269623 | 0.1130782e-1 | |
| | 0.001 | **0.6827548320** | **0.6356852e-3** | **0.6827548320** | **0.6356852e-3** | **0.6832214582** | **0.1102311e-2** | **0.6832214582** | **0.1102311e-2** | |
| $t = 0.8$ | 0.1 | 0.7251013208 | 0.5833989e-1 | 0.7251013208 | 0.5833989e-1 | 0.7790689712 | 0.1123075 | 0.7790689712 | 0.1123075 | **0.6667614350** |
| | 0.01 | 0.6712269841 | 0.4465549e-2 | 0.6712269841 | 0.4465549e-2 | 0.6757718415 | 0.9010407e-2 | 0.6757718415 | 0.9010407e-2 | |
| | 0.001 | **0.6671945142** | **0.4330792e-3** | **0.6671945142** | **0.4330792e-3** | **0.6676400806** | **0.8786456e-3** | **0.6676400806** | **0.8786456e-3** | |
| $t = 1.0$ | 0.1 | 0.7283406140 | 0.4088942e-1 | 0.7283406140 | 0.4088942e-1 | 0.7819157314 | 0.4088942e-1 | 0.7819157314 | 0.4088942e-1 | **0.6874511974** |
| | 0.01 | 0.6903427960 | 0.2891599e-2 | 0.6903427960 | 0.2891599e-2 | 0.6951540023 | 0.2891599e-2 | 0.6951540023 | 0.2891599e-2 | |
| | 0.001 | **0.6877284390** | **0.2772416e-3** | **0.6877284390** | **0.2772416e-3** | **0.6882018981** | **0.2772416e-3** | **0.6882018981** | **0.2772416e-3** | |

**Table 3.6.** Two-step LPI and NPI solution and convergence study of example 3.2 at $\alpha(t) = \tanh(t+1)$ and $\gamma = 1$.

| | | Power-law kernel type two-step LPI | Mittag-Leffler kernel type two-step LPI | Power-law kernel type two-step NPI | Mittag-Leffler kernel type two-step NPI |
|---|---|---|---|---|---|
| $t$ | Step Size $h$ | Solution $n = 1000$ | Solution $n = 1000$ | Solution $n = 1000$ | Solution $n = 1000$ |
| $t = 0.2$ | 0.1 | 0.8121581091 | 0.8121581091 | 0.8121581091 | 0.8121581091 |
| | 0.01 | 0.8094486710 | 0.7579481704 | 0.8163962400 | 0.7619801076 |
| | 0.001 | **0.7990741897** | **0.7549301225** | **0.7997207889** | **0.7553286477** |
| $t = 0.4$ | 0.1 | 0.8088309579 | 0.7675832962 | 0.9338088620 | 0.8555415799 |
| | 0.01 | 0.7323928212 | 0.7328451814 | 0.7370577196 | 0.7361664895 |
| | 0.001 | **0.7258555155** | **0.7298970143** | **0.7263050089** | **0.7302183121** |
| $t = 0.6$ | 0.1 | 0.7591309209 | 0.7548862513 | 0.8266875515 | 0.8036193604 |
| | 0.01 | 0.6956515496 | 0.7205591709 | 0.6995907552 | 0.7236754177 |
| | 0.001 | **0.6909663529** | **0.7176133157** | **0.6912926700** | **0.7178741260** |
| $t = 0.8$ | 0.1 | 0.7381265742 | 0.7588650876 | 0.7814094572 | 0.7895177604 |
| | 0.01 | 0.6901192338 | 0.7272541060 | 0.6939270367 | 0.7304368391 |
| | 0.001 | **0.6865478155** | **0.7245583082** | **0.6870612980** | **0.7249942975** |
| $t = 1.0$ | 0.1 | 0.7527541369 | 0.7879371521 | 0.7946725039 | 0.8190497757 |
| | 0.01 | 0.7173120559 | 0.7632997051 | 0.7214299442 | 0.7668586811 |
| | 0.001 | **0.7152365621** | **0.7617387577** | **0.7154489147** | **0.7619592330** |

**Table 3.7.** Two-step LPI and NPI solution and convergence study of example 3.2 at $\alpha(t) = \frac{0.01}{1+e^{-\frac{t}{4}}} + 0.99$ and $\gamma = 1$.

| $t$ | Step Size $h$ | Power-law kernel type two-step LPI Solution $n = 1000$ | Mittag-Leffler kernel type two-step LPI Solution $n = 1000$ | Power-law kernel type two-step NPI Solution $n = 1000$ | Mittag-Leffler kernel type two-step NPI Solution $n = 1000$ |
|---|---|---|---|---|---|
| $t = 0.2$ | 0.1 | 0.8986288065 | 0.8986288065 | 0.8986288065 | 0.8986288065 |
| | 0.01 | 0.8496133039 | 0.8476446094 | 0.8569770536 | 0.8548926745 |
| | 0.001 | **0.8371677543** | **0.8354840345** | **0.8378701646** | **0.8361762685** |
| $t = 0.4$ | 0.1 | 0.8497514723 | 0.8471753646 | 0.9449040012 | 0.9420017512 |
| | 0.01 | 0.7445921577 | 0.7444935872 | 0.7501952225 | 0.7500392656 |
| | 0.001 | **0.7361341191** | **0.7361870596** | **0.7366723015** | **0.7367202922** |
| $t = 0.6$ | 0.1 | 0.7633583095 | 0.7629348665 | 0.8282101180 | 0.8269610941 |
| | 0.01 | 0.6884305286 | 0.6894092437 | 0.6931919865 | 0.6941390804 |
| | 0.001 | **0.6825286727** | **0.6835999885** | **0.6830632247** | **0.6841322419** |
| $t = 0.8$ | 0.1 | 0.7251196304 | 0.7259011819 | 0.7787938902 | 0.7790209553 |
| | 0.01 | 0.6717155114 | 0.6735085611 | 0.6762332730 | 0.6780084698 |
| | 0.001 | **0.6675858806** | **0.6694587601** | **0.6681814625** | **0.6700456480** |
| $t = 1.0$ | 0.1 | 0.7290783894 | 0.7308182140 | 0.7823240038 | 0.7836921774 |
| | 0.01 | 0.6916238419 | 0.6943189644 | 0.6964114398 | 0.6991002755 |
| | 0.001 | **0.6888126740** | **0.6915857012** | **0.6895446248** | **0.6923137534** |

**Table 3.8.** Three-step LPI and NPI solution and convergence study of example 3.1 at $\alpha(t) = 1$ and $\gamma = 1$.

| | | Power-law kernel type three-step LPI | | Mittag-Leffler kernel type three-step LPI | | Power-law kernel type three-step NPI | | Mittag-Leffler kernel type three-step NPI | | |
|---|---|---|---|---|---|---|---|---|---|---|
| $t$ | Step Size $h$ | Solution $n = 1000$ | Absolute error | Solution $n = 1000$ | Absolute error | Solution $n = 1000$ | Absolute error | Solution $n = 1000$ | Absolute error | Exact solution |
| $t = 0.2$ | 0.1 | 0.3678794412 | 0.7406671e-2 | 0.3678794412 | 0.7406671e-2 | 0.3678794412 | 0.7406671e-2 | 0.3678794412 | 0.7406671e-2 | **0.3752861123** |
| | 0.01 | 0.3744846933 | 0.801419e-3 | 0.3744846933 | 0.801419e-3 | 0.3743959582 | 0.8901541e-3 | 0.3743959582 | 0.8901541e-3 | |
| | 0.001 | **0.3752109953** | **0.75117e-4** | **0.3752109953** | **0.75117e-4** | **0.3752109886** | **0.751237e-4** | **0.3752109886** | **0.751237e-4** | |
| $t = 0.4$ | 0.1 | 0.3771306213 | 0.2096582e-1 | 0.3771306213 | 0.2096582e-1 | 0.3789337083 | 0.1916273e-1 | 0.3789337083 | 0.1916273e-1 | **0.3980964390** |
| | 0.01 | 0.3964880728 | 0.1608366e-2 | 0.3964880728 | 0.1608366e-2 | 0.3964476478 | 0.1648791e-2 | 0.3964476478 | 0.1648791e-2 | |
| | 0.001 | **0.3979408305** | **0.1556085e-3** | **0.3979408305** | **0.1556085e-3** | **0.3979477205** | **0.1487185e-3** | **0.3979477205** | **0.1487185e-3** | |
| $t = 0.6$ | 0.1 | 0.4075181182 | 0.3056982e-1 | 0.4075181182 | 0.3056982e-1 | 0.3975738821 | 0.4051406e-1 | 0.3975738821 | 0.4051406e-1 | **0.4380879389** |
| | 0.01 | 0.4355520061 | 0.2535933e-2 | 0.4355520061 | 0.2535933e-2 | 0.4356743486 | 0.241359e-2 | 0.4356743486 | 0.241359e-2 | |
| | 0.001 | **0.4378399507** | **0.2479882e-3** | **0.4378399507** | **0.2479882e-3** | **0.4378668538** | **0.2210851e-3** | **0.4378668538** | **0.2210851e-3** | |
| $t = 0.8$ | 0.1 | 0.4560961354 | 0.4212726e-1 | 0.4560961354 | 0.4212726e-1 | 0.4454744433 | 0.5274896e-1 | 0.4454744433 | 0.5274896e-1 | **0.4982233994** |
| | 0.01 | 0.4945802695 | 0.364313e-2 | 0.4945802695 | 0.364313e-2 | 0.4950665945 | 0.3156805e-2 | 0.4950665945 | 0.3156805e-2 | |
| | 0.001 | **0.4978653014** | **0.358098e-3** | **0.4978653014** | **0.358098e-3** | **0.4979342690** | **0.2891304e-3** | **0.4979342690** | **0.2891304e-3** | |
| $t = 1.0$ | 0.1 | 0.5262821323 | 0.5628998e-1 | 0.5262821323 | 0.5628998e-1 | 0.5156338556 | 0.6693826e-1 | 0.5156338556 | 0.6693826e-1 | **0.5825721108** |
| | 0.01 | 0.5775903363 | 0.4981774e-2 | 0.5775903363 | 0.4981774e-2 | 0.5787499127 | 0.3822198e-2 | 0.5787499127 | 0.3822198e-2 | |
| | 0.001 | **0.5820810921** | **0.4910187e-3** | **0.5820810921** | **0.4910187e-3** | **0.5822251673** | **0.3469435e-3** | **0.5822251673** | **0.3469435e-3** | |

**Table 3.9.** Three-step LPI and NPI solution and convergence study of example 3.1 at $\alpha(t) = \tanh(t+1)$ and $\gamma = 1$.

| | | Power-law kernel type three-step LPI | Mittag-Leffler kernel type three-step LPI | Power-law kernel type three-step NPI | Mittag-Leffler kernel type three-step NPI |
|---|---|---|---|---|---|
| $t$ | Step Size $h$ | Solution $n = 1000$ | Solution $n = 1000$ | Solution $n = 1000$ | Solution $n = 1000$ |
| $t = 0.2$ | 0.1 | 0.3678794412 | 0.3678794412 | 0.3678794412 | 0.3678794412 |
| | 0.01 | 0.3782355771 | 0.3909700934 | 0.3781403923 | 0.3909105816 |
| | 0.001 | **0.3791021373** | **0.3927505541** | **0.3791025596** | **0.3927556017** |
| $t = 0.4$ | 0.1 | 0.3827278787 | 0.3952160909 | 0.3857091408 | 0.3857091408 |
| | 0.01 | 0.4040747802 | 0.4234645844 | 0.4040918074 | 0.4235749822 |
| | 0.001 | **0.4055185880** | **0.4253328882** | **0.4055386313** | **0.4253613385** |
| $t = 0.6$ | 0.1 | 0.4177603435 | 0.4367597680 | 0.4064095771 | 0.4256024501 |
| | 0.01 | 0.4455577409 | 0.4679681740 | 0.4458084352 | 0.4683799533 |
| | 0.001 | **0.4473318739** | **0.4699405080** | **0.4477873321** | **0.4704141447** |
| $t = 0.8$ | 0.1 | 0.4688099244 | 0.4908528415 | 0.4579667300 | 0.4801469489 |
| | 0.01 | 0.5061216890 | 0.5303365250 | 0.5068124815 | 0.5312657815 |
| | 0.001 | **0.5127658757** | **0.5367268768** | **0.5093445844** | **0.5337055235** |
| $t = 1.0$ | 0.1 | 0.5406031631 | 0.5646843061 | 0.5306146608 | 0.5554259518 |
| | 0.01 | 0.5903383896 | 0.6163798877 | 0.5917885372 | 0.6181701462 |
| | 0.001 | **0.5877338336** | **0.6133411914** | **0.5948269532** | **0.6210221807** |

**Table 3.10.** Three-step LPI and NPI solution and convergence study of example 3.1 at $\alpha(t) = \frac{0.01}{1 + e^{-\frac{t}{4}}} + 0.99$ and $\gamma = 1$.

| | | Power-law kernel type three-step LPI | Mittag-Leffler kernel type three-step LPI | Power-law kernel type three-step NPI | Mittag-Leffler kernel type three-step NPI |
|---|---|---|---|---|---|
| $t$ | Step Size $h$ | Solution $n = 1000$ | Solution $n = 1000$ | Solution $n = 1000$ | Solution $n = 1000$ |
| $t = 0.2$ | 0.1 | 0.3678794412 | 0.3678794412 | 0.3678794412 | 0.3678794412 |
| | 0.01 | 0.3745700765 | 0.3748883350 | 0.3744809537 | 0.3747997452 |
| | 0.001 | **0.3753028398** | **0.3756538084** | **0.3753034142** | **0.3756541046** |
| $t = 0.4$ | 0.1 | 0.3772608436 | 0.3776023894 | 0.3790831062 | 0.3790831062 |
| | 0.01 | 0.3967569347 | 0.3974334194 | 0.3967180202 | 0.3973968136 |
| | 0.001 | **0.3982328295** | **0.3989439778** | **0.3982265028** | **0.3989392847** |
| $t = 0.6$ | 0.1 | 0.4078913741 | 0.4085805051 | 0.3978940139 | 0.3985881403 |
| | 0.01 | 0.4360682473 | 0.4371597076 | 0.4361959698 | 0.4372934983 |
| | 0.001 | **0.4388387033** | **0.4397997677** | **0.4383955955** | **0.4395319432** |
| $t = 0.8$ | 0.1 | 0.4567633393 | 0.4578630584 | 0.4461214764 | 0.4472256015 |
| | 0.01 | 0.4954112372 | 0.4970293182 | 0.4959111051 | 0.4975411818 |
| | 0.001 | **0.5012878154** | **0.5037299591** | **0.4987819384** | **0.5004576736** |
| $t = 1.0$ | 0.1 | 0.5273156104 | 0.5289401938 | 0.5166979800 | 0.5183506148 |
| | 0.01 | 0.5788253488 | 0.5811369222 | 0.5800076839 | 0.5823415445 |
| | 0.001 | **0.5820870698** | **0.5851680338** | **0.5834784602** | **0.5858654256** |

**Table 3.11.** Three-step LPI and NPI solution and convergence study of example 3.2 at $\alpha(t) = 1$ and $\gamma = 1$.

| | | Power-law kernel type three-step LPI | | Mittag-Leffler kernel type three-step LPI | | Power-law kernel type three-step NPI | | Mittag-Leffler kernel type three-step NPI | | |
|---|---|---|---|---|---|---|---|---|---|---|
| $t$ | Step Size $h$ | Solution $n = 1000$ | Absolute error | Solution $n = 1000$ | Absolute error | Solution $n = 1000$ | Absolute error | Solution $n = 1000$ | Absolute error | Exact solution |
| $t = 0.2$ | 0.1 | 0.9 | 0.6297621e-1 | 0.9 | 0.6297621e-1 | 0.9 | 0.6297621e-1 | 0.9 | 0.6297621e-1 | **0.8370237914** |
| | 0.01 | 0.8581709175 | 0.2114713e-1 | 0.8581709175 | 0.2114713e-1 | 0.8650972961 | 0.280735e-1 | 0.8650972961 | 0.280735e-1 | |
| | 0.001 | **0.8390770038** | **0.2053212e-2** | **0.8390770038** | **0.2053212e-2** | **0.8397410469** | **0.2717255e-2** | **0.8397410469** | **0.2717255e-2** | |
| $t = 0.4$ | 0.1 | 0.9452992451 | 0.2092897 | 0.9452992451 | 0.2092897 | 0.9452992452 | 0.2092897 | 0.9452992452 | 0.2092897 | **0.7360095376** |
| | 0.01 | 0.7510614302 | 0.1505189e-1 | 0.7510614302 | 0.1505189e-1 | 0.7554644268 | 0.1945489e-1 | 0.7554644268 | 0.1945489e-1 | |
| | 0.001 | **0.7374747484** | **0.1465211e-2** | **0.7374747484** | **0.1465211e-2** | **0.7378714275** | **0.186189e-2** | **0.7378714275** | **0.186189e-2** | |
| $t = 0.6$ | 0.1 | 0.8287662467 | 0.1466471 | 0.8287662467 | 0.1466471 | 0.8287662466 | 0.1466471 | 0.8287662466 | 0.1466471 | **0.6821191468** |
| | 0.01 | 0.6934269625 | 0.1130782e-1 | 0.6934269625 | 0.1130782e-1 | 0.6955625823 | 0.1344344e-1 | 0.6955625823 | 0.1344344e-1 | |
| | 0.001 | **0.6832214583** | **0.1102311e-2** | **0.6832214583** | **0.1102311e-2** | **0.6833840576** | **0.1264911e-2** | **0.6833840576** | **0.1264911e-2** | |
| $t = 0.8$ | 0.1 | 0.7790689710 | 0.1123075 | 0.7790689710 | 0.1123075 | 0.7790689712 | 0.1123075 | 0.7790689712 | 0.1123075 | **0.6667614350** |
| | 0.01 | 0.6757718420 | 0.9010407e-2 | 0.6757718420 | 0.9010407e-2 | 0.6757287257 | 0.8967291e-2 | 0.6757287257 | 0.8967291e-2 | |
| | 0.001 | **0.6676400804** | **0.8786454e-3** | **0.6676400804** | **0.8786454e-3** | **0.6675805509** | **0.8191159e-3** | **0.6675805509** | **0.8191159e-3** | |
| $t = 1.0$ | 0.1 | 0.7819157311 | 0.9446453e-1 | 0.7819157311 | 0.9446453e-1 | 0.7819157314 | 0.9446453e-1 | 0.7819157314 | 0.9446453e-1 | **0.6874511974** |
| | 0.01 | 0.6951540028 | 0.7702805e-2 | 0.6951540028 | 0.7702805e-2 | 0.6928576718 | 0.5406474e-2 | 0.6928576718 | 0.5406474e-2 | |
| | 0.001 | **0.6882018981** | **0.7507007e-3** | **0.6882018981** | **0.7507007e-3** | **0.6879133703** | **0.4621729e-3** | **0.6879133703** | **0.4621729e-3** | |

**Table 3.12.** Three-step LPI and NPI solution and convergence study of example 3.2 at $\alpha(t) = \tanh(t + 1)$ and $\gamma = 1$.

| $t$ | Step Size $h$ | Power-law kernel type three-step LPI Solution $n = 1000$ | Mittag-Leffler kernel type three-step LPI Solution $n = 1000$ | Power-law kernel type three-step NPI Solution $n = 1000$ | Mittag-Leffler kernel type three-step NPI Solution $n = 1000$ |
|---|---|---|---|---|---|
| $t = 0.2$ | 0.1 | 0.8121581091 | 0.8121581091 | 0.8121581091 | 0.8121581091 |
| | 0.01 | 0.8163961856 | 0.7619800654 | 0.8142572667 | 0.7593816658 |
| | 0.001 | **0.7996588132** | **0.7552774017** | **0.7980926580** | **0.7539403959** |
| $t = 0.4$ | 0.1 | 0.9338088622 | 0.8555415799 | 0.6314689090 | 0.6314689090 |
| | 0.01 | 0.7370580540 | 0.7361668245 | 0.7251547830 | 0.7241848459 |
| | 0.001 | **0.7264567858** | **0.7303729235** | **0.7231263372** | **0.7278415870** |
| $t = 0.6$ | 0.1 | 0.8266875534 | 0.8036193624 | 0.7872314094 | 0.7454730864 |
| | 0.01 | 0.6995817960 | 0.7236836694 | 0.6797542959 | 0.7033005342 |
| | 0.001 | **0.7001188614** | **0.7292521853** | **0.6874436999** | **0.7156242166** |
| $t = 0.8$ | 0.1 | 0.7814094486 | 0.7895177532 | 0.6885164814 | 0.6972323783 |
| | 0.01 | 0.6938971663 | 0.7304112127 | 0.6649085937 | 0.7001802493 |
| | 0.001 | **0.6470578260** | **0.6943876338** | **0.6814512509** | **0.7107264268** |
| $t = 1.0$ | 0.1 | 0.7946724921 | 0.8190497646 | 0.7012944749 | 0.7214400441 |
| | 0.01 | 0.7214186187 | 0.7668715164 | 0.6798476263 | 0.7230207352 |
| | 0.001 | **0.8064360150** | **0.8008972963** | **0.7028905804** | **0.7501401811** |

**Table 3.13.** Three-step LPI and NPI solution and convergence study of example 3.2 at $\alpha(t) = \frac{0.01}{1 + e^{-\frac{t}{4}}} + 0.99$ and $\gamma = 1$.

| $t$ | Step Size $h$ | Power-law kernel type three-step LPI Solution $n = 1000$ | Mittag-Leffler kernel type three-step LPI Solution $n = 1000$ | Power-law kernel type three-step NPI Solution $n = 1000$ | Mittag-Leffler kernel type three-step NPI Solution $n = 1000$ |
|---|---|---|---|---|---|
| $t = 0.2$ | 0.1 | 0.8986288065 | 0.8986288065 | 0.8986288065 | 0.8986288065 |
| | 0.01 | 0.8569770419 | 0.8548926791 | 0.8638569696 | 0.8616885923 |
| | 0.001 | **0.8379406940** | **0.8362519262** | **0.8385220817** | **0.8368189459** |
| $t = 0.4$ | 0.1 | 0.9449040014 | 0.9420017514 | 0.7558961698 | 0.7558961698 |
| | 0.01 | 0.7501974702 | 0.7500409526 | 0.7544278787 | 0.7542096823 |
| | 0.001 | **0.7363375258** | **0.7362412677** | **0.7370283706** | **0.7370847203** |
| $t = 0.6$ | 0.1 | 0.8282101179 | 0.8269610942 | 0.8578055928 | 0.8557348600 |
| | 0.01 | 0.6931814072 | 0.6941282884 | 0.6950578153 | 0.6959636834 |
| | 0.001 | **0.6710704329** | **0.6812999199** | **0.6832372012** | **0.6842875210** |
| $t = 0.8$ | 0.1 | 0.7787938858 | 0.7790209509 | 0.7844179209 | 0.7844169889 |
| | 0.01 | 0.6762261769 | 0.6779915489 | 0.6758151188 | 0.6775651779 |
| | 0.001 | **0.6385922459** | **0.6354133850** | **0.6679173641** | **0.6699305189** |
| $t = 1.0$ | 0.1 | 0.7823239974 | 0.7836921751 | 0.7891275147 | 0.7902845371 |
| | 0.01 | 0.6964298598 | 0.6990843893 | 0.6936097566 | 0.6962812328 |
| | 0.001 | **0.6477029183** | **0.6733959290** | **0.6888389641** | **0.6922400847** |

## 3.8 Conclusion

In this chapter, the Adams–Bashforth approach has been implemented to find the solution of two variable-order fractal-fractional differential equations. We have successfully applied two-step and three-step NPI and LPI to variable-order fractal-fractional differential equations defined in the Caputo and Atangana–Baleanu–Caputo senses. The accuracy of the proposed techniques has been tested by comparing the present solutions with the exact solutions at integer-order cases which are in exact agreement. The convergence analysis for variable and integer-order differential equations has been effectively conducted by reducing step lengths $h$, as seen in various tables. The included tables show that one can obtain high accuracy results by reducing the step length $h$. Three steps NPI and LPI with the Adams–Bashforth approach offer a better solution than two steps NPI and LPI. The approach outlined in this paper is precise and reliable, and it can be used to compute and analyze numerical solutions for many other real-world models. This approach is easy to apply, and by replacing the NPI and LPI with other known polynomials, we can construct a new area within the broad field of numerical analysis.

## Conflicts of interest

The authors state that they have no conflict of interest in publishing this paper.

## Acknowledgement

Mr Rajarama Mohan Jena acknowledges the Department of Science and Technology, Government of India, for providing INSPIRE Fellowship (IF170207).

## References

[1] Jena R M, Chakraverty S and Jena S K 2020 Analysis of the dynamics of phytoplankton nutrient and whooping cough models with nonsingular kernel arising in the biological system *Chaos, Solitons Fractals* **141** 110373

[2] Jena R M, Chakraverty S and Baleanu D 2020 SIR epidemic model of childhood diseases through fractional operators with Mittag-Leffler and exponential kernels *Math. Comput. Simul.* **182** 514–34

[3] Jena R M, Chakraverty S, Baleanu D and Jena S K 2020 Analysis of time-fractional dynamical model of romantic and interpersonal relationships with non-singular kernels: a comparative study *Math. Methods Appl. Sci.* **44** 2183–99

[4] Jena R M, Chakraverty S, Rezazadeh H and Ganji D D 2020 On the solution of time-fractional dynamical model of Brusselator reaction-diffusion system arising in chemical reactions *Math. Methods Appl. Sci.* **43** 3903–13

[5] Srivastava H M, Jena R M, Chakraverty S and Jena S K 2020 Dynamic response analysis of fractionally-damped generalized Bagley–Torvik equation subject to external loads *Russ. J. Math. Phys.* **27** 254–68

[6] Jena R M, Chakraverty S and Baleanu D 2020 A novel analytical technique for the solution of time-fractional ivancevic option pricing model *Physica A* **550** 124380

[7] Jena R M, Chakraverty S and Baleanu D 2020 Solitary wave solution for a generalized Hirota–Satsuma coupled KdV and MKdV equations: a semi-analytical approach *Alexandria Eng. J.* **59** 2877–89

[8] Jena R M, Chakraverty S and Yavuz M 2020 Two-hybrid techniques coupled with an integral transformation for caputo time-fractional Navier–Stokes equations *Prog. Fractional Differentiation Applications* **6** 201–13

[9] Bonyah E 2018 Chaos in a 5-D hyperchaotic system with four wings in the light of non-local and non-singular fractional derivatives *Chaos, Solitons Fractals* **116** 316–31

[10] Owolabi K M 2019 Mathematical modelling and analysis of love dynamics: a fractional approach *Physica A* **525** 849–65

[11] Araz S I 2020 Numerical analysis of a new Volterra integro-differential equation involving fractal-fractional operators *Chaos, Solitons Fractals* **130** 1109396

[12] Toufik M and Atangana A 2017 New numerical approximation of fractional derivative with non-local and non-singular kernel: application to chaotic models *European Phys. J.* **132** 444

[13] Werner W 1984 Polynomial interpolation: Lagrange versus Newton *Mathematics and Computation* **43** 205–17

[14] Krogh T F 1970 Efficient algorithms for polynomial interpolation and numerical differentiation *Mathematics and Computation* **24** 185–90

[15] Yang Y and Gordon S P 2015 Visualizing and understanding the components of Lagrange and Newton interpolation *Problems, Resources, and Issues in Mathematics Undergraduate Studies* **26** 39–52

[16] Dimitrov D K and Philipps G M 1994 A note on convergence of Newton interpolating polynomials *J. Comput. Appl. Math.* **51** 127–30

[17] Srivastava R B and Shukla S 2012 *Numerical Accuracies of Lagrange's and Newton Polynomial Interpolation: Numerical Accuracies of Interpolation Formulas* (Mauritius: LAP LAMBERT Academic Publishing)

[18] Rosenfeld J A and Dixon W E 2017 Approximating the caputo fractional derivative through the Mittag-Leffler reproducing kernel Hilbert space and the kernelized Adams–Bashforth–Moulton method *SIAM J. Numer. Anal.* **55** 1201–17

[19] Diethelm K, Ford N J and Freed A D 2002 A predictor-corrector approach for the numerical solution of fractional differential equations *Nonlinear Dyn.* **29** 3–22

[20] Diethelm K, Ford N J and Freed A D 2004 Detailed error analysis for a fractional Adams method *Numer. Algorithms* **36** 31–52

[21] Gnitchogna R and Atangana A 2017 New two step Laplace Adam–Bashforth method for integer a noninteger order partial differential equations *Numer. Methods Partial Differ. Equ.* **34** 1739–58

[22] Zhao L and Deng W 2014 Jacobian-predictor-corrector approach for fractional differential equations *Adv. Comput. Math.* **40** 137–65

[23] Samko S G, Kilbas A A and Marichev O I 1993 *Fractional Integrals and Derivatives: Theory and Applications* (Yverdon: Gordon and Breach)

[24] Caputo M and Fabrizio M 2015 A new definition of fractional derivative without singular kernel *Progr. Fract. Differ. Appl.* **1** 73–85

[25] Atangana A and Baleanu D 2016 New fractional derivatives with nonlocal and non-singular kernel: theory and application to heat transfer model *Therm. Sci.* **20** 763–9

[26] Atangana A 2017 Fractal-fractional differentiation and integration: connecting fractal calculus and fractional calculus to predict complex system *Chaos, Solitons Fractals* **102** 396–406

[27] Atangana A, Akgül A and Owolabi K M 2020 Analysis of fractal fractional differential equations *Alexandria Eng. J.* **1** 1117–34

[28] Atangana A and Qureshi S 2019 Modeling attractors of chaotic dynamical systems with fractal-fractional operators *Chaos, Solitons Fractals* **123** 320–37

[29] Gómez-Aguilar J F and Atangana A 2020 New chaotic attractors: application of fractal-fractional differentiation and integration *Math. Meth. Appl. Sci.* **44** 3036–65

[30] Solís-Pérez J E, Gómez-Aguilar J F and Atangana A 2018 Novel numerical method for solving variable-order fractional differential equations with power, exponential and Mittag-Leffler laws *Chaos, Solitons Fractals* **114** 175–85

**IOP** Publishing

Polynomial Paradigms
Trends and applications in science and engineering
**S Chakraverty**

# Chapter 4

# Polynomial-based numerical methods for singularly perturbed differential equation on layer-adapted meshes

**Jugal Mohapatra and Subal Ranjan Sahu**

## Symbols

| | |
|---|---|
| $\varepsilon$ | parameter |
| $\delta$, $\eta$ | Delay parameters |
| $o(\cdot)$ | Landau symbol (little-oh) |
| $N$ | Number of mesh-intervals |
| $x$, $x_i$, $x_j$ | Spatial variables |
| $h_i$, $h_j$, $H$ | Mesh size |
| $\|\cdot\|$ | Norm |
| $\Omega$ | Domain |
| $C(\Omega)$ | Continuous functions on [0, 1] |
| $\bar{\Omega}^N$ | Discrete spatial domains |
| $L_\varepsilon$, $L$ | Differential operators |
| $L^N_{\text{mu}}$, $L^N_{\text{cs}}$, $L^N$ | Discreet differential operators |
| $D^+D^-$, $D^+$, $D^-$ | Difference operators |
| $E_N$ | Maximum pointwise errors |
| $R_N$ | Rate of convergence |
| $\aleph^{\pm}$ | Barrier function |
| $\omega(x)$, $\chi(x)$ | Mesh generating functions |
| SPDE | Singularly perturbed differential equations |
| SPDDE | Singularly perturbed differential-difference equations |

This chapter describes a polynomial-based numerical technique for singularly perturbed differential-difference equations. At first, the original differential-difference problem is reduced to an analogous singularly perturbed boundary value

doi:10.1088/978-0-7503-5067-9ch4

problem via Taylor series expansion. The reduced problem is then solved using a polynomial-based numerical approximation on Shishkin type meshes. Here, the numerical approximation comprises a midpoint upwind approximation in the region of coarse mesh and a cubic spline approximation in the region of fine mesh. A thorough analysis of the proposed methods is presented and certain numerical simulations are carried out to support the theoretical findings.

## 4.1 Introduction

A differential equation whose highest order coefficient is associated with a small parameter ($\varepsilon$) is generally known as a singularly perturbed differential equation (SPDE). Such SPDEs widely arise in the fields of scientific disciplines such as fluid dynamics, fluid mechanics, plasma dynamics, aerodynamics, oceanography and many more domains of the world of fluid motion [6, 12]. If these types of SPDEs involve some positive and negative shifts then the differential equations are known as singularly perturbed differential-difference equations (SPDDEs) [9]. These SPDDEs generally occur in the modeling of the first exit time problems. When the limiting value of $\varepsilon$ approaches to zero, the solution of SPDEs behaves non uniformly. The solutions of such kinds of SPDEs have multi-scale characters and they basically occur in a thin region of the domain where they vary rapidly. This region is generally known as the layer region. When classical numerical methods are applied to such types of SPDEs, numerous difficulties arise, demanding special attention to these problems. Generally, to tackle these types of problems, there are two types of approaches used by the researchers over the last few decades, namely, the fitted operator method and fitted mesh method [16, 9, 12]. In the fitted operator method basically the standard finite difference operator is replaced by a finite difference operator, which reflects the layer behavior of the differential operator. On the other hand, the fitted mesh method of the standard numerical method involves a special kind of mesh that adapts the layers phenomena.

In this chapter, we have considered the following second-order SPDDE of the form:

$$\begin{cases} L_\varepsilon z = \varepsilon z''(x) + p_1(x)z'(x) + p_2(x)z(x-\delta) - p_3(x)z(x) & x \in \Omega, \\ \qquad + p_4(x)z(x+\eta) = f(x), & \\ z(x) = \vartheta(x), & -\delta \leqslant x \leqslant 0, \\ z(x) = \phi(x), & 1 \leqslant x \leqslant 1+\eta, \end{cases} \tag{4.1}$$

where the diffusion coefficient $\varepsilon$ is sufficiently small i.e., $0 < \varepsilon \ll 1$. The parameters $\delta = o(\varepsilon)$ and $\eta = o(\varepsilon)$, are the positive and negative shift respectively. Assume that the coefficients $p_1(x)$, $p_2(x)$, $p_3(x)$, $p_4(x)$ and $f(x)$ are regular, bounded and independent of $\varepsilon$ on $\bar{\Omega} = [0, 1]$. We now apply the following Taylor series approximation in the vicinity of $x$ for the approximation of the negative and positive shift terms of (4.1),

$$\begin{cases} z(x-\delta) \approx z(x) - \delta z'(x) + o(\delta)^2, \\ z(x+\eta) \approx z(x) + \eta z'(x) + o(\eta)^2. \end{cases} \tag{4.2}$$

By omitting the second-order terms from (4.2) and plugging them in (4.1), we get

$$\begin{cases} Lz = \varepsilon z''(x) + a(x)z'(x) - b(x)z(x) = f(x), \\ z(0) = \vartheta(0), \quad z(1) = \phi(1). \end{cases} \tag{4.3}$$

Here, $a(x) = p_1(x) - \delta p_2(x) + \eta p_4(x)$, $b(x) = p_3(x) - p_2(x) - p_4(x)$. The layer behavior of (4.3) generally depends on the convective term $a(x)$, that is, if $a(x) > 0$, $a(x) < 0$ and $a(x) = 0$, $\forall x \in \Omega$ then the layer will appear at the left-hand side, right-hand side and interior of the domain respectively. In this chapter, we have considered $a(x) \geqslant \beta > 0$ and $b(x) \geqslant 0$ for any constant $\beta$ so that, the solution of the problem (4.3) forms a boundary layer near $x = 0$.

During the last few decades, SPDDEs have been studied extensively by many researchers. Here, we have accounted the studies of the researchers on various kinds of numerical approaches for SPDDEs. These types of SPDDEs have been broadly researched due to their multiple applications. Lange and Miura [9, 9] adopted singular perturbation approach and Laplace transformation to treat the SPDDEs numerically. Kadalbajoo and Sharma [12] developed a parameter uniform numerical method to solve the SPDDES of mixed type. Similarly, Rao and Chakravarthy [15] established a finite difference scheme by constructing some special type of mesh to handle the positive and the negative shifts. Using grid equidistribution, an uniformly convergent method was established by Mohapatra and Natesan [12]. Duressa *et al* [6] presented a domain decomposition method to tackle the original problem.

We consider a second-order SPDDE and numerically address it on several layer-adapted meshes such as the Shishkin mesh, Bakhvalov–Shishkin mesh (B–S) and Vulanovi'c mesh (V). Initially, via Taylor series expansion, the model problem is reduced to an equivalent singularly perturbed boundary value problem. Most of numerical methods for SPDDEs that are available in the literature are of first order convergent. To design a numerical scheme of higher order is our primary focus in this chapter. In [15], the authors have developed an approximation that solves singularly perturbed reaction-diffusion problems on Shishkin mesh using both the conventional finite difference approximation and the cubic spline approximation. The spline-based numerical approximation for SPDDEs were devised by Priyadharshini *et al* [16]. The cubic spline approximation only holds the discrete supremum norm in the fine mesh part and the cubic spline method's solution is unstable in the coarse mesh part. Therefore, we establish a numerical scheme which comprises of a midpoint upwind approximation in the coarse mesh region and a cubic spline approximation in the fine mesh region. The proposed scheme is thoroughly investigated, and it is established that it is second-order uniform convergent with respect to the perturbation parameter.

The following is how the chapter is organized: the properties of the solution and its derivatives are presented in section 4.2. The construction of a spline-based numerical scheme on various types of Shishkin type meshes is explained in section 4.3. Section 4.4 presents error bounds for the proposed method. We looked at some numerical test problems in section 4.5, and the results are discussed there. Throughout this chapter, $C$ has been treated as a generic positive constant that is independent of the perturbation parameter $\varepsilon$, mesh size, and mesh points. The supremum norm is defined as follows $\| H(x)\| = \max_{x\in\bar{\Omega}} | H(x)|$, for every continuous function $H(x)$.

## 4.2 Derivative bounds and solution decomposition

**Lemma 4.1.** Let us assume a function $z \in C^2(\Omega) \cap C(\bar{\Omega})$ with $z(0) \geqslant 0$, $z(1) \geqslant 0$ and $Lz \leqslant 0$, $\forall x \in \Omega$, then $z(x) \geqslant 0$, $\forall x \in \bar{\Omega}$.

Proof. We shall prove this lemma by the method of contradiction. Let us assume that there exists a point $x^* \in \bar{\Omega}$ such that $z(x^*) = \min(z(x))$ and $z(x^*) < 0$. This implies $z'(x^*) = 0$, $z''(x^*) > 0$. Also,

$$Lz = \varepsilon z''(x^*) + a(x^*)z'(x^*) - b(x^*)z(x^*) = \varepsilon z''(x^*) - b(x^*)z(x^*) \geqslant 0.$$

This is in direct opposition to our presumptions. Thus, $z(x) \geqslant 0$, $\forall x \in \bar{\Omega}$.

**Lemma 4.2.** Let $z$ be the solution of (4.3), then we can estimate:

$$\| z \| \leqslant \|f\|\beta^{-1} + \max\left\{z(0), z(1)\right\}, \tag{4.4}$$

for all $x \in \Omega$.

Proof. Let us choose $\aleph^{\pm}(x) = \|f\|\beta^{-1} \pm z(x) + \max\{z(0), z(1)\}$ as a barrier function. It is evident that $\aleph^{\pm}(x) \geqslant 0$ at the points $x = 0, 1$ and

$$\begin{aligned} L\aleph^{\pm}(x) &= \varepsilon(\aleph^{\pm}(x))'' + a(x)(\aleph^{\pm}(x))' - b(x)(\aleph^{\pm}(x)) \\ &= b(x)\left[\frac{\|f\|}{\beta} + \max\{|z(0)|, |z(1)|\}\right] \pm Lz(x) \\ &\geqslant b(x)\max\{|z(0)|, |z(1)|\} + |f| \pm f(x) \\ &\geqslant 0. \end{aligned}$$

Now, by applying lemma 4.1, one can get the required estimate.

**Lemma 4.3.** We have the following bound for the solution $z$ of (4.3):

$$| z^{(l)}(x)| \leqslant C(1 + \varepsilon^{-l}e^{(-\beta x/\varepsilon)}), \tag{4.5}$$

where $0 \leqslant l \leqslant 4$.

Proof. From the mean value theorem, there exists a $\varsigma \in (0, \varepsilon)$ such that $z'(\varsigma) = \frac{z(0) - z(1)}{\varepsilon}$. Thus, we have

$$\varepsilon z'(\varsigma) < 2\|z\|. \tag{4.6}$$

Integrating (4.3) from 0 to $\zeta$, we get

$$\varepsilon z'(\varsigma) - \varepsilon z'(0) = \int_0^{\varsigma} (f(\varkappa) - a(\varkappa)z'(\varkappa) + b(\varkappa))d\varkappa. \tag{4.7}$$

$$\left| \int_0^{\varsigma} (f(\varkappa) - a(\varkappa)z'(\varkappa) + b(\varkappa)z(\varkappa))d\varkappa \right| \leqslant \|f\||\varsigma| + \int_0^{\varsigma} (a(\varkappa)y'(\varkappa))d\varkappa + \|b\|\|z\||\zeta|. \tag{4.8}$$

Since $\int_0^{\varsigma} (a(\varkappa)z'(\varkappa))d\varkappa = a(\varkappa)z(\varkappa)|_0^{\varsigma} - \int_0^{\zeta} (a'(\varkappa)z(\varkappa))d\varkappa$. Therefore,

$$\left| \int_0^{\varsigma} (a(\varkappa)z'(\varkappa))d\varkappa \right| \leqslant 2\|a\|\|z\| - \|a'\|\|z\|\|\varkappa\|. \tag{4.9}$$

Now, combining the inequalities (4.8) and (4.9), we have

$$\left| \int_0^{\varsigma} (f(\varkappa) - a(\varkappa)z'(\varkappa) + b(\varkappa)y(\varkappa))d\varkappa \right| \leqslant \|f\| + (2\|a\| + \|a'\|\|b\|)\|z\|. \tag{4.10}$$

From (4.6) and (4.7), we obtain $|z'(0)| \leqslant C\varepsilon^{-1}$. For $\varsigma = x \in \Omega$, we can have $|z'(x)| \leqslant C\varepsilon^{-1}$. Now using the estimates of $z$ and $z'$, one can have the required estimate.

Now, for the batter analysis, the solution $z$ of (4.3) is decomposed into the regular $r$ and singular component $s$ such that $z = r + s$. The asymptotic expansion for $r$ can be written as: $r(x) = r_0 + \varepsilon r_1 + \varepsilon^2 r_2 + \varepsilon^3 r_3$, where $r_0$, $r_1$, $r_2$, and $r_3$ are the solutions of following equations:

$$a(x)r_0' - b(x)r_0 = f(x), \qquad r_0(0) = z(0), \tag{4.11}$$

$$a(x)r_1' - b(x)r_1 = -r_0''(x), \quad r_1(0) = 0, \tag{4.12}$$

$$\varepsilon r_2'' + a(x)r_2' - b(x)r_2 = -r_1''(x), \quad r_2(0) = 0, \quad r_2(1) = 0, \tag{4.13}$$

$$\varepsilon r_3'' + a(x)r_3' - b(x)r_3 = -r_2''(x), \quad r_3(0) = 0, \quad r_3(1) = 0. \tag{4.14}$$

Also, $r$ and $s$ are the solutions of the following boundary value problems:

$$Lr(x) = f(x), \quad r(0) = z(0), \quad r(1) = r_0(1) + \varepsilon r_1(1) + \varepsilon^2 r_2(1) + \varepsilon^3 r_3(1), \tag{4.15}$$

$$Ls(x) = 0, \quad s(0) = 0, \quad s(1) = z(1) - r(1). \tag{4.16}$$

**Lemma 4.4.** For $0 \leqslant l \leqslant 4$, the smooth and the layer component of $z(x)$ meets the following requirements

$$|r^{(l)}| \leqslant C(\mathrm{e}^{(-\beta x/\varepsilon)}\varepsilon^{(3-l)} + 1) \tag{4.17}$$

$$|s^{(l)}| \leqslant C\mathrm{e}^{(-\beta x/\varepsilon)}\varepsilon^{-l}. \tag{4.18}$$

Proof. From the decomposition of $z$, we get $|r_0^{(l)}| \leqslant C, \quad$ for $0 \leqslant l \leqslant 4$ and $|r_1^{(l)}| \leqslant C, \quad$ for $0 \leqslant l \leqslant 4$. From lemma 4.3, we can easily obtain

$$|r_2^{(l)}(x)| \leqslant C(\mathrm{e}^{(-\beta x/\varepsilon)}\varepsilon^{(2-l)} + 1)$$

and

$$|r_3^{(l)}(x)| \leqslant C(\mathrm{e}^{(-\beta x/\varepsilon)}\varepsilon^{(3-l)} + 1)$$

for $0 \leqslant l \leqslant 4$. Now, for $s$, if we choose $\aleph^{\pm}(x) = |s(1)|\mathrm{e}^{(-\beta x/\varepsilon)} \pm s(x)$. as the barrier function, then clearly, $\aleph^{\pm}(x) \geqslant 0$, at the end points of $\Omega$ and $L\aleph^{\pm}(0) \geqslant 0$. Thus, by applying lemma 4.1, we obtain

$$|s(x)| \leqslant C\mathrm{e}^{(-\beta x/\varepsilon)}.$$

The singular component $s(x)$ is also can be rewritten as $s(x) = s(0)\nu(x) + s(1)(1 - \nu(x))$ where $0 \leqslant \nu(x) \leqslant 1$ and $\nu$ is the solution of the following

$$L\nu(x) = 0, \quad \nu(0) = 0, \quad \nu(1) = 0. \tag{4.19}$$

Therefore $s' = (s(0) - s(1))\nu'$. So

$$|s'(x)| \leqslant C|\nu'(x)| \leqslant C\varepsilon^{-1}\mathrm{e}^{(-\beta x/\varepsilon)}. \tag{4.20}$$

As a result, the derivatives of $s$ can be calculated directly from the bounds of $s$ and $s'$. Hence,

$$|s^{l}(x)| \leqslant C\mathrm{e}^{(-\beta x/\varepsilon)}\varepsilon^{-l}, \tag{4.21}$$

for $0 \leqslant l \leqslant 4$.

## 4.3 The discrete problem

### 4.3.1 Construction of non-uniform meshes

In the neighborhood of $x = 0$, the solution of (4.3) contains an exponential boundary layer. So, it is necessary to use layer-adapted meshes in order to achieve $\varepsilon$-uniform convergence of the numerical scheme. In the area of SPDE, the most commonly-studied meshes are the Shishkin type meshes because of their simplicity. These types of meshes are now being explored for (4.1). Let $m \in (0, 1)$ and $\kappa > 0$ be mesh parameters. Now set the transition parameter $\tau$ by

$$\tau = \min\left\{m, \frac{\kappa\varepsilon}{\beta}\ln N\right\},$$

which divided $\Omega^N$ into two intervals, that is $[0, \tau]$ and $[\tau, 1]$. The intervals $[0, \tau]$ and $[\tau, 1]$ contain $mN$ and $(1 - m)N$ number of sub intervals respectively. In most cases, the transition parameter is chosen in such a manner that the exponential term $\exp(-\beta x/\varepsilon)$ must be less than the $N^{-\kappa}$. The parameter $\kappa$ is usually set to the same value as the numerical method's order or large enough to accommodate the

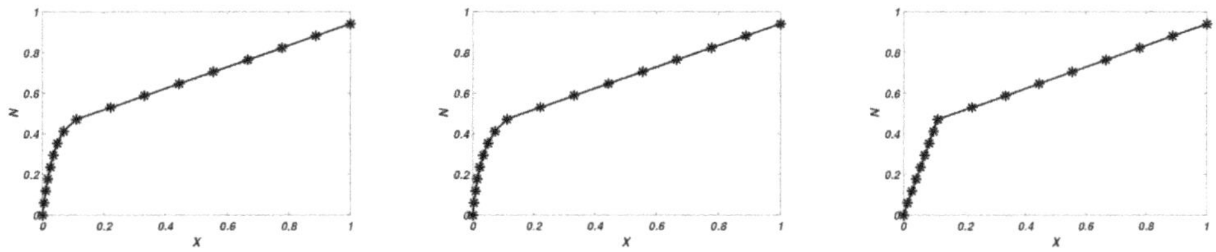

*(a) B-S mesh* *(b) V mesh* (c) S mesh

**Figure 4.1.** Mesh construction with $N = 16$, $\varepsilon = 0.01$.

convergence analysis. For problem (4.1), we have considered $m = 1/2$ and $\kappa = 2$. Basically, this type of mesh is developed by a mesh generating function $\omega$ with $\omega(0) = 0$ and $\omega(1/2) = \ln N$. The $\omega$ is monotonic increasing and continuously differentiable function. Define the mesh point

$$x_i = \begin{cases} \frac{\kappa\varepsilon}{\beta}\omega(t_i), & \text{for } t_i = \frac{i}{N}, \quad i = 0, 1, \dots, N/2, \\ 1 - \left(1 - \frac{\kappa\varepsilon}{\beta}\ln N\right)\frac{2(N-i)}{N}, & i = N/2 + 1, \dots, N. \end{cases}$$

To describe the layer-adapted meshes more clearly, consider the characterizing function $\chi$ which is closely related to the generating function $\omega$. That is, $\omega = -\ln\chi$ with $\chi(0) = 1$ and $\chi(1/2) = N^{-1}$. The $\chi$ is calculated as follows for various layer-adapted meshes:

$$\chi(t) = \begin{cases} \{e^{-2t(\ln N)}, & \text{(Smesh)} \\ 1 - 2t - N^{-1}t, & \text{(B–Smesh)} \\ e^{-\frac{t}{t_1 - t}}, \quad \text{with} \quad t_1 = 0.5(1 + (\ln N)^{-1}) & \text{(Vmesh).} \end{cases} \tag{4.22}$$

We introduce the following notation on the non-uniform mesh $\Omega^N$: for $i = 0, 1, \dots, N/2 - 1$ $h_i = x_{i+1} - x_i$, for $i = N/2 + 1, \dots, N$, $h_i = 2(1 - \tau)N^{-1} = H$ and $\hbar_i = (h_i + h_{i+1})$. It is clear that, $h_i \leqslant h_{i+1}$ for $i = 0, 1, \dots, N/2 - 1$, and $N^{-1} \leqslant H \leqslant 2N^{-1}$. We assume $\varepsilon \leqslant N^{-1}$, throughout the chapter as the problem is dominated by the convective term (figure 4.1).

### 4.3.2 Cubic spline-based numerical method

Let's define $P_3(x)$ as an interpolating cubic spline function that satisfies the following criteria:

(i) On each sub-interval $[x_{i-1}, x_i]$, $i = 1, \dots, N$, $P_3(x)$ is a third degree polynomial.
(ii) $P_3(x) \in C^2[0, 1]$.
(iii) $P_3(x) = Z(x_i)$, $i = 0, 1, \dots, N$.

Here, $Z(x)$ is a discrete function defined on the points $x = x_0, x_1, \dots, x_N$. Then, for $x_{i-1} \leqslant x \leqslant x_i$, $i = 1, \dots, N$, we can write $P_3(x)$ as:

$$P_3(x) = \frac{(x_i - x)^3}{6h_i}Q_{i-1} + \frac{(x - x_{i-1})^3}{6h_i}Q_i + \left(Z(x_{i-1}) - \frac{h_i^2}{6}Q_{i-1}\right)\left(\frac{(x_i - x)}{h_i}\right),$$

where $Q_i = P_3''(x_i)$. When the conditions of a cubic spline are applied, one can be observed that

$$\frac{h_i}{6}Q_{i-1} + \frac{\hbar_i}{3}Q_i + \frac{h_{i+1}}{6}Q_{i+1} = \frac{Z(x_{i+1}) - Z(x_i)}{h_{i+1}} - \frac{Z(x_i) - Z(x_{i-1})}{h_i} \tag{4.23}$$

for $i = 1, \dots, N - 1$. One can get the first and second-order estimates of $z(x)$ via Taylor series expansion of $Z$ in the vicinity of $x_i$,

$$Z(x_{i+1}) \approx Z(x_i) + h_{i+1}Z'(x_i) + \frac{h_{i+1}^2}{2}Z''(x_i) \tag{4.24}$$

$$Z(x_{i-1}) \approx Z(x_i) - h_iZ'(x_i) + \frac{h_i^2}{2}Z''(x_i). \tag{4.25}$$

Now by solving (4.25) and (4.24), we obtain $Z'(x_i) \approx \frac{1}{h_ih_{i+1}(\hbar_i)}(-h_{i+1}^2Z(x_{i-1}) + (h_{i+1}^2 - h_i^2)Z(x_i) + h_i^2Z(x_{i+1}))$,

Similarly, we have $Z''(x_i) \approx \frac{1}{h_ih_{i+1}(\hbar_i)}(h_{i+1}^2Y(x_{i-1}) + (h_i + h_i^2 + 1)Z(x_i) + h_iZ(x_{i+1}))$.

Plugging the above approximations in $Z'(x_{i+1}) \approx Z'(x_i) + h_{i+1}Z''(x_i)$ and $Z'(x_{i-1}) \approx Z'(x_i) - h_iZ''(x_i)$, we get

$$Z'_j(x_{i+1}) \approx \frac{1}{h_ih_{i+1}(\hbar_i)}(h_{i+1}^2Z(x_{i-1}) - (h_i + h_{i+1})^2Z(x_i) + (h_i^2 + 2h_ih_{i+1})Z(x_{i+1})),$$

$$Z'_j(x_{i-1}) \approx \frac{1}{h_ih_{i+1}(\hbar_i)}(-(h_{i+1}^2 + 2h_ih_{i+1})Z(x_{i-1}) - (h_i^2 + h_{i+1}^2)^2Z(x_i) + (h_i^2)Z(x_{i+1})).$$

For $j = i, i \pm 1$, consider the following estimate for the problem (4.1): $\varepsilon Q_j + a(x_j)Z'(x_j) - b(x_j)Z(x_j) = f(x_j)$. Using the above, we can get the following system of linear equations from (4.23):

$$L_{cs}^NZ_i = K_i^-Z_i - K_i^cZ_i + K_i^+Z_i = E_i^N, \quad i = 1, \dots, N - 1, \tag{4.26}$$

where $L_{cs}^N$ is a cubic spline operator.

### 4.3.3 Spline-based hybrid scheme

In this section, the spline-based hybrid numerical approximation, which comprises a midpoint upwind approximation in the coarse mesh region and a cubic spline approximation in the fine mesh region, is discussed. Define the discrete problem for (4.3) as:

$$L^N Z_j = \begin{cases} L_{cs}^N Z_j, & \text{for} \quad 0 < j \leqslant N/2, \\ L_{mu}^N Z_j, & \text{for} \quad N/2 < j \leqslant N-1. \end{cases} \tag{4.27}$$

Here, the midpoint upwind scheme [16, 14] $L_{mu}^N$ is given by:

$$L_{mu}^N Z_j = \begin{cases} \varepsilon D^+ D^- Z_j + a_{j+1/2} D^+ Z_j - b_{j+1/2} Z_{j+1/2} = f_{j+1/2}, \\ Z(0) = z(0), \quad Z(1) = z(1), \end{cases} \tag{4.28}$$

where $\varrho_{j+1/2} = \varrho((x_j + x_{j+1})/2)$ for any $\varrho$. Then, the numerical discretization of (4.3) is given by the system of equations $AZ = B$, where here,

$$A = \begin{pmatrix} 1 & 0 & 0 & 0 & & 0 & 0 & 0 & 0 \\ A_1^- & A_1^c & A_1^+ & 0 & \cdots & 0 & 0 & 0 & 0 \\ 0 & A_2^- & A_2^c & A_2^+ & \cdots & 0 & 0 & 0 & 0 \\ \vdots & \vdots & \vdots & \vdots & \ddots & \vdots & \vdots & \vdots & \vdots \\ 0 & 0 & 0 & 0 & \cdots & A_{N-2}^- & A_{N-2}^c & A_{N-2}^+ & 0 \\ 0 & 0 & 0 & 0 & \cdots & 0 & A_{N-1}^- & A_{N-1}^c & A_{N-1}^+ \\ 0 & 0 & 0 & 0 & & 0 & 0 & 0 & 1 \end{pmatrix},$$

$$Z = \begin{pmatrix} Z_0 \\ Z_1 \\ Z_2 \\ \vdots \\ Z_{N-2} \\ Z_{N-1} \\ Z_N \end{pmatrix}, \quad B = \begin{pmatrix} z(0) \\ F_1 \\ F_2 \\ \vdots \\ F_{N-2} \\ F_{N-1} \\ z(1) \end{pmatrix}$$

$$A_j^- = \begin{cases} a_{j-1}\dfrac{(2h_j + h_{j+1})}{2\hbar_j} + b_{j-1}\dfrac{h_j}{2} + a_j\dfrac{h_{j+1}}{h_j} - a_{j+1}\dfrac{h_{j+1}^2}{2h_j(\hbar_j)} - \dfrac{3\varepsilon}{h_j}, & \text{for } 1 \leqslant j \leqslant N/2, \\ \dfrac{3\varepsilon}{h_j(\hbar_j)}, & \text{for } N/2 < j \leqslant N-1, \end{cases}$$

$$A_j^c = \begin{cases} a_{j-1}\dfrac{\hbar_j}{2(h_{j+1})} + b_j(\hbar_j) + a_j\dfrac{(h_{j+1}^2 - h_j^2)}{h_j h_{j+1}} - a_{j+1}\dfrac{\hbar_j}{2h_j} - \dfrac{3\varepsilon(\hbar_j)}{h_j h_{j+1}}, & \text{for } 1 \leqslant j \leqslant N/2 \\ \dfrac{2\varepsilon}{(\hbar_j)h_{j+1}} + \dfrac{2\varepsilon}{h_j(\hbar_j)} + \dfrac{a_j}{h_{j+1}} - \dfrac{b_j}{2}, & \text{for } N/2 < j \leqslant N-1, \end{cases}$$

$$A_j^+ = \begin{cases} a_{j-1}\dfrac{h_j^2}{2(\hbar_j)h_{j+1}} - a_j\dfrac{h_j}{h_{j+1}} - a_{j+1}\dfrac{(h_j + 2h_{j+1})}{2(\hbar_j)} + b_{j+1}\dfrac{h_{j+1}}{2} - \dfrac{3\varepsilon}{h_j}, & \text{for } 1 \leqslant j \leqslant N/2, \\ \dfrac{2\varepsilon}{(\hbar_j)h_{j+1}} + \dfrac{a_j}{h_{j+1}} + \dfrac{b_j}{2}, & \text{for } N/2 < j \leqslant N-1, \end{cases}$$

$$F_j^N = \begin{cases} -\dfrac{1}{2}h_j f_{j-1} - (\hbar_j) - \dfrac{1}{2}h_{j+1}f_{j+1}, & \text{for } j \leqslant N/2, \\ \dfrac{1}{2}(f_j + f_{j+1}), & \text{for } j \geqslant N/2. \end{cases}$$

**Lemma 4.5.** Suppose $Z(x_j)$ be the mesh function that meets the requirements $Z(x_0) \geqslant 0$, $Z(x_N) \geqslant 0$ and $L^N Z(x_j) \leqslant 0$, $\forall Z_j \in \Omega^N$, then $Z(x_j) \geqslant 0\ \forall x_j \in \Omega^N$.

Proof. The $(N+1) \times (N+1)$ matrix $A$ associated with $L^N$, is an $M$-matrix. Hence, the matrix is irreducible and has positive inverse. So the solution exist for the system $L^N Z_j = f_j$ for $0 \leqslant j \leqslant N$. Now, by applying the logic presented in [16], one can obtain the required outcomes.

## 4.4 Error analysis

The error for the proposed scheme on S-mesh and B–S mesh is examined in this section.

### 4.4.1 Error estimate on S-mesh

**Lemma 4.6.** The following estimate holds for the regular component $r$ of the solution $z$ of (4.3):

$$\left|(r - R)(x_j)\right| \leqslant CN^{-2} \tag{4.29}$$

Proof. The truncation error analysis of regular component gives,

$$\begin{aligned} \left| L^N(r - R)(x_j)\right| &= \left| L^N r(x_j) - L^N R(x_j)\right| \\ &= \left| L^N r(x_j) - f \right| \\ &= \left|(L^N - L)r(x_j)\right| \\ &\leqslant \begin{cases} \varepsilon h_j^2 \parallel r^{(4)} \parallel, & j \leqslant N/2, \\ \varepsilon h_j \parallel r^{(3)} \parallel + Ch_j^2(\parallel r^{(3)} \parallel + \parallel r^{(2)} \parallel), & j \geqslant N/2. \end{cases} \end{aligned}$$

Now by the help of lemma 4.4, we get

$$\left| L^N(r - R)(x_j)\right| \leqslant CN^{-2}.$$

Let us consider $\aleph^{\pm}(x_j) = \frac{C}{\beta}N^{-2} \pm (r - R)(x_j)$ as a barrier function. Then one can see that, $\aleph^{\pm}(x_j) \geqslant 0$ at $x_0$ and $x_N$. Also, for $1 \leqslant j \leqslant N/2$, $L^N\aleph^{\pm}(x_j) = \frac{C}{N^2} \pm \frac{C}{\beta N^2}\left(A_j^- + A_j^c + A_j^+\right)(x_j)$ and for $N/2 + 1 \leqslant j < N$, $L^N\aleph^{\pm}(x_j) = CN^{-2} \pm \frac{C}{\beta N^2}\left(a_{j+1/2} - b_{j+1/2}\right)$ are nonnegative. Now, from the application of lemma 4.5, we get the desired result.

**Lemma 4.7.** On the S-mesh, the following bounds holds for the component $s$ of $z$:

$$\left|(s - S)(x_j)\right| \leqslant CN^{-2}\ln^3 N \tag{4.30}$$

Proof. As we have,

$$\begin{aligned}\left| L^N(s - S)(x_j)\right| &= \left| L^N s(x_j) - L^N S(x_j)\right| \\ &= \left| L^N s(x_j) - 0\right| \\ &= \left|(L^N - L)s(x_j)\right|.\end{aligned}$$

Now, we divide the proof into two cases.

Case-1: For $x_j \in [0, \tau]$, the truncation error is estimated as

$$\left| L^N(s - S)(x_j)\right| \leqslant \varepsilon h_j^2 \parallel s^{(4)} \parallel.$$

From lemma 4.4, the bound for the singular component is used to get

$$\left| L^N(s - S)(x_j)\right| \leqslant CN^{-2}\tau^2\varepsilon^{-3}. \tag{4.31}$$

Now, considering $\aleph^{\pm}(x_j) = CN^{-2} + CN^{-2}\tau^2\varepsilon^{-3}(x_j - \tau) \pm \left\|(s - S)(x_j)\right\|$ as a barrier function. Then we can see that, $\aleph^{\pm}(x) > 0$ at $x = 0$ and $x = \tau$. Also, $L^N\aleph^{\pm}(x_j) \geqslant 0$. Thus, from lemma 4.5, we get

$$\left|(s - S)(x_j)\right| \leqslant CN^{-2}\tau^3\varepsilon^{-3} \leqslant CN^{-2}(\ln N)^3. \tag{4.32}$$

Case-2: For $x_j \in [\tau, 1]$, as we have $|Z(x_{N+1})| \leqslant C$ and $|S(x_{N+1})| \leqslant C\sqrt{\varepsilon}$. So by applying the claims given in [12], we obtain $|S(x_j)| \leqslant |S(x_{N+1})|N^{-2} \leqslant CN^{-2}\sqrt{\varepsilon}$ and $|(S - s)(x_j)| \leqslant |S(x_j)| + |s(x_j)| \leqslant CN^{-2}\sqrt{\varepsilon}$. Hence from all the above results, one can obtain the desired outcome that is $|(S - s)(x_j)| \leqslant CN^{-2}\ln^3 N$.

**Theorem 4.1.** We have the following error estimate on the S-mesh::

$$\left| z_j - Z_j \right| \leqslant CN^{-2}\ln^3 N. \tag{4.33}$$

Proof. Using the triangular inequality, $\left| z_j - Z_j \right| \leqslant \left| r_j - R_j \right| + \left| s_j - S_j \right|$. Now using the results of lemmas 4.1 and 4.2, one can get the required estimate.

### 4.4.2 Error estimate on B–S mesh

**Lemma 4.8.** The following estimate holds for the regular component $r$ of the solution $z$ of (4.3):

$$\left|(r - R)(x_j)\right| \leqslant CN^{-2}. \tag{4.34}$$

Proof. As,

$$\begin{aligned}\left| L^N(r - R)(x_j)\right| &= \left|(L^N - L_\varepsilon)r(x_j)\right| \\ &\leqslant \begin{cases} \varepsilon h_j^2 \parallel r^{(4)} \parallel, & j \leqslant N/2, \\ Ch_j^2(\parallel r^{(3)} \parallel + \parallel r^{(2)} \parallel) + \varepsilon h_j \parallel r^{(3)} \parallel, & j \geqslant N/2. \end{cases} \\ &\leqslant CN^{-2}.\end{aligned}$$

Now, by selecting an appropriate barrier function and using the lemma 4.5, the desired outcome can be obtained.

**Lemma 4.9.** For the singular component $s$ of $z$, we have the following estimate:

$$\left|(s - S)(x_j)\right| \leqslant CN^{-2} \tag{4.35}$$

Proof. We have

$$\left| L^N(s - S)(x_j)\right| \leqslant \varepsilon h_j \parallel s^{(3)} \parallel + Ch_j^2(\parallel s^{(3)} \parallel + \parallel s^{(2)} \parallel) \leqslant CN^{-2}.$$

Now, we have considered the proof for two cases. Case-1: for $x_j \in [0, \tau]$, we have the following truncation error bound

$$\left| L^N(s - S)(x_j)\right| \leqslant \varepsilon h_j^2 \parallel s^{(4)} \parallel \leqslant CN^{-2}.$$

Case-2: For $x_j \in [\tau, 1]$, then we have

$$\left| L^N(s - S)(x_j)\right| \leqslant \varepsilon h_j \parallel s^{(3)} \parallel + Ch_j^2(\parallel s^{(3)} \parallel + \parallel s^{(2)} \parallel) \leqslant CN^{-2}.$$

By considering

$$\aleph^{\pm}(x_j) = CN^{-2} \pm \left\|(s - S)(x_j)\right\|$$

as the barrier function, one can observe that $\aleph^{\pm}(x_0) \geqslant 0$, $\aleph^{\pm}(x_1) \geqslant 0$ and $L^N\aleph^{\pm}(x_j) \geqslant 0$, $\forall\, 1 \leqslant j \leqslant N - 1$. Thus, using lemma 4.5, one can complete the proof.

**Theorem 4.2.** Let $z_j$ represent the continuous solution and $Z_j$ represent the numerical solution derived using the proposed scheme (4.25). On the B–S mesh, we have the following bound:

$$\left| z_j - Z_j \right| \leqslant CN^{-2}. \tag{4.36}$$

Proof. From the triangular inequality, we have $\left| z_j - Z_j \right| \leqslant \left| r_j - R_j \right| + \left| s_j - S_j \right|$. Now, one can obtain the proof of theorem 4.6 from lemmas 4.8 and 4.9.

**Remark 4.1.** One can estimate the error on V mesh for the numerical solution using a similar approach as in the case of B–S mesh, that is:

$$\left| z_j - Z_j \right| \leqslant CN^{-2}. \tag{4.37}$$

## 4.5 Numerical results

To demonstrate the theoretical findings, we explore three test problems of the kind (4.1) in this section. We employ the spline-based hybrid technique on several types of layer-adapted meshes for comparison.

**Example 4.1.** Following is the test problem where only the delay term $\delta$ is involved.

$$\begin{cases} \varepsilon z''(x) + z'(x) + 2z(x - \delta) - 3z(x) = 0, \\ z(0) = 1, \quad z(1) = 1. \end{cases} \tag{4.38}$$

**Example 4.2.** Following is the test problem where only the advanced term $\eta$ is involved.

$$\begin{cases} \varepsilon z''(x) + z'(x) - 3z(x) + 2z(x + \eta) = 0, \\ z(0) = 1, \quad z(1) = 1. \end{cases} \tag{4.39}$$

**Example 4.3.** Following is the test problem where both the delay term $\delta$ and advanced term $\eta$ are involved.

$$\{\varepsilon z''(x) + z'(x) - 2z(x - \delta) - 5z(x) + y(x + \eta) = 0,\ z(0) = 1,\ z(1) = 1. \tag{4.40}$$

The solution of the above examples takes the form

$$z(x) = C_1 e^{m_1 x} + C_2 e^{m_2 x}$$

where $m_1 = \dfrac{-a(x) + \sqrt{(a(x))^2 + 4\varepsilon b(x)}}{2\varepsilon}$, $m_2 = \dfrac{-a(x) - \sqrt{(a(x))^2 + 4\varepsilon b(x)}}{2\varepsilon}$, $C_1 = \dfrac{e^{m_2} - 1}{e^{m_2} - e^{m_1}}$, $C_2 = \dfrac{1 - e^{m_2}}{e^{m_2} - e^{m_1}}$. For every value of $N$, the maximum pointwise error $E^N$ is given by: $E^N = \max\limits_{0 < x \leqslant N} \left\| z_j - Z_j \right\|$, here, $z_i$ and $Z_i$, are respectively the solutions of (4.3) and (4.27). Correspondingly the rate of convergence $R^N$ is given as: $R^N = \log_2\left(\frac{E^N}{E^{2N}}\right)$.

The $E^N$ and $R^N$ attained by the proposed method on the various meshes mentioned above for example 4.5.1 are shown in table 4.1. Tables 4.2 and 4.3 show similar results for examples 4.2 and 4.3 respectively. In comparison to standard S-mesh, the employment of B–S mesh and V mesh produces better acceptable results, as shown in the tables. In table 4.4, we have compared our

**Table 4.1.** $E^N$ and $R^N$ are obtained by using the scheme (4.27) for example 4.1 with $\delta = 0.5\varepsilon$.

| | $\varepsilon = 1 \times 10^{-3}$ | | | $\varepsilon = 1 \times 10^{-8}$ | | |
|---|---|---|---|---|---|---|
| $N$ | S mesh | B–S mesh | V mesh | S mesh | B–S mesh | V mesh |
| 32 | $4.722 \times 10^{-2}$ | $1.030 \times 10^{-2}$ | $8.888 \times 10^{-3}$ | $4.722 \times 10^{-2}$ | $1.032 \times 10^{-2}$ | $8.902 \times 10^{-3}$ |
| | 1.577 | 2.000 | 1.923 | 1.578 | 2.000 | 1.924 |
| 64 | $1.581 \times 10^{-2}$ | $2.576 \times 10^{-3}$ | $2.342 \times 10^{-3}$ | $1.581 \times 10^{-2}$ | $2.578 \times 10^{-3}$ | $2.344 \times 10^{-3}$ |
| | 1.633 | 1.975 | 1.925 | 1.633 | 1.976 | 1.926 |
| 128 | $5.098 \times 10^{-3}$ | $6.522 \times 10^{-4}$ | $6.169 \times 10^{-4}$ | $5.096 \times 10^{-3}$ | $6.520 \times 10^{-4}$ | $6.167 \times 10^{-4}$ |
| | 1.655 | 1.988 | 1.935 | 1.656 | 1.991 | 1.938 |
| 256 | $1.618 \times 10^{-3}$ | $1.651 \times 10^{-4}$ | $1.612 \times 10^{-4}$ | $1.617 \times 10^{-3}$ | $1.646 \times 10^{-4}$ | $1.608 \times 10^{-4}$ |
| | 1.655 | 1.987 | 1.943 | 1.655 | 1.987 | 1.943 |
| 512 | $5.135 \times 10^{-4}$ | $4.614 \times 10^{-5}$ | $4.190 \times 10^{-5}$ | $5.131 \times 10^{-4}$ | $4.131 \times 10^{-5}$ | $4.158 \times 10^{-5}$ |
| | 1.697 | 1.982 | 1.946 | 1.698 | 1.997 | 1.960 |
| 1024 | $1.582 \times 10^{-4}$ | $1.053 \times 10^{-5}$ | $1.087 \times 10^{-5}$ | $1.580 \times 10^{-4}$ | $1.034 \times 10^{-5}$ | $1.069 \times 10^{-5}$ |
| | 1.724 | 1.967 | 1.940 | 1.726 | 1.998 | 1.966 |

**Table 4.2.** $E^N$ and $R^N$ are obtained by using the scheme (4.27) for example 4.2 with $\eta = 0.5\varepsilon$.

| | $\varepsilon = 1 \times 10^{-3}$ | | | $\varepsilon = 1 \times 10^{-8}$ | | |
|---|---|---|---|---|---|---|
| $N$ | S mesh | B–S mesh | V mesh | S mesh | B–S mesh | V mesh |
| 32 | $4.730 \times 10^{-2}$ | $1.031 \times 10^{-2}$ | $8.904 \times 10^{-3}$ | $4.722 \times 10^{-2}$ | $1.032 \times 10^{-2}$ | $8.902 \times 10^{-3}$ |
| | 1.577 | 1.999 | 1.923 | 1.578 | 2.000 | 1.924 |
| 64 | $1.585 \times 10^{-2}$ | $2.580 \times 10^{-3}$ | $2.346 \times 10^{-3}$ | $1.581 \times 10^{-2}$ | $2.579 \times 10^{-3}$ | $2.345 \times 10^{-3}$ |
| | 1.634 | 1.975 | 1.925 | 1.633 | 1.977 | 1.926 |
| 128 | $5.107 \times 10^{-3}$ | $6.560 \times 10^{-4}$ | $6.177 \times 10^{-4}$ | $5.096 \times 10^{-3}$ | $6.550 \times 10^{-4}$ | $6.167 \times 10^{-4}$ |
| | 1.655 | 1.988 | 1.935 | 1.656 | 1.992 | 1.939 |
| 256 | $1.621 \times 10^{-3}$ | $1.653 \times 10^{-4}$ | $1.614 \times 10^{-4}$ | $1.617 \times 10^{-3}$ | $1.646 \times 10^{-4}$ | $1.608 \times 10^{-4}$ |
| | 1.655 | 1.988 | 1.944 | 1.656 | 1.994 | 1.951 |
| 512 | $5.143 \times 10^{-4}$ | $4.169 \times 10^{-5}$ | $4.196 \times 10^{-5}$ | $5.130 \times 10^{-4}$ | $4.131 \times 10^{-5}$ | 4.158e-5 |
| | 1.697 | 1.982 | 1.946 | 1.698 | 1.997 | 1.960 |
| 1024 | $1.585 \times 10^{-4}$ | $1.055 \times 10^{-5}$ | $1.088 \times 10^{-5}$ | $1.580 \times 10^{-4}$ | $1.034 \times 10^{-5}$ | $1.068 \times 10^{-5}$ |
| | 1.724 | 1.967 | 1.940 | 1.726 | 1.999 | 1.967 |

results with the results of [9]. Figure 4.2 portrays the exact and the approximate solutions of examples 4.1 and 4.3. The error $E^N$ for examples 4.2 and 4.3, are displayed as a function of $\varepsilon$ and $N$ in figure 4.3. It can be seen that as the number of intervals $N$ increases, the error decreases for all values of $\varepsilon$. The log-log plots of the $E^N$ are displayed in figure 4.4 for examples 4.1–4.3, to reflect the rate of convergence.

**Table 4.3.** $E^N$ and $R^N$ are obtained by using the scheme (4.27) for example 4.3 with $\delta = \eta = 0.5\varepsilon$.

| | $\varepsilon = 1 \times 10^{-3}$ | | | $\varepsilon = 1 \times 10^{-8}$ | | |
|---|---|---|---|---|---|---|
| $N$ | S mesh | B–S mesh | V mesh | S mesh | B–S mesh | V mesh |
| 32 | $7.509 \times 10^{-2}$ | $1.630 \times 10^{-2}$ | $1.405 \times 10^{-2}$ | $7.450 \times 10^{-2}$ | $1.627 \times 10^{-2}$ | $1.403 \times 10^{-2}$ |
| | 1.573 | 1.997 | 1.925 | 1.578 | 2.000 | 1.924 |
| 64 | $2.522 \times 10^{-2}$ | $4.084 \times 10^{-3}$ | $3.701 \times 10^{-3}$ | $2.495 \times 10^{-2}$ | $4.067 \times 10^{-3}$ | $3.698 \times 10^{-3}$ |
| | 1.635 | 1.977 | 1.925 | 1.633 | 1.977 | 1.926 |
| 128 | $8.166 \times 10^{-3}$ | $1.037 \times 10^{-3}$ | $9.754 \times 10^{-4}$ | $8.041 \times 10^{-3}$ | $1.034 \times 10^{-3}$ | $9.727 \times 10^{-4}$ |
| | 1.654 | 1.988 | 1.936 | 1.656 | 1.992 | 1.939 |
| 256 | $2.578 \times 10^{-3}$ | $2.613 \times 10^{-4}$ | $2.546 \times 10^{-4}$ | $2.550 \times 10^{-3}$ | $2.595 \times 10^{-4}$ | $2.534 \times 10^{-4}$ |
| | 1.657 | 1.987 | 1.944 | 1.656 | 1.994 | 1.951 |
| 512 | $8.176 \times 10^{-4}$ | $6.590 \times 10^{-5}$ | $6.614 \times 10^{-5}$ | $8.095 \times 10^{-4}$ | $6.515 \times 10^{-5}$ | $6.557 \times 10^{-5}$ |
| | 1.698 | 1.983 | 1.947 | 1.698 | 1.997 | 1.960 |
| 1024 | $2.524 \times 10^{-4}$ | $1.666 \times 10^{-5}$ | $1.715 \times 10^{-5}$ | $2.493 \times 10^{-4}$ | $1.631 \times 10^{-5}$ | $1.685 \times 10^{-5}$ |
| | 1.724 | 1.967 | 1.941 | 1.726 | 1.999 | 1.966 |

**Table 4.4.** Comparison of $E^N$ for example 4.1.

| $N$ | | $\varepsilon = 10^{-3}$ | | $\varepsilon=10^{-5}$ | | $\varepsilon = 10^{-8}$ | |
|---|---|---|---|---|---|---|---|
| | | Result in [9] | Our result | Result in [9] | Our result | Result in [9] | Our result |
| 8 | $E^N$ | 0.187 2710 | 0.199 9335 | 0.217 3719 | 0.206 3460 | 0.206 3460 | 0.217 5910 |
| 16 | $E^N$ | 0.106 9782 | 0.044 9201 | 0.120 4841 | 0.045 205 | 0.120 6323 | 0.045 0913 |
| 32 | $E^N$ | 0.059 0411 | 0.010 2530 | 0.069 8994 | 0.010 3211 | 0.070 0229 | 0.010 3218 |
| 64 | $E^N$ | 0.030 7968 | 0.002 5783 | 0.037 2137 | 0.002 5791 | 0.037 2876 | 0.002 5791 |
| 128 | $E^N$ | 0.015 6796 | 0.000 6626 | 0.019 3277 | 0.000 6551 | 0.019 3712 | 0.000 6550 |
| 256 | $E^N$ | 0.007 9907 | 0.000 5097 | 0.009 8423 | 0.000 1647 | 0.009 8657 | 0.000 1646 |

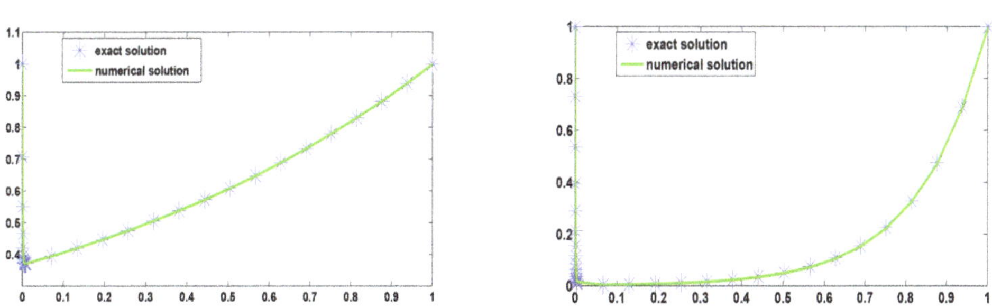


(a) For Example 1.5.1 (b) For Example 1.5.3

**Figure 4.2.** Exact solution and approximate solution for $N = 32,\quad \varepsilon = 10^{-3}$.

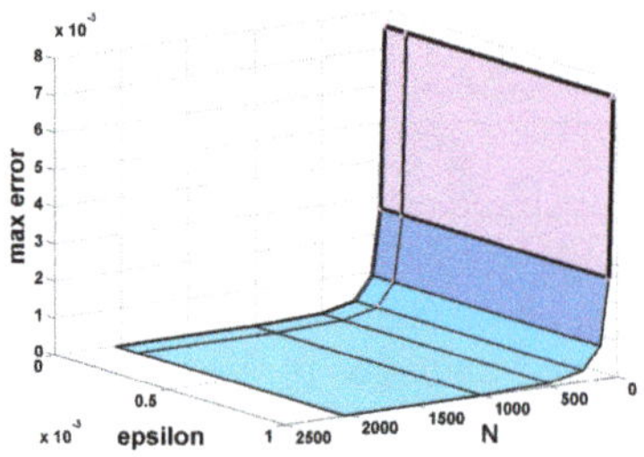


(a) For Example 1.5.2

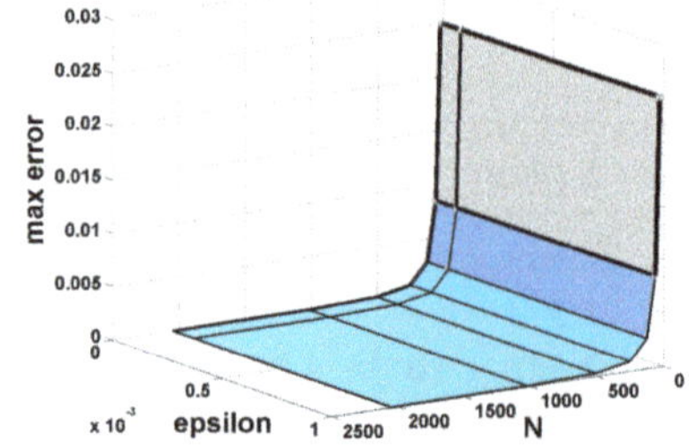


(b) For Example 1.5.3

**Figure 4.3.** $E^N$ as a function of $N$ and $\varepsilon$.

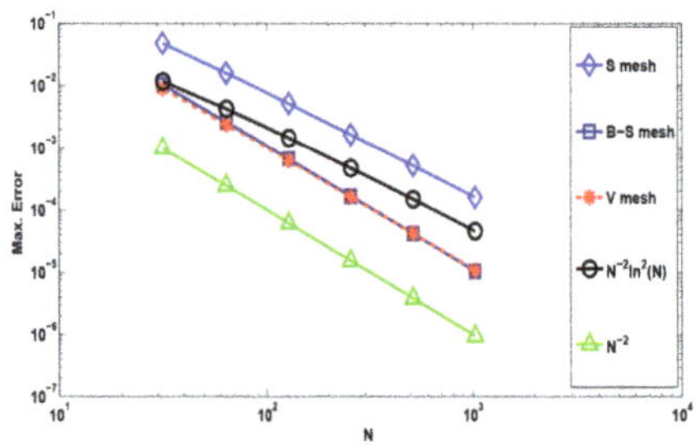


(a) For Example 1.5.1

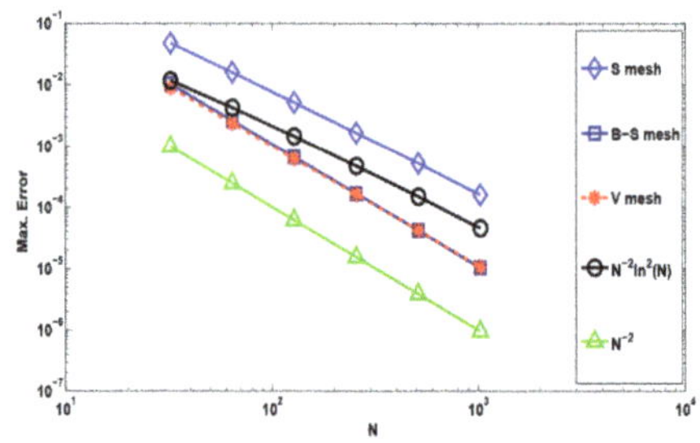


(b) For Example 1.5.2

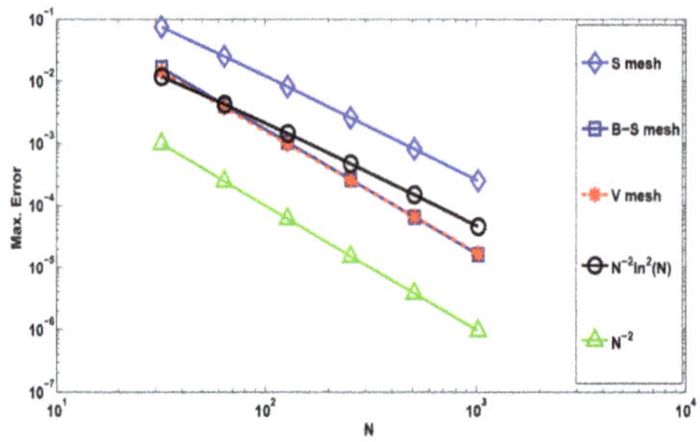


(c) For Example 1.5.3

**Figure 4.4.** Log-log plot of $E^N$.

## 4.6 Conclusion

In this chapter, we have considered a second-order SPDDE. Then, via Taylor series expansion, the SPDDE is reduced to an equivalent SPDE. The problem is then solved using a spline-based hybrid scheme on several Shishkin type meshes such as B–S mesh and V mesh. To validate the efficacy of the proposed method, a thorough examination of the numerical solution is performed. The proposed method's error is shown to be second-order convergent regardless of the parameter $\varepsilon$. When the B–S mesh and V mesh results are compared to the S mesh results, it can be concluded that the B–S mesh and V mesh are more accurate than the standard S-mesh.

## References


[1] Duressa G and Reddy Y N 2015 Domain decomposition method for singularly perturbed differential difference equations with layer behavior *Int. J. Eng. Appl. Sci.* **7** 86–102

[2] Farrell P A, Hegarty A F, Miller J J H, O'Riordan E and Shishkin G I 2000 *Robust Computational Techniques for Boundary Layers* (New York: Chapman and Hall/CRC Press)

[3] Kadalbajoo M K and Sharma K K 2004 $\varepsilon$-uniform fitted mesh method for singularly perturbed differential-difference equations: mixed type of shifts with layer behavior, *Int. J. Comput. Math.* **81** 49–62

[4] Kadalbajoo M K and Sharma K K 2005 Numerical treatment of a mathematical model arising from a model of neuronal variability, *J. Math. Anal. Appl.* **307** 606–27

[5] Kellogg R B and Tsan A 1978 Analysis of some difference approximations for a singular perturbation problem without turning points, *Math. Comp.* **32** 1025–39

[6] Kevorkian J and Cole J D 2013 *Perturbation Methods in Applied Mathematics* (New York: Springer)

[7] Lange C G and Miura R M 1982 Singular perturbation analysis of boundary value problems for differential-difference equation, *SIAM J. Appl. Math.* **42** 502–31

[8] Lange C G and Miura R M 1994 Singular perturbation analysis of boundary-value problems for differential-difference equations, V. small shifts with layer behavior, *SIAM J. Appl. Math.* **54** 249–72

[9] Miller J J H, O'Riordan E and Shishkin G I 2012 *Fitted Numerical Methods for Singular Perturbation Problems* (Singapore: World Scientific)

[10] Mohapatra J and Natesan S 2011 Uniformly convergent numerical method for singularly perturbed differential-difference equation using grid equidistribution, *Int. J. Numer. Method Biomed. Eng.* **27** 1427–45

[11] Natesan S and Bawa R K 2007 Second-order numerical scheme for singularly perturbed reaction-diffusion Robin problems, *J. Numer. Anal. Ind. Appl. Math.* **2** 177–92

[12] O'Malley R E 1974 *Introduction to Singular Perturbations* (New York: Academic)

[13] Priyadharshini R M, Ramanujam N and Valanarasu T 2010 Hybrid difference schemes for singularly perturbed problem of mixed type with discontinuous source term, *J. Appl. Math. Info.* **28** 1035–54

[14] Ramesh V P and Kadalbajoo M K 2008 Upwind and midpoint upwind difference methods for time-dependent differential difference equations with layer behaviour, *Appl. Math. Comp.* **202** 453–71

[15] Rao R N and Chakravarthy P P 2013 A finite difference method for singularly perturbed differential-difference equations arising from a model of neuronal variability, *J. Taibah Univ. Sci.* **7** 128–36

[16] Stynes M and Roos H G 1997 The midpoint upwind scheme, *Appl. Numer. Math.* **23** 361–74

# Chapter 5

# Modelling the impact of preventive and treatment-based control interventions on the transmission dynamics of Leptospirosis disease

**G N Nkem, E A Bakare, S Hoskova-Mayerova and O S Obabiyi**

Leptospirosis is one of the most common and widespread zoonotic infections in the world. It is a worldwide public health problem and the greatest concern for humid tropical and subtropical regions. There are few works on the modelling of Leptospirosis till now but none of them have considered the influence of pesticides/rodent control and treatment with antibiotics (doxycycline) on the dynamics of transmission of Leptospirosis. In this chapter, we developed an ordinary differential equation deterministic model to predict the impact of preventive and treatment-based control interventions on the transmission dynamics of Leptospirosis. Qualitative analysis of model were carried out by investigating the existence and positivity of the solutions. The next generation matrix technique was used to obtain the epidemiological threshold quantity called the reproduction number $R_0$. Equilibrium analysis of the model was performed using the fixed point theory to determine the long term behaviour of the disease. Stability analysis of the equilibrium solutions were performed using the Jacobian matrix approach and the Lyapunov function theory. The centre manifold theory was applied to explore the local asymptotic stability of the endemic equilibrium. Sensitivity and elasticity analysis were used to ascertain the reliability of the model prediction and the choice of the best intervention strategies. Polynomial based solutions was used to solve the related system of differential equations It was established that the model was mathematically well-posed and epidemiologically meaningful. Several other mathematical analyses were performed and our theoretical findings were validated using numerical simulations and were found to be in good agreement with the polynomial based solutions. It was established that the model was mathematically well-posed and epidemiologically meaningful. We found the region where the model is epidemiologically feasible; and stability analysis of the disease-free equilibrium via the reproduction number $\mathcal{R}_0$

doi:10.1088/978-0-7503-5067-9ch5

obtained using the next generation matrix technique were established. The model results show that the disease-free equilibrium is asymptotically stable when $\mathcal{R}_0 < 1$ and unstable when $\mathcal{R}_0 > 1$. The existence of a unique endemic equilibrium and the global asymptotic stability of the disease-free equilibrium under certain conditions were determined. The centre manifold theory revealed that the model exhibited a backward bifurcation. The elasticity analysis predicted that in order to eliminate Leptospirosis, the best strategy to employ is to apply all the available control interventions but in areas with little resources, treatment with antibiotics and pesticides/rodent control will be the best options. A stochastic based differential equation (SDE) model was developed based on the existing ODE model, in order to study the variabilities in the population. Results of the SDE and its deterministic model counterpart was compared using numerical simulation. Furthermore, various numerical simulations were carried out with the use of the typical computer algebra system called Matlab to verify the analytic results and investigate the possible behaviour of the formulated model. The developed model predicted the control and/or elimination of Leptospirosis through the use of treatment-based and preventive control interventions.

## 5.1 Introduction

Endemic zoonotic diseases affect, impoverish and attack not only people's health but also their livelihoods. They remain neglected in most endemic countries because of a lack of information and awareness about the extent of the problem. An absence of suitable diagnostic tools and sustainable strategies for prevention and control worsens the problem. In Africa, zoonotic infections are directly responsible for both human illness and death, and indirectly impact human well-being as a result of reduced livestock productivity and food insecurity.

Leptospirosis is a bacterial disease that affects humans and animals that is caused by bacteria of the genus *Leptospira*. It is one of the most common and widespread zoonotic infections in the world and is recognised as a neglected disease by the World Health Organization [1]. It is particularly common in the tropical areas where people and animals live in closed contact, and warm and humid conditions favour environmental survival and transmission of the pathogen. Leptospirosis is estimated to cause 1.03 million cases; where Occupational exposure probably accounts for 30% to 50% cases, and 59 000 deaths each year worldwide [2]. The incidence of Leptospirosis, which is in part a rodent-borne disease, depends on environmental and climatic conditions that influence rodent population dynamics, size and behaviour. The Intergovernmental Panel on Climate Change (IPCC) has suggested that the predicted increase in heavy rainfall in the 21st century could increase the risk of Leptospirosis through contamination of flood waters of run-off by rodent populations. In the past decade alone, Leptospirosis has proven to be an endemic disease with endemic potential. Recent reported outbreaks in Guyana, India, Kenya, Tanzania, Lao People's Democratic Republic, New Caledonia, Nicaragua, the Philippines and Thailand have highlighted the strong link between Leptospirosis and extreme weather events. Transmission of Leptospirosis can either be direct or indirect. Direct transmission occurs when Leptospires from tissues, body fluids or urine of acutely infected or asymptomatic carrier animals enter a new host and establish infection. The presence of *Leptospires* in genital tracts, as well as trans-placental transmission, has been demonstrated in animals [3]. Indirect transmission occurs when an animal or human being acquires Leptospirosis from environmental *Leptospires*, originating in the urine of carrier animals. The most common portal of entry into new host usually occurs when the pathogen gains entry via the skin through small abrasions or other breaches of the surface integument. They may also enter directly into the bloodstream or lymphatic system via the conjunctiva, the genital tract, the nasopharyngeal mucosa, and the lungs following inhalation of aerosols [4]. Humans are accidental hosts and acquire infection due to several occupational and recreational hazards associated with Leptospirosis. Agricultural workers, veterinarians, workers in the fishing industry, plumbers, meat handlers and slaughterhouse workers, coal miners, military troops, milkers, sewer workers and rodent control workers are at increased risk [5, 6]. Outbreaks associated with recreational exposure to water have been reported from several countries [7]. Direct transmission between humans is not indicated. They are not proven to be important epidemiological source of transmission although the excretion of *Leptospires* in human urine occurs for months after recovery has been recorded [8, 9]. In human, symptoms are gen*rally*flu-like, but the disease can result in liver damage and renal failure.

Mathematical modelling has played a substantial role to understanding the epidemiology of an infectious disease. It is used to model and simulate a real-world problem, which results in the understanding of the problem in a faster, safer and cost effective manner than using empirical studies [10]. These models provide us with the qualitative descriptions of the complicated, nonlinear process of disease transmission and help us to obtain insight into the dynamics of disease with decisions for public health policy. Understanding the transmission characteristics of infectious diseases in communities, regions, and countries can lead to better approaches to decrease the transmission of these diseases. Many mathematical models have been proposed to represent the compartmental dynamics of both human and vector population.

Recently, there have been various mathematical studies so far on Leptospirosis transmission dynamics [6, 11–19] but none of them have considered including in their models the removal or reduction of the vector through pesticide control; and treatment for the class of exposed and infectious human population to the best of our knowledge.

In this chapter, a compartmental model, for both human and vector population is used to examine the dynamical interaction of Leptospirosis, a vector borne disease. We present some of the pertinent aspects of Leptospirosis that are particularly important to the development and use of epidemiology model, keenly focusing on the basic disease progression in humans, the demographics of Leptospirosis, and efforts at controlling the infection.

The aim here, is to study the dynamics of the transmission of Leptospirosis and to predict the impact of preventive and treatment-based control intervention on the transmission of Leptospirosis. We applied the ordinary differential equation (ODE) deterministic model and the stochastic differential equation (SDE) model to study the dynamics of transmission of Leptospirosis in the presence of the two control interventions.

This chapter is organised as follows: in section 5.2, we give a full description of the model formulation. In section 5.3, we discussed the qualitative analysis of the model and the derivation of the basic reproduction number. We show the bifurcation analysis in section 5.4. Section 5.5 is devoted to the sensitivity and elasticity analysis of the parameters in the model. In section 5.6, the intervention strategies for the Leptospirosis disease is discussed. We presents some computer simulations on the reproduction number $R_0$, numerical results and discussion in section 5.7. In section 5.8, we give the concluding remark and finally, the references are presented.

## 5.2 Model formulation

In this section, we present the model formulation of the Leptospirosis disease with their interaction with human population considering the following assumptions:

1. New-borns are not immunised and are therefore vulnerable instantly.
2. Humans can be infected by infected vectors but not by infected humans.
3. Infected humans can infect susceptible vectors.
4. All compartments of vector and human populations are well-mixed.
5. Treatment such as doxycycline or penicillin for the exposed individuals.
6. Treatment such as intravenous antibiotics for the infectious individuals.
7. Infected individuals and vectors are also infectious.
8. The vectors remain infectious for life and have no recovered class.

The human population at any time $t$, denoted by $N_{\mathrm{h}}(t)$, is subdivided into four compartments; $S_{\mathrm{h}}(t)$—susceptible individuals, $E_{\mathrm{h}}(t)$—exposed individuals, $I_{\mathrm{h}}(t)$—infectious individuals (those who have been infected) and $R_{\mathrm{h}}(t)$—recovered individuals (those who recovered from infection or removed). Hence, we have

$$N_{\mathrm{h}}(t) = S_{\mathrm{h}}(t) + E_{\mathrm{h}}(t) + I_{\mathrm{h}}(t) + R_{\mathrm{h}}(t). \tag{5.1}$$

The population of the susceptible is increased by recruitment of individuals (which is assumed susceptible), at the rate $\lambda_{\mathrm{h}}$. The susceptible population decreased by the effective contact with the infectious vectors (in the $I_{\mathrm{v}}$). The population of susceptible is further decreased with the natural death rate $\mu_{\mathrm{h}}$, while increased with the number of individuals that have lost immunity at a rate of $\gamma R_{\mathrm{h}}$, the treatment rate and proportion of effective treatment of exposed individuals and infectious individuals ($\alpha_1 \epsilon E_{\mathrm{h}}$ and $\alpha_2 \rho I_{\mathrm{h}}$). So, we can write the rate of change of susceptible population given as

$$\frac{dS_{\mathrm{h}}}{dt} = \lambda_{\mathrm{h}} - \frac{\beta_{\mathrm{h}} S_{\mathrm{h}} I_{\mathrm{v}}}{N_{\mathrm{h}}} + \alpha_1 \epsilon E_{\mathrm{h}} + \alpha_2 \rho I_{\mathrm{h}} - \mu_{\mathrm{h}} S_{\mathrm{h}} + \gamma R_{\mathrm{h}}. \tag{5.2}$$

The population of the exposed individuals is generated by the infection of susceptible individuals (at a rate $\frac{\beta_{\mathrm{h}} S_{\mathrm{h}} I_{\mathrm{v}}}{N_{\mathrm{h}}}$). The exposed class for individuals decreased by the natural death rate $\mu_h$, treatment rate and proportion of effective treatment of exposed individuals $\alpha_1 \epsilon E_h$, and progression rate of the exposed class for individuals to infectious individuals $\sigma$. So, we can express their rate of change for exposed individuals as follows

$$\frac{dE_h}{dt} = \frac{\beta_h S_h I_v}{N_h} - \alpha_1 \epsilon E_h - \mu_h E_h - \sigma E_h. \tag{5.3}$$

The population of the infected individuals is generated by the progression rate of the exposed class for individuals to infectious individuals $\sigma E_h$. The population of infected individuals is decreased by the natural mortality rate $\mu_h$, disease-induced death rate $\delta_h$, the treatment rate and proportion of effective treatment of infectious individuals $\alpha_2 \rho I_h$ and from the recovery rate of infection $\omega$. So, we can write their rate of change for infected population as

$$\frac{dI_h}{dt} = \sigma E_h - \alpha_2 \rho I_h - \delta_h I_h - \mu_h I_h - \omega I_h. \tag{5.4}$$

The population of recovered individuals is generated by the recovery $\omega$ (from infectious class), while decreased by natural death rate $\mu_h$ and $\gamma$ (those individuals who lost immunity). Thus, their rate of change can be expressed as follows:

$$\frac{dR_h}{dt} = \omega I_h - \mu_h R_h - \gamma R_h. \tag{5.5}$$

The vector population, denoted by $N_v(t)$, is subdivided into three classes: $S_v(t)$—susceptible vector, $E_v(t)$—exposed class for vector and $I_v(t)$—infectious vector. The vectors remain infectious for life and have no recovered class. Thus, the total size of the vector population at any time $t$ is given by

$$N_v(t) = S_v(t) + E_v(t) + I_v(t). \tag{5.6}$$

The population of susceptible vector is increased by the recruitment of the vector at a rate $\lambda_v$. This population is decreased by following the effective contact with infectious human $I_h$ (at a rate of $\frac{\beta_v S_v I_h}{N_h}$). Also, the population of susceptible vector is decreased by the natural death rate $\mu_v$ and $q$ (failure in use of pesticide). Thus, the rate of change for the susceptible vector is given by

$$\frac{dS_v}{dt} = \lambda_v - \frac{\beta_v S_v I_h}{N_h} - \mu_v S_v - q S_v. \tag{5.7}$$

The exposed vector population is generated by the infection of susceptible vector, while decreased by natural death $\mu_v$, the failure rate in the use of pesticide $q$ and progression rate of exposed class for vector to infectious vector $\xi_v$. So, we can write the rate of change of the exposed class for vector population as:

$$\frac{dE_v}{dt} = \frac{\beta_v S_v I_h}{N_h} - \mu_v E_v - \xi_v E_v - q E_v. \tag{5.8}$$

The population of the infected vector is generated by the progression rate of exposed class for the vector $\xi_v$, while decreased by the natural death $\mu_v$, disease related death $\delta_v$, and $q$ (the failure rate in the use of pesticide). Their rate of change can be written as:

$$\frac{dI_v}{dt} = \xi_v E_v - \mu_v I_v - \delta_v I_v - q I_v. \tag{5.9}$$

The flow diagram in figure 5.1 is used to describe the model.

The complete transfer flow diagram of human and vector population interaction is shown in figure 5.1. Thus, the system for the transmission dynamics of

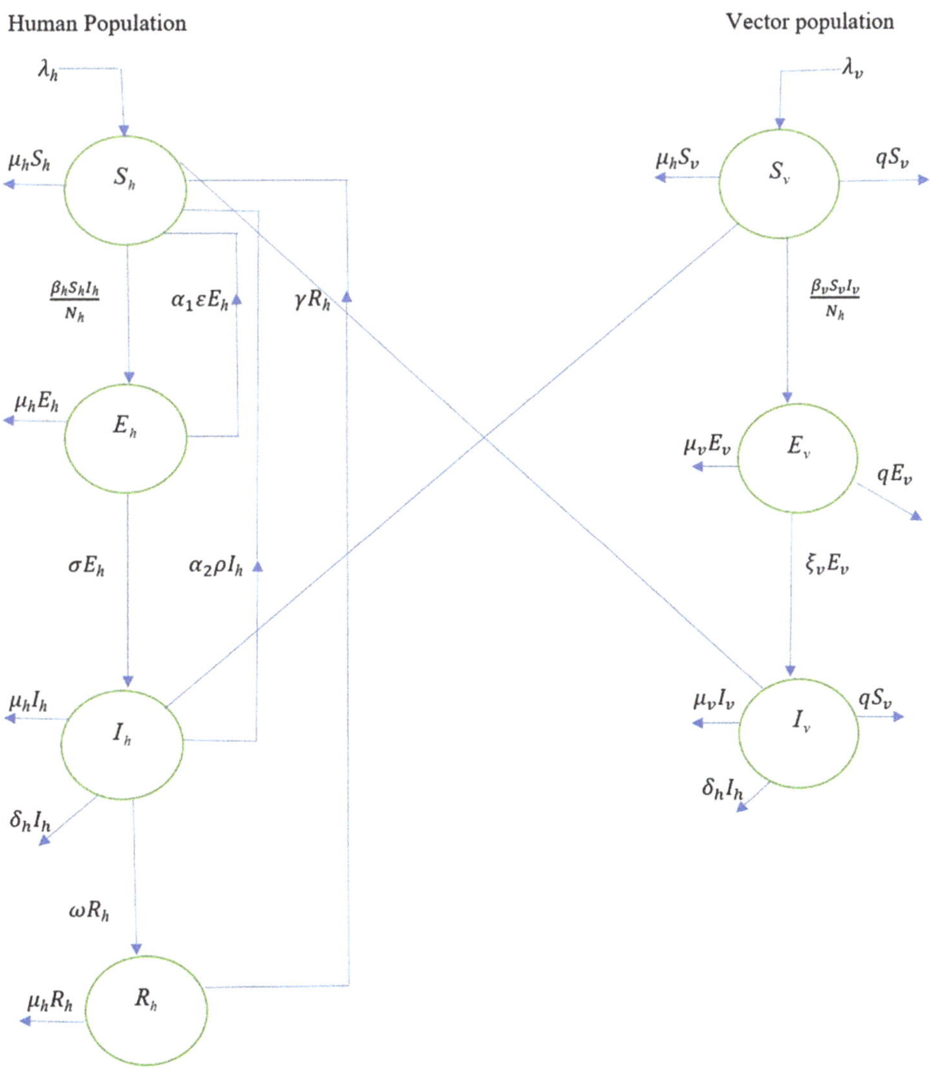


**Figure 5.1.** The flowchart diagram of human and rodent interaction.

Leptospirosis disease is given by the following nonlinear system of differential equations:

$$
\begin{aligned}
\frac{dS_h}{dt} &= \lambda_h - \frac{\beta_h S_h I_v}{N_h} + \alpha_1 \epsilon E_h + \alpha_2 \rho I_h - \mu_h S_h + \gamma R_h \\
\frac{dE_h}{dt} &= \frac{\beta_h S_h I_v}{N_h} - \alpha_1 \epsilon E_h - \mu_h E_h - \sigma E_h \\
\frac{dI_h}{dt} &= \sigma E_h - \alpha_2 \rho I_h - \delta_h I_h - \mu_h I_h - \omega I_h \\
\frac{dR_h}{dt} &= \omega I_h - \mu_h R_h - \gamma R_h \\
\frac{dS_v}{dt} &= \lambda_v - \frac{\beta_v S_v I_h}{N_h} - \mu_v S_v - q S_v \\
\frac{dE_v}{dt} &= \frac{\beta_v S_v I_h}{N_h} - \mu_v E_v - \xi_v E_v - q E_v \\
\frac{dI_v}{dt} &= \xi_v E_v - \mu_v I_v - \delta_v I_v - q I_v
\end{aligned}
\tag{5.10}
$$

with initial conditions, $S_h(0) \geqslant 0$, $E_h(0) \geqslant 0$, $I_h(0) \geqslant 0$, $R_h(0) \geqslant 0$, $S_v(0) \geqslant 0$, $E_v(0) \geqslant 0$ and $I_v(0) \geqslant 0$.

### 5.2.1 Notations and meanings

State variables and parameters used for the Leptospirosis model are described in tables 5.1 and 5.2.

## 5.3 Model qualitative analysis

Here, we provide the following results which guarantee that the Leptospirosis model governed by system (10) is epidemiologically and mathematically well-posed in a feasible region $\Omega$ given by:

$$\Omega = \Omega_h \bigcup \Omega_v \subset \mathbb{R}_+^4 \times \mathbb{R}_+^3$$

**Table 5.1.** Description of state variables.

| | |
|---|---|
| $S_h(t)$ | Number of susceptible humans to Leptospirosis infection at time $t$ |
| $E_h(t)$ | Number of exposed individuals to Leptospirosis infection at time $t$ |
| $I_h(t)$ | Number of infectious humans at time $t$ |
| $R_h(t)$ | Number of recovered humans at time $t$ |
| $S_v(t)$ | Number of susceptible vectors at time $t$ |
| $E_v(t)$ | Number of exposed vectors at time $t$ |
| $I_v(t)$ | Number of infectious vectors at time $t$ |

**Table 5.2.** Description of model parameters.

| | |
|---|---|
| $\lambda_h$ | Recruitment rate for susceptible humans |
| $\beta_h$ | Transmission rate between susceptible humans and infected vectors |
| $\mu_h$ | Natural death rate for humans population |
| $\delta_h$ | Disease-induced death rate of human |
| $\alpha_1$ | Treatment rate of exposed individuals |
| $\alpha_2$ | Treatment rate of infectious human |
| $\epsilon$ | Proportion of effective treatment of exposed individuals |
| $\rho$ | Proportion of effective treatment of infectious human |
| $\gamma$ | Loss of immunity in humans |
| $\omega$ | Rate of recovery |
| $\sigma$ | Progression rate of exposed individuals to infectious human |
| $\lambda_v$ | Recruitment rate for vector |
| $\mu_v$ | Natural death rate for vectors population |
| $\xi_v$ | Progression rate of exposed vectors to infectious vector |
| $\beta_v$ | Transmission rate between susceptible vector and infected vectors |
| $q$ | Use of pesticide |
| $\delta_v$ | Disease-induced death rate of vector |

where

$$\Omega_h = \left\{(S_h(t),\, E_h(t),\, I_h(t),\, R_h(t)) \in \mathbb{R}^4_+ \colon N_h(t) \leqslant \frac{\lambda_h}{\mu_h}\right\}$$

and

$$\Omega_v = \left\{(S_v(t),\, E_v(t),\, I_v(t)) \in \mathbb{R}^3_+ \colon N_v(t) \leqslant \frac{\lambda_v}{\mu_v + q}\right\}.$$

It implies that;

$$\Omega = \left\{(S_h(t),\, E_h(t),\, I_h(t),\, R_h(t),\, S_v(t),\, E_v(t),\, I_v(t)) \in \mathbb{R}^7_+ \colon N_h(t) \leqslant \frac{\lambda_h}{\mu_h},\, N_v(t) \leqslant \frac{\lambda_v}{\mu_v + q}\right\}$$

which is attracting and positively invariant with respect to the flow of system (5.10).

### 5.3.1 Existence and positivity of solutions

**Theorem 5.1.** There exist a domain $\Omega$ as shown above in which the solution set $\{(S_h(t),\, E_h(t),\, I_h(t),\, R_h(t),\, S_v(t),\, E_v(t),\, I_v(t))\}$ with the given nonnegative initial conditions is contained and bounded for all $t \geqslant 0$.

Proof. We obtain the total dynamics of the human population by adding the first, second, third and fourth equations of the system (5.10), we have:

$$N_h(t) = S_h(t) + E_h(t) + I_h(t) + R_h(t).$$

Differentiating with respect to time $t$ and simplifying, we obtain:

$$\frac{dN_h}{dt} + \mu_h N_h \leqslant \lambda_h. \tag{5.11}$$

The above equation (5.11) is a first order linear differential inequality, which on solving we have:

$$N_h(t) \leqslant \frac{\lambda_h}{\mu_h}(1 - \exp(-\mu_h t)) + N_h(0)\exp(-\mu_h t). \tag{5.12}$$

Similarly, by adding the last three equations of system (5.10) for the vector subclasses, we get the total dynamics of the vector population given by:

$$\frac{dN_v}{dt} + N_v(\mu_v + q) \leqslant \lambda_v. \tag{5.13}$$

Solving the above differential inequality, we obtain:

$$N_v(t) \leqslant \frac{\lambda_v}{(\mu_v + q)}(1 - \exp(-\mu_v t)) + N_v(0)\exp(-\mu_v t). \tag{5.14}$$

Consequently, taking the limits as $t \longrightarrow \infty$ gives $N_h(t) \leqslant \frac{\lambda_h}{\mu_h}$ and $N_v(t) \leqslant \frac{\lambda_v}{(\mu_v + q)}$. So $N_h(t)$ is a bounded function of time, together with $N_v(t)$ which is also a bounded function of time. We can say that $\Omega$ is bounded and at limiting equilibrium $\lim_{t \longrightarrow \infty} N_h(t) = \frac{\lambda_h}{\mu_h}$. Besides, any sum or difference of variables in $\Omega$ with positive initial values will remain in $\Omega$ or in a neighbourhood of $\Omega$. Thus $\Omega$ is a positively invariant and is attracting with respect to the flow of system (5.10).

We shall proceed by showing that the solutions of system (5.10) are nonnegative in $\Omega$ for any time $t \geqslant 0$, since the model represents humans and vectors populations.

**Theorem 5.2.** The existing solutions of all the equations in system (5.10) with nonnegative initial conditions are all positive in $\Omega$ for all $t \geqslant 0$.

Proof. From the first equation of system (5.10), we have:

$$\frac{dS_h}{dt} = \lambda_h - \frac{\beta_h S_h I_v}{N_h} + \alpha_1 \epsilon E_h + \alpha_2 \rho I_h - \mu_h S_h + \gamma R_h \quad \forall t \geqslant 0$$

which implies

$$\frac{dS_h}{dt} \geqslant -S_h\left(\frac{\beta_h I_v}{N_h} + \mu_h\right). \tag{5.15}$$

Solving the above equation (5.15) using separable variable method yields;

$$S_h \geqslant S_h(0)\exp\left(\int_0^t \frac{\beta_h I_v}{N_h} dt\right)\exp(-\mu_h t) \tag{5.16}$$

which is positive given that $S_h(0)$ is also positive.

The remaining equations of system (5.10) yields

$$
\begin{aligned}
E_h &\geqslant E_h(0)\exp[-(\alpha_1\epsilon + \mu_h + \sigma)]t \\
I_h &\geqslant I_h(0)\exp[-(\alpha_2\rho + \mu_h + \omega + \delta_h)]t \\
R_h &\geqslant R_h(0)\exp[-(\mu_h + \gamma)]t \\
S_v &\geqslant S_v(0)\exp\left[\int_0^t \left(\frac{\beta_v I_h}{N_h}\right)dt\right]\exp[-(\mu_v + q)]t \\
E_v &\geqslant E_v(0)\exp[-(\mu_v + \xi_v + q)]t \\
&\text{And} \\
I_v &\geqslant I_v(0)\exp[-(\mu_v + \delta_h + q)]t.
\end{aligned}
$$

So, $E_h$, $I_h$, $R_h$, $S_v$, $E_v$ and $I_v$ are also all positive for nonnegative initial conditions. Thus, all the state variables are positive and this completes the proof.

### 5.3.2 Existence and stability of equilibrium points

Our model has two steady states: the disease-free equilibrium (DFE), which describes the total absence of Leptospirosis in the studied population, and the endemic equilibrium (EE), which exists at any positive prevalence of Leptospirosis in the population. It is important to note that there are no trivial equilibrium points as the recruitment terms $\lambda_h$ and $\lambda_v$ are not zero. This implies that $(S_h^*, E_h^*, I_h^*, R_h^*, S_v^*, E_v^*, I_v^*, ) \neq (0, 0, 0, 0, 0, 0, 0)$ and the population will not be extinct.

### 5.3.3 Existence of disease-free equilibrium point

At the disease-free equilibrium point $E_0$, we set all the infectious compartment to zero, which implies that $E_h^* = 0$, $I_h^* = 0$, $R_h^* = 0$, $E_v^* = 0$, $I_h^* = 0$, and $(S_h^*, E_h^*, I_h^*, R_h^*, S_v^*, E_v^*, I_v^*, ) = (S_h^*, 0, 0, 0, S_v^*, 0, 0)$ for the Leptospirosis model (5.10). Solving the first and fifth equation of system (5.10) yields, $S_h^* = \frac{\lambda_h}{\mu_h}$ and $S_v^* = \frac{\lambda_v}{(\mu_v + q)}$ respectively. Thus, we obtain:

$$
E_0 = \left(\frac{\lambda_h}{\mu_h}, 0, 0, 0\frac{\lambda_v}{\mu_v + q}, 0, 0\right).
$$

### 5.3.4 Reproduction number

One of the most important concerns about any infectious disease is its ability to invade a population. Many epidemiological models have a disease-free equilibrium at which the population remains in the absence of disease. These models usually have a threshold parameter, known as the basic reproduction number, $\mathcal{R}_0$, such that if $\mathcal{R}_0 < 1$, then the DFE is locally asymptotically stable, and the disease cannot invade the population, but if $\mathcal{R}_0 > 1$, then the DFE is unstable and invasion is always possible (the disease will become endemic). This reproduction number $\mathcal{R}_0$ is typically defined by (5.6) as: the average number of secondary cases produced by a 'typical' infected (assumed infectious) individual during his/her entire life as infectious (infectious period) when introduced in a population of susceptible.

$\mathcal{R}_0$ is often found using the next generation matrix technique extensively discussed in (5.6) and (5.21).

The infectious compartment model can be expressed in the form

$$\frac{dx_i}{dt} = \mathcal{F}_i(x) - \mathcal{V}_i(x)$$

where $\mathcal{F}_i$ are the new infections and $\mathcal{V}_i$ are the transfers of infections from one compartment to another. $\mathcal{V}_i(x) = \mathcal{V}_i^-(x) - \mathcal{V}_i^+(x)$, $x \in \mathbb{R}_+^7$, and $i = 1, 2, \ldots, m$ is the number of infectious compartment. It is assumed that each function is continuously differentiable at least twice in each variable (5.21). Evaluating at the disease-free equilibrium $E_0$, we then have, $F = \left[\frac{\partial \mathcal{F}_i}{\partial x_j}(E_0)\right]$ and $V = \left[\frac{\partial \mathcal{V}_i}{\partial x_j}(E_0)\right]$ with $i \leqslant 1$ and $j \leqslant m$ such that $\mathcal{R}_0$ is the maximum or dominant eigenvalue of the matrix formed from $FV^{-1}$.

In the case of our model, we have:

$$\mathcal{F} = \begin{pmatrix} \dfrac{\beta_h S_h I_v}{N_h} \\ 0 \\ \dfrac{\beta_v S_v I_h}{N_h} \\ 0 \end{pmatrix} \quad \text{and} \quad \mathcal{V} = \begin{pmatrix} \alpha_1 \epsilon E_h + \mu_h E_h + \sigma E_h \\ \alpha_2 \rho I_h + \mu_h I_h + \delta_h I_h + \omega I_h - \sigma E_h \\ \mu_v E_v + \xi_v E_v + q E_v \\ \mu_v I_v + \delta_v I_v + q I_v - \xi_v E_v \end{pmatrix}.$$

Finding the derivatives of $\mathcal{F}$ and $\mathcal{V}$ at the disease-free equilibrium point $E_0$ gives $F$ and $V$ respectively. Where

$$F = \begin{pmatrix} 0 & 0 & 0 & \beta_h \\ 0 & 0 & 0 & 0 \\ 0 & \beta_v & 0 & 0 \\ 0 & 0 & 0 & 0 \end{pmatrix}$$

and

$$V = \begin{pmatrix} Q_1 & 0 & 0 & 0 \\ -\sigma & Q_2 & 0 & 0 \\ 0 & 0 & Q_3 & 0 \\ 0 & 0 & -\xi_v & Q_4 \end{pmatrix}$$

so that,

$$V^{-1} = \frac{1}{Q_1 Q_2 Q_3 Q_4} \begin{pmatrix} Q_2 Q_3 Q_4 & 0 & 0 & 0 \\ \sigma Q_3 Q_4 & Q_1 Q_2 Q_3 & 0 & 0 \\ 0 & 0 & Q_1 Q_2 Q_4 & 0 \\ 0 & 0 & \xi_v Q_1 Q_2 & Q_1 Q_2 Q_3 \end{pmatrix}.$$

Therefore

$$FV^{-1} = \begin{pmatrix} 0 & 0 & \frac{\beta_h \xi_v}{Q_3 Q_4} & \frac{\beta_h}{Q_4} \\ 0 & 0 & 0 & 0 \\ \frac{\beta_v \sigma}{Q_1 Q_2} & \frac{\beta_v}{Q_2} & 0 & 0 \\ 0 & 0 & 0 & 0 \end{pmatrix}.$$

The basic reproduction number, $\mathcal{R}_0 = \rho(FV^{-1})$, is the spectral radius of the product $FV^{-1}$. Hence for the model (5.10), we arrive at

$$R_0 = \sqrt{\frac{\beta_h \beta_v \sigma \xi_v}{Q_1 Q_2 Q_3 Q_4}} \tag{5.17}$$

where

$$Q_1 = \alpha_1 \epsilon + \mu_h + \sigma, \quad Q_2 = \alpha_2 \rho + \mu_h + \delta_h + \omega.$$
$$Q_3 = \mu_v + \xi_v + q, \quad Q_4 = \mu_v + \delta_v + q.$$

In (5.17), $\frac{1}{Q_2}$ is the average duration of the infectious period of the human, $\frac{\beta_h}{Q_1}$ is the probability that an individual in $E_h$ moves to $I_h$ (that is, an individual will survive the exposed state to become infectious). $\frac{1}{Q_3}$ is the state to become infectious, $\frac{\beta_v}{Q_3}$ is the probability that a vector in $E_v$ moves to $I_v$ (that is, a vector will survive the exposed state to become infectious) and $\frac{1}{Q_4}$ is the average duration of the infectious period of the vector.

Let the basic reproduction number, $R_0$, be written as

$$R_0 = \sqrt{R_h R_v} \tag{5.18}$$

where $R_h = \frac{\beta_h \sigma}{Q_1 Q_2}$ and $R_v = \frac{\beta_v \xi_v}{Q_3 Q_4}$. Here, $R_h$ describes the number of humans that one infectious vector infects over its expected infection period in a completely susceptible humans population, while $R_v$ is the number of vectors infected by one infectious human during the period of infectiousness in a completely susceptible vectors population.

Also, the computational value of $R_0 = 0.002\,28$ by applying the parameters in table 5.3. Considering the parameters for our model from table 5.3, the value of $R_0$ appears to be far less than 1 implying that the Leptospirosis will actually die out in the population.

### 5.3.5 Stability of disease-free equilibrium point

We analyse the stability of the equilibrium point using the basic reproduction number obtained for the model (5.10) in the following result.

**Table 5.3.** Elasticities of $R_0$.

| Parameter | Elasticity |
|---|---|
| $\beta_h$ | +0.832 2 |
| $\mu_h$ | −0.740 05 |
| $\delta_h$ | −0.047 24 |
| $\alpha_1$ | −0.003 1884 |
| $\alpha_2$ | −0.460 62 |
| $\varepsilon$ | −2.104 33 |
| $\rho$ | −0.460 33 |
| $\omega$ | −0.318 89 |
| $\sigma$ | +0.245 51 |
| $\mu_v$ | −0.687 5 |
| $\xi_v$ | +0.555 2 |
| $\beta_v$ | +0.832 17 |
| $q$ | −0.988 52 |
| $\delta_v$ | −0.006 8775 |

**Theorem 5.3.** The disease-free state, $E_0$, is locally asymptotically stable if $R_0 < 1$ and unstable if $R_0 > 1$.

Proof. In other to prove the local stability of the DFE, we want to show that the Jacobian matrix, $J(E_0)$, of the system (5.10) at the disease-free equilibrium has negative eigenvalues. The Jacobian matrix of the system (5.10) evaluated at the disease-free equilibrium point $E_0$, is obtained as

$$J(E_0) = \begin{pmatrix} -\mu_h & \alpha_1\epsilon & \alpha_2\rho & \gamma & 0 & 0 & \beta_h \\ 0 & -Q_1 & 0 & 0 & 0 & 0 & \beta_h \\ 0 & \sigma & -Q_2 & 0 & 0 & 0 & 0 \\ 0 & 0 & \omega & -Q_6 & 0 & 0 & 0 \\ 0 & 0 & \beta_v & 0 & -Q_5 & 0 & 0 \\ 0 & 0 & \beta_v & 0 & 0 & -Q_3 & 0 \\ 0 & 0 & 0 & 0 & 0 & \xi_v & -Q_4 \end{pmatrix}$$

where

$$Q_1 = \alpha_1\epsilon + \mu_h + \sigma, \quad Q_2 = \alpha_2\rho + \mu_h + \delta_h + \omega.$$
$$Q_3 = \mu_v + \xi_v + q, \quad Q_4 = \mu_v + \delta_v + q.$$
$$Q_5 = \mu_v + q, \quad Q_6 = \mu_h + \gamma.$$

From the above Jacobian matrix, we observe that the first and fifth columns contain only the diagonal terms which form the two negative eigenvalues, $-\mu_h$ and $-Q_5 = -(\mu_v + q)$. The other five eigenvalues can be obtained from the sub-matrix

$J_1(E_0)$, formed by eliminating the first and fifth rows and columns of $J(E_0)$. Hence, we have

$$J_1(E_0) = \begin{pmatrix} -Q_1 & 0 & 0 & 0 & \beta_h \\ \sigma & -Q_2 & 0 & 0 & 0 \\ 0 & \omega & -Q_6 & 0 & 0 \\ 0 & \beta_v & 0 & -Q_3 & 0 \\ 0 & 0 & 0 & \xi_v & -Q_4 \end{pmatrix}.$$

Again, the third column of $J_1(E_0)$ contains only the diagonal term which forms a negative eigenvalue, $-Q_6 = -(\mu_h + \gamma)$. The remaining four eigenvalues are obtained from the sub-matrix,

$$J_2(E_0) = \begin{pmatrix} -Q_1 & 0 & 0 & \beta_h \\ \sigma & -Q_2 & 0 & 0 \\ 0 & \beta_v & -Q_3 & 0 \\ 0 & 0 & \xi_v & -Q_4 \end{pmatrix}.$$

The eigenvalues of the matrix $J_2(E_0)$ are the roots of the characteristic equation

$$(Q_1 + \lambda)(Q_2 + \lambda)(Q_3 + \lambda)(Q_4 + \lambda) - (\beta_h\beta_v\sigma\xi_v) = 0. \tag{5.19}$$

From equation (5.18) after some simplifications, we obtain:

$$A_4\lambda^4 + A_3\lambda^3 + A_2\lambda^2 + A_1\lambda + A_0 = 0, \tag{5.20}$$

where

$$\begin{aligned} A_4 &= 1 \\ A_3 &= Q_1 + Q_2 + Q_3 + Q_4 \\ A_2 &= Q_1Q_2 + Q_3Q_4 + (Q_1 + Q_2)(Q_3 + Q_4) \\ A_1 &= Q_1Q_2(Q_3 + Q_4) + Q_3Q_4(Q_1 + Q_2) \\ A_0 &= Q_1Q_2Q_3Q_4 - \beta_h\beta_v\sigma\xi_v. \end{aligned} \tag{5.21}$$

From equation (5.23), we can further re-write $A_0$ in terms of the basic reproduction number, $\mathcal{R}_0$.

$$A_0 = Q_1Q_2Q_3Q_4\,(1 - R_0^2). \tag{5.22}$$

Now, we employ the Routh Hurwitz criterion in [20] which states that all roots of the polynomial (5.20) have negative real parts if and only if the coefficients $A_i$ are positive and matrices $H_i > 0$, for $i = 0, 1, 2, 3, 4$. From equation (5.21), it is easy to see that $A_1 > 0$, $A_2 > 0$, $A_3 > 0$, $A_4 > 0$. Also, the Hurwitz matrices for the polynomial (5.20) are found to be positive. That is,

$$H_1 = A_3 > 0, \quad H_2 = \begin{vmatrix} A_3 & A_4 \\ A_1 & A_2 \end{vmatrix} > 0$$

$$H_3 = \begin{vmatrix} A_3 & A_4 & 0 \\ A_1 & A_2 & A_3 \\ 0 & A_0 & A_1 \end{vmatrix} > 0, \quad H_4 = \begin{vmatrix} A_3 & A_4 & 0 & 0 \\ A_1 & A_2 & A_3 & A_4 \\ 0 & A_0 & A_1 & A_2 \\ 0 & 0 & 0 & A_0 \end{vmatrix} > 0.$$

Therefore, all the eigenvalues of the Jacobian matrix $J(E_0)$ have negative real parts when $\mathcal{R}_0 < 1$ and the disease-free equilibrium point is locally asymptotically stable and unstable when $\mathcal{R}_0 > 1$.

However, when $\mathcal{R}_0 > 1$, we see that $A_0 < 0$ and by Descartes' rule of signs [16], there is exactly one sign change in the sequence $A_4 < 0$, $A_3 < 0$, $A_2 < 0$, $A_1 < 0$, $A_0 < 0$, of coefficients of the polynomial (5.24). So, there is one eigenvalue with positive real part and the disease-free equilibrium point is unstable.

### 5.3.6 Existence of endemic equilibrium point

In this subsection, we show the existence of endemic equilibrium point ($E_e$) for the system (5.10). The endemic equilibrium point is a nonnegative steady state solution where the Leptospirosis disease persists or cannot be totally eradicated in the population.

**Theorem 5.4.** The Leptospirosis system (5.10) has no endemic equilibrium point when $R_0 < 1$ and a unique endemic equilibrium exists when $R_0 > 1$.

Proof. Let $\theta_h^{**} = \frac{\beta_h I_v}{N_h}$ and $\theta_v^{**} = \frac{\beta_v I_h}{N_h}$ be the force of infection in humans and the rodents. Also, let $E_e = (S_h^{**}, E_h^{**}, I_h^{**}, R_h^{**}, S_v^{**}, E_v^{**}, I_v^{**})$ be a nontrivial equilibrium of the system (5.10); i.e., all components of $E_e$ are positive. If we set all the differential equation in (5.10) to be equal to zero, we get:

$$\begin{aligned}
S_h^{**} &= \frac{Q_1Q_2Q_6\lambda_h}{Q_1Q_2Q_6(\theta_h^{**} + \mu_h) - Q_2Q_6\alpha_1\epsilon\theta_h^{**} - Q_6\alpha_2\rho\sigma\theta_h^{**} - \gamma\omega\sigma\theta_h^{**}} \\
E_h^{**} &= \frac{Q_2Q_6\lambda_h\theta_h^{**}}{Q_1Q_2Q_6(\theta_h^{**} + \mu_h) - Q_2Q_6\alpha_1\epsilon\theta_h^{**} - Q_6\alpha_2\rho\sigma\theta_h^{**} - \gamma\omega\sigma\theta_h^{**}} \\
I_h^{**} &= \frac{Q_6\lambda_h\sigma\theta_h^{**}}{Q_1Q_2Q_6(\theta_h^{**} + \mu_h) - Q_2Q_6\alpha_1\epsilon\theta_h^{**} - Q_6\alpha_2\rho\sigma\theta_h^{**} - \gamma\omega\sigma\theta_h^{**}} \\
R_h^{**} &= \frac{\lambda_h\sigma\omega\theta_h^{**}}{Q_1Q_2Q_6(\theta_h^{**} + \mu_h) - Q_2Q_6\alpha_1\epsilon\theta_h^{**} - Q_6\alpha_2\rho\sigma\theta_h^{**} - \gamma\omega\sigma\theta_h^{**}} \\
S_v^{**} &= \frac{\lambda_v}{(\theta_v^{**} + Q_5)} \\
E_v^{**} &= \frac{\lambda_v\theta_v^{**}}{Q_3(\theta_v^{**} + Q_5)} \\
I_v^{**} &= \frac{\xi_v\lambda_v\theta_v^{**}}{Q_3Q_4(\theta_v^{**} + Q_5)}
\end{aligned} \tag{5.23}$$

and $\theta_h^{**}$ is a positive solution of an equation given by

$$k_0\theta_h^{**} + a_0 = 0, \tag{5.24}$$

where

$$k_0 = Q_3Q_4Q_6c\beta_v\lambda_h\sigma + Q_1Q_2Q_3Q_4Q_6\lambda_v - Q_2Q_3Q_4Q_6\alpha_1\epsilon\lambda_v - Q_3Q_4Q_6\alpha_2\rho\sigma\lambda_v - Q_3Q_4\gamma\omega\sigma\lambda_v$$

$$a_0 = Q_1Q_2Q_3Q_4Q_6\lambda_v\mu_h(1 - R_0^2),$$

$$Q_1 = \alpha_1\epsilon + \mu_h + \sigma, \quad Q_2 = \alpha_2\rho + \mu_h + \delta_h + \omega.$$

$$Q_3 = \mu_v + \xi_v + q, \quad Q_4 = \mu_v + \delta_v + q.$$

$$Q_5 = \mu_v + q, \quad Q_6 = \mu_h + \gamma$$

and

$$R_0^2 = \frac{\beta_h\beta_v\sigma\xi_v}{Q_1Q_2Q_3Q_4}.$$

The positive solution of equation (5.24) depends upon $k_0$ and $a_0$ respectively. When $R_0$ is less than one ($R_0 < 1$), the constant $a_0$ is positive ($a_0 > 0$) and when $R_0$ is greater than one ($R_0 > 1$), the constant $a_0$ is negative ($a_0 < 0$).

For $k_0 > 0$, when $R_0 < 1$, one can see that $a_0 > 0$ and it implies that the system (5.10) has no positive solution.

However, when $R_0 > 1$, then $a_0 < 0$, there exist a unique endemic equilibrium given by $\theta_h^{**} = \frac{a_0}{k_0}$. This complete the proof.

### 5.3.7 Global stability

To show that the system (5.10) is globally asymptotically stable, we use the Lyapunov functional theory to solve for the disease-free equilibrium. We present the globally stability of the disease-free equilibrium.

**Theorem 5.5.** The disease-free equilibrium of the system (5.10) is globally asymptotically stable if $\sigma(\delta_h + \mu_h + \omega) > \frac{\sigma\gamma\omega}{Q_6}$ and $E_h\sigma\alpha_1\epsilon \geqslant Q_3Q_4I_v$, otherwise unstable.

Proof. For the global stability of the disease-free we define the Lyapunov function in the following form

$$P(t) = U_1S_h + U_2E_h + U_3I_h + U_4R_h + U_5S_v + U_6E_v + U_7I_v, \tag{5.25}$$

where $U_i(i = 1\ldots7)$ are positive constant to be chosen later.

Taking the derivative with respect to time $t$ of the above equation (5.24) we have:

$$P'(t) = U_1S'_h + U_2E'_h + U_3I'_h + U_4R'_h + U_5S'_v + U_6E'_v + U_7I'_v.$$

Now using the system (5.10), we get:

$$P'(t) = U_1\left[\lambda_h - \frac{\beta_h S_h I_v}{N_h} + \alpha_1\epsilon E_h + \alpha_2\rho I_h - \mu_h S_h + \gamma R_h\right]$$
$$+ U_2\left[\frac{\beta_h S_h I_v}{N_h} - \alpha_1\epsilon E_h - \mu_h E_h - \sigma E_h\right]$$
$$+ U_3[\sigma E_h - \alpha_2\rho I_h - \delta_h I_h - \mu_h I_h - \omega I_h] + U_4[\omega I_h - \mu_h R_h - \gamma R_h]$$
$$+ U_5\left[\lambda_v - \frac{\beta_v S_v I_h}{N_h} - \mu_v S_v - qS_v\right] + U_6\left[\frac{\beta_v S_v I_h}{N_h} - \mu_v E_v - \xi_v E_v - qE_v\right]$$
$$+ U_7[\xi_v E_v - \mu_v I_v - \delta_v I_v - qI_v],$$

where (′) denotes the derivative with respect to time and after some simplification we obtain

$$P'(t) = E_v[U_7\xi_v - U_6 Q_3] + \frac{\beta_h S_h I_v}{N_h}[U_2 - U_1] + \frac{\beta_v S_v I_h}{N_h}[U_6 - U_5]$$
$$+ E_h[U_3\sigma - U_2 Q_1] + R_h[U_1\gamma - U_4 Q_6]$$
$$+ I_h[U_1\alpha_2\rho - U_2 Q_2 + U_4\omega] + E_h U_1\alpha_1\epsilon - U_7 I_v Q_4$$
$$+ U_1[\lambda_h - \mu_h S_h] + U_5[\lambda_v - S_v Q_5],$$

where

$$Q_1 = \alpha_1\epsilon + \mu_h + \sigma, \quad Q_2 = \alpha_2\rho + \mu_h + \delta_h + \omega.$$
$$Q_3 = \mu_v + \xi_v + q, \quad Q_4 = \mu_v + \delta_v + q.$$
$$Q_5 = \mu_v + q, \quad Q_6 = \mu_h + \gamma.$$

Now choosing the constants, $U_1 = U_2 = \sigma$, $U_3 = Q_1$, $U_4 = \frac{\sigma\gamma}{Q_6}$, $U_5 = U_6 = \xi_v$, $U_7 = Q_3$, $\lambda_h = S_h^0\mu_h$, $\lambda_v = (\mu_v + q)S_v^0$, $\mu_v = (\mu_v + q) = Q_5$, and after some rearrangement, we get

$$P'(t) = -\mu_h\sigma(S_h - S_h^0) - \xi_v Q_5(S_v - S_v^0) - I_h\left[\sigma(\delta_h + \mu_h + \omega) - \frac{\sigma\gamma\omega}{Q_6}\right]$$
$$+ E_h\sigma\alpha_1\epsilon - Q_3 Q_4 I_v.$$

For $S_h = S_h^0$ and $S_v = S_v^0$, we have

$$P'(t) = -\left[I_h\left[\sigma(\delta_h + \mu_h + \omega) - \frac{\sigma\gamma\omega}{Q_6}\right] - E_h\sigma\alpha_1\epsilon + Q_3 Q_4 I_v\right].$$

Then $P'(t)$ is negative when $\sigma(\delta_h + \mu_h + \omega) > \frac{\sigma\gamma\omega}{Q_6}$, and $E_h\sigma\alpha_1\epsilon \geqslant Q_3 Q_4 I_v$ and $P'(t) = 0$, for $E_h = I_h = R_h = E_v = I_v = 0$.

Thus by Lassalle's principle [18] the disease-free equilibrium of the system (5.10) is globally asymptotically stable on $\Omega$.

## 5.4 Bifurcation analysis for the Leptospirosis model

The eigenvalues obtained from the Jacobian matrix computation at each endemic equilibrium is used to determine the stability of the endemic equilibrium. The centre manifold theory [21] is applied in this section to explore the local asymptotic stability of the endemic equilibrium because of the complexity of finding the eigenvalues of the Jacobian. We made the following change of variables in order to apply the centre manifold theory such that: $S_h = x_1$, $E_h = x_2$, $I_h = x_3$, $R_h = x_4$, $S_v = x_5$, $E_v = x_6$, $I_v = x_7$, where $\underset{2}{X} = (x_1, x_2, x_3, x_4, x_5, x_6, x_7)^T$, $\underset{2}{X}$ is the vector notation for $x_1, x_2, x_3, \dots, x_7$. Hence, the Leptospirosis model is rewritten in the following vectorial pattern: $\frac{dX}{dt} = H = (h_1, h_2, h_3, h_4, h_5, h_6, h_7)^T$, such that:

$$
\begin{aligned}
\frac{dx_1}{dt} &= \Lambda_h - \frac{\beta_h x_1 x_7}{x_1 + x_2 + x_3 + x_4} + \gamma x_4 - \mu_h x_1 = h_1 \\
\frac{dx_2}{dt} &= \frac{\beta_h x_1 x_7}{x_1 + x_2 + x_3 + x_4} - (\alpha_1 \epsilon - \sigma + \mu_h) x_2 = h_2 \\
\frac{dx_3}{dt} &= \sigma x_2 - (\alpha_2 \rho + \omega + \delta_h + \mu_h) x_3 = h_3 \\
\frac{dx_4}{dt} &= \omega x_3 - (\gamma + \mu_h) x_4 = h_4 \\
\frac{dx_5}{dt} &= \Lambda_v + \frac{\beta_v x_5 x_3}{x_1 + x_2 + x_3 + x_4} - (q + \mu_h) x_5 = h_5 \\
\frac{dx_6}{dt} &= \frac{\beta_v x_5 x_3}{x_1 + x_2 + x_3 + x_4} - (\xi_v + q + \mu_v) x_6 = h_6 \\
\frac{dx_7}{dt} &= \xi_v x_6 - (\mu_v + \delta_v + q) x_7 = h_7.
\end{aligned}
\tag{5.26}
$$

At the equilibrium state $\Omega$, the Jacobian of the system (5.26) is given by

$$
J_\Omega = \begin{pmatrix}
-\mu_h & \alpha_1\epsilon & \alpha_2\rho & \gamma & 0 & 0 & -\beta_h \\
0 & -Q_1 & 0 & 0 & 0 & 0 & \beta_h \\
0 & \sigma & -Q_2 & 0 & 0 & 0 & 0 \\
0 & 0 & 0 & \omega & -Q_6 & 0 & 0 \\
0 & 0 & -\dfrac{\beta_v \Lambda_v \mu_h}{(q + \mu_v)\Lambda_h} & 0 & -Q_5 & 0 & 0 \\
0 & 0 & \dfrac{\beta_v \Lambda_v \mu_h}{(q + \mu_v)\Lambda_h} & 0 & 0 & -Q_3 & 0 \\
0 & 0 & 0 & 0 & 0 & \xi_v & -Q_4
\end{pmatrix}
\begin{pmatrix} J_1 \\ J_2 \\ J_3 \\ J_4 \\ J_5 \\ J_6 \\ J_7 \end{pmatrix}
= \begin{pmatrix} 0 \\ 0 \\ 0 \\ 0 \\ 0 \\ 0 \\ 0 \end{pmatrix}
\tag{5.27}
$$

$Q_1 = \alpha_1\epsilon + \mu_h + \sigma$, $Q_2 = \alpha_2\rho + \mu_h + \delta_h$, $Q_3 = \mu_v + \xi_v + q$, $Q_4 = \mu_v + \delta_v + q$, $Q_5 = \mu_v + q$, $Q_6 = \mu_h + \gamma$. Next, we calculate the left and right eigenvector which are associated with zero as a single eigenvalue of (5.27). If $J = (J_1, J_2, J_3, J_4, J_5, J_6, J_7)$ are to be the right eigenvector associated with the zero eigenvalue, then the equation below are obtained

$$
\begin{aligned}
&-\mu_h J_1 + \alpha_1 \epsilon J_2 + \alpha_2 \rho J_3 + \gamma J_4 - \beta_h J_7 = 0\\
&-Q_1 J_2 + \beta_h J_7 = 0\\
&\sigma J_2 - Q_2 J_3 = 0\\
&\omega J_3 - Q_6 J_4 = 0\\
&-\beta_v J_3 - Q_5 J_5 = 0\\
&\beta_v J_3 - Q_3 J_6 = 0\\
&\xi_v J_6 - Q_4 J_7 = 0.
\end{aligned}
$$

If we find the solution of the system (5.26) we obtain

$$
\begin{aligned}
J_1 &= \frac{-\beta_h \xi_v \beta_v \Lambda_v \mu_h \sigma Q_2 Q_6 + \alpha_1 \epsilon Q_3 Q_4 (q + \mu_v) \Lambda_h Q_2 Q_6 + \alpha_2 \rho \sigma Q_3 Q_4 (q + \mu_v) \Lambda_h + \gamma \omega \sigma Q_3 Q_4 (q + \mu_v) \Lambda_h}{Q_2 Q_3 Q_4 Q_6 \mu_h (q + \mu_v) \Lambda_h}\\
J_2 &= J_2 > 0\\
J_3 &= \frac{\sigma}{Q_2} > 0\\
J_4 &= \frac{\omega \sigma}{Q_2 Q_6} > 0\\
J_5 &= \frac{-\beta_v \Lambda_v \mu_h \sigma}{Q_5 (q + \mu_v) Q_2} < 0\\
J_6 &= \frac{\beta_v \Lambda_v \mu_h \sigma}{Q_3 (q + \mu_v) \Lambda_h Q_2} > 0\\
J_7 &= \frac{\xi_v \beta_v \Lambda_v \mu_h \sigma}{Q_3 Q_4 (q + \mu_h) \Lambda_h} > 0.
\end{aligned}
$$

Suppose $\beta_v$ represents a bifurcation parameter. The system is linearised at the disease-free equilibrium point $J_\Omega$, such that $R_0 = 1$ yields

$$\beta_v^* = \frac{Q_1 Q_2 Q_3 Q_4}{\beta_h \sigma \xi_v}.$$

The model (5.10) with $\beta_v$ as bifurcation parameter has a simple zero eigenvalues. The model equation (5.26) is analysed by the centre manifold theory near $\beta_v^*$, the Jacobian matrix near $\beta_v^*$ has a left eigenvector associated with the zero eigenvalue given by $(g_1, g_2, g_3, g_4, g_5, g_6, g_7)$ for the system (5.11). The left eigenvector given by $g = (g_1, g_2, g_3, g_4, g_5, g_6, g_7)$ is solved and satisfying $g.\, h = 1$, we take the transpose of (5.27) to obtain

$$
J_\Omega^T = \begin{pmatrix}
-\mu_h & 0 & 0 & 0 & 0 & 0 & 0\\
\alpha_1 \epsilon & -Q_1 & \sigma & 0 & 0 & 0 & 0\\
\alpha_2 \rho & 0 & -Q_2 & \omega & \dfrac{\beta_v \Lambda_v \mu_h}{(q + \mu_v)\Lambda_h} & \dfrac{\beta_v \Lambda_v \mu_h}{(q + \mu_v)\Lambda_h} & 0\\
\gamma & 0 & 0 & -Q_6 & 0 & 0 & 0\\
0 & 0 & 0 & 0 & -Q_5 & 0 & 0\\
0 & 0 & 0 & 0 & 0 & -Q_3 & \xi_v\\
-\beta_h & \beta_h & 0 & 0 & 0 & 0 & -Q_4
\end{pmatrix}
\begin{pmatrix} g_1\\ g_2\\ g_3\\ g_4\\ g_5\\ g_6\\ g_7 \end{pmatrix}
= \begin{pmatrix} 0\\ 0\\ 0\\ 0\\ 0\\ 0\\ 0 \end{pmatrix}. \tag{5.28}
$$

Therefore, we obtain the following systems of equation

$$g_1 = 0, \quad g_2 = \frac{\sigma\beta_v \xi_v \Lambda_v \beta_h \mu_h g_2}{Q_2 Q_4 \Lambda_h (q + \mu_v)} > 0,$$

$$g_3 = \frac{\beta_v \xi_v \Lambda_v \beta_h \mu_h g_2}{Q_2 Q_4 \Lambda_h (q + \mu_v)} > 0,$$

$$g_4 = 0, \quad g_5 = 0, \quad g_6 = \frac{\xi_v^2 \beta_h \sigma \beta_v \Lambda_v \beta_h \mu_h g_2}{Q_2 Q_4^2 \Lambda_h (q + \mu_v)} > 0.$$

Applying the Castillo-Chavez and Song theorem to prove the local stability of the endemic equilibrium point near $R_0 = 1$. We used the Castillo-Chavez and Song bifurcation at the critical value of a parameter. Using the centre manifold theorem, we compute the values of $a$ and $b$, since $g_1 = g_4 = g_5 = 0$ for $k = 1, 4, 5$ then the values of $k$ at 2, 3, 6 and 7 will be investigated. The disease-free equilibrium point is given by

$$\begin{aligned} a &= \Sigma_{k,\, i,\, j=1}^{n} g_k J_i J_j \frac{\partial^2 h_k(0,\,0)}{\partial x_i \partial x_j} \\ b &= \Sigma_{k,\, i,\, j=1}^{n} g_k J_i \frac{\partial^2 h_k(0,\,0)}{\partial x_i \partial \beta_h}. \end{aligned} \tag{5.29}$$

We discussed that all the second-order partial derivatives at $E_0$ and $\beta_h^*$ are equal to zero except the ones below:

$$\frac{\partial^2 h_1}{\partial x_1 \partial x_7} = \frac{\partial_1^h}{\partial x_7 \partial x_1} = \frac{\beta_h}{N_h} = \frac{-\beta_h \mu_h}{\Lambda_h}$$

$$\frac{\partial^2 h_2}{\partial x_1 \partial x_7} = \frac{\partial_2^h}{\partial x_7 \partial x_1} = \frac{\beta_h}{N_h} = \frac{-\beta_h \mu_h}{\Lambda_h}$$

$$\frac{\partial^2 h_5}{\partial x_3 \partial x_5} = \frac{-\beta_v \mu_h}{\Lambda_h}$$

$$\frac{\partial^2 h_6}{\partial x_3 \partial x_5} = \frac{\beta_v \mu_h}{\Lambda_h}$$

$$\frac{\partial^2 h_1}{\partial x_7 \partial \beta_h} = \frac{\partial^2 h_1}{\partial \beta_h \partial x_7} = -1,$$

$$\frac{\partial^2 h_1}{\partial x_7 \partial \beta_h} = \frac{\partial^2 h_1}{\partial \beta_h \partial x_7} = 1.$$

We obtain the nature of bifurcation at $R_0 = 1$ by the signs of the bifurcation coefficients $a$ and $b$, which was obtained from the above partial derivatives given, respectively by

$$a = \Sigma_{k,\, i,\, j=1}^{n} g_k J_i J_j \frac{\partial^2 h_k(E_0,\, \beta_h^*)}{\partial x_i \partial x_j}$$

$$a = \frac{\sigma\beta_v\xi_v\Lambda_v\beta_h\mu_h g_2}{Q_2Q_4\Lambda_h(q+\mu_v)}\left[\frac{\beta_h\mu_h - Z}{\Lambda_h}\frac{-\beta_v\Lambda_v\mu_h\sigma}{Q_5(q+\mu_v)Q_2} + \frac{\beta_h\mu_h}{\Lambda_h}(-Z)\left(\frac{-\beta_v\Lambda_v\mu_h\sigma}{Q_5(q+\mu_v)Q_2}\right)\right]$$

$$a > 0 \quad b = \Sigma_{k,\, i=1}^{7} g_k J_i \frac{\partial^2 h_k(E_0, \beta_h^*)}{\partial x_i \partial \beta_h}$$

$$g_1\left[J_7\frac{\partial^2 h_1(E_0, \beta_h^*)}{\partial x_7 \partial \beta_h}\right] + g_2\left[J_7\frac{\partial^2 h_2(E_0, \beta_h^*)}{\partial x_7 \partial \beta_h}\right]$$

$$= 0 + \frac{\sigma\beta_v\xi_v\Lambda_v\beta_h\mu_h g_2}{Q_2Q_4\Lambda_h(q+\mu_v)}\left[\frac{\xi_v\beta_v\Lambda_v\mu_h\sigma J_2}{Q_3Q_4(q+\mu_v)\Lambda_h \times 1}\right]$$

$b > 0$, it reveals that the Leptospirosis model (5.26) shows a backward bifurcation if the threshold parameter the crosses unity. This implies the co-existence of disease-free and endemic equilibrium at $R_0$ at little bit less than 1. Therefore, we obtain the following.

Proposition: The Leptospirosis model given by (5.26) exhibits the phenomenon of backward bifurcation at $E_0$ and $R_0 = 1$.

**Remark 5.1.** It implies that the reduction of $R_0$ below 1 is not enough but necessary to eliminate Leptospirosis from the environment. Eliminating Leptospirosis may also depend on the location and on whom the disease is found.

## 5.5 Sensitivity and elasticity analysis of the parameters in the model

The need to apply sensitivity and elasticity analysis of the parameters in the model become important when we need to ascertain the reliability of any predictions and it can help in determining how best we can reduce human mortality and morbidity due to Leptospirosis by investigating the relative significance of the many factors responsible for its transmission and prevalence. With sensitivity analysis we can determine the relative change in a static quantity when the parameter changes, Hence, the sensitivity of the static quantity $R_0$ (the reproduction number) with respect to any of the parameter is given by

$$\Gamma_{\Lambda_h}^{R_0} := \frac{\partial R_0}{\partial \Lambda_h}.$$

While we define the elasticity of $R_0$ with respect to the parameter $\Lambda_h$ as given by

$$\Gamma_{\Lambda_h}^{R_0} := \frac{\partial R_0}{\partial \Lambda_h} \cdot \frac{\partial \Lambda_h}{\partial R_0}.$$

Therefore, the sensitivity or elasticity of $R_0$ with respect to $\Lambda_h$ is positive when $R_0$ is increasing with respect to $\Lambda_h$ but negative when $R_0$ is decreasing with respect to $\Lambda_h$. We provide in table 5.4 the sensitivity analysis of the parameters in the model

The most sensitive parameters are $\beta_h$ and $\beta_v$ which implies that the rate of transmission among susceptible human, infected *Mastomys* rat and susceptible *Mastomys* rat has a very strong influence on the rate of change of the reproduction number $R_0$. The positive sign of the sensitivity index of $R_0$ to the model parameters implies that an increase or decrease in the value of each of the parameters (that are positive) will evolve to increase or decrease in the $R_0$ of Leptospirosis. For instance, $\Gamma_{\beta_h}^{R_0} = 0.832\ 17$ implies that increasing or reducing the transmission rate between susceptible *Mastomys* rat and infected *Mastomys* rat by 8% increases or decreases the $R_0$ by 8%. The negative sign of the sensitivity index of the $R_0$ to the model parameters indicates that an increase or decrease in the value of each of the parameters (that are negative) lead to an increase or decrease in the $R_0$ of Leptospirosis. Hence, sensitivity analysis of the Leptospirosis model gives a very strong insight into the relative significance of different factors responsible for the transmission and prevalence of Leptospirosis. It is therefore a useful technique that can be used to guide policy makers and National control program on Leptospirosis in finding the optimal intervention strategies to control and eliminate the spread of Leptospirosis. Figure 5.2 shows that increase in $\beta_h$ (transmission rate between susceptible humans and infected rodents) leads to increase in the reproduction number, $R_0$. This also leads to an increase in the diseases induced death rate, $\delta_h$. This implies that transmissibility is a major factor in the spread of the disease. Figure 5.3

**Table 5.4.** Elasticities of $I_h^{**}$.

| Parameter | Elasticity |
|---|---|
| $\beta_h$ | −0.005 9279 |
| $\beta_v$ | −1.0 |
| $\mu_h$ | −1.000 143 875 |
| $\delta_h$ | −0.000 015 88 |
| $\alpha_1$ | +0.000 024 59 |
| $\alpha_2$ | =0.000 062 51 |
| $\varepsilon$ | +0.000 024 628 159 35 |
| $\rho$ | +0.000 000 2596 |
| $\omega$ | −0.000 053 542 |
| $\sigma$ | +0.179 7 |
| $\mu_v$ | +0.842 7 |
| $\Lambda_v$ | −0.000 012 312 |
| $\Lambda_h$ | +0.002 263 34 |
| $q$ | +0.168 6 |
| $\gamma$ | −0.000 000 065 84 |
| $\delta_v$ | +0.000 048 991 |

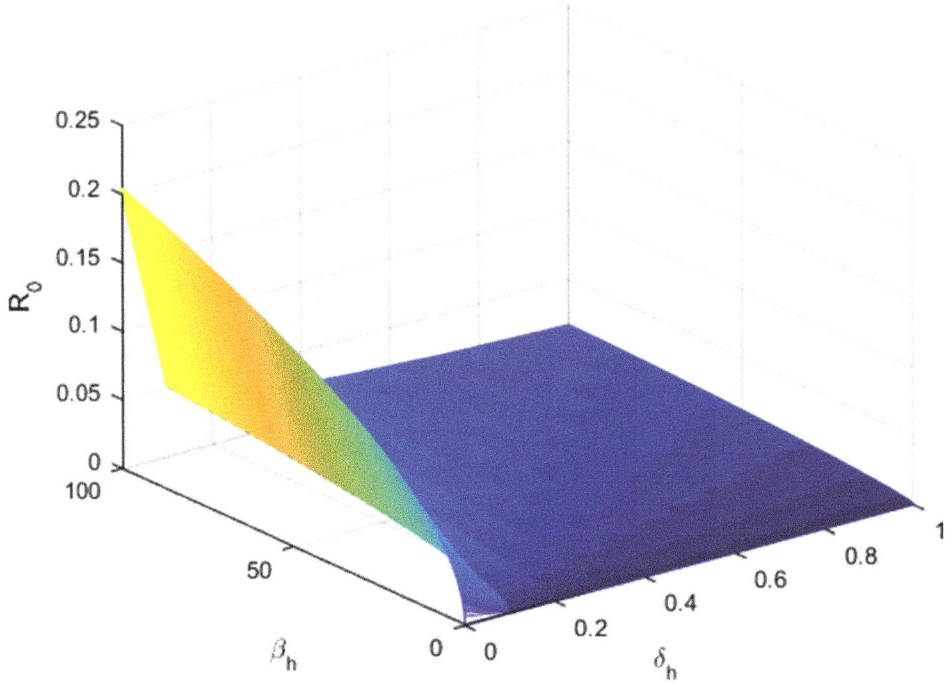


**Figure 5.2.** Simulation showing the effect of varying $\beta_h$ and $\delta_h$ on the reproduction number $R_0$.

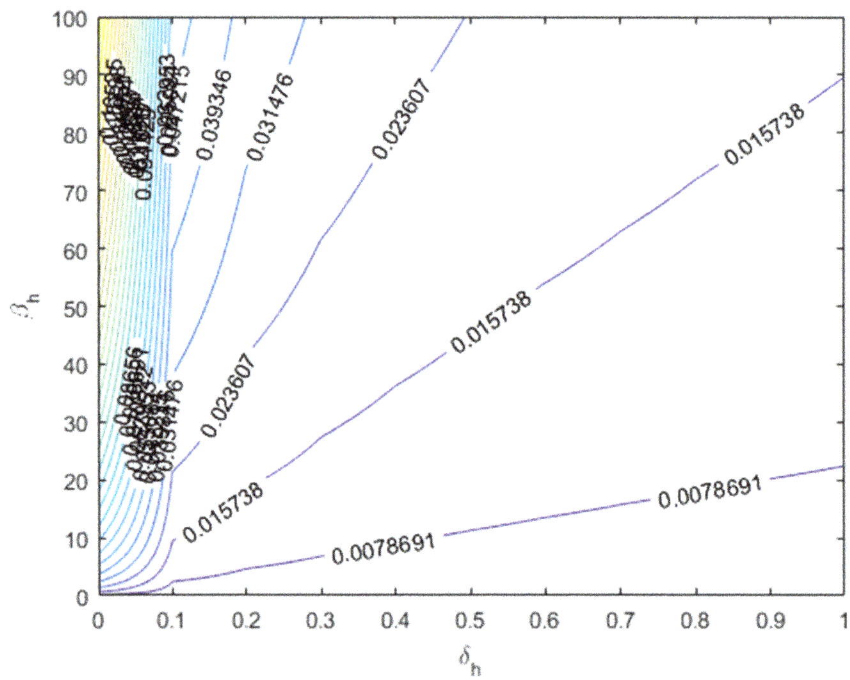


**Figure 5.3.** Simulation showing the effect of varying $\beta_h$ and $\delta_h$.

depicts the disease-induced mortality rate, $\delta_h$ increases as the transmission rate $\beta_h$ increases over time. This means that the more the spread of the disease the more the possibility of a large proportion of people dying due to the disease.

## 5.6 Determining intervention strategies for Leptospirosis diseases

Leptospirosis is a disease caused by bacteria and it affects human and animals. There are preventive and treatment regimen that are capable of reducing the morbidity and mortality of the disease in human if used at the appropriate time. Below we list some of the known intervention that can reduce the morbidity or mortality of the disease.

- Prevention with wearing protective clothing for people at occupational risk and avoidance of swimming in water that may be contaminated.
- Vaccination of the rodents.
- Culling with pesticides and rodent control.
- Treatment with antibiotics (doxycycline) and proper diagnostics.

It is important to develop various scenarios for the intervention strategies and combination of these strategies are applied in different forms in various places. In the interest of the local national control program on Leptospirosis, we offer some combined intervention package that can reduce the morbidity and mortality of the disease. Our aim here is to minimise the number of infected and exposed humans and the number of infected rodents. Hence, we define the elasticity of $I_H^{**}$ with respect to $\alpha_2$ as: $\coprod_{\alpha_2}^{I_h^{**}} = \frac{\partial I_h^{**}}{\partial \alpha_2} \cdot \frac{\alpha_2}{I_h^{**}}$. From our observation of table 5.5, we discover that some interventions, for instance, the use of prevention with wearing protective clothing for people at occupational risk and avoidance of swimming in water that may be contaminated, has more effectiveness in reducing the prevalence of Leptospirosis in humans than other interventions. Here, we investigate those parameters each interventions would influence in order to compare and contrast among the available interventions. For instance, prevention with wearing protective clothing for people at occupational risk and avoidance of swimming in water that may be contaminated affect $\beta_h$, treatment with doxycycline affects $\gamma$, $\omega$, $\alpha_1$, $\alpha_2$, $\epsilon$, $\rho$, and $\delta_h$, culling with pesticides and rodent control affect $q$, $\mu_v$, $\delta_h$, and $\beta_v$, while vaccination of the rodents affects $\beta_v$ and $\delta_v$. Hence, we provide the overall effects of each intervention in table 5.5. Table 5.6 suggests that prevention with wearing protective clothing for

**Table 5.5.** List of intervention strategies and their efficacy.

| Intervention | Affected parameter | Total efficacy | Rank |
|---|---|---|---|
| Treatment with antibiotics (doxycycline) | $\gamma$, $\omega$, $\alpha_1$, $\alpha_2$, $\epsilon$, $\rho$, $\delta_h$ | 0.000 082 520 04 | 4 |
| Culling with pesticides and rodent control | $q$, $\mu_v$, $\delta_h$, $\beta_v$ | 0.011 284 12 | 3 |
| Vaccination of the rodents | $\beta_v$, $\delta_v$ | 0.999 951 009 | 2 |
| Prevention with wearing protective clothing | $\beta_h$ | 1.0 | 1 |

**Table 5.6.** Numerical values of parameters used in the simulation.

| Parameters | Description | Value | Source |
|---|---|---|---|
| $\lambda_h$ | Recruitment rate for susceptible humans | $1.2d^{-1}$ | [13] |
| $\beta_h$ | Transmission rate between susceptible humans and infected vectors | $0.008\,9$ d$^{-1}$ | [11] |
| $\mu_h$ | Natural death rate for humans population | $4.6 \times 10^{-5}$ d$^{-1}$ | [13] |
| $\delta_h$ | Disease-induced death rate of human | $8 \times 10^{-4}$ d$^{-1}$ | [12] |
| $\alpha_1$ | Treatment rate of exposed individuals | $0.002$ d$^{-1}$ | Assumed |
| $\alpha_2$ | Treatment rate of infectious human | $0.006$ d$^{-1}$ | Assumed |
| $\epsilon$ | Proportion of effective treatment of exposed individuals | $0.000\,1$ d$^{-1}$ | Assumed |
| $\rho$ | Proportion of effective treatment of infectious human | $0.65$ d$^{-1}$ | Assumed |
| $\gamma$ | Loss of immunity in humans | $2.85 \times 10^{-3}$ | [4] |
| $\omega$ | Rate of recovery | $2.7 \times 10^{-3}$ d$^{-1}$ | [11] |
| $\sigma$ | Progression rate of exposed individuals to infectious individuals | $2.0 \times 10^{-6}$ d$^{-1}$ | [24] |
| $\lambda_v$ | Recruitment rate for vector | $1.3$ d$^{-1}$ | [13] |
| $\mu_v$ | Natural death rate for vectors population | $1.0 \times 10^{-1}$ d$^{-1}$ | [12] |
| $\xi_v$ | Progression rate of exposed vectors to infectious vectors | $1.0 \times 10^{-2}$ d$^{-1}$ | [11] |
| $\beta_v$ | Transmission rate between susceptible vectors and infected vectors | $0.007\,9$ d$^{-1}$ | [12] |
| $q$ | Use of pesticide dimensionless | 0.98 Estimated | |
| $\delta_v$ | Disease-induced death rate of vector | $1.0 \times 10^{-3}$ d$^{-1}$ | [18] |

people at occupational risk and avoidance of swimming in water that may be contaminated is the most effective strategy, while vaccination of the rodents is the second most efficient followed by culling with pesticides and rodent control effective and lastly treatment with antibiotics (doxycycline).

## 5.7 Stochastic model of the transmission dynamics of Leptospirosis

Here, we say that $(\Omega, \mathfrak{F}, \mathbb{P})$ represents a complete probability space with a filtration $\{\mathfrak{F}_t\}_{t\geqslant 0}$ which satisfies the criteria that it is increasing and right continuous while $\mathfrak{F}_0$ contain all $\mathbb{P}$-null sets. In this case we introduce randomness into the model (5.10) by considering that the white noise depends on the size of the corresponding population. It means we use the corresponding of the form $\tau_i(S_h(t), E_h(t), I_h(t), R_h(t), S_v(t), E_v(t), I_v(t))dW(t)$ where $\tau$ denotes the intensity of the random perturbation, $i \in [1, 7]$ and $W(t)_{t\geqslant 0}$ is a one-dimensional Brownian motion defined on a complete probability space $(\Omega, \mathfrak{F}, (\mathfrak{F}_t)_{t\geqslant 0}, \mathbb{P})$. Thus, the stochastic model of Leptospirosis is illustrated by the stochastic differential equations (SDE):

$$
\begin{aligned}
dS_h &= \left(\lambda_h - \frac{\beta_h S_h I_v}{N_h} + \alpha_1 \epsilon E_h + \alpha_2 \rho I_h - \mu_h S_h + \gamma R_h\right)dt \\
&\quad + \tau_1 S_h(t) dW(t) \\
dE_h &= \left(\frac{\beta_h S_h I_v}{N_h} - \alpha_1 \epsilon E_h - \mu_h E_h - \sigma E_h\right)dt + \tau_2 E_h(t) dW(t) \\
dI_h &= \left(\sigma E_h - \alpha_2 \rho I_h - \delta_h I_h - \mu_h I_h - \omega I_h\right)dt + \tau_3 I_h(t) dW(t) \\
dR_h &= \left(\omega I_h - \mu_h R_h - \gamma R_h\right)dt + \tau_1 R_h(t) dW(t) \\
dS_v &= \left(\lambda_v - \frac{\beta_v S_v I_h}{N_h} - \mu_v S_v - q S_v\right)dt + \tau_4 S_v(t) dW(t) \\
dE_v &= \left(\frac{\beta_v S_v I_h}{N_h} - \mu_v E_v - \xi_v E_v - q E_v\right)dt + \tau_1 E_v(t) dW(t) \\
dI_v &= \left(\xi_v E_v - \mu_v I_v - \delta_v I_v - q I_v\right)dt + \tau_1 I_v(t) dW(t).
\end{aligned}
\tag{5.30}
$$

Suppose $X(t) = \begin{pmatrix} S_h(t) \\ E_h(t) \\ I_h(t) \\ R_h(t) \\ S_v(t) \\ E_v(t) \\ I_v(t) \end{pmatrix}$, then we can re-write the system (5.30) in the form of a single SDEs of the form $dX(t) = G(X(t), t)dt + H(X(t), t)dW(t)$, where the function $G: \mathbb{R}_+^7 \times \mathbb{R}^+ \to \mathbb{R}^7$ is defined by

$$
G = \begin{pmatrix}
\lambda_h - \dfrac{\beta_h S_h I_v}{N_h} + \alpha_1 \epsilon E_h + \alpha_2 \rho I_h - \mu_h S_h + \gamma R_h \\
\dfrac{\beta_h S_h I_v}{N_h} - \alpha_1 \epsilon E_h - \mu_h E_h - \sigma E_h \\
\sigma E_h - \alpha_2 \rho I_h - \delta_h I_h - \mu_h I_h - \omega I_h \\
\omega I_h - \mu_h R_h - \gamma R_h \\
\lambda_v - \dfrac{\beta_v S_v I_h}{N_h} - \mu_v S_v - q S_v \\
\dfrac{\beta_v S_v I_h}{N_h} - \mu_v E_v - \xi_v E_v - q E_v \\
\xi_v E_v - \mu_v I_v - \delta_v I_v - q I_v
\end{pmatrix}
$$

and the function $H: \mathbb{R}_+^7 \times \mathbb{R}^+ \to \mathbb{R}^7$ is given by

$$H = \begin{pmatrix} \tau_1 S_h(t) \\ \tau_2 E_h(t) \\ \tau_3 I_h(t) \\ \tau_4 R_h(t) \\ \tau_5 S_v(t) \\ \tau_6 E_v(t) \\ \tau_7 I_v(t) \end{pmatrix}.$$

### 5.7.1 Positivity of the solution for the SDE model

**Theorem 5.6.** If $(S_h(0), E_h(0), I_h(0), R_h(0), S_v(0), E_v(0), I_v(0)) \in \Delta$, then model (5.30) allows a unique solution $(S_h^*, E_h^*, I_h^*, R_h^*, S_v^*, E_v^*, I_v^*)$ on $t \geqslant 0$ and the solution persists in $\Delta$ with probability 1.

Proof. Whenever $(S_h(0), E_h(0), I_h(0), R_h(0), S_v(0), E_v(0), I_v(0)) \in \Delta$ then the total human population reveals that the equation

$$\frac{dN_h^*}{dt} = \lambda_h - \mu_h N_h^* - \delta_h I^* \leqslant \lambda_h - \mu N_h^* - \delta_h N_h^*$$

&

and the total vector population is given

$$\frac{dN_v^*}{dt} = \lambda_v - (\mu_v + q)N_v^*.$$

If $(S_h^*, E_h^*, I_h^*, R_h^*, S_v^*, E_v^*, I_v^*) \in \mathbb{R}_+^7$ a.s. (a.s.) for all $t \in [0, T]$, then we obtain

$$dN_h^* \leqslant (\lambda_h - (\mu_h - \delta_h)N_h^*)dr$$

a.s.

$$dN_v^* \leqslant (\lambda_v - (\mu_v + q)N_v^*)dr.$$

Hence, by the comparison theorem, we have that,

$$N_h^* \leqslant \frac{\lambda_h}{(\mu_h + \delta_h)} + \left(N_h^*(0) - \frac{\lambda_h}{\mu_h + \delta_h}\right)e^{-(\mu_h+\delta_h)t} < \frac{\lambda_h}{(\mu_h + \delta_h)}$$

a.s. $\forall\, t \in [0, T]$

$$N_v^* \leqslant \frac{\lambda_v}{(\mu_v + q)} + \left(N_v^*(0) - \frac{\lambda_v}{(\mu_v + q)}\right)e^{-(\mu_v+q)t} < \frac{\lambda_v}{(\mu_v + q)}$$

a.s. $\forall\, t \in [0, T]$.

Then, $N_h^* \leqslant \frac{\lambda_h}{(\mu_h + \delta_h)}$ & $N_v^* \leqslant \frac{\lambda_v}{(\mu_v + q)}$, so we obtain $(S_h^*, E_h^*, I_h^*, R_h^*, S_v^*, E_v^*, I_v^*) \in \Delta\,\forall$, $t \in [0, T]$. □

We have in [22] a theorem and proof of existence and uniqueness of solutions of SDEs

**Theorem 5.7.** Suppose $T > 0, G(X(t), t)$: $\mathbb{R}^2 \times \mathbb{R}^2 \to \mathbb{R}^7$ and $H(X(t), t)$: $\mathbb{R}^7 \times \mathbb{R}^+ \to \mathbb{R}^7$ are measurable functions and the constants $K > 0,\ M > 0$ such that

1. $|G(t, x)| + |H(x, t)| \leqslant K(1 + |x|)\forall x \in \mathbb{R}, t \in [0, T]$
2. $|G(t, x) - G(t, y)| + |H(t, x) - H(t, y)| \leqslant M|x - y| \forall x, y \in \mathbb{R}, t \in [0, T]$

then $dX(t) = G(X(t), t)dt + H(X(t), t)dW(t), X(0) = X_0$ has a unique solution.

## 5.8 Homotopy analysis approach

This is one of the most powerful analytic approach for solving nonlinear problems, which has a convenient means to control and adjust the convergence region and ratio of approximation series, when it is crucial. In order to explain the fundamental concept of HAM, we consider the following nonlinear ordinary differential equation of the form.

$$Z[v(t)] = 0. \tag{5.31}$$

Such that $Z$ is a nonlinear operator, $t$ represent the time, an independent variable $v(t)$ is an unknown function. We derived the zero-order definition equation by applying an embedding parameter $p \in [0, 1]$ [23] given by

$$(1 - p)L[\varphi(t; p) - v_0(t)] = ph_1 H(t)Z(t; q) \tag{5.32}$$

such that $h_1 \neq 0$ is an auxiliary parameter, $H(t)$ is an auxiliary function, $L$ is auxiliary linear operator, $v_0(t)$ is an initial guess of $v(t)$, $\varphi(t, p)$ is an unknown function. In the HAM based solution, we have freedom to select $h_1$, $H(t)$, and $L(t)$. it can easily be shown that $p = 0$, we get

$$\varphi(t; 0) = v_0(t) \tag{5.33}$$

and when $p = 1$, we obtain

$$\varphi(t; 1) = v(t). \tag{5.34}$$

Therefore, as $p$ increases from 0 to 1, the solution $\varphi(t; p)$ varies arbitrary for the initial approximation $v_0(t)$ to the exact solution $v(t)$, such a kind of condition variation is called definition in topology. By expanding $\varphi(t; p)$ by Taylor series with respect to $p$, we have

$$\varphi(t; p) = v_0(t) + \Sigma_{m=1}^{\infty} v_m p^m \tag{5.35}$$

$$\varphi(0; p) = 0, \quad \left.\frac{\varphi(t; p)}{\partial t}\right|_{t=0} = 0, \quad \left.\frac{\partial\varphi(t; p)}{\partial t}\right|_{t=+\infty} = 0, \tag{5.36}$$

such that

$$v_m(t) = \frac{1}{m!}\frac{\partial^m \varphi(t; p)}{\partial p^m}\Bigg|\, p = 0. \tag{5.37}$$

(5.37) is the definition derivative. If the auxiliary nonlinear operator $Z$, the initial approximation $V_0(t)$, the auxiliary parameter $h_1$ and the auxiliary function $H(x)$ are chosen cautiously such that

1. The solution $\varphi(t; p)$ of the zero-order deformation equation (5.37) exists for all $p \in [0, 1]$.
2. The deformation derivative (5.37) exists for all $m = 1, 2, ....$
3. The series (5.37) converge at $p = 1$.

Therefore, we have the series solution as;

$$v(t) = v_0(t) + \sum_{m=1}^{\infty} v_m(t) \tag{5.38}$$

which should be one of the solution of the output of equation (5.38) as proved by Liao [23]. In short, we define the vector

$$\vec{v}_n(t) = \{v_0(t), v_1(t), \dots, v_n(t)\}. \tag{5.39}$$

By differentiating the zero-order equations (5.39) $n$-times with respect to the embedding parameter $p$, then by setting $p = 0$ and finally,dividing the resulting equation by $n!$, we have the $n$th order deformation equation given as

$$L[v_n(t) - \lambda_n v_{n-1}(t)] = h_1 H(t) P_n(\vec{v}_{n-1}(t)) \tag{5.40}$$

subject to initial condition

$$v_n(0) = 0 \tag{5.41}$$

where

$$P_n(\vec{v}_{n-1}(t)) = \frac{1}{(n-1)!}\frac{\partial^{n-1} A[(t; p)]}{\partial p^{n-1}}\Bigg|\, p = 0 \tag{5.42}$$

and then

$$\lambda_n = \begin{cases} 0 & \text{if } n \leqslant 1, \\ 1 & \text{if } n > 1 \end{cases}. \tag{5.43}$$

According to the definition (5.38), $v_n(t)$ for $n \geqslant 1$ can be obtained for equation (5.40) which for uncoupled linear first order differential equation when solved subject to the initial condition (5.41). Using the symbol computer software such as MAPLE, MATHEMATICA.

## 5.9 Solution of the Leptospirosis model by HAM

We consider the first equation of the model equation (5.31) to solve the model equation (5.10) by HAM and choose the linear operator

$$Z[S_h(t;\, p)] = \frac{dS_h(t;\, p)}{dt} \tag{5.44}$$

with the property that

$$Z[\theta_1] = 0 \tag{5.45}$$

where $\theta_1$ is a constant of integration. The inverse operator $Z^{-1}$ is given by

$$Z^{-1}(\cdot) = \int_0^t (\cdot)dt. \tag{5.46}$$

Suppose the nonlinear operator is defined as;

$$Z[S_h(t;\, p)] = \frac{dS_h(t;\, p)}{dt}\left[\lambda_h + \gamma R_h(t,\, p) - \frac{\beta_h S_h(t;\, p) I_v(t;\, p)}{N_h} + \alpha_1 \epsilon E_h(t;\, p) + \alpha_2 \rho I_h(t;\, p) - \mu_h S_h(t;\, p)\right]. \tag{5.47}$$

By constructing the zero-order definition equation

$$(1 - p)Z[S_h(t;\, p) - S_{h,0}(t;\, p)] = qh_1 H(t) Z[s_h(t;\, p)] \tag{5.48}$$

we have the following:

1. if $p = 0$, then $S_h(t;\, 0) = S_{h,0}(t)$
2. if $p = 1$, then $S_h(t;\, 1) = S_h(t)$.

Hence, we have the $n$th order definition equation as;

$$Z[S_{h,n}(t) - \Lambda S_{h,(n-1)(t)}] = h_1 H(t) \mathcal{P}(S_{n,(n-1)(t)}), \quad n \geqslant 1 \tag{5.49}$$

where,

$$\begin{aligned} P_n\big(S_{h,(n-1)}(t)\big) = {} & \frac{dS_{h,(n-1)}(t)}{dt} - \lambda_h + \gamma R_h(t,\, p) \\ & - \frac{\beta_h S_h(t;\, p) I_v(t;\, p)}{N_h} + \alpha_1 \epsilon E_h(t;\, p) + \alpha_2 \rho I_h(t;\, p) - \mu_h S_h(t;\, p). \end{aligned} \tag{5.50}$$

The solution of the $n$th order definition equation (19) for $n \geqslant 1$ and using $h_1 = -0.01$ and $H(t) = 1$, we have that

$$\begin{aligned} S_{h,n}(t) = {} & \lambda_n S_{h,(n-1)}(t) - 0.01 \int_\infty^t \frac{dS_{h,(n-1)}(t)}{dt} - \lambda_h + \gamma R_{h,(n-1)}(t) \\ & - \frac{\beta_h S_{h,(n-1)}(t) I_{v,(n-1)}(t)}{N_h} + \alpha_1 \epsilon E_{h,(n-1)}(t) + \alpha_2 \rho I_{h,(n-1)}(t) \\ & - \mu_h S_{h,(n-1)}(t). \end{aligned} \tag{5.51}$$

According to the earlier step, we obtain,

$$E_{h,n}(t) = \lambda_n E_{h,(n-1)}(t) - 0.01 \int_{\infty}^{t} \left[ \frac{dE_{h,(n-1)}(t)}{dt} - \frac{\beta_{\text{h}} S_{h,(n-1)}(t) I_{v,(n-1)}(t)}{N_{\text{h}}} + \left(\alpha_1 \epsilon + \mu_{\text{h}} + \sigma\right) E_{h,(n-1)} \right] dt, \quad n \geqslant 1 \tag{5.52}$$

$$I_{h,n}(t) = \lambda_n I_{h,(n-1)}(t) - 0.01 \int_{\infty}^{t} \left[ \frac{dI_{h,(n-1)}}{dt} - \sigma E_{h,(n-1)} + \left(\alpha_2 \rho + \delta_{\text{h}} + \mu_{\text{h}} + \omega\right) I_{h,(n-1)} \right] dt, \quad n \geqslant 1 \tag{5.53}$$

$$R_{h,n}(t) = \lambda_n R_{h,(n-1)}(t) - 0.01 \int_{\infty}^{t} \left[ \frac{dR_{h,(n-1)}(t)}{dt} - \omega I_{h,(n-1)} + \left(\mu_{\text{h}} + \gamma\right) R_{h,(n-1)} \right] dt, \quad n \geqslant 1 \tag{5.54}$$

$$S_{v,n}(t) = \lambda_{\text{v}} S_{v,(n-1)}(t) - 0.01 \int_{\infty}^{t} \left[ \frac{dS_{v,(n-1)}(t)}{dt} - \lambda_{\text{v}} + \frac{\beta_{\text{v}} S_{v,(n-1)} I_{h,(n-1)}}{N_{\text{h}}} + \left(\mu_{\text{v}} + q\right) S_{v,(n-1)} \right] dt, \quad n \geqslant 1 \tag{5.55}$$

$$E_{v,n}(t) = \lambda_{\text{v}} E_{v,(n-1)}(t) - 0.01 \int_{\infty}^{t} \left[ \frac{dE_{v,(n-1)}(t)}{dt} - \frac{\beta_{\text{v}} S_{v,(n-1)} I_{h,(n-1)}}{N_{\text{h}}} + \left(\mu_{\text{v}} + \xi_{\text{v}} + q\right) E_{v,(n-1)} \right] dt, \quad n \geqslant 1 \tag{5.56}$$

$$I_{v,n}(t) = \lambda_{\text{v}} I_{v,(n-1)}(t) - 0.01 \int_{\infty}^{t} \left[ \frac{dI_{v,(n-1)}(t)}{dt} - \xi_{\text{v}} E_{v,(n-1)} + \left(\mu_{\text{v}} + \delta_{\text{v}} + q\right) I_{v,(n-1)} \right] dt, \quad n \geqslant 1. \tag{5.57}$$

The 7th to 9th approximations for $S_{\text{h}}$, $E_{\text{h}}$, $I_{\text{h}}$, $R_{\text{h}}$, $S_{\text{v}}$, $E_{\text{v}}$, $I_{\text{v}}$ are obtained and presented below:

7th approximations:
$S_{h,7}(t) = -0.029\,901\,248\,18t + 0.006\,234\,005\,238t^2 - 0.000\,056\,319\,414\,30t^3$
$+2.036\,090\,016 \times 10^{-7}t^4 - 3.312\,127\,875 \times 10^{(-10)}t^5 + 2.394\,358\,877 \times 10^{(-13)}t^6$
$-6.181\,115\,670 \times 10^{(-17)}t^7$

$E_{h,7}(t) = 0.389\,089\,1309t - 0.005\,192\,317\,924t^2 + 0.000\,045\,055\,452\,46t^3$
$- 1.730\,666\,287 \times 10^{(-7)}t^4 + 3.091\,276\,122 \times 10^{(-10)}t^5 - 2.394\,327\,946 \times 10^{(-13)}t^6$
$+ 6.181\,128\,326 \times 10^{(-17)}t^7$

$I_{h,7}(t) = -0.175\,162\,3953t + 9.403\,515\,451 \times 10^{(-6)}t^2 - 2.936\,674\,411 \times 10^{-10}t^3$
$+ 2.504\,908\,307 \times 10^{(-13)}t^4 - 3.716\,145\,882 \times 10^{(-16)}t^5 + 2.981\,369\,729 \times 10^{(-19)}t^5$
$- 1.151\,651\,672 \times 10^{(-22)}t^7$

$R_{h,7}(t) = 0.045\,304\,024\,79t - 0.000\,018\,020\,687\,03t^2 + 6.893\,651\,875 \times 10^{(-10)}t^3$
$- 1.116\,842\,104 \times 10^{(-14)}t^4 + 1.112\,442\,078 \times 10^{(-18)}t^5 - 5.668\,317\,898 \times 10^{(-22)}t^6$
$+ 1.776\,841\,471 \times 10^{(-27)}t^7$

$S_{v,7}(t) = 700 - 14.929\,129\,40t + 0.078\,950\,855\,38t^2 - 0.000\,717\,648\,1170t^3$
$+ 3.945\,569\,024 \times 10^{(-6)}t^4 - 8.624\,930\,367 \times 10^{(-9)}t^5 + 7.837\,029\,617 \times 10^{(-12)}t^6$
$- 2.438\,610\,939 \times 10^{(-15)}t^7$

$E_{v,7}(t) = -2.892\,697\,760t + 0.095\,526\,136\,38t^2 - 0.000\,876\,461\,5107t^3$
$+ 3.216\,644\,093 \times 10^{(-6)}t^4 - 5.312\,335\,888 \times 10^{(-9)}t^5 + 3.899\,290\,310 \times 10^{(-12)}t^6$
$- 1.022\,180\,480 \times 10^{(-15)}t^7$

$I_{v,7}(t) = -4.042\,715\,761t + 0.131\,552\,9928t^2 - 0.001\,189\,002\,711t^3$
$+ 4.298\,139\,089 \times 10^{(-6)}t^4 - 6.991\,093\,820 \times 10^{(-9)}t^5 + 5.053\,366\,372 \times 10^{(-12)}t^6$
$- 1.304\,400\,227 \times 10^{(-15)}t^7.$

8th approximations:
$S_{h,8}(t) = -0.017\,602\,235\,70t + 0.007\,200\,265\,408t^2 - 0.000\,078\,058\,709\,68t^3$
$+ 3.527\,525\,950 \times 10^{-7}t^4 - 7.651\,015\,391 \times 10^{(-10)}t^5 + 8.296\,453\,504 \times 10^{(-13)}t^6$
$- 4.283\,513\,160 \times 10^{(-16)}t^7 + 8.292\,613\,999 \times 10^{(-20)}t^8$

$E_{h,8}(t) = 0.385\,198\,2396t - 0.006\,168\,493\,380t^2 + 0.000\,066\,907\,096\,66t^3$
$- 3.225\,145\,934 \times 10^{(-7)}t^4 + 7.432\,351\,672 \times 10^{(-10)}t^5 - 8.296\,417\,284 \times 10^{(-13)}t^6$
$+ 4.283\,521\,936 \times 10^{(-16)}t^7 - 8.292\,630\,999 \times 10^{(-20)}t^8$

$I_{h,8}(t) = -0.173\,410\,7713t + 0.000\,012\,475\,928\,24t^2 - 4.385\,332\,005 \times 10(-10)t^3$
$+4.759\,142\,675 \times 10(-13)t^4 - 1.061\,974\,002 \times 10^{(-15)}t^5 + 1.327\,817\,478 \times 10^{(-18)}t^6$
$- 7.996\,451\,752 \times 10^{(-22)}t^7 + 1.550\,480\,349 \times 10^{(-25)}t^8$

$R_{h,8}(t) = 0.045\,603\,037\,27t - 0.000\,083\,381\,434\,08t^2 + 5.641\,420\,995 \times 10(-7)t^3$
$- 2.036\,108\,157 \times 10(-9)t^4 + 3.312\,130\,404 \times 10^{(-12)}t^5 - 2.394\,361\,120 \times 10^{(-15)}t^6$
$+ 6.181\,127\,211 \times 10^{(-19)}t^7 - 3.886\,888\,714 \times 10^{(-28)}t^8$

$S_{v,8}(t) = 700 - 22.326\,838\,11t + 0.158\,782\,1050t^2 - 0.000\,994\,694\,8390t^3$
$+ 5.843\,763\,253 \times 10^{(-6)}t^4 - 1.706\,111\,016 \times 10^{(-8)}t^5 + 2.328\,353\,398 \times 10^{(-11)}t^6$
$- 1.450\,564\,195 \times 10^{(-14)}t^7 + 3.292\,124\,768 \times 10^{(-18)}t^8$

$E_{v,8}(t) = -2.863\,770\,782t + 0.110\,332\,6184t^2 - 0.001\,214\,775\,068t^3$
$+ 5.572\,835\,266 \times 10^{(-6)}t^4 - 1.227\,149\,665 \times 10^{(-8)}t^5 + 1.351\,104\,094 \times 10^{(-11)}t^6$
$- 7.083\,710\,729 \times 10^{(-15)}t^7 + 1.392\,720\,904 \times 10^{(-18)}t^8$

$I_{v,8}(t) = -4.002\,288\,603t + 0.151\,943\,7067t^2 - 0.001\,647\,957\,763t^3$
$+ 7.446\,525\,989 \times 10^{(-6)}t^4 - 1.614\,942\,671 \times 10^{(-8)}t^5 + 1.750\,991\,447 \times 10^{(-11)}t^6$
$- 9.039\,493\,574 \times 10^{(-15)}t^7 + 1.749\,793\,550 \times 10^{(-18)}t^8.$

9th approximations:

$S_{h,9}(t) = -0.005\,426\,213\,34t + 0.008\,146\,584\,930t^2 - 0.000\,103\,038\,0920t^3$
$+ 5.587\,641\,237 \times 10^{(-7)}t^4 - 1.514\,912\,655 \times 10^{(-9)}t^5 + 2.190\,279\,456 \times 10^{(-12)}t^6$
$- 1.696\,280\,959 \times 10^{(-15)}t^7 + 6.567\,772\,308 \times 10^{(-19)}t^8 - 9.888\,146\,126 \times 10^{(-23)}t^9$

$E_{h,9}(t) = 0.381\,346\,2572t - 0.007\,124\,626\,095t^2 + 0.000\,091\,997\,092\,76t^3$
$- 5.288\,235\,682 \times 10^{(-7)}t^4 + 1.493\,251\,978 \times 10^{(-9)}t^5 - 2.190\,261\,186 \times 10^{(-12)}t^6$
$+ 1.696\,274\,687 \times 10^{(-15)}t^7 - 6.567\,763\,751 \times 10^{(-19)}t^8 + 9.888\,166\,399 \times 10^{(-23)}t^9$

$I_{h,9}(t) = -0.171\,676\,6636t + 0.000\,015\,485\,952\,42t^2 - 6.254\,397\,473 \times 10^{(-10)}t^3$
$+ 8.096\,494\,665 \times 10^{(-13)}t^4 - 2.344\,849\,688 \times 10^{(-15)}t^5 + 3.798\,381\,173 \times 10^{(-18)}t^6$
$- 3.168\,903\,304 \times 10^{(-21)}t^7 + 1.227\,987\,437 \times 10^{(-24)}t^8 - 1.849\,027\,761 \times 10^{(-28)}t^9$

$R_{h,9}(t) = 0.045\,147\,006\,90t - 0.000\,085\,548\,997\,13t^2 + 5.594\,178\,706 \times 10^{(-7)}t^3$
$- 2.019\,834\,424 \times 10^{(-9)}t^4 + 3.290\,804\,808 \times 10^{(-12)}t^5 - 2.386\,408\,837 \times 10^{(-15)}t^6$
$+ 6.218\,425\,295 \times 10^{(-19)}t^7 - 2.240\,651\,654 \times 10^{(-24)}t^8 + 4.663\,948\,191 \times 10^{(-31)}t^9$

$S_{v,9}(t) = 700 - 29.650\,569\,73t + 0.277\,762\,6346t^2 - 0.001\,556\,363\,633t^3$
$+ 8.471\,001\,691 \times 10^{(-6)}t^4 - 2.951\,302\,769 \times 10^{(-8)}t^5 + 5.376\,069\,693 \times 10^{(-11)}t^6$
$- 5.028\,375\,226 \times 10^{(-14)}t^7 + 2.284\,182\,016 \times 10^{(-17)}t^8 - 3.950\,549\,722 \times 10^{(-21)}t^9$

$E_{v,9}(t) = -2.835\,133\,074t + 0.124\,833\,4181t^2 - 0.001\,603\,502\,333t^3$
$+ 8.827\,368\,968 \times 10^{(-6)}t^4 - 2.429\,756\,256 \times 10^{(-8)}t^5 + 3.566\,914\,944 \times 10^{(-11)}t^6$
$- 2.805\,149\,449 \times 10^{(-14)}t^7 + 1.103\,034\,956 \times 10^{(-17)}t^8 - 1.686\,739\,760 \times 10^{(-21)}t^9$

$I_{v,9}(t) = -3.962\,265\,717t + 0.171\,913\,4510t^2 - 0.002\,175\,304\,254t^3$
$+ 0.000\,011\,795\,297\,21t^4 - 3.197\,586\,492 \times 10^{(-8)}t^5 + 4.622\,617\,415 \times 10^{(-11)}t^6$
$- 3.579\,639\,453 \times 10^{(-14)}t^7 + 1.385\,836\,492 \times 10^{(-17)}t^8 - 2.086\,221\,799 \times 10^{(-21)}t^9.$

## 5.10 Numerical results and discussions

We execute the numerical analysis of the model using the parameters obtained from different literatures. Table 5.7 shows the details of the parameters and their values.

## 5.11 Numerical simulations

The numerical simulations are conducted using ODE 45 in MATLAB, and the results are given in figures 5.4–5.10 illustrating the system's behaviour for the different values of the model's parameters. Here, we study numerically the behaviour of our system (5.10) using some of the parameter values compatible with Leptospirosis as given in table 5.6, and by considering initial conditions, $S_h(0) = 8000$, $E_h(0) = 4000$, $I_h(0) = 5000$, $R_h(0) = 3000$, $S_v(0) = 700$, $E_v(0) = 400$ and $I_v(0) = 300$. We present here, the time evolution of the human and rodents population with an initially entirely susceptible human and rodents population and a single infected and infectious rodent, numerical simulations on the effect of

**Table 5.7.** Numerical values of parameters used in the simulations.

| Parameter | Symbol | Value |
|---|---|---|
| Disease-induced death rate | $\psi$ | 0.05 |
| Probability of human getting infected | $\beta_1$ | 0.833 3 |
| Probability of mosquito getting infected | $\beta_2$ | 0.09 |
| Per capita birth rate of mosquitoes | $b_v$ | 1000 |
| Natural death rate of humans | $d_h$ | 0.000 04 |
| Progression of human from the exposed to the infectious state | $\alpha_h$ | 0.058 8 |
| Progression of mosquitoes from the exposed to the infectious state | $\alpha_v$ | |
| Per capita biting rate of mosquitoes | $\epsilon_v$ | 0.2 |
| Contact rate of vector per human per unit time | $\phi$ | 0.502 |
| Per capita biting rate of humans | $\epsilon_h$ | 0.2 |
| Human spontaneous recovery | $\theta$ | 0.005 |
| Natural death rate of mosquitoes | $d_v$ | 0.142 9 |
| Rate of loss of immunity from humans | $\gamma$ | 0.790 2 |
| Per capita birth rate of humans | $b_h$ | 100 |

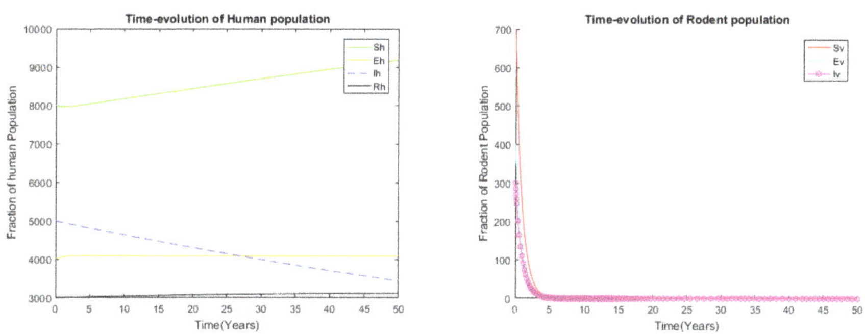


(a) (b)

**Figure 5.4** Simulation showing the time evolution of the human and rodents population using numerical method.

treatment on the exposed and infected human population, numerical simulation on effect of pesticides and rodent control, numerical simulations on the effect of effect of variation of transmissibility rate on the susceptible human population, numerical simulations of the infected human and rodents on the force of infection (FOI) and lastly the numerical simulation of the reproduction number against some important model parameters. Figures 5.4(a) and (b) show the time evolution of the human and rodents population respectively. Figures 5.7(a) and (b) show the plots on the effect of treatment on the exposed and infected humans. Figures 5.5(a) and (b) show the time evolution of the human and rodents population using homotopy analysis method.

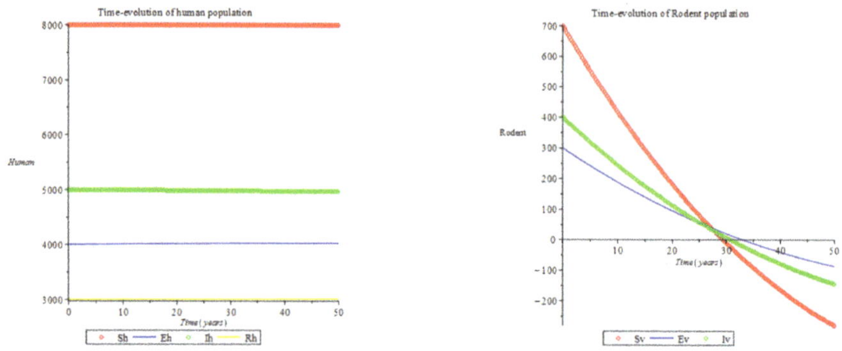


(a) (b)

**Figure 5.5** Simulation showing the time evolution of the human and rodents population using HAM.

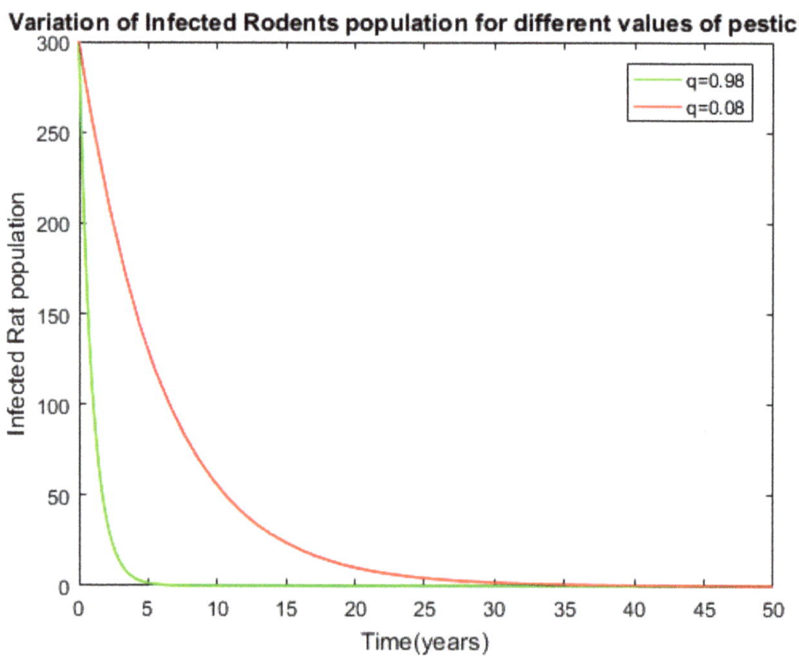


**Figure 5.6** Simulation showing the effect of varying the use of pesticides and rodent control.

Figure 5.6 shows the plot on the effect of pesticides and rodent control on the infected rodents. Figure 5.8 shows the plot effect of variation of transmissibility rate on the susceptible human population. Figures 5.9(a) and (b) show the plot on the variation of infected human and rodents against the force of infection (FOI). Figure 5.10 shows the plots of the reproduction number against some important model parameters.

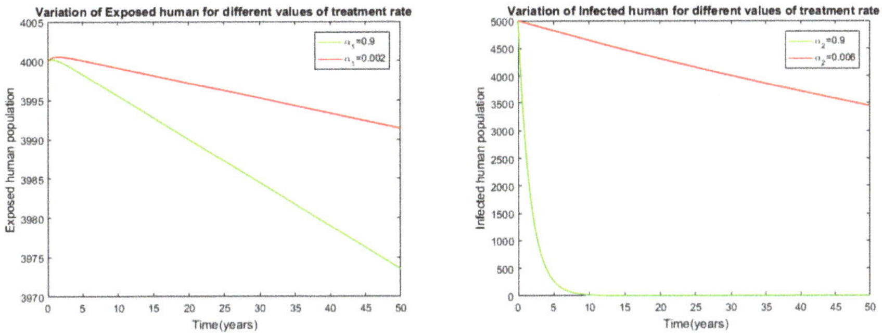


(a) (b)

**Figure 5.7** Simulation showing effect of treatment on the exposed and infected human.

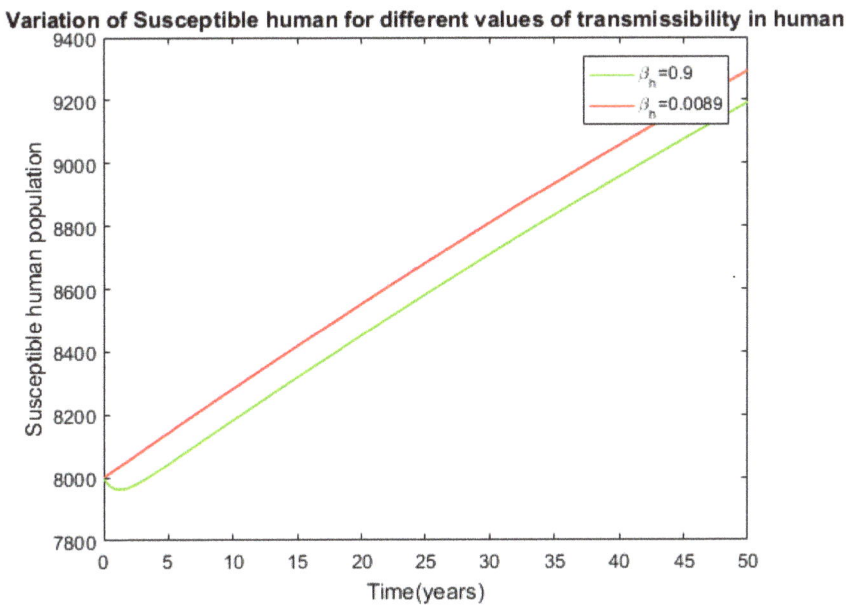


**Figure 5.8** Simulation on the effect of varying transmissibility rate on the susceptible human population.

### 5.11.1 Numerical simulations on the time evolution of the human and rodents population

Figure 5.4 shows the simulation showing the time evolution of the human and rodents population, which includes figures 5.5(a) and (b). In figure 5.4(a), the susceptible human population increases as time evolves which implies that there is constant recruitment of the susceptible in the population (due to birth and loss of immunity) which gives the opportunity for continuous and progressive low-level transmission of Leptospirosis in the population. This may also lead to some rapid outbreaks at any point in time. In order words, the disease is endemic in the population. It was also observed in figure 5.4(a) that the population of the exposed

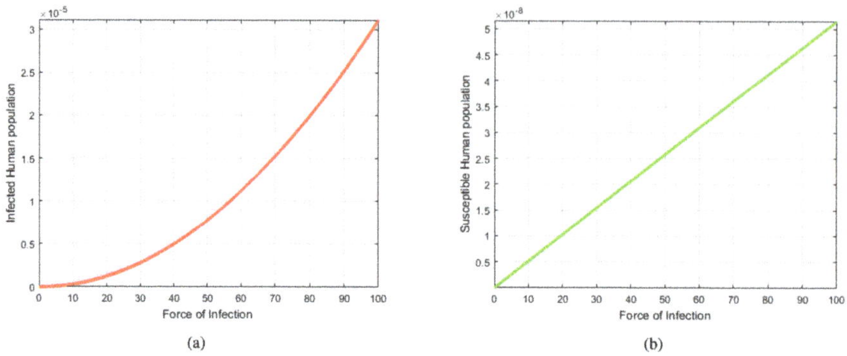


**Figure 5.9** Simulation showing the variation of infected and susceptible humans against the FOI.

human decreases as time evolves due to the influence of treatment with doxycycline to a level. In the same figure 5.4(a), we also observed that the population of the recovered human increases over time due to treatment of the exposed and infected human population who recovered from the disease. In figure 5.4(b), the susceptible, exposed and infected rodent population decreases over time due to the introduction of pesticides and rodent control into the population of rodents.

### 5.11.2 Numerical simulation on the effect of pesticides and rodent control

It is observed from figure 5.6 that the variation of infected rodents at $q = 0.08$ decreases at a slower rate compare to when $q = 0.98$. This shows that when the rate at which the pesticides and rodent control is 98% successful the population of the infected rodents go into extinction earlier and as rapidly as possible compared to when the rate of success is lower.

### 5.11.3 Numerical simulations on the effect of treatment on the exposed and infected human population

We observed from figure 5.7(a) that the variation of the exposed human at different values of the treatment rate of the exposed showed different behaviours. The simulation showed that at $\alpha_1 = 0.9$, the exposed human population depletes a lot and faster while at $\alpha_1 = 0.002$, the fraction of the exposed human population decreases a little. This shows that when the rate at which the treatment of exposed human population is 90% successful, the population of the exposed human decreases to a larger extent than when the treatment of the exposed human population is 0.2%. In figure 5.7(b), the variation of the infected human population at two different values of the treatment rate of the infected human was captured. The simulation showed that at $\alpha_2 = 0.9$, the infected human population decreases drastically and faster while at $\alpha_2 = 0.006$, the fraction of the infected human population has little effect on the dynamics of infected human in the long run. This depicts that when the success of the treatment rate is very effective, the infected human population decreases drastically as time evolves compared to when the treatment rate is at a much lower rate.

### 5.11.4 Numerical simulations on the effect of variation of transmissibility rate on the susceptible human population

Figure 5.8 show the simulation on the variation of susceptible human at different values of transmissibility. It is observed that the higher the transmissibility the lower the number of susceptible human in the population over time and the lower the transmissibility the higher the number of susceptible human in the population over time.

### 5.11.5 Numerical simulations on the variation of infected human and rodents against the force of infection (FOI)

In figure 5.9(a), we observed that the infected human population increases as the FOI increases over time. This implies that the presence of infected human population has a great role to play in the spread of an infection in a population. The FOI, which is the per capita rate at which susceptible human is infected. Hence, FOI increases as the prevalence of infection in the population increases. In figure 5.9(b), it is observed that the population of the susceptible human increases as the FOI increases. This means that more susceptible humans are contracting infection in the population.

### 5.11.6 Numerical simulation of the reproduction number against some important model parameters

Here, we investigate the effect of different parameter values on $R_0$. It is observed from figure 5.10(a) that as $R_0$ increases, there is also an increase in transmission rate between susceptible humans and infected *mastomys* rat ($\beta_h$), at the same time we also observed from figure 5.10(b)that as $R_0$ increases, there is also an increase in transmission rate between susceptible *mastomys* rat and infected *mastomys* rat ($\beta_v$), which implies that transmissibility plays a significant role in the spread of Leptospirosis and infectious diseases in general. We also observed from figure 5.10(c) that there is a drastic decrease in $R_0$ as the rate of recovery increases ($\omega$), which implies that the more people get recovered the lower the reproduction number and the lower the spread of Leptospirosis in the population. Finally, in figure 5.10(d), it is observed that as $q$ (culling with pesticides and rodent control) increases the reproduction number $R_0$ decreases drastically, which also means that there is a great effect of the intervention on the spread of the disease. When the number of *mastomys* rats are reduced and controlled then the rate of Leptospirosis spread will decrease drastically.

### 5.11.7 Numerical simulation of the SDE model

Here, we simulate the SDE model (5.30) by applying the Milstein method given the set of parameters in table 5.6. We obtained the numerical results of the stochastic model for 200 runs which was shown and compared with the mean of 200 runs of stochastic model simulation with the results of the corresponding deterministic model given in figures 5.11 and 5.12, which presented the time series solution of all

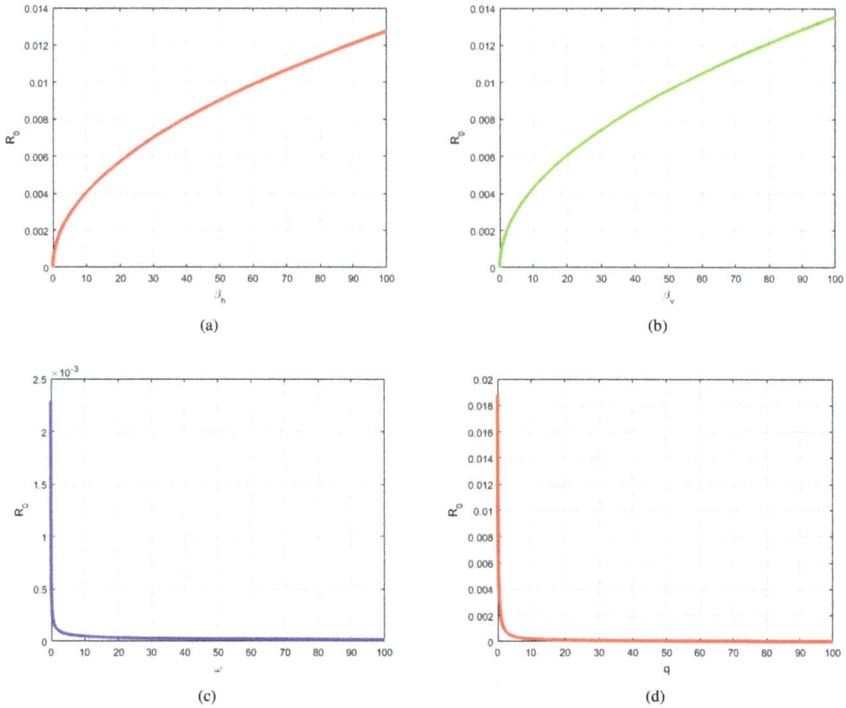


**Figure 5.10** Simulation showing the effect of different parameter values on $R_0$.

the variables of the model (table 5.7). We also observed from figures 5.11(a) and (b), 5.12(a) and (b) that the results of the stochastic model simulation is little lower than the simulation results of their corresponding deterministic model.

## 5.12 Conclusion

In this paper, we have formulated and analysed a compartmental model for Leptospirosis transmission in human and vector populations with treatment rate of exposed and infectious human population, proportion of effective treatment of exposed, and infectious human population and the use of pesticide on the rodent population. The human population was divided into four compartments: susceptible, exposed, infectious and recovered, while the rodent population was divided into three compartments: susceptible, exposed and infectious; since vectors remain infectious for life and have no recovered class.

We found the region where the model is epidemiologically feasible; and stability analysis of the disease-free equilibrium via the reproduction number $\mathcal{R}_0$ obtained using the next generation matrix technique were established. It was established that the disease-free equilibrium of the model is globally asymptotically stable. The model results showed that the disease-free equilibrium is asymptotically stable when $\mathcal{R}_0 < 1$ and unstable when $\mathcal{R}_0 > 1$. The existence of a unique endemic equilibrium and the global asymptotic stability of the disease-free equilibrium under certain

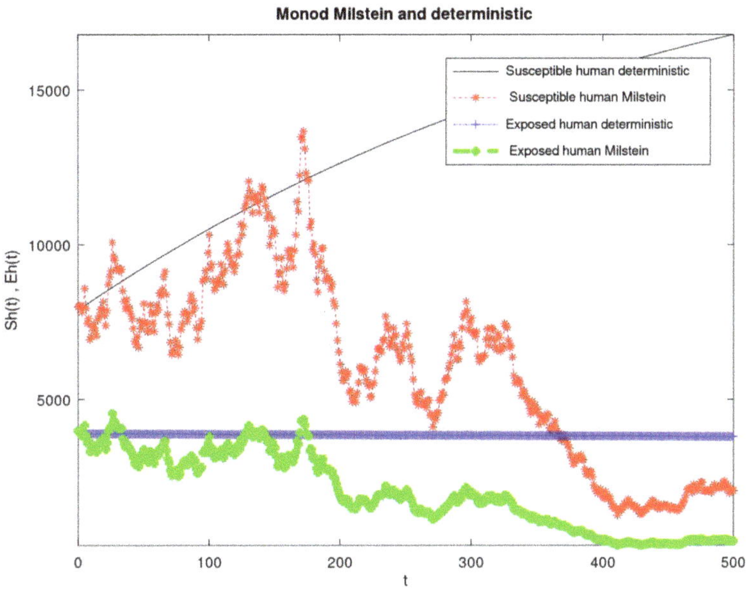


(a)

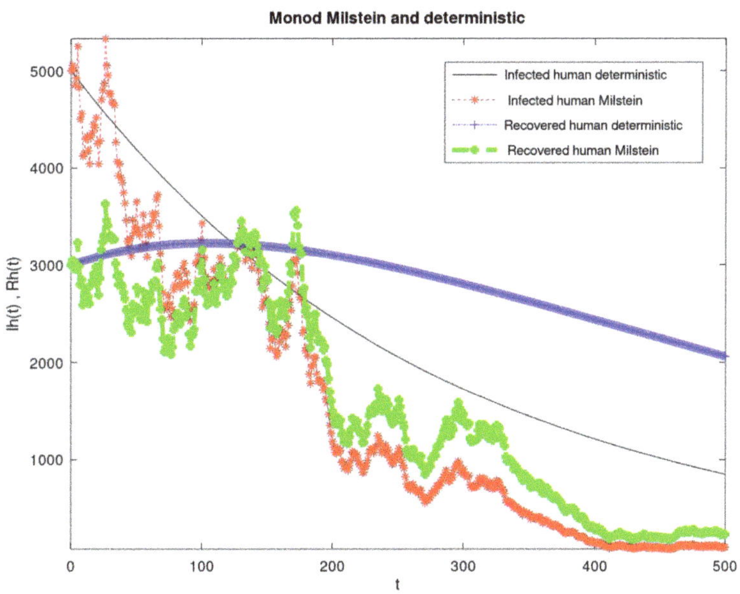


(b)

**Figure 5.11** Simulation showing the dynamic behaviour of the susceptible human, exposed human population, infected human, recovered human population.

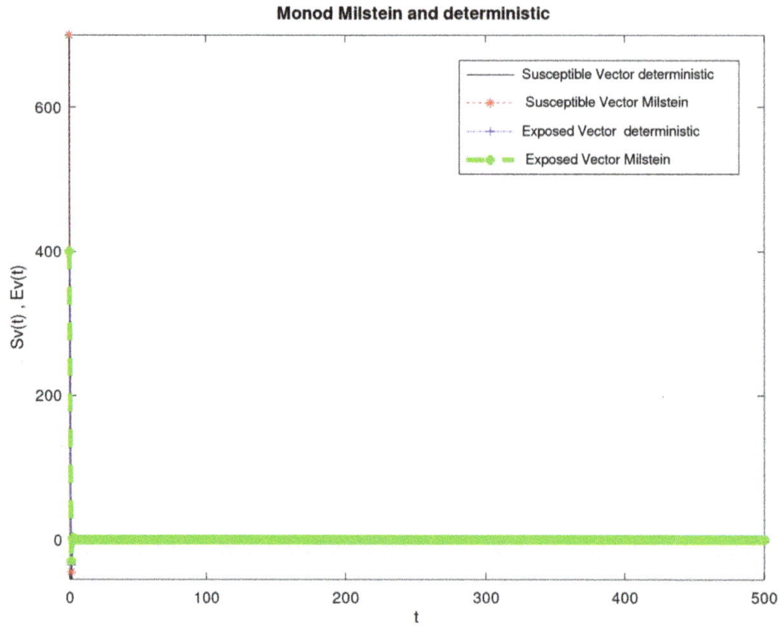


(a)

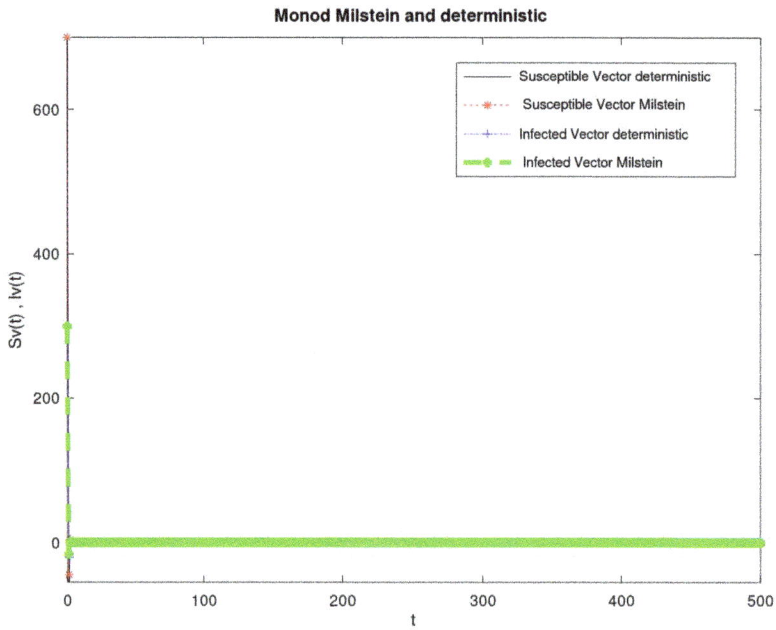


(b)

**Figure 5.12** Simulation showing the dynamic behaviour of the susceptible vector, exposed vector population and infected vector population.

conditions were determined. It was observed that the centre manifold theory showed that the model exhibited a backward bifurcation. The elasticity analysis predicted that in order to eliminate Leptospirosis, the best strategy to employ is to apply all the available multiple control interventions but in areas with little resources, treatment with antibiotics and pesticides/rodent control will be the best options.

We applied the homotopy analysis method to approximately solve the system of nonlinear equations of Leptospirosis disease dynamics, specifically. We discovered that from our results the strength and efficiency of homotopy analysis method in providing solution to nonlinear problems is very powerful and dependable. Moreover, the homotopy analysis method established a convenient means of modifying and controlling the convergence space and rate of approximation series when necessary.

We also studied numerically the behaviour of our system (5.10) using some of the parameter values compatible with Leptospirosis as given above in table 5.6 and found that there is a decrease of Leptospirosis disease in the population. Also, the computational value of $R_0 = 0.002\,28$ which is far less than 1, implies that the Leptospirosis will actually die out and not remain endemic in the population. This shows that the preventive and the treatment-based control interventions predicted a decrease in the transmission of Leptospirosis in the population over time. It is our opinion from this result that if the local National Control program on diseases can implement the progressive and active use of this preventive and treatment-based control interventions Leptospirosis will be eliminated in the population and mathematical modelling will become an integral part of decision and policy making in the global and public health decision making.

Finally, the deterministic model (5.10) was converted to stochastic model in (5.30) were compared with the corresponding deterministic model. It was observed that the level of infected human population than the result from their corresponding deterministic model simulation.

## References

[1] World Health Organization 2010 *Report of the First Meeting of the Leptospirosis Burden Epidemiology Reference Group*

[2] Costa F, Hagan J E, Calcagno J, Kane M, Torgerson P, Martinez-Silveira M S, Stein C, Abela-Ridder B and Ko A I 2015 Global morbidity and mortality of Leptospirosis: a systematic review. *PLoS Negl. Trop. Dis.* **9** e0003898

[3] Solomon F *et al* 1994 *Leptospira and Leptospirosis* (Boca Raton, FL: CRC Press)

[4] Ellis W A, Brien J J, Neill S D and Bryson D G 1986 Bovine Leptospirosis: experimental serovar hardjo infection *Vet. Microbiol.* **11** 293–9

[5] Carole A B and Paul K 1988 Human-to-human transmission of *Leptospira interrogans* by milk *J. Infect. Dis.* **158** 246–7

[6] Pongsuumpun P, Miami T and Kongnuy R 2008 Age structural transmission model for Leptospirosis *Proc. 3rd Int. Symp. on Biomedical Engineering (Changsha, China, 8-10 June 2008)* pp 411–6

[7] Anderson R M and May R M 1982 *Population Biology of Infectious Diseases: Report of the Dahlem Workshop on Population Biology of Infectious Disease Agents* (Berlin: Springer) pp 14–9

[8] Bal A E, Gravekamp C, Hartskeerl R A, De Meza-Brewster J, Korver H and Terpstra W J 1994 Detection of leptospires in urine by PCR for early diagnosis of Leptospirosis *J. Clin. Microbiol.* **32** 1894–8
[9] Bharti A R *et al* 2003 Leptospirosis: a zoonotic disease of global importance *Lancet Infect. Dis.* **3** 757–71
[10] Bakare E A and Snehahish C 2017 Homotopy analysis based solution of a malaria transmission dynamics model *Int. J. Ecol. Econ. Stat.* **38** 120–43
[11] Holt J, Davis S and Leirs H 2006 A model of Leptospirosis infection in an African rodent to determine risk to humans: seasonal fluctuations and the impact of rodent control *Acta Trop.* **99** 218–25
[12] Khan M A, Islam S, Ullah M, Khan S A, Zaman G and Saddiq S F 2013 Analytical solution of the Leptospirosis epidemic model by homotopy perturbation method *Res. J. Recent Sci.* **2** 66–71
[13] Khan M A, Islam S and Khan S A 2014 Mathematical modeling towards the dynamical interaction of Leptospirosis *Appl. Math. Inf. Sci.* **8** 1049
[14] Khan M A, Saddiq S F, Islam S, Khan I and Ching D L C 2014 Epidemic model of Leptospirosis containing fractional order *Abstr. Appl. Anal.* **2014** 317201
[15] Khan M A, Saddiq S F, Islam S, Khan I and Shafie S 2016 Dynamic behavior of Leptospirosis disease with saturated incidence rate *Int. J. Appl. Comput. Math.* **2** 435–52
[16] Pimpunchat B, Wake G C, Modchang C, Triampo W and Babylon A M 2013 Mathematical model of Leptospirosis: linearized solutions and stability analysis *Appl. Math.* **4** 77
[17] Pongsumpun P 2012 Mathematical model for the transmission of Leptospirosis in juvennile and adults humans *Int. J. Math. Comput. Sci.* **6** 1639
[18] LaSalle J P 1976 *The Stability of Dynamical Systems* vol 25 (Philadelphia, PA: SIAM)
[19] Castillo-Chavez C and Song B 2004 Dynamical models of tuberculosis and their applications *Math. Biosci. Eng.* **1** 361–404
[20] Tangkanakul W, Tharmaphornpil P, Plikaytis B D, Bragg S, Poonsuksombat D, Choomkasien P, Kingnate D and Ashford D A 2000 Risk factors associated with Leptospirosis in Northeastern Thailand *Am. J. Trop. Med. Hyg.* **63** 204–8
[21] Zaman G, Khan M A, Islam S, Chohan M I and Jung I H 2012 Modeling dynamical interactions between Leptospirosis infected vector and human population *Appl. Math. Sci.* **6** 1287–302
[22] Oksendal B 2000 *Stochastic Differential equations. An Introduction with Applications* 5th edn (New York: Springer)
[23] Liao S J 1992 The proposed homotopy analysis method for the solutions of nonlinear problems *PhD Thesis* Shanghai Jiao Tong University, Shanghai, China
[24] Ideris S H, Malim M R and Shaadan N 2019 A stochastic model for relative risk estimation of Leptospirosis in Malaysia *J. Phys.: Conf. Ser.* **1366** 012118

**IOP** Publishing

Polynomial Paradigms
Trends and applications in science and engineering
**S Chakraverty**

# Chapter 6

# Polynomials based semi-analytical methods for the solutions of fractional order Volterra–Fredholm integro differential equations

**Jugal Mohapatra and Abhilipsa Panda**

## Symbols

| | |
|---|---|
| $\mu$ | Order of derivative |
| $\mathcal{B}$ | Bernstein polynomials |
| $\mathcal{T}$ | Chebyshev polynomials |
| $\mathbb{D}^{\mu}$ | Fractional derivative |
| $\mathbb{I}^{\mu}$ | Fractional integral |
| $z$ | Unknown function |
| $\mathcal{K}$ | Kernel function |
| $\mathcal{N}$ | Nonlinear functions |
| $\mathcal{A}_n$ | Adomian polynomials |
| $\mathcal{T}(x)$ | Chebyshev polynomials |
| $\mathcal{B}(x)$ | Bernstein polynomials |
| $\Omega$ | Domain |
| $\mathbb{N}$ | Natural numbers |
| $\mathbb{R}$ | Real numbers |
| $\mathbb{R}^+$ | Positive real numbers |
| $C([0, 1])$ | Continuous functions on [0, 1] |
| $B([0, 1])$ | Banach space of continuous functions on [0, 1] |
| $f_C(x)$ | Approximation of $f(x)$ using Chebyshev polynomials |
| $f_B(x)$ | Approximation of $f(x)$ using Bernstein polynomials |
| $f_T(x)$ | Approximation of $f(x)$ using Taylor's polynomials |
| $\mathbb{E}_n^{\infty}$ | Absolute pointwise errors |
| $C_1$, $C_2$ | Lipschitz constants |
| IDE | Integro-differential equations |
| FIDE | Fractional integro-differential equations |

 

| | |
|---|---|
| ADM | Adomian decomposition method |
| VIM | Variational iteration method |
| CP | Chebyshev polynomials |
| BP | Bernstein polynomials |

This work introduces the series solution for the linear and nonlinear fractional order Volterra–Fredholm integro-differential equations using Chebyshev and Bernstein polynomials in the Adomian decomposition method. The convergence and uniqueness of the solution is proved. Reliability and efficiency of the proposed technique is depicted using the tabular datas and the plots. Comparison shows that the proposed method is highly productive and has a wide applicability.

## 6.1 Introduction

The considerable interest in integro-differential equations (IDEs) have mainly arisen due to its major applications in the theory of mechanical engineering [1], elasticity [2] and several others. The well-known mathematician Niels Henrik Abel obtained the famous integral equation of the first kind with kernel function $\mathcal{K}(x, t) = (x - t)^{-\mu}$, for $\mu = 1/2$ by solving the mechanical problem of *Tautochrone* as described in [3], which he then generalized for $0 < \mu < 1$. The theory given by Abel in [3] further paved the way for researchers to look deep into the idea of fractional order integro-differential equations (FIDEs). The wide application of FIDEs for electromagnetic waves in dielectric media and unsteady aerodynamics have generated great interest to explore more in this field. Several analytical and numerical techniques have been introduced to obtain the solutions of FIDEs with the singular and nonsingular kernels. Aghajani *et al* established the existence of solutions of FIDEs in [4]. The operational tau approximation method based on orthogonal polynomials was implemented on a class of FIDEs by Vanani and Aminataei in [5]. Heydari and Hooshmandasl [6] used the Chebyshev wavelet method to solve the nonlinear FIDEs on a large interval by converting the fractional differential and integral parts of the FIDE to some operational matrices. Then, they obtained the solution by solving a set of algebraic equations. Higher order FIDEs, such as the fourth-order FIDEs were solved by Amer *et al* [7] using the ADM and VIM, where the solution was given by an infinite convergent series. Also, quite a few approximated techniques described in [8, 9] have been discussed in the past to solve the linear and nonlinear FIDEs in [7, 10].

But all the model problems solved have considered the source term as a polynomial function which is comparatively easier to approximate. So, we propose a new modification of ADM for obtaining the solution of a class of FIDE where the source function is not a polynomial one. The general way of ADM was first introduced by G Adomian [11] to solve the linear and nonlinear problems. Gradually, ADM was improvised using the Chebyshev [12], Legendre [13] and Bernstein polynomials [14]. These modified techniques are used to solve a class of ordinary and partial differential equations where the source function is exponential, trigonometric, or hyperbolic functions rather than the polynomial one. The approximation of functions by polynomials is extremely important as different scientific experiments rely on them, such as the study of statistics in population dynamics [15], temperatures, and also in the approximation theory [16]. Moreover, polynomials are the best mathematical techniques to approximate as they can be characterized, figured, separated, and incorporated effortlessly. Orthogonal polynomials such as the Chebyshev polynomials have been widely used in approximating functions in a wide variety of problems. These are the eigen functions of singular Sturm–Liouville problems. It is well known that these eigen functions allow the approximation of functions in $C^{\infty}[a, b]$, where the truncation error approaches zero faster than any function used in the approximation as described in [17]. Gottlieb in [18] described this effect as the 'spectral accuracy'. For more information, one may refer [19, 20]. In this chapter, we are using the first kind orthogonal Chebyshev polynomials $\{\mathcal{T}_k\}_{k=0}^{\infty}$ given as

$$\sqrt{1 - x^2}\mathcal{T}'(x) + \frac{k^2}{\sqrt{1 - x^2}}\mathcal{T}_k(x) = 0.$$

Also, we have used the Bernstein polynomials for the modification of ADM. These polynomials approximate the function with a few number of terms in comparison to the approximations done using the Taylor series. They are utilized in the fields of connected arithmetic and material science as well as computer-assisted geometric outlines. They are also used in conjunction with other techniques like the Galerkin and collocation methods to solve some differential and integral problems. In addition, the proposed modified ADM is numerically performed through Matlab programming. The obtained results show the advantage of using the modified ADM over the classical ADM.

The chapter is structured as follows: section 6.2 describes some of the definitions and properties. In section 6.3, the model problem is defined and a detailed study on the proposed methods are given in section 6.4. Section 6.5 shows the convergence analysis and the error bounds. Numerical tests are performed in section 6.6 with the demonstration of graphs and tables. Finally, section 6.7 draws the concluding remarks.

## 6.2 Some definitions and properties

**Definition 6.1.** The Bernstein basis polynomials of degree $m$ over the interval [0, 1] are defined as:

$$\mathcal{B}_{i,m}(x) = \binom{m}{i} x^i(1 - x)^{m-i}.$$

**Definition 6.2.** The Riemann–Liouville fractional integral of order $\mu > 0$ for a function $f(x)$ is defined as:

$$\mathbb{I}^{\mu}f(x) = \frac{1}{\Gamma(\mu)} \int_0^x (x - \tau)^{\mu-1}f(\tau)d\tau.$$

**Definition 6.3.** The Caputo derivative of order $\mu \in \mathbb{R}^+$ for a function $f(x)$ is defined as:

$$\mathbb{D}^{\mu}f(x) = \begin{cases} \dfrac{1}{\Gamma(n-\mu)} \displaystyle\int_0^x (x-\tau)^{n-\mu-1} f(\tau) d\tau, & n-1 < \mu < n, \\ \dfrac{d^n f(x)}{dx^n}, & \mu = n, \quad n \in \mathrm{N}. \end{cases} \tag{6.1}$$

Some of the important properties of fractional derivatives and integrals are discussed as follows:

- $\mathbb{D}^{\mu}\mathbb{I}^{\mu}f(x) = f(x)$ and $\mathbb{I}^{\mu}\mathbb{D}^{\mu}f(x) = f(x) - f(0+)$.
- Linearity property is sustained while defining the derivative in Caputo sense, given as:

$$\mathbb{D}^{\mu}(\psi_1 m(x) + \psi_2 n(x)) = \psi_1 \mathbb{D}^{\mu} m(x) + \psi_2 \mathbb{D}^{\mu} n(x).$$

- For $0 < \mu \leqslant 1$ and $\vartheta \in \mathbb{R}$, $\mathbb{I}^{\mu}x^{\vartheta} = \frac{\Gamma(\vartheta+1)}{\Gamma(\mu+\vartheta+1)}$, $\mu > 0$, $\vartheta > -1$, $x > 0$.

## 6.3 Model problem

Consider the Volterra–Fredholm FIDE of order $\mu$:

$$\begin{cases} \mathbb{D}^{\mu}z(x) + a(x)z(x) = f(x) + \displaystyle\int_0^x \mathcal{K}_1(x, s)\mathcal{N}_1(z(s)) \\ \qquad\qquad ds + \displaystyle\int_0^1 \mathcal{K}_2(x, s)\mathcal{N}_2(z(s))ds, \\ z(0) = z_0. \end{cases} \tag{6.2}$$

Here $0 < \mu \leqslant 1$, $x \in [0, 1]$, and the fractional order derivative $\mathbb{D}^{\mu}$ is defined in Caputo sense which is assumed to be invertible. The source function $f(x)$, $a(x)$, and the nonlinear operators denoted by $\mathcal{N}_1$ and $\mathcal{N}_2$ are continuous functions on [0, 1]. $\mathcal{K}_1(x, s)$ and $\mathcal{K}_2(x, s)$ are smooth kernel functions defined on ([0, 1] × [0, 1]). The given initial condition is symbolized as $z_0$. Throughout this chapter, for a function $f(x)$, defined on $\Omega = [0, 1]$, we define $\|f(x)\|_{\infty} = \max_{x\in\Omega}|f(x)|$ and $C$ is defined as a generic constant, independent of $\mu$.

## 6.4 Methodology

### 6.4.1 Adomian decomposition method (ADM)

A brief description of the modified ADM is discussed in this section. Consider the FIDE (6.2). The nonlinear operator is approximated using the Adomian polynomials $\mathcal{A}_n$. One may refer [9, 11] for the formula of $\mathcal{A}_n$. The solution $z$ is represented as a series solution given by $z = \sum_{n=0}^{\infty} z_n$. Operating the inverse operator $\mathbb{I}^{\mu}$ on both sides of (6.2), we get

$$z(x) = \mathbb{I}^{\mu}(f(x)) - \mathbb{I}^{\mu}\left(a(x)z_0(x) + \int_0^x \mathcal{K}_1(x, s)\mathcal{N}_1(z(s))ds + \int_0^1 \mathcal{K}_2(x, s)\mathcal{N}_2(z(s))ds\right). \quad (6.3)$$

Following the classical ADM, the recurrence relation for the solution of (6.3) is obtained as:

$$\begin{cases} z_0 = \mathbb{I}^{\mu}(f(x)) + z(0^+), \\ z_1 = \mathbb{I}^{\mu}\left(-a(x)z_0(x) + \int_0^x \mathcal{K}_1(x, s)\mathcal{A}_1(z_0(s))ds + \int_0^1 \mathcal{K}_2(x, s)\mathcal{A}_2(z_0(s))ds\right), \\ z_2 = \mathbb{I}^{\mu}\left(-a(x)z_1(x) + \int_0^x \mathcal{K}_1(x, s)\mathcal{A}_1(z_1(s))ds + \int_0^1 \mathcal{K}_2(x, s)\mathcal{A}_2(z_1(s))ds\right), \\ \vdots \end{cases} \quad (6.4)$$

Finally, we calculate the solution as $z = \sum_{n=0}^{\infty} z_n$, if $\sum_{n=0}^{\infty} z_n$ converges.

### 6.4.2 ADM based on Chebyshev polynomials (ADM-CP)

In usual algorithm of ADM, the approximation of $f(x)$ is done using the Taylor's series expansion as $f(x) = \sum_{i=0}^{n} \frac{f^n(0)}{n!} x^n$ for an arbitrary $\mathbb{N}$. Hosseini [12] modified the ADM by expanding $f(x)$ using the Chebyshev polynomial approximation,

$$f_C(x) \approx \sum_{i=0}^{n} C_n \mathcal{T}_n(x), \quad (6.5)$$

where $\mathcal{T}_n(x)$ is the first kind orthogonal Chebyshev polynomial. Some of the Chebyshev polynomials are noted below.

$$\begin{cases} \mathcal{T}_0(x) = 1, \\ \mathcal{T}_1(x) = x, \\ \mathcal{T}_2(x) = 2x^2 - 1, \\ \mathcal{T}_3(x) = 4x^3 - 3x, \\ \vdots \\ \mathcal{T}_{n+1}(x) = 2x\mathcal{T}_n - \mathcal{T}_{n-1}, \quad n \geqslant 1. \end{cases} \quad (6.6)$$

Using (6.4) and (6.5), the following approximations for the solution of (6.2) are obtained as

$$\begin{cases} z_0 = \mathbb{I}^{\mu}(C_0\mathcal{T}_0(x) + C_1\mathcal{T}_1(x) + C_2\mathcal{T}_2(x)\ldots C_n\mathcal{T}_n(x)) + z(0^+), \\ z_1 = \mathbb{I}^{\mu}\left(-a(x)z_0(x) + \int_0^x \mathcal{K}_1(x, s)\mathcal{A}_1(z_0(s))ds + \int_0^1 \mathcal{K}_2(x, s)\mathcal{A}_2(z_0(s))ds\right), \\ z_2 = \mathbb{I}^{\mu}\left(-a(x)z_1(x) + \int_0^x \mathcal{K}_1(x, s)\mathcal{A}_1(z_1(s))ds + \int_0^1 \mathcal{K}_2(x, s)\mathcal{A}_2(z_1(s))ds\right), \\ \vdots \end{cases} \quad (6.7)$$

This work will prove that the approximated solution obtained by (6.7) is more reliable than any other existing methods. In addition, One may also approximate using the following algorithm as described in [12]

$$\begin{cases} z_0 = \mathbb{I}^{\mu}(C_0\mathcal{T}_0(x)) + z(0^+), \\ z_1 = \mathbb{I}^{\mu}\left( C_1\mathcal{T}_1(x) - a(x)z_0(x) + \int_0^x \mathcal{K}_1(x, s)\mathcal{A}_1(z_0(s)) \right. \\ \qquad \left. ds + \int_0^1 \mathcal{K}_2(x, s)\mathcal{A}_2(z_0(s))ds \right), \\ z_2 = \mathbb{I}^{\mu}\left( C_2\mathcal{T}_2(x) - a(x)z_1(x) + \int_0^x \mathcal{K}_1(x, s)\mathcal{A}_1(z_1(s)) \right. \\ \qquad \left. ds + \int_0^1 \mathcal{K}_2(x, s)\mathcal{A}_2(z_1(s))ds \right), \\ \vdots \end{cases} \tag{6.8}$$

Now (6.5) can also be written in the standard form as

$$f(x) \approx p_0 + p_1x + p_2x^2 + \cdots p_rx^r,$$

$$\begin{bmatrix} p_0 \\ p_1 \\ p_2 \\ \vdots \\ p_r \end{bmatrix} = \begin{bmatrix} 1 & 0 & -1 & 0 & 1 & 0\ldots \\ 0 & 1 & 0 & -3 & 0 & 5\ldots \\ 0 & 0 & 2 & 0 & -8 & 0\ldots \\ 0 & 0 & 0 & 4 & 0 & -20\ldots \\ & & & & & \vdots \end{bmatrix} \begin{bmatrix} C_0 \\ C_1 \\ C_2 \\ \vdots \\ C_r \end{bmatrix},$$

and

$$\begin{cases} z_0 = \mathbb{I}^{\mu}(p_0) + z(0^+), \\ z_1 = \mathbb{I}^{\mu}\left( p_1 - a(x)z_0(x) + \int_0^x \mathcal{K}_1(x, s)\mathcal{A}_1(z_0(s))ds + \int_0^1 \mathcal{K}_2(x, s)\mathcal{A}_2(z_0(s))ds \right), \\ z_2 = \mathbb{I}^{\mu}\left( p_2 - a(x)z_1(x) + \int_0^x \mathcal{K}_1(x, s)\mathcal{A}_1(z_1(s))ds + \int_0^1 \mathcal{K}_2(x, s)\mathcal{A}_2(z_1(s))ds \right), \\ \vdots \end{cases}$$

Finally using (6.8), the series solution $z(x)$ is obtained as follows:

$$z(x) = z_0(x) + z_1(x) + z_2(x)\ldots \tag{6.9}$$

### 6.4.3 ADM based on Bernstein polynomials (ADM-BP)

In this segment, in order to improve the accuracy and reliability of ADM, the source function is expressed in the form of the Bernstein polynomial approximation,

$$f_B(x) = \sum_{i=0}^{n} D_i\mathcal{B}_i(x), \tag{6.10}$$

where $\mathcal{B}_i(x)$ are the Bernstein polynomials. Using (6.3) and (6.10), the approximated solution for FIDE (6.2) is obtained as follows:

$$\begin{cases} z_0 = \mathbb{I}^{\mu}(D_0\mathcal{B}_0(x) + D_1\mathcal{B}_1(x) + D_2\mathcal{B}_2(x)\ldots D_n\mathcal{B}_n(x)) + z(0^+), \\ z_1 = \mathbb{I}^{\mu}\left(-a(x)z_0(x) + \int_0^x \mathcal{K}_1(x, s)\mathcal{A}_1(z_0(s))ds + \int_0^1 \mathcal{K}_2(x, s)\right. \\ \qquad \left.\mathcal{A}_2(z_0(s))ds\right), \\ z_2 = \mathbb{I}^{\mu}\left(-a(x)z_1(x) + \int_0^x \mathcal{K}_1(x, s)\mathcal{A}_1(z_1(s))ds + \int_0^1 \mathcal{K}_2(x, s)\right. \\ \qquad \left.\mathcal{A}_2(z_1(s))ds\right), \\ \vdots \end{cases} \tag{6.11}$$

The Bernstein polynomials of degree $m$ are obtained as

$$\mathcal{B}_m f(x) = \sum_{i=0}^{m} \binom{m}{i} x^i (1 - x)^{(m-i)} f\left(\frac{i}{m}\right).$$

For each function $f$: $[0, 1] \to \mathrm{R}$, we have $\lim_{m\to\infty}\mathcal{B}_m f(x) = f(x)$. Finally using (6.11), the solution is obtained as

$$z(x) = z_0(x) + z_1(x) + z_2(x)\ldots \tag{6.12}$$

## 6.5 Analysis of the proposed methods

### 6.5.1 Existence and uniqueness of the solution

In this segment, some of the hypothesis are stated which will be further used in the analysis.

**H1** Consider two Lipschitz constants $(C_1, C_2) > 0$ such that $\mathcal{N}_1(z(x))$ and $\mathcal{N}_2(z(x))$ satisfy the Lipschitz condition given as

$$\begin{cases} \|\mathcal{N}_1(z_1(x)) - \mathcal{N}_1(z_2(x))\| \leqslant C_1\|z_1 - z_2\|, \\ \|\mathcal{N}_2(z_1(x)) - \mathcal{N}_2(z_2(x))\| \leqslant C_2\|z_1 - z_2\|. \end{cases}$$

**H2** Consider $\mathcal{Q} = \{(x, t) \in \mathbb{R} \times \mathbb{R}: 0 \leqslant t \leqslant x \leqslant 1\}$ and $\mathcal{K}_1^*, \ \mathcal{K}_2^* \in C(\mathcal{Q}, \mathbb{R}^+)$, such that

$$\mathcal{K}_1^* = \sup_{x\in[0,1]} \int_0^x |\mathcal{K}_1(x, s)|dt < \infty, \quad \mathcal{K}_2^* = \sup_{x\in[0,1]} \int_0^x |\mathcal{K}_2(x, s)|dt < \infty.$$

**Theorem 6.1.** Assuming that **(H1)** and **(H2)** hold, if $\left(\frac{\mathcal{K}_1^*C_1+\mathcal{K}_2^*C_2}{\Gamma(\mu+1)}\right) < 1$, then there exists a unique solution $z(x) \in C[0, 1]$ for (6.2).
Proof. The proof of the above theorem is well explained in theorem 7 of [21]. Here we provide the outline of the proof in very few lines. Applying $\mathbb{I}^\mu$ on both sides of (6.2) we get, $z(x) = Tz(x)$, where

$$(Tz)(x) = z_0 + \mathbb{I}^\mu\left[-a(x)z(x) + f(x) + \int_0^x \mathcal{K}_1(x, s)\mathcal{N}_1(z(s))ds - \int_0^1 \mathcal{K}_2(x, s)\mathcal{N}_2(z(s))ds\right].$$

Since, we know $z_1(x)$, $z_2(x)$, $a(x) \in C[0, 1]$, so

$$\begin{aligned}|(Tz_1)(x) - (Tz_2)(x)| &\leqslant \frac{1}{\Gamma(\mu)}\int_0^x (x-s)^{\mu-1}\left[a(x)|z_1(x) - z_2(x)| + \int_0^t |\mathcal{K}_1(x, s)\right. \\ &\quad \left.|(|\mathcal{N}_1(z_1(s)) - \mathcal{N}_1(z_2(s))|)ds + \int_0^1 |\mathcal{K}_2(x, s)|(|\mathcal{N}_2(z_1(s)) - \mathcal{N}_2(z_2(s))|)ds\right]dt, \\ &\leqslant \frac{\mathcal{K}_1^*}{\Gamma(\mu)}\int_0^x (x-s)^{\mu-1}\left[\int_0^t \mathcal{N}_1|z_1(s) - z_2(s)|ds\right]dt + \frac{\mathcal{K}_2^*}{\Gamma(\mu)}\int_0^x (x-s)^{\mu-1} \\ &\quad \left[\int_0^1 \mathcal{N}_2|z_1(s) - z_2(s)|ds\right]dt, \\ &\leqslant \frac{\mathcal{K}_1^*C_1 + \mathcal{K}_2^*C_2}{\Gamma(\mu+1)}|z_1 - z_2|.\end{aligned}$$

As we know that,

$$\frac{\mathcal{K}_1^*C_1 + \mathcal{K}_2^*C_2}{\Gamma(\mu+1)} < 1,$$

so, $\|T(z_1(x)) - T(z_2(x))\| \leqslant \|z_1 - z_2\|$. This proves that $T$ is a contraction mapping in Banach space. So, we can conclude that (6.2) has a unique solution using the Banach contraction principle.

**Theorem 6.2.** Suppose $B([0, 1], \|\cdot\|)$ is the Banach space of all continuous functions on $\Omega$. Then $z = \sum_{i=0}^{\infty} z_i(x)$ uniformly converges to the exact solution on [0, 1].

Proof. As proved in [10], consider $|z_1(x)| < \infty$ for all $x \in [0, 1]$. The sequence of partial sum of the series is denoted as $s_p$. Let $s_p$ and $s_q$ be arbitrary partial sums with $p \geqslant q$. We need to prove that $s_p = \sum_{i=0}^{p} z_i(x)$ is a Cauchy sequence in $C([0, 1], \|\cdot\|)$.

$$
\begin{aligned}
\left\|s_p - s_q\right\|_\infty &= \max_{x\in[0,1]}|s_p - s_q| \\
&= \max_{x\in[0,1]}\left|\sum_{i=0}^{p} z_i(x) - \sum_{i=0}^{q} z_i(x)\right| = \left|\sum_{i=p+1}^{q} z_i(x)\right| \\
&= \max_{x\in[0,1]}\left(\left|\int_0^t \mathcal{K}_1(x, s)(\mathcal{N}_1(s_{p-1}) - \mathcal{N}_1(s_{q-1}))ds\right.\right. \\
&\quad \left.\left. + \int_0^1 \mathcal{K}_2(x, s)(\mathcal{N}_2(s_{p-1}) - \mathcal{N}_2(s_{q-1}))ds\right| dt\right) \\
&\leqslant \max_{x\in[0,1]}\left(\frac{1}{\Gamma(\mu)}\left|\int_0^t \mathcal{K}_1(x, s)(\mathcal{N}_1(s_{p-1}) - \mathcal{N}_1(s_{q-1}))ds\right.\right. \\
&\quad \left.\left. + \int_0^1 \mathcal{K}_2(x, s)(\mathcal{N}_2(s_{p-1}) - \mathcal{N}_2(s_{q-1}))ds\right| dt\right) \\
&\leqslant \frac{1}{\Gamma(\mu + 1)}\left[\mathcal{K}_1^* C_1\left\|s_{p-1} - s_{q-1}\right\|_\infty + \mathcal{K}_2^* C_2\left\|s_{p-1} - s_{q-1}\right\|_\infty\right] \\
&= \left(\frac{\mathcal{K}_1^* C_1 + \mathcal{K}_2^* C_2}{\Gamma(\mu + 1)}\right)\|s_{p-1} - s_{q-1}\|_\infty = \gamma_1\|s_{p-1} - s_{q-1}\|_\infty
\end{aligned}
$$

where $\gamma = \left(\frac{\mathcal{K}_1^* C_1 + \mathcal{K}_2^* C_2}{\Gamma(\mu + 1)}\right)$. Also for $p = q + 1$,

$$
\begin{aligned}
\left\|s_p - s_q\right\|_\infty &\leqslant \left\|s_{q+1} - s_q\right\|_\infty + \left|s_{q+2} - s_{q+1}\right|_\infty + \cdots + \left|s_p - s_{p-1}\right|_\infty \\
&\leqslant [\gamma_1^q + \gamma_1^{q+1} + \cdots + \gamma_1^{p-1}]\left\|s_1 - s_0\right\|_\infty \\
&\leqslant \gamma_1^q\left(\frac{1 - \gamma_1^{p-q}}{1 - \gamma_1}\right)\|z_1\|_\infty .
\end{aligned}
$$

Since, $0 < \gamma_1 < 1$, we have $(1 - \gamma_1^{p-q}) < 1$, then

$$
\|s_p - s_q\|_\infty \leqslant \frac{\gamma_1^q}{1 - \gamma_1}\|z_1\|_\infty .
$$

As $\|z_1(x)\| < \infty$ and $m \to \infty$, we get $\left\|s_p - s_q\right\|_\infty \to 0$. Hence, it can be concluded that $s_p$ is a Cauchy sequence in $B[0, 1]$ and $z = \lim_{n\to\infty} z_n$. Thus, the series is proved to be convergent by Weierstrass $M$-test.

### 6.5.2 Error bound

The exact solution for (6.2) is given by $z(x) = \lim_{N\to\infty}\Psi_N$ and the numerical solution can be obtained by truncating the series (6.9) and (6.12) up to a finite number of terms If $\Psi_N$ gives the $N$ terms approximated solution then, the absolute pointwise error bound depends on the partial sum $\sum_{n=0}^{N-1} z_n(x)$ which is bounded by $\frac{\mathcal{M}\gamma_1^N}{1-\gamma_1}$. $\gamma_1$ is defined in theorem 6.2 which satisfies $0 < \gamma_1 < 1$ and $z_0 \leqslant \mathcal{M}$.

## 6.6 Numerical experiments

This section consists of two numerical example which clearly depicts the efficiency of the modified technique.

**Example 6.1.** Consider the following model

$$\mathbb{D}^{\mu} z(x) = f(x) + \int_0^x \mathcal{K}_1(x, s) z(s) ds + \int_0^1 \mathcal{K}_2(x, s) z(s) ds, \tag{6.13}$$

with the initial condition $z(0) = 1$. Here $\mathcal{K}_1(x, t) = 1$, $\mathcal{K}_2(x, t) = 2t - 1$ and $f(x) = \exp(x) - 1 + x^{1-\mu} E_{1,2-\mu}(x)$. The exact solution is $z(x) = \exp(x)$.

First approximating $f(x)$ using the Chebyshev polynomials, we get,

$$f_C(x) = \sum_{i=0}^{5} C_i T_i(2x - 1), \quad 0 \leqslant x \leqslant 1,$$

where

$$C_0 = \frac{1}{\pi} \int_{-1}^{1} \frac{f(0.5x + 0.5) T_0(x)}{\sqrt{1 - x^2}} dx$$

and

$$C_i = \frac{2}{\pi} \int_{-1}^{1} \frac{f(0.5x + 0.5) T_i(x)}{\sqrt{1 - x^2}} dx, \quad i = 1, 2, \dots, 6.$$

It implies that

$$\begin{aligned} f_C(x) \approx{} & \frac{x^{1-\mu}}{\Gamma(2-\mu)} + \frac{x^{2-\mu}}{\Gamma(3-\mu)} + \frac{x^{3-\mu}}{\Gamma(4-\mu)} + \frac{x^{4-\mu}}{\Gamma(5-\mu)} + \frac{x^{5-\mu}}{\Gamma(6-\mu)} \\ & + \frac{x^{6-\mu}}{\Gamma(7-\mu)} - 1.000\,2x \\ & - 0.499\,197x^2 - 0.166\,489x^3 - 0.043\,7939x^4 - 0.008\,686\,82x^5 \\ & - 0.000\,04. \end{aligned} \tag{6.14}$$

Applying (6.7) we obtain the two term approximated solution as

$$\begin{aligned}
z(x) &= z_0(x) + z_1(x)\\
&= 1 + x + \frac{x^2}{2!} + \frac{x^3}{3!} + \frac{x^4}{4!} + \frac{x^5}{5!} + \frac{x^6}{6!} - \frac{1.000\,2(x^{1+\mu})}{\Gamma(2+\mu)} - 0.499\,197\frac{(x^{2+\mu})}{\Gamma(3+\mu)}\\
&- 0.166\,489\frac{(x^{3+\mu})}{\Gamma(4+\mu)} - 0.043\,7939\frac{(x^{4+\mu})}{\Gamma(5+\mu)} - 0.008\,6862\frac{(x^{5+\mu})}{\Gamma(6+\mu)}\\
&- 0.000\,04\frac{(x^{\mu})}{\Gamma(1+\mu)} + 0.000\,198\,413\frac{x^{7+\mu}\Gamma(8)}{\Gamma(8+\mu)} + 0.001\,388\,89\frac{x^{6+\mu}\Gamma(7)}{\Gamma(7+\mu)}\\
&+ 0.008\,333\,33\frac{x^{5+\mu}\Gamma(6)}{\Gamma(6+\mu)} + 0.041\,6667\frac{x^{4+\mu}\Gamma(5)}{\Gamma(5+\mu)} + 0.166\,667\frac{x^{3+\mu}\Gamma(4)}{\Gamma(4+\mu)}\\
&+ 0.5\frac{x^{2+\mu}\Gamma(3)}{\Gamma(3+\mu)} + \frac{x^{1+\mu}\Gamma(2)}{\Gamma(2+\mu)} - 1.000\,2\frac{x^{2+2\mu}\Gamma(3+\mu)}{\Gamma(3+2\mu)(2+\mu)} - 0.499\,197\frac{x^{3+2\mu}\Gamma(4+\mu)}{\Gamma(4+2\mu)(3+\mu)}\\
&- 0.166\,489\frac{x^{4+2\mu}\Gamma(5+\mu)}{\Gamma(5+2\mu)(4+\mu)} - 0.043\,7939\frac{x^{5+2\mu}\Gamma(6+\mu)}{\Gamma(6+2\mu)(5+\mu)}\\
&- 0.008\,6862\frac{x^{6+2\mu}\Gamma(7+\mu)}{\Gamma(7+2\mu)(6+\mu)} - 0.000\,04\frac{x^{1+2\mu}\Gamma(2+\mu)}{\Gamma(2+2\mu)(1+\mu)}.
\end{aligned}$$

Simplifying the problem using Bernstein polynomials $\sum_{i=0}^{n} \mathcal{D}_i \mathcal{B}_i(x)$ with $i = 5$ gives the approximation for $f(x)$ as

$$\begin{aligned}
f_B(x) \approx 1 &+ \frac{x^{1-\mu}}{\Gamma(2-\mu)} + \frac{x^{2-\mu}}{\Gamma(3-\mu)} + \frac{x^{3-\mu}}{\Gamma(4-\mu)} + \frac{x^{4-\mu}}{\Gamma(5-\mu)} + \frac{x^{5-\mu}}{\Gamma(6-\mu)} + \frac{x^{6-\mu}}{\Gamma(7-\mu)}\\
&- 1.107\,013\,79x - 0.490\,191\,813x^2 - 0.108\,529\,819x^3 - 0.012\,014\,4007x^4\\
&- 0.000\,532\,004\,29x^5.
\end{aligned}$$

Following the steps for series solution using ADM given in (6.11), we obtain the two term approximated solution as

$$\begin{aligned}
z(x) &= z_0(x) + z_1(x)\\
&= 1 + x + \frac{x^2}{2!} + \frac{x^3}{3!} + \frac{x^4}{4!} + \frac{x^5}{5!} + \frac{x^6}{6!} - \frac{1.107\,013\,79(x^{1+\mu})}{\Gamma(2+\mu)} - 0.490\,191\,813\frac{(x^{2+\mu})}{\Gamma(3+\mu)}\\
&- 0.108\,529\,819\frac{(x^{3+\mu})}{\Gamma(4+\mu)} - 0.012\,014\,4007\frac{(x^{4+\mu})}{\Gamma(5+\mu)} - 0.000\,532\,004\,29\frac{(x^{5+\mu})}{\Gamma(6+\mu)}\\
&+ 0.000\,198\,413\frac{x^{7+\mu}\Gamma(8)}{\Gamma(8+\mu)} + 0.001\,388\,89\frac{x^{6+\mu}\Gamma(7)}{\Gamma(7+\mu)} + 0.008\,333\,33\frac{x^{5+\mu}\Gamma(6)}{\Gamma(6+\mu)}\\
&+ 0.041\,6667\frac{x^{4+\mu}\Gamma(5)}{\Gamma(5+\mu)} + 0.166\,667\frac{x^{3+\mu}\Gamma(4)}{\Gamma(4+\mu)} + 0.5\frac{x^{2+\mu}\Gamma(3)}{\Gamma(3+\mu)} + \frac{x^{1+\mu}\Gamma(2)}{\Gamma(2+\mu)}\\
&- 1.107\,013\,79\frac{x^{2+2\mu}\Gamma(3+\mu)}{\Gamma(3+2\mu)(2+\mu)} - 0.490\,191\,813\frac{x^{3+2\mu}\Gamma(4+\mu)}{\Gamma(4+2\mu)(3+\mu)}\\
&- 0.108\,529\,819\frac{x^{4+2\mu}\Gamma(5+\mu)}{\Gamma(5+2\mu)(4+\mu)} - 0.012\,014\,4007\frac{x^{5+2\mu}\Gamma(6+\mu)}{\Gamma(6+2\mu)(5+\mu)}\\
&- 0.000\,532\,004\,29\frac{x^{6+2\mu}\Gamma(7+\mu)}{\Gamma(7+2\mu)(6+\mu)}.
\end{aligned}$$

Finally, using the classical iterative technique and approximating using the Taylor's polynomial for $f(x)$, we get

$$f_T(x) \approx 1 + \frac{x^{1-\mu}}{\Gamma(2-\mu)} + \frac{x^{2-\mu}}{\Gamma(3-\mu)} + \frac{x^{3-\mu}}{\Gamma(4-\mu)} + \frac{x^{4-\mu}}{\Gamma(5-\mu)} + \frac{x^{5-\mu}}{\Gamma(6-\mu)} + \frac{x^{6-\mu}}{\Gamma(7-\mu)} - x - \frac{x^2}{2!} - \frac{x^3}{3!} - \frac{x^4}{4!} - \frac{x^5}{5!},$$

Recursively, using the scheme for ADM, the solution is obtained as

$$\begin{aligned}
z(x) &= z_0(x) + z_1(x) \\
&= 1 + x + \frac{x^2}{2!} + \frac{x^3}{3!} + \frac{x^4}{4!} + \frac{x^5}{5!} + \frac{x^6}{6!} - \frac{(x^{1+\mu})}{\Gamma(2+\mu)} - \frac{(x^{2+\mu})}{\Gamma(3+\mu)} - \frac{(x^{3+\mu})}{\Gamma(4+\mu)} \\
&\quad - \frac{(x^{4+\mu})}{\Gamma(5+\mu)} - \frac{(x^{5+\mu})}{\Gamma(6+\mu)} + 0.000\,198\,413\frac{x^{7+\mu}\Gamma(8)}{\Gamma(8+\mu)} + 0.001\,388\,89\frac{x^{6+\mu}\Gamma(7)}{\Gamma(7+\mu)} \\
&\quad + 0.008\,333\,33\frac{x^{5+\mu}\Gamma(6)}{\Gamma(6+\mu)} + 0.041\,6\,667\frac{x^{4+\mu}\Gamma(5)}{\Gamma(5+\mu)} + 0.166\,667\frac{x^{3+\mu}\Gamma(4)}{\Gamma(4+\mu)} \\
&\quad + 0.5\frac{x^{2+\mu}\Gamma(3)}{\Gamma(3+\mu)} + \frac{x^{1+\mu}\Gamma(2)}{\Gamma(2+\mu)} - \frac{x^{2+2\mu}\Gamma(3+\mu)}{\Gamma(3+2\mu)(2+\mu)} - \frac{x^{3+2\mu}\Gamma(4+\mu)}{\Gamma(4+2\mu)(3+\mu)} \\
&\quad - \frac{x^{4+2\mu}\Gamma(5+\mu)}{\Gamma(5+2\mu)(4+\mu)} - \frac{x^{5+2\mu}\Gamma(6+\mu)}{\Gamma(6+2\mu)(5+\mu)} - \frac{x^{6+2\mu}\Gamma(7+\mu)}{\Gamma(7+2\mu)(6+\mu)}.
\end{aligned}$$

Figure 6.1 shows the error plot at $\mu = 0.25$ using the two term expansion of the modified ADM technique and the classical ADM. One can observe the robustness of ADM-BP and ADM-CP rather than approximating by the general ADM as the decrement in error is more in case of our proposed techniques as compared to the classical technique. Similarly, figure 6.2 depicts the comparison of $\mathbb{E}_2^\infty$ between all the three techniques. The error in case of ADM-CP and ADM-BP is minimal as compared to general ADM which makes it efficient for use when the source term in the model problem is any function rather than a polynomial function. Tables 6.1 and 6.2

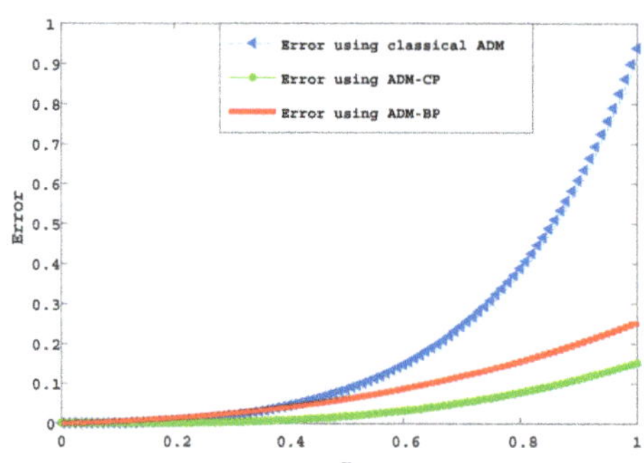


(a) $\mu = 0.25$

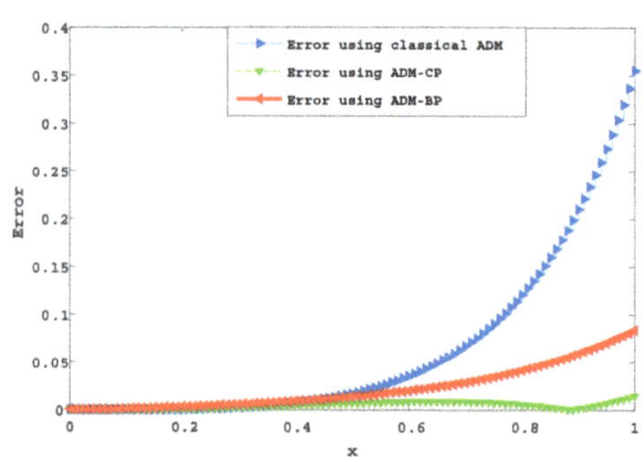


(b) $\mu = 0.75$

**Figure 6.1.** Error plots for example 6.1.

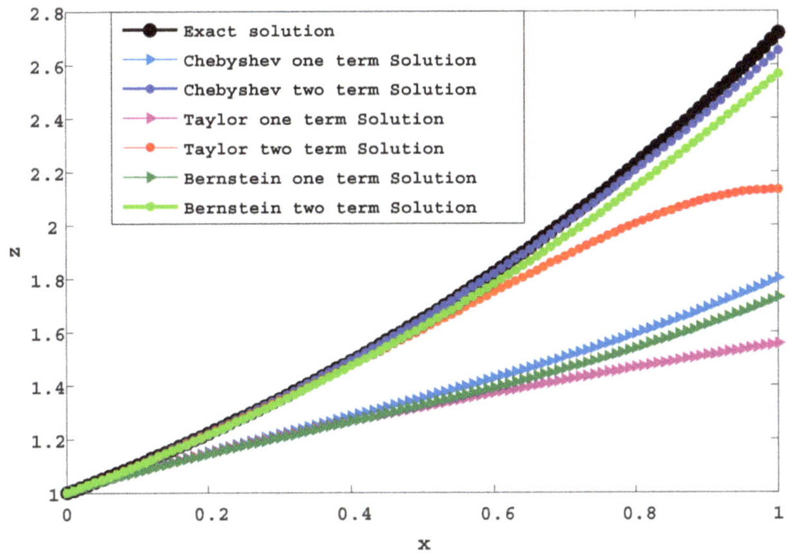


**Figure 6.2.** Solution plot at $\mu = 0.5$.

**Table 6.1.** Absolute point wise errors with $\mu = 0.01$ for example 6.1.

| $x$ | $E_2^\infty$ using ADM-BP | $E_2^\infty$ using ADM-CP | $E_2^\infty$ using classical ADM |
|---|---|---|---|
| 0.2 | $3.2232 \times 10^{-2}$ | $9.6072 \times 10^{-3}$ | $2.1724 \times 10^{-2}$ |
| 0.4 | $8.8292 \times 10^{-2}$ | $3.9878 \times 10^{-2}$ | $1.0156 \times 10^{-1}$ |
| 0.6 | $1.6813 \times 10^{-1}$ | $9.2759 \times 10^{-2}$ | $2.7822 \times 10^{-1}$ |
| 0.8 | $2.7053 \times 10^{-1}$ | $1.7013 \times 10^{-1}$ | $6.4637 \times 10^{-1}$ |
| 1.0 | $3.9279 \times 10^{-1}$ | $2.7399 \times 10^{-1}$ | 1.4248 |

**Table 6.2.** Absolute point wise errors with $\mu = 0.95$ for example 6.1.

| $x$ | $E_2^\infty$ using ADM-BP | $E_2^\infty$ using ADM-CP | $E_2^\infty$ using classical ADM |
|---|---|---|---|
| 0.2 | $1.7740 \times 10^{-3}$ | $6.4018 \times 10^{-4}$ | $1.9533 \times 10^{-4}$ |
| 0.4 | $5.3420 \times 10^{-3}$ | $4.0840 \times 10^{-3}$ | $3.3104 \times 10^{-3}$ |
| 0.6 | $1.0931 \times 10^{-2}$ | $1.0136 \times 10^{-2}$ | $1.8948 \times 10^{-2}$ |
| 0.8 | $2.2620 \times 10^{-2}$ | $1.4696 \times 10^{-2}$ | $7.2710 \times 10^{-2}$ |
| 1.0 | $4.9022 \times 10^{-2}$ | $8.5299 \times 10^{-3}$ | $2.3222 \times 10^{-1}$ |

give the error for $x \in [0, 1]$ at $\mu = 0.01$ and $\mu = 0.95$ respectively. Though at some points close to zero the error in classical ADM seems to be less than our proposed methods, but at rest all of the node points the proposed methods prove to be more accurate and efficient which clearly shows their reliability (figure 6.2).

**Example 6.2.** Consider the following nonlinear Volterra–Fredholm FIDE

$$\mathbb{D}^{0.75}z(x) + \frac{x^2e^x}{5}z(x) = \frac{6}{\Gamma(2.25)}x^{2.25} + \int_0^x e^x s z(s)ds + \int_0^1 (4 - s^{-3})z(s)ds \tag{6.15}$$

with the initial condition $z(0) = 0$. The exact solution is $z(x) = x^3$.

Here, the source function is in the form of a polynomial function. Approximating $f(x)$ using the Chebyshev polynomials and then applying the recursive algorithm, we obtain the series solution

$$f_C(x) = \sum_{i=0}^{4} C_i T_i(2x - 1), \quad 0 \leqslant x \leqslant 1,$$

where

$$C_0 = \frac{1}{\pi}\int_{-1}^{1} \frac{f(0.5x + 0.5)T_0(x)}{\sqrt{1 - x^2}}dx$$

and

$$C_i = \frac{2}{\pi}\int_{-1}^{1} \frac{f(0.5x + 0.5)T_i(x)}{\sqrt{1 - x^2}}dx, \quad i = 1, 2, \dots, 6.$$

It implies that

$$f_C(x) \approx -0.229\,3888x^4 + 0.976\,5696x^3 + 1.666\,256x^2 - 0.061\,0748x + 0.001\,0169.$$

Applying (6.7) we obtain the approximated solution

$$z(x) = -0.229\,3888x^{4.75}\frac{\Gamma(5)}{\Gamma(5.75)} + 0.976\,5696x^{3.75}\frac{\Gamma(4)}{\Gamma(4.75)} + 1.666\,256x^{2.75}\frac{\Gamma(3)}{\Gamma(3.75)} - 0.061\,0748x^{1.75}\frac{\Gamma(2)}{\Gamma(2.75)} + 0.001\,0169x^{0.75}\frac{\Gamma(1)}{\Gamma(1.75)}.$$

Further, $f(x)$ is approximated using the Bernstein polynomial $\sum_{i=0}^{n} D_i B_i(x)$ with $i = 10$ and then applying the recursive algorithm for ADM, we obtain the series solution

$$f_B(x) = \sum_{i=0}^{10} D_i B_i(x), \quad 0 \leqslant x \leqslant 1,$$

such that

$$f_B(x) \approx -0.000\,569\,441\,9493x^{10} + 0.006\,482\,451\,697x^9 - 0.034\,08x^8 + 0.110\,17x^7 - 0.247\,127x^6 + 0.416\,952\,685x^5 - 0.584\,450\,992x^4 + 0.911\,942\,5917x^3 + 1.641\,949\,867x^2 + 0.132\,354\,172x.$$

Substituting $f_B(x)$ in (6.11) we obtain the approximated solution which converges to the exact solution as shown in figure 6.3. Here the error is calculated using

$$\mathbb{E}_n^{\infty} = |z(x) - \sum_{i=0}^{n} z_i(x)|.$$

Hamoud and Ghadle in [10] solved this example using the classical ADM and obtained the exact solution in the first iteration itself. So, one can observe that, since the source term is already a polynomial function (i.e. in the Taylor series expansion), the proposed techniques (ADM-BP and ADM-CP) doesn't contribute much in decreasing the error while obtaining the solution in comparison to [10]. Table 6.3 shows the pointwise error obtained after the first term series solution using ADM-BP and ADM-CP. Though the error is less, but the still proposed method is not effective for such model problems. Hence, we can conclude that the proposed techniques are best suitable for the model problems where the source term is any other function except the polynomials.

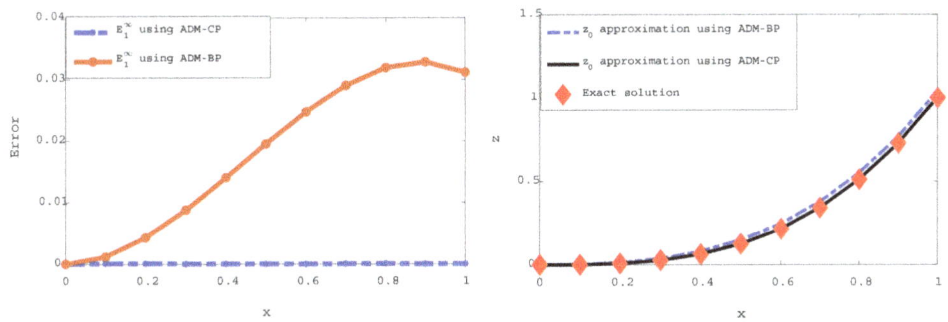


(a) Error plot

(b) Solution plot for $\mu = 0.75$

**Figure 6.3.** Solutions for example 6.2.

**Table 6.3.** Absolute point wise errors with $\mu = 0.75$ for example 6.2.

| $x$ | $\mathbb{E}_1^{\infty}$ using ADM-BP | $\mathbb{E}_1^{\infty}$ using ADM-CP |
|---|---|---|
| 0.1 | $1.5167 \times 10^{-3}$ | $7.7086 \times 10^{-5}$ |
| 0.2 | $5.4081 \times 10^{-3}$ | $1.1525 \times 10^{-4}$ |
| 0.3 | $1.0989 \times 10^{-2}$ | $4.4583 \times 10^{-5}$ |
| 0.4 | $1.7565 \times 10^{-2}$ | $2.3456 \times 10^{-5}$ |
| 0.5 | $2.4440 \times 10^{-2}$ | $2.9868 \times 10^{-5}$ |
| 0.6 | $3.0919 \times 10^{-2}$ | $1.6814 \times 10^{-5}$ |
| 0.7 | $3.6305 \times 10^{-2}$ | $6.7225 \times 10^{-5}$ |
| 0.8 | $3.9902 \times 10^{-2}$ | $7.1810 \times 10^{-5}$ |
| 0.9 | $4.1014 \times 10^{-2}$ | $2.6809 \times 10^{-5}$ |
| 1 | $3.8945 \times 10^{-2}$ | $2.2190 \times 10^{-5}$ |

## 6.7 Conclusion

The chapter focuses on new polynomials based analytical methods for finding the solutions of a class of fractional order Volterra–Fredholm IDE. Bernstein and Chebyshev polynomials have been used to modify the ADM in order to make it more reliable and obtain the solutions with less computational cost and greater accuracy. The convergence and uniqueness of the solution is proved. Efficiency of the proposed techniques are depicted using the tabular data and plots.

## References


[1] Chang Y K and Nieto J J 2009 Existence of solutions for impulsive neutral integro-differential inclusions with nonlocal initial conditions via fractional operators *Numer. Funct. Anal. Optim.* **30** 227–44

[2] Tarasov V E 2009 Fractional integro-differential equations for electromagnetic waves in dielectric media *Theor. Math. Phys.* **158** 355–9

[3] Podlubny I, Magin R L and Trymorush I 2017 Niels Henrik Abel and the birth of fractional calculus *Fract. Calc. Appl. Anal.* **20** 1068–75

[4] Aghajani A, Jalilian Y and Trujillo J J 2012 On the existence of solutions of fractional integro-differential equations *Fract. Calc. Appl. Anal.* **15** 44–69

[5] Vanani S K and Aminataei A 2011 Operational tau approximation for a general class of fractional integro-differential equations *Comput. Appl. Math.* **30** 655–74

[6] Heydari M H, Hooshmandasl M R, Maalek Ghaini F M and Li M 2013 Chebyshev wavelets method for solution of nonlinear fractional integro-differential equations in a large interval *Adv. Math. Phys.* **2013** 482083

[7] Amer S M, Saleh M H, Mohamed M A and Abdelrhman N S 2013 Variational iteration method and adomian decomposition method for fourth-order fractional integro-differential equations *Int. J. Comput. Appl.* **80** 7–14

[8] Das P, Rana S and Ramos H 2019 A perturbation based approach for solving fractional-order Volterra-Fredholm integro-differential equations and its convergence analysis *Int. J. Comput. Math.* **97** 1994–2014

[9] Panda A, Santra S and Mohapatra J 2021 Adomian decomposition and homotopy perturbation method for the solution of time fractional partial integro-differential equations *J. Appl. Math. Comput.* **68** 2065–2082

[10] Hamoud A A and Ghadle K P 2018 Modified Laplace decomposition method for fractional Volterra-Fredholm integro-differential equations *J. Math. Model.* **6** 91–104

[11] Adomian G 1988 A review of the decomposition method in applied mathematics *J. Math. Anal. Appl.* **135** 501–44

[12] Hosseini M M 2006 Adomian decomposition method with Chebyshev polynomials *Appl. Math. Comput.* **175** 1685–93

[13] Liu Y 2009 Adomian decomposition method with orthogonal polynomials: Legendre polynomials *Math. Comput. Model.* **49** 1268–73

[14] Qasim A F and AL-Rawi E S 2018 Adomian decomposition method with modified Bernstein polynomials for solving ordinary and partial differential equations *J. Appl. Math.* **2018** 1803107

[15] Solari H G and Natiello M A 2014 Linear processes in stochastic population dynamics: theory and application to insect development *Sci. World J.* **2014** 873624

[16] de Boor C, Gout C, Kunoth A and Rabut C 2008 Multivariate approximation: theory and applications *An overview. Numer. Algorithms* **48** 1–9

[17] Canuto C, Hussaini M Y, Quarteroni A and Zang T A 2007 *Spectral Methods: Evolution to Complex Geometries and Applications to Fluid Dynamics* (Heidelberg: Spinger)

[18] Gottlieb D and Orszag S A 1997 *Numerical Analysis of Spectral Methods: Theory and Applications* (Philadelphia, PA: CBMS-NSF, Society for Industrial and Applied Mathematics)

[19] Babolian E and Hosseini M M 2002 A modified spectral method for numerical solution of ordinary differential equations with non-analytic solution *Appl. Math. Comput.* **132** 341–51

[20] Hosseini M M 2005 Numerical solution of ordinary differential equations with impulse solution *Appl. Math. Comput.* **163** 373–81

[21] Hamoud A A, Ghadle K P, Bani Issa M S H and Giniswamy G 2018 Existence and uniqueness theorems for fractional Volterra-Fredholm integro-differential equations *Int. J. Appl. Math.* **31** 333–48

# Chapter 7

# Comparing different polynomials-based shape functions in the Rayleigh–Ritz method for investigating dynamical characteristics of nanobeam

**Subrat Kumar Jena, Dineshkumar Harursampath, Vinyas Mahesh and Sathiskumar A Ponnusami**

In this chapter, a numerically efficient method, namely the Rayleigh–Ritz method is employed to investigate the dynamical characteristics of nanobeam embedded in the Winkler–Pasternak elastic foundation. Orthogonal polynomials such as Chebyshev polynomials, Legendre polynomials, and Hermite polynomials are used as shape functions in the Rayleigh–Ritz method. The main advantages of these polynomials are orthogonality, making the technique more computationally efficient, and avoiding ill-conditioning for higher values of polynomials. The governing differential equations of the present model are obtained by incorporating Hamilton's principle, and the frequency parameters are computed for classical boundary conditions. Additionally, a comparative study has been carried out to justify the effectiveness and convergence of the present model or method for different polynomials.

## 7.1 Introduction

Lord Rayleigh proposed the Rayleigh method in 1877 to solve vibration problems by considering mode shapes while calculating the fundamental frequency of continuous systems. Moreover, in 1908, Ritz explained his method for handling boundary value problems and eigenvalue problems. The Rayleigh–Ritz approach, an approximate energy-based strategy, has been used to solid mechanic problems throughout the century. This technique approximates the displacement function as a linear combination of trial functions. To determine the maximum kinetic energy and maximum potential energy of a structure, the displacement function must fulfil geometrical or essential boundary conditions (displacement or slope). The efficiency of this strategy

doi:10.1088/2053-2563/ac9580CH007  

is determined by the trial functions chosen. In the Rayleigh–Ritz technique, the trial function is a basic algebraic polynomial, however in the Hermite–Ritz and shifted Chebyshev polynomials-based Rayleigh–Ritz methods, the trial functions are Hermite polynomials and shifted Chebyshev polynomials, respectively. Maximum kinetic energy and maximum potential energy are equalised to determine the Rayleigh quotient. Eigenvalue problems are formulated by differentiating the Rayleigh quotient of the displacement function with respect to the unknown coefficients; hence, any eigenvalue problem solver can be used to solve them.

Ozkose and Bagdatli (2010) investigated the natural frequencies of three-supported beam using the Hamilton, Rayleigh, and Rayleigh–Ritz methods. Zhu (2011) utilised the Rayleigh–Ritz method to investigate flapwise and chordwise bending vibration analysis of rotating pretwisted beams based on Timoshenko beam theory. Behera and Chakraverty (2014) investigated the free vibration of Euler–Bernoulli and Timoshenko nanobeams by employing simple polynomials and orthonormal polynomials as trial functions in the Rayleigh–Ritz method based on Eringen's nonlocal elasticity theory. Chakraverty and Behera (2015) investigated the vibration and buckling of Euler–Bernoulli nanobeams immersed in an elastic medium by employing boundary characteristic orthogonal polynomials as form functions in the Rayleigh–Ritz method. Fakher and Hosseini-Hashemi (2017) studied the static bending and free vibration characteristics of Euler nanobeams based on nonlocal strain gradient elasticity, taking into consideration the differential form of nonlocal strain gradient, the integral nonlocal strain gradient with satisfying higher order boundary conditions, and the basic form of integral nonlocal strain gradient without satisfying higher order boundary conditions, utilising polynomial and combination of polynomials. Kumar (2018) reviewed the Rayleigh–Ritz approach for linear dynamic, static, and buckling behaviour of beams, shells, and plates. Jena *et al* (2019) investigated propagation of uncertainty in the vibration characteristics of Euler–Bernoulli nanobeam using a triangular fuzzy number and Rayleigh–Ritz technique. Using the Rayleigh–Ritz method, Abdullah *et al* (2020) examined the torsional vibrations of carbon nanotubes/nanorods put in an elastic medium and subjected to heat loads.

In this chapter, orthogonal polynomials such as Chebyshev polynomials, Legendre polynomials, and Hermite polynomials are used as shape functions in the Rayleigh–Ritz method to demonstrate the effectiveness over algebraic polynomials in terms of time to investigate the dynamical properties of a nanobeam embedded in a Winkler–Pasternak elastic foundation. A comparison analysis has been performed to illustrate the efficacy and convergence of the existing model for various polynomials with regard to time. Validation of frequency parameters has been accomplished using the existing work.

## 7.2 Preliminaries

In this section, basic preliminaries of different orthogonal polynomials such as Chebyshev polynomials, Legendre polynomials, and Hermite polynomials are discussed with respect to the undertaken problem related to the dynamical characteristics of nanobeam.

### 7.2.1 Chebyshev polynomials

Chebyshev polynomials of the first kind ($T_n(x)$), as shown in figure 7.1, are a sequence of orthogonal polynomials with $x \in [-1, 1]$ and few terms of the sequence with recurrence relations are defined as (Jena *et al* 2021)

$$T_0(x) = 1$$
$$T_1(x) = x$$
$$T_2(x) = 2x^2 - 1$$
$$T_n(x) = 2xT_{n-1}(x) - T_{n-2}(x), \qquad n = 2, 3, \ldots \tag{7.1}$$

The main advantages of using Chebyshev polynomials over algebraic polynomials $(1, x, x^2, x^3, \ldots x^n)$ are the orthogonal properties of Chebyshev polynomials which reduce the computational effort, and for the higher value of $n$ $(n > 10)$, the system avoids ill-conditioning. Since the domain of the problem lies in $[0, \quad 1]$, the Chebyshev polynomials must be reduced to shifted Chebyshev polynomials of the first kind $(T_n^*(x))$ with $x \in [0, 1]$. This is achieved by transforming $x \mapsto 2x - 1$ and there exists one-to-one correspondence between $[0, 1]$ and $[-1, 1]$. Accordingly, the first few terms of shifted Chebyshev polynomials of the first kind $(T_n^*(x))$ where $T_n^*(x) = T_n(2x - 1)$, can be written as

$$T_0^*(x) = 1$$
$$T_1^*(x) = 2x - 1$$
$$T_2^*(x) = 8x^2 - 8x + 1$$
$$T_n^*(x) = 2(2x - 1)T_{n-1}^*(x) - T_{n-2}^*(x), \qquad n = 2, 3, \ldots \tag{7.2}$$

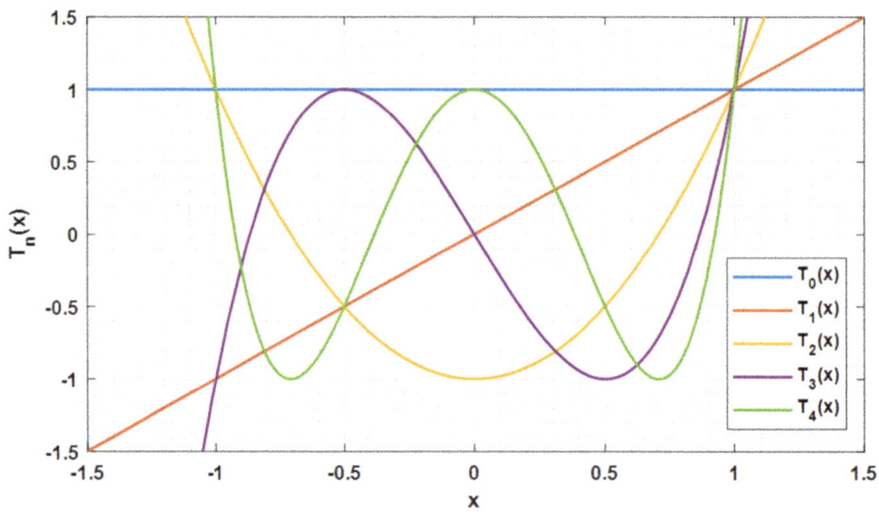


**Figure 7.1.** First five Chebyshev polynomials of the first kind.

### 7.2.2 Legendre polynomials

Legendre polynomials($P_n(x)$) are orthogonal polynomials with $x \in [-1, 1]$ and have a vast number of mathematical properties. A few terms of the sequence are defined as

$$P_0(x) = 1$$

$$P_1(x) = x$$

$$P_2(x) = \frac{1}{2}(3x^2 - 1)$$

$$P_n(x) = \frac{1}{n}[(2n - 1)xP_{n-1}(x) - (n - 1)P_{n-2}(x)], \qquad n = 2, 3, \dots \tag{7.3}$$

The orthogonal properties of Legendre polynomials, as shown in figure 7.2, which reduce computational effort and prevent ill-conditioning, are the main advantages of adopting Legendre polynomials over algebraic polynomials. As the domain of the problem lies in [0, 1], the Legendre polynomials must be reduced to shifted Legendre polynomials. The shifted Legendre polynomials are defined as $P_n^*(x) = P_n(2x - 1)$. The affine transformation $x \mapsto 2x - 1$ bijectively maps the interval [0, 1] to the interval $[-1, 1]$, implying that the polynomials $P_n^*(x)$ are orthogonal on [0, 1]. The first few terms of shifted Legendre polynomials ($P_n^*(x)$) with recurrence relations can be written as

$$P_0^*(x) = 1$$

$$P_1^*(x) = 2x - 1$$

$$P_2^*(x) = 6x^2 - 6x + 1$$

$$P_n^*(x) = \frac{1}{n}[(2n - 1)(2x - 1)P_{n-1}^*(x) - (n - 1)P_{n-2}^*(x)], \qquad n = 2, 3, \dots \tag{7.4}$$

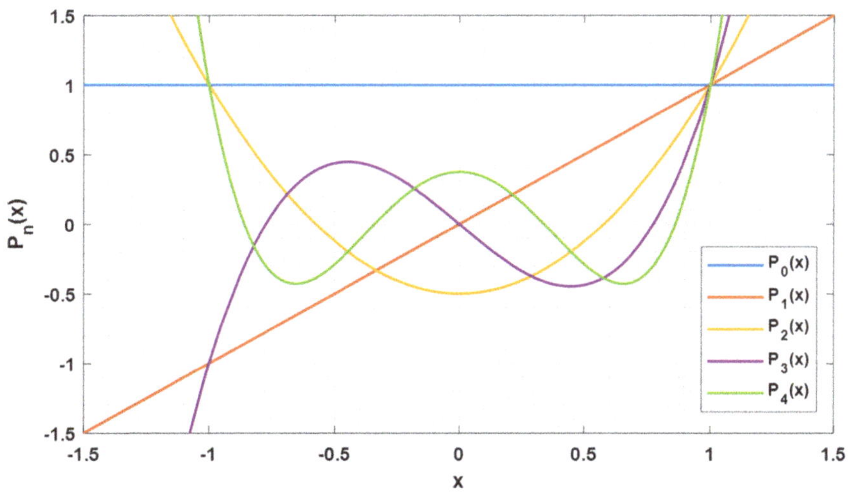


**Figure 7.2.** First five Legendre polynomials.

### 7.2.3 Hermite polynomials

Hermite polynomials ($H_n(x)$) are sets of orthogonal polynomials with respect to the weight function $e^{-x^2}$defined over the domain $(-\infty, \infty)$, i.e.,

$$\int_{-\infty}^{\infty} e^{-x^2} H_m(x)\, H_n(x)\, \mathrm{d}x = \begin{cases} \sqrt{\pi}\, 2^n n!, & n = m \\ 0, & n \neq m \end{cases}. \tag{7.5}$$

The first few terms of Hermite polynomials with recurrence relations can be expressed as (Jena *et al* 2020)

$$H_0(x) = 1$$

$$H_1(x) = 2x$$

$$H_2(x) = 4x^2 - 2$$

$$H_n(x) = 2x\, H_{n-1}(x) - 2(n-1)\, H_{n-2}(x) \text{ and } H'_n(x) = 2n\, H_{n-1}(x). \tag{7.6}$$

The Hermite polynomials, as shown in figure 7.3, are chosen as shape functions as the Hermite polynomials are orthogonal polynomials, which reduce computation time. In contrast to other orthogonal polynomials such as Chebyshev polynomials, Legendre polynomials, etc, the domain is $(-\infty, \ \infty)$ provides flexibility in the limit Lagrangian energy function. It aids in preventing ill-conditioning of the matrix for higher polynomial values.

## 7.3 Governing equations of motion for the proposed model

According to Euler–Bernoulli beam theory, the displacement field can be given as (Reddy 2007)

$$u_1(x, z, t) = -z\left(\frac{\partial w}{\partial x}\right) u_2(x, z, t) = 0\ u_3(x, z, t) = w\,(x, t). \tag{7.7}$$

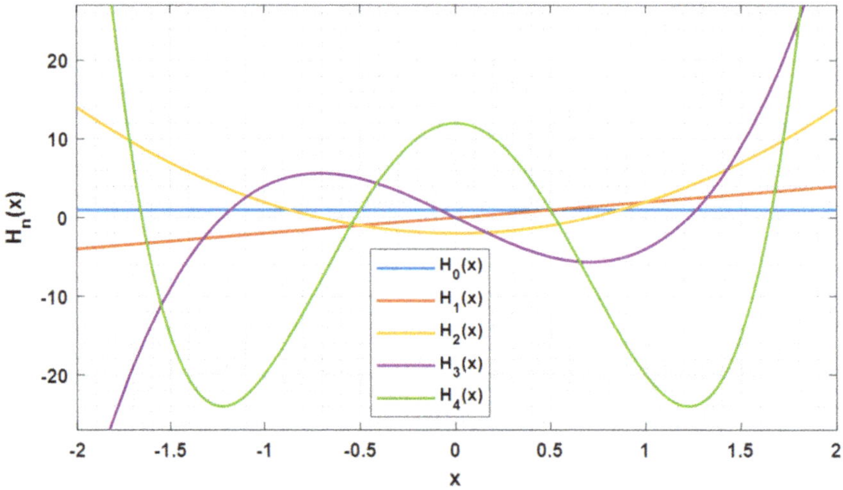


**Figure 7.3.** First five Hermite polynomials.

Here $w\ (x,\ t)$ represents the transverse displacements on the mid-plane of the nanobeam. The strain-displacement relation of the nanobeam is stated as

$$\varepsilon_{xx} = \frac{\partial u_1(x,\, z,\, t)}{\partial x} = -z\frac{\partial^2 w\ (x,\ t)}{\partial x^2}\ . \tag{7.8}$$

The strain energy ($S$) of the nanobeam is expressed as (Jena and Chakraverty 2021)

$$\begin{aligned} S &= \frac{1}{2}\int_0^L \int_A (\sigma_{xx}\varepsilon_{xx})\ \mathrm{d}A\ \mathrm{d}x \\ &= \frac{1}{2}\int_0^L \int_A \left[\sigma_{xx}\left(-z\frac{\partial^2 w\ (x,\ t)}{\partial x^2}\right)\right] \mathrm{d}A\ \mathrm{d}x \\ &= \frac{1}{2}\int_0^L \left[-M\left(\frac{\partial^2 w\ (x,\ t)}{\partial x^2}\right)\right] \mathrm{d}x. \end{aligned} \tag{7.9}$$

From equation (7.9), the bending moment is defined as $M = \int_A (z\sigma_{xx})\mathrm{d}A$. Next, the variation in strain energy ($\delta S$) can be written as (Jena and Chakraverty 2021)

$$\begin{aligned} \delta S &= \int_0^L \int_A (\sigma_{xx}\delta\varepsilon_{xx})\ \mathrm{d}A\ \mathrm{d}x \\ &= \int_0^L \int_A \left[\sigma_{xx}\left(-z\frac{\partial^2 \delta w\ (x,\ t)}{\partial x^2}\right)\right] \mathrm{d}A\ \mathrm{d}x \\ &= \int_0^L \left[-\left(\frac{\partial^2 M}{\partial x^2}\right)\right] \delta w\ \mathrm{d}x. \end{aligned} \tag{7.10}$$

The kinetic energy ($T$) of the nanobeam is defined as (Jena and Chakraverty 2021)

$$\begin{aligned} T &= \frac{1}{2}\int_0^L \int_A \rho\left[\left(\frac{\partial u_1}{\partial t}\right)^2 + \left(\frac{\partial u_2}{\partial t}\right)^2 + \left(\frac{\partial u_3}{\partial t}\right)^2\right] \mathrm{d}A\ \mathrm{d}x \\ &= \frac{1}{2}\int_0^L \int_A \rho\left[\left(-z\left(\frac{\partial^2 w}{\partial x \partial t}\right)\right)^2 + \left(\frac{\partial w}{\partial t}\right)^2\right] \mathrm{d}A\ \mathrm{d}x \\ &= \frac{1}{2}\int_0^L \left[I_0\left(\frac{\partial w}{\partial t}\right)^2\right] \mathrm{d}x. \end{aligned} \tag{7.11}$$

In which $I_0 = \int_A \rho \, dA = \rho A$ is the mass moment of inertia. Variation in kinetic energy ($\delta T$) from equation (7.11) can be expressed as (Jena and Chakraverty 2021)

$$\delta T = \int_0^L \left[ -I_0 \left( \frac{\partial^2 w}{\partial t^2} \right) \right] \delta w \, \mathrm{d}x. \tag{7.12}$$

Further, the work done ($W$) due to external loads Winkler–Pasternak elastic foundation can be obtained as (Jena and Chakraverty 2021)

$$W = \frac{1}{2} \int_0^L \left[ k_w w^2 - k_p w \left( \frac{\partial^2 w}{\partial x^2} \right) \right] \mathrm{d}x, \tag{7.13}$$

where $k_w$ is the Winkler modulus, $k_p$ is the shear modulus. The virtual work done ($\delta W$) may be obtained from equation (7.13) as (Jena and Chakraverty 2021)

$$\delta W = \int_0^L \left[ k_w w - k_p \left( \frac{\partial^2 w}{\partial x^2} \right) \right] \delta w \, \mathrm{d}x. \tag{7.14}$$

Assuming the motion of the nanobeam as sinusoidal, i.e., $w = w_0(x) \sin \omega t$, where $\omega$ is the natural frequency of vibration, using Hamilton's principle $\delta \prod = \int_0^t \delta \, (T - (U + W)) \, \mathrm{d}t$, and collecting the co-efficient $\delta w$, the governing equation of motion may be obtained as

$$\frac{\partial^2 M}{\partial x^2} - k_w w + k_p \left( \frac{\mathrm{d}^2 w_0}{\mathrm{d}x^2} \right) + \rho A \omega^2 w_0 = 0. \tag{7.15}$$

From Eringen's nonlocal elasticity theory (Eringen 1983), we have

$$\left( 1 - (e_0 a)^2 \frac{\partial^2}{\partial x^2} \right) \sigma_{xx} = E \varepsilon_{xx} \tag{7.16}$$

where $\sigma_{xx}$ is the normal stress, $E$ is Young's modulus, $\varepsilon_{xx}$ is the classical strain, and $e_0 a$ is the nonlocal parameter. Now, multiplying equation(7.16) by $z \, \mathrm{d}A$ and integrating over the area, we may get

$$\left( 1 - (e_0 a)^2 \frac{\partial^2}{\partial x^2} \right) M = -EI \frac{\partial^2 w_0}{\partial x^2}. \tag{7.17}$$

Substituting equation (7.15) in equation (7.17), we have

$$M = -EI \frac{\mathrm{d}^2 w_0}{\mathrm{d}x^2} + (e_0 a)^2 \left( k_w w - k_p \left( \frac{\mathrm{d}^2 w_0}{\mathrm{d}x^2} \right) - \rho A \omega^2 w_0 \right). \tag{7.18}$$

Equating maximum kinetic and potential energies, we get

$$\int_0^L \begin{bmatrix} EI\left(\frac{d^2w_0}{dx^2}\right)^2 - (e_0a)^2k_w w_0\left(\frac{d^2w_0}{dx^2}\right) + (e_0a)^2 \\ k_p\left(\frac{d^2w_0}{dx^2}\right)^2 + k_w w_0^2 - k_p w_0\left(\frac{d^2w_0}{dx^2}\right) \end{bmatrix} dx = \int_0^L \begin{bmatrix} (\rho A\omega^2)w_0^2 \\ -(e_0a)^2(\rho A\omega^2)w_0\left(\frac{d^2w_0}{dx^2}\right) \end{bmatrix} dx. \tag{7.19}$$

Let's introduce the following non-dimensional parameters for simplicity and convenience in mathematical formulation as

$$X = \frac{x}{L},\ W = \frac{w_0}{L},\ \lambda^2 = \frac{\rho A\omega^2 L^4}{EI},\ \alpha = \frac{e_0 a}{L},\ K_w = \frac{k_w L^4}{EI},\ K_p = \frac{k_p L^2}{EI}. \tag{7.20}$$

Using the above non-dimensional parameters i.e., equation (7.20) in equation (7.19), we have

$$\int_0^1 \begin{bmatrix} \left(\frac{d^2W}{dX^2}\right)^2 - \alpha^2K_wW\left(\frac{d^2W}{dX^2}\right) + \alpha^2 \\ K_p\left(\frac{d^2W}{dX^2}\right)^2 + K_wW^2 - K_pW\left(\frac{d^2W}{dX^2}\right) \end{bmatrix} dX = \lambda^2\int_0^1 \left[W^2 - \alpha^2W\left(\frac{d^2W}{dX^2}\right)\right] dX. \tag{7.21}$$

## 7.4 Solution procedures

In this section, orthogonal polynomials such as Chebyshev polynomials, Legendre polynomials, and Hermite polynomials are utilised as shape functions in the Rayleigh–Ritz method.

### 7.4.1 Application of different polynomials in Rayleigh–Ritz method

The procedures of Rayleigh–Ritz method along with algebraic polynomials, Chebyshev polynomials, Legendre polynomials, and Hermite polynomials have been described to solve the governing equations of motion for the proposed model. The transverse displacement $W(X)$ of the nanobeam for the proposed model can be written as (Jena *et al* 2020)

$$W(X) = X^k(1 - X)^l\sum_{i=1}^{n} d_iP(i - 1,\ X). \tag{7.22}$$

Here $d_i$ are unknown coefficients, $P(n,\ X)$ is the $n$th term of shape function which could be algebraic polynomials, Chebyshev polynomials, Legendre

**Table 7.1.** $k$ and $l$ for different boundary conditions (BC) (Jena *et al* 2020).

| BC | $k$ | $l$ |
|---|---|---|
| P-P | 1 | 1 |
| C-P | 2 | 1 |
| C-C | 2 | 2 |

polynomials, and Hermite polynomials. $X^k(1 - X)^l$ is the admissible functions with exponents $k$, and $l$. For different boundary conditions $k$, and $l$ possess different values, as shown in table 7.1.

Substituting equation (7.22) into equation (7.21) and minimising $\lambda^2$ with respect to the unknown coefficients $d_i$, $i = 1, 2, 3, \ldots, n$, yields the following generalised eigenvalue problem

$$[K]\,\{X\} = \lambda^2\,[M]\,\{X\} \tag{7.23}$$

where $\{X\} = [c_1, c_2, c_3, \ldots c_n]^T$, $[K]$ and $[M]$ represent the stiffness and mass matrices, respectively.

## 7.5 Numerical results and discussions

Rayleigh–Ritz method has been used with different shape functions to solve the titled problem. The effective properties are acquired as the Young's modulus $E$=1 TPa, thickness of the SWCNT $h$=0.34 nm, mass density $\rho$=1370 kg m$^{-3}$, and Poisson's ratio $\nu = 0.19$ from Zhen and Fang (2010).

### 7.5.1 Validation

To validate the current findings, a comparison study has been undertaken with the results of Wang *et al* (2007). Table 7.2 illustrates the first four frequency parameters of the P-P, C-P, and C-C boundary conditions that are taken into account with $K_w = 0$, and $K_p = 0$. The small scale parameters are set to 0, 0.1, 0.3, 0.5, and 0.7 in this instance. The results are found to be substantially similar to those found by Wang *et al* (2007).

### 7.5.2 Convergence

A convergence study has been conducted to assess the efficacy and effectiveness of the proposed method as well as to compute the results. The convergence of the first five frequency parameters $(\sqrt{\lambda})$ of the P-P, C-P, and C-C boundary conditions is investigated in this study, as can be seen in table 7.3, taking values of the parameters as $\alpha = 0.1$, $K_w = 100$, and $K_p = 1$. These results demonstrate that $n = 6$ is adequate for the convergence of first mode frequencies, whereas $n = 8$, $n = 10$, $n = 12$, and $n = 14$ are required for the second, third, fourth, and fifth modes of frequency parameters, respectively, and this holds true for all the undertaken boundary conditions. It's also worth noting that different shape functions, such as algebraic,

**Table 7.2.** Comparisons of frequency parameters($\sqrt{\lambda}$) with [A] (Wang *et al* 2007).

(a) P-P boundary condition

| | $\sqrt{\lambda}_1$ | | $\sqrt{\lambda}_2$ | | $\sqrt{\lambda}_3$ | | $\sqrt{\lambda}_4$ | |
|---|---|---|---|---|---|---|---|---|
| $\alpha = e_0a/L$ | Present | [A] | Present | [A] | Present | [A] | Present | [A] |
| 0 | 3.1416 | 3.1416 | 6.2832 | 6.2832 | 9.4248 | 9.4248 | 1.566 | 1.566 |
| 0.1 | 3.0685 | 3.0685 | 5.7817 | 5.7817 | 8.0400 | 8.0400 | 9.9161 | 9.9161 |
| 0.3 | 2.6800 | 2.6800 | 4.3013 | 4.3013 | 5.4422 | 5.4422 | 6.3630 | 6.3630 |
| 0.5 | 2.3022 | 2.3022 | 3.4604 | 3.4604 | 4.2941 | 4.2941 | 4.9820 | 4.9820 |
| 0.7 | 2.0212 | 2.0212 | 2.9585 | 2.9585 | 3.6485 | 3.6485 | 4.2234 | 4.2234 |

(b) C-P boundary condition

| | $\sqrt{\lambda}_1$ | | $\sqrt{\lambda}_2$ | | $\sqrt{\lambda}_3$ | | $\sqrt{\lambda}_4$ | |
|---|---|---|---|---|---|---|---|---|
| $\alpha = e_0a/L$ | Present | [A] | Present | [A] | Present | [A] | Present | [A] |
| 0 | 3.9266 | 3.9266 | 7.0686 | 7.0686 | 10.2102 | 10.2102 | 13.3518 | 13.3518 |
| 0.1 | 3.8209 | 3.8209 | 6.4649 | 6.4649 | 8.6517 | 8.6517 | 10.469 | 10.469 |
| 0.3 | 3.2828 | 3.2828 | 4.7668 | 4.7668 | 5.8371 | 5.8371 | 6.7143 | 6.7143 |
| 0.5 | 2.7899 | 2.7899 | 3.8325 | 3.8325 | 4.6105 | 4.6105 | 5.2632 | 5.2632 |
| 0.7 | 2.4364 | 2.4364 | 3.2776 | 3.2776 | 3.9201 | 3.9201 | 4.4644 | 4.4644 |

(c) C-C boundary condition

| | $\sqrt{\lambda}_1$ | | $\sqrt{\lambda}_2$ | | $\sqrt{\lambda}_3$ | | $\sqrt{\lambda}_4$ | |
|---|---|---|---|---|---|---|---|---|
| $\alpha = e_0a/L$ | Present | [A] | Present | [A] | Present | [A] | Present | [A] |
| 0 | 4.7300 | 4.7300 | 7.8532 | 7.8532 | 10.9956 | 10.9956 | 14.1372 | 14.1372 |
| 0.1 | 4.5945 | 4.5945 | 7.1402 | 7.1402 | 9.2583 | 9.2583 | 11.016 | 11.016 |
| 0.3 | 3.9184 | 3.9184 | 5.1963 | 5.1963 | 6.2317 | 6.2317 | 7.0482 | 7.0482 |
| 0.5 | 3.3153 | 3.3153 | 4.1561 | 4.1561 | 4.9328 | 4.9328 | 5.5213 | 5.5213 |
| 0.7 | 2.8893 | 2.8893 | 3.5462 | 3.5462 | 4.1996 | 4.1996 | 4.6816 | 4.6816 |

Chebyshev, Legendre, and Hermite polynomials, require the same number of terms and follow the same pattern of convergence. As such, $n = 14$ is considered for the sake of computation throughout the study.

### 7.5.3 Comparisons of shape functions with respect to convergence

In this subsection, comparisons among all the considered shape functions, such as algebraic, Chebyshev, Legendre, and Hermite polynomials, are carried out in terms of the convergence of the P-P boundary condition and the time period required to attain the results. In this regards the first four frequency parameters of P-P boundary

**Table 7.3.** Convergence of frequency parameters with $\alpha = 0.1$, $K_w = 100$, and $K_p = 1$.

(a) P-P boundary condition

| $n$ | $\sqrt{\lambda_1}$ | $\sqrt{\lambda_2}$ | $\sqrt{\lambda_3}$ | $\sqrt{\lambda_4}$ | $\sqrt{\lambda_5}$ |
|---|---|---|---|---|---|
| 2 | 3.8481 | 6.6212 | | | |
| 4 | 3.7539 | 5.9651 | 9.6943 | 12.6711 | |
| 6 | 3.7537 | 5.9543 | 8.1895 | 10.1546 | 15.5292 |
| 8 | 3.7537 | 5.9542 | 8.1302 | 9.9874 | 11.9300 |
| 10 | 3.7537 | 5.9542 | 8.1293 | 9.9817 | 11.5892 |
| 12 | 3.7537 | 5.9542 | 8.1293 | 9.9816 | 11.5681 |
| 14 | 3.7537 | 5.9542 | 8.1293 | 9.9816 | 11.5676 |
| 16 | 3.7537 | 5.9542 | 8.1293 | 9.9816 | 11.5676 |

(b) C-P boundary condition

| $n$ | $\sqrt{\lambda_1}$ | $\sqrt{\lambda_2}$ | $\sqrt{\lambda_3}$ | $\sqrt{\lambda_4}$ | $\sqrt{\lambda_5}$ |
|---|---|---|---|---|---|
| 2 | 4.2515 | 7.8790 | | | |
| 4 | 4.2480 | 6.6260 | 8.9615 | 14.0180 | |
| 6 | 4.2479 | 6.5975 | 8.7378 | 10.8114 | 13.0720 |
| 8 | 4.2479 | 6.5972 | 8.7296 | 10.5434 | 12.1637 |
| 10 | 4.2479 | 6.5972 | 8.7295 | 10.5299 | 12.0766 |
| 12 | 4.2479 | 6.5972 | 8.7295 | 10.5297 | 12.0723 |
| 14 | 4.2479 | 6.5972 | 8.7295 | 10.5297 | 12.0722 |
| 16 | 4.2479 | 6.5972 | 8.7295 | 10.5297 | 12.0722 |

(c) C-C boundary condition

| $n$ | $\sqrt{\lambda_1}$ | $\sqrt{\lambda_2}$ | $\sqrt{\lambda_3}$ | $\sqrt{\lambda_4}$ | $\sqrt{\lambda_5}$ |
|---|---|---|---|---|---|
| 2 | 4.8759 | 7.3435 | | | |
| 4 | 4.8669 | 7.2477 | 9.6601 | 11.8013 | |
| 6 | 4.8669 | 7.2462 | 9.3396 | 11.1197 | 13.8433 |
| 8 | 4.8669 | 7.2462 | 9.3280 | 11.0746 | 12.6949 |
| 10 | 4.8669 | 7.2462 | 9.3278 | 11.0734 | 12.5785 |
| 12 | 4.8669 | 7.2462 | 9.3278 | 11.0733 | 12.5726 |
| 14 | 4.8669 | 7.2462 | 9.3278 | 11.0733 | 12.5725 |
| 16 | 4.8669 | 7.2462 | 9.3278 | 11.0733 | 12.5725 |

conditions are taken into consideration, as indicated in table 7.4, with $\alpha = 0.3$, $K_w = 100$, and $K_p = 2$. The main advantages of using Chebyshev, Legendre, and Hermite polynomials as shape functions over algebraic polynomials are orthogonality, making the technique more computationally efficient, and avoiding ill-conditioning for higher values of polynomials. It can be concluded from these tabular results that for lower values of $n$, all polynomials behave similarly in terms of time. However, as we keep increasing the values of $n$, time required by the

**Table 7.4.** Convergence of frequency parameters with $\alpha = 0.3$, $K_w = 100$, and $K_p = 2$.

(a) Fist mode of P-P boundary condition

| $n$ | Algebraic Polynomials | | Chebyshev Polynomials | | Legendre Polynomials | | Hermite Polynomials | |
|---|---|---|---|---|---|---|---|---|
| | $\sqrt{\lambda_1}$ | Time (s) | $\sqrt{\lambda_1}$ | Time (s) | $\sqrt{\lambda_1}$ | Time (s) | $\sqrt{\lambda_1}$ | Time (s) |
| 2 | 3.6883 | 3.2535 | 3.6883 | 3.4192 | 3.6883 | 3.1005 | 3.6883 | 3.2030 |
| 3 | 3.6181 | 3.5302 | 3.6181 | 3.4224 | 3.6181 | 3.1630 | 3.6181 | 3.4658 |
| 4 | 3.6181 | 3.6954 | 3.6181 | 3.7549 | 3.6181 | 3.6129 | 3.6181 | 3.6834 |
| 5 | 3.6179 | 3.9400 | 3.6179 | 3.8584 | 3.6179 | 3.6940 | 3.6179 | 3.7219 |
| 6 | 3.6179 | 4.1138 | 3.6179 | 4.0996 | 3.6179 | 3.9324 | 3.6179 | 4.0569 |
| 7 | 3.6179 | 4.1739 | 3.6179 | 4.1269 | 3.6179 | 3.9841 | 3.6179 | 4.1565 |
| 8 | 3.6179 | 4.2472 | 3.6179 | 4.2262 | 3.6179 | 4.1278 | 3.6179 | 4.2234 |

(b) Second mode of P-P boundary condition

| $n$ | Algebraic Polynomials | | Chebyshev Polynomials | | Legendre Polynomials | | Hermite Polynomials | |
|---|---|---|---|---|---|---|---|---|
| | $\sqrt{\lambda_2}$ | Time (s) | $\sqrt{\lambda_2}$ | Time (s) | $\sqrt{\lambda_2}$ | Time (s) | $\sqrt{\lambda_2}$ | Time (s) |
| 2 | 5.2151 | 3.5728 | 5.2151 | 3.5459 | 5.2151 | 3.2836 | 5.2151 | 3.5516 |
| 3 | 5.2151 | 3.6137 | 5.2151 | 3.5885 | 5.2151 | 3.4210 | 5.2151 | 3.5921 |
| 4 | 4.7852 | 3.7561 | 4.7852 | 3.6367 | 4.7852 | 3.5422 | 4.7852 | 3.6843 |
| 5 | 4.7852 | 3.8313 | 4.7852 | 3.7436 | 4.7852 | 3.6842 | 4.7852 | 3.7693 |
| 6 | 4.7782 | 3.9862 | 4.7782 | 3.8210 | 4.7782 | 3.7198 | 4.7782 | 3.8778 |
| 7 | 4.7782 | 4.0488 | 4.7782 | 3.9139 | 4.7782 | 3.8456 | 4.7782 | 3.9972 |
| 8 | 4.7782 | 4.1114 | 4.7782 | 3.9977 | 4.7782 | 3.9176 | 4.7782 | 4.0919 |

(c) Third mode of P-P boundary condition

| $n$ | Algebraic Polynomials | | Chebyshev Polynomials | | Legendre Polynomials | | Hermite Polynomials | |
|---|---|---|---|---|---|---|---|---|
| | $\sqrt{\lambda_3}$ | Time (s) | $\sqrt{\lambda_3}$ | Time (s) | $\sqrt{\lambda_3}$ | Time (s) | $\sqrt{\lambda_3}$ | Time (s) |
| 3 | 6.7949 | 3.5470 | 6.7949 | 3.0585 | 6.7949 | 3.1939 | 6.7949 | 3.0359 |
| 4 | 6.7949 | 3.7499 | 6.7949 | 3.1584 | 6.7949 | 3.3407 | 6.7949 | 3.1442 |
| 5 | 5.8656 | 3.9028 | 5.8656 | 3.5225 | 5.8656 | 3.6515 | 5.8656 | 3.4637 |
| 6 | 5.8656 | 3.9921 | 5.8656 | 3.7065 | 5.8656 | 3.7734 | 5.8656 | 3.6696 |
| 7 | 5.8301 | 4.5790 | 5.8301 | 3.8701 | 5.8301 | 3.9309 | 5.8301 | 3.8330 |
| 8 | 5.8301 | 4.7113 | 5.8301 | 4.2347 | 5.8301 | 4.4501 | 5.8301 | 4.1807 |
| 9 | 5.8295 | 5.0638 | 5.8295 | 4.5374 | 5.8295 | 4.6557 | 5.8295 | 4.4359 |
| 10 | 5.8295 | 5.4470 | 5.8295 | 5.0800 | 5.8295 | 5.1266 | 5.8295 | 4.8842 |
| 11 | 5.8295 | 5.7828 | 5.8295 | 5.2710 | 5.8295 | 5.4539 | 5.8295 | 5.1758 |

| (d) Fourth mode of P-P boundary condition | | | | | | | | |
|---|---|---|---|---|---|---|---|---|
| $n$ | Algebraic Polynomials | | Chebyshev Polynomials | | Legendre Polynomials | | Hermite Polynomials | |
| | $\sqrt{\lambda_4}$ | Time (s) | $\sqrt{\lambda_4}$ | Time (s) | $\sqrt{\lambda_4}$ | Time (s) | $\sqrt{\lambda_4}$ | Time (s) |
| 4 | 8.3425 | 3.5354 | 8.3425 | 3.4245 | 8.3425 | 3.4347 | 8.3425 | 3.4572 |
| 5 | 8.3425 | 3.7092 | 8.3425 | 3.6324 | 8.3425 | 3.4960 | 8.3425 | 3.8244 |
| 6 | 6.8335 | 3.8387 | 6.8335 | 3.7739 | 6.8335 | 3.7894 | 6.8335 | 3.8633 |
| 7 | 6.8335 | 4.3183 | 6.8335 | 4.2208 | 6.8335 | 4.2890 | 6.8335 | 4.3989 |
| 8 | 6.7365 | 4.3755 | 6.7365 | 4.2074 | 6.7365 | 4.2307 | 6.7365 | 4.3688 |
| 9 | 6.7365 | 4.8668 | 6.7365 | 4.3046 | 6.7365 | 4.7444 | 6.7365 | 4.8565 |
| 10 | 6.7330 | 4.9806 | 6.7330 | 4.8038 | 6.7330 | 4.9333 | 6.7330 | 4.9792 |
| 11 | 6.7330 | 5.2498 | 6.7330 | 5.0967 | 6.7330 | 5.2320 | 6.7330 | 5.3207 |
| 12 | 6.7330 | 5.5772 | 6.7330 | 5.1289 | 6.7330 | 5.5050 | 6.7330 | 5.6721 |
| 13 | 6.7330 | 5.8718 | 6.7330 | 5.3501 | 6.7330 | 5.5581 | 6.7330 | 5.7281 |

orthogonal polynomial will be lesser than that of algebraic polynomials and this trend will be more remarkable if the problem was related to dynamical analysis of two- or three-dimensional structures i.e., plates or shells. As a result, one could conclude that orthogonal polynomials are better suited as shape functions in the Rayleigh–Ritz method for dealing with complicated structural dynamics problems.

## 7.6 Conclusion

The vibration characteristics of nanobeam embedded in Winkler–Pasternak elastic foundation are investigated using the Rayleigh–Ritz method in this chapter. In the Rayleigh–Ritz method, orthogonal polynomials such as Chebyshev polynomials, Legendre polynomials, and Hermite polynomials are utilised as shape functions. The orthogonality of these polynomials makes the technique more computationally efficient, and it avoids ill-conditioning for higher polynomial values. In addition, validations of present results are carried out with existing literature and a comparative study has been conducted to demonstrate the efficacy and convergence of the current model for various polynomials with respect to time. $n = 6$ is sufficient for the convergence of first mode frequencies, but $n = 8$, $n = 10$, $n = 12$, and $n = 14$ are needed for the second, third, fourth, and fifth modes of frequency parameters, respectively, and this holds true for all boundary conditions considered. In terms of time, all polynomials behave identically for smaller values of $n$. However, when n values are increased, the time required by orthogonal polynomials reduces relative to algebraic polynomials, making it more suitable for shape functions in the Rayleigh–Ritz method.

## References

Abdullah S S, Hosseini-Hashemi S, Hussein N A and Nazemnezhad R 2020 Temperature change effect on torsional vibration of nanorods embedded in an elastic medium using Rayleigh–Ritz method *J. Brazilian Soc. Mechanical Sci. Eng.* **42** 588

Behera L and Chakraverty S 2014 Free vibration of Euler and Timoshenko nanobeams using boundary characteristic orthogonal polynomials *Appl. Nanosci.* **4** 347–58

Chakraverty S and Behera L 2015 Vibration and buckling analyses of nanobeams embedded in an elastic medium *Chin. Phys. B* **24** 097305

Eringen A C 1983 On differential equations of nonlocal elasticity and solutions of screw dislocation and surface waves *J. Appl. Phys.* **54** 4703–10

Fakher M and Hosseini-Hashemi S 2017 Bending and free vibration analysis of nanobeams by differential and integral forms of nonlocal strain gradient with Rayleigh–Ritz method *Mater. Res. Express* **4** 125025

Jena S K and Chakraverty S 2021 Vibration analysis of nanostructural members using the Hermite–Ritz method *Static and Dynamic Behaviors of Nano Scaled Structural Problems* (Melville, NY: AIP Publishing) pp 8.1–8.26

Jena S K, Chakraverty S and Jena R M 2019 Propagation of uncertainty in free vibration of Euler–Bernoulli nanobeam *J. Brazilian Soc. Mechanical Sci. Eng.* **41** 436

Jena S K, Chakraverty S and Malikan M 2021 Application of shifted Chebyshev polynomial-based Rayleigh–Ritz method and Navier's technique for vibration analysis of a functionally graded porous beam embedded in Kerr foundation *Eng. Comput.* **37** 3569–89

Jena S K, Chakraverty S, Malikan M and Sedighi H 2020 Implementation of Hermite–Ritz method and Navier's technique for vibration of functionally graded porous nanobeam embedded in Winkler–Pasternak elastic foundation using bi-Helmholtz nonlocal elasticity *J. Mechanics Mater. Structures* **15** 405–34

Kumar Y 2018 The Rayleigh–Ritz method for linear dynamic, static and buckling behavior of beams, shells and plates: a literature review *J. Vib. Control* **24** 1205–27

Özköse H and Bağdatli S M 2010 The comparison of Hamilton method with Rayleigh's and Rayleigh–Ritz methods for the natural frequencies of three-supported beam *Int. J. Recent Res. Appl. Studies* **2010** 279–87

Reddy J 2007 Nonlocal theories for bending, buckling and vibration of beams *Int. J. Eng. Sci.* **45** 288–307

Wang C M, Zhang Y Y and He X Q 2007 Vibration of nonlocal Timoshenko beams *Nanotechnology* **18** 105401

Zhen Y and Fang B 2010 Thermal–mechanical and nonlocal elastic vibration of single-walled carbon nanotubes conveying fluid *Comput. Mater. Sci.* **49** 276–82

Zhu T L 2011 The vibrations of pre-twisted rotating Timoshenko beams by the Rayleigh–Ritz method *Comput. Mech.* **47** 395–408

# Chapter 8

# Application of polynomial functions in analyzing anti-plane wave profiles in a functionally graded piezoelectric–viscoelastic–poroelastic structure with buffer layer

**A K Singh and Sonam Singh**

This chapter illustrates the propagation of the anti-plane wave in a model consisting of three layers coated with a thin film. The first and third layers are composed of piezoelectric–viscoelastic (PV) and piezoelectric–poroelastic (PP) materials, respectively. The second layer lying between these piezo-composite layers is an integration of these two materials and is hence named as the buffer layer. The properties of this buffer material are a mix-up of PV and PP materials defined in terms of their volume fractions with the help of a gradient parameter in such a way that at the interfaces, this buffer becomes identical to the adjoining layers. The incorporation of volume fraction makes it difficult to solve the mechanics of the buffer by conventional methods, as the properties of the material become arbitrary functions of depth. Thus, to solve the governing equations of the buffer, the material parameters and the displacement and potential functions are considered as polynomials of infinite order, which are solved using the power series approach. The outcomes of this analysis are portrayed graphically and discussed in detail.

## 8.1 Introduction

In this era of developed materials which satisfies the need of the hour, there exist some non-centrosymmetric material which produces electric response and also accumulates electrical energy when they are mechanically disturbed and reversibly, their crystalline structure gets disturbed when subjected to an electric field. These materials are piezoelectric materials that are highly applicable in various scientific and engineering applications, such as sound detection, Love wave bio-sensors, SAW

doi:10.1088/978-0-7503-5067-9ch8  

devices, microbalances, and piezoelectric motors. Due to their material properties, some countries are using these materials for large-scale constructions such as roads, subways, etc. Some small machines like watches, earphones, headphones, speakers, printers, lighters, keyboards, phones also utilize these materials. A deep understanding of these materials and various wave propagation aspects in these materials may be found in the notable works of [1–6].

Despite numerous applications, there are some drawbacks such as brittle nature and the inability to bear high deformation essential for some applications, such as in parts of space crafts, aerospace, naval engineering, etc., have demoted the utilization of these materials. Therefore, the revival of this monolithic material may be done by improving their properties by adding a second phase of similar or different (i.e., piezoelectric or non-piezoelectric) material to these materials and making them composite or grading these materials with some other materials and making them functionally graded. This kind of tailoring or grading can only fulfill the ever-increasing requirements of engineering applications. Some of these composite materials adopted in the present work are piezoelectric–viscoelastic (PV) and piezoelectric–poroelastic (PP) materials. These materials offer lower acoustic impedance, higher hydrostatic coefficients, higher piezoelectric sensitivity, and lower density and stiffness than monolithic piezoceramics. In their works, [7–11] have explored these materials in brief. Wave propagation in composite PP material was discussed in the works of [12, 13]. The works of [14–20] discussed the characterization, development, and wave propagation aspects of different types of improved smart composite materials. This work has also considered the functionally graded materials (FGMs). These are a class of engineered materials and have spatially inhomogeneous microstructural properties. Aeronautical engineering, ceramic engines, nuclear fusion, electronic devices, coatings with anti-corrosion/wear properties, thin optical films, case hardened steel, gas turbine thermal barrier coatings, and bio-materials are some examples and areas of utilization of these graded materials. The works of [21, 22] discussed the characterization, development, and wave aspects of these improved composite materials.

When an FGM is considered, its material properties are not constant, but they are mostly defined with the help of volume fraction and constant material parameters of two different materials and gradient parameters. Thus the material properties of these FGMs are variable in form, and as a result, the governing equations related to these are ordinary differential equations (ODEs) with variable coefficients. The conventional method that solves an ODE with constant coefficients cannot give a solution in this case. To solve these ODEs with variable coefficient some higher methods are adopted such as Wentzel–Kramers–Brillouin (WKB) [10, 23, 24], analytical methods based on special function [25–28], perturbation technique [29–31], etc. These methods can be applied when the gradient of the material is considered in some special form or in some specific condition, for example, when the properties are defined as a special function of depth (linear, quadratic or exponential, etc) or in the case of high frequency or for small gradient coefficient.

The limitations of these methods bound the researchers to consider the gradient in a limited manner. Thus, to choose the gradient materials in the desired form and solve such ODEs, the power series technique was introduced. In this situation, the material parameters, the displacements and potential functions are chosen in terms of a polynomial function of infinite order. A reduced case for such consideration may be linear, quadratic, exponential, or trigonometric function form of gradient material. The governing equations in such cases are solved with the help of a convergent Taylor's series. The power series method is more stable than explicit alternatives, resulting in greater processing efficiency. It is precise as hybrid strategies like Crank Nicolson, but it's easier to put into practice. The power series method considering material properties and solutions of governing equations in terms of a polynomial function has been employed by many researchers such as [32–34] in the recent past.

For the mass loading process, in a layered model, a thin film is coated on the free surface of the model. The thickness of the coated film is considered to be infinitesimally small so that only the thickness and density and not the other material properties of the coated film have any impact on the model. This impact is mathematically represented by changing the stress-free condition of the guiding layer of the model. In physical terms, the prime sensing principal of acoustic sensors of Love wave devices largely depends on the mass loading. The entrapment of acoustic energy in the sensing device depends on the propagation speed of the wave. The difference in the properties of layer and half-space plays a vital role in keeping the wave energy near the surface. Thus any kind of loading at the surface of the guiding layer significantly affects the entrapment of acoustic energy in the sensing devices. Given this point, the study of mass loading sensitivity became an essential aspect of analyzing the efficiency of the considered model. Some previous works exploring the mass loading are [35–37].

In general, the bonding between two solids in a layered structure is not perfect due to impurities, micro-defects, other physical and chemical reasons, etc. This imperfect nature of bonding causes disruption in the continuous transfer of the displacement and potential. A buffer layer may be considered to avoid these discontinuities at the interface of two solids. The properties of the guiding layer and half-space material and their volume fractions are used to represent the material characteristics of the buffer layer. The fundamental reason for considering such a buffer layer at the layer interface is to reduce inter-laminar stress and other discontinuities at the laminate interface due to material mismatch across the interface. In their works [32, 38] has made some progress in this area, with the authors taking into consideration monolithic piezoelectric material.

In the present work, a layered structure composed of a PV guiding layer bonded to PP half-space with the help of a buffer layer is considered. The top of the PV layer is coated with a thin film to analyze the model's mass loading sensitivity. The expressions for the mechanical displacement and electric potential of the PV layer and PP half-space are derived by the separation of variable techniques. As the displacement and potential of the buffer layer are considered a polynomial function, they are derived using the power series method. The presence of viscosity in the guiding layer introduces loss. A complex equation relating the phase velocity with the complex form of wave number is obtained using these expressions in the considered boundary conditions. The dispersion and attenuation relation is derived from separating the real and imaginary parts of this complex equation. The impact of guiding and buffer layer width on dispersion and attenuation of the anti-plane wave is portrayed. The impact of the coated film is also demonstrated. The electromechanical coupling (EMC) is plotted, showing the impact of the guiding layer, buffer layer, and coated film width. The mass loading sensitivity is also discussed in both electrically open and short cases.

## 8.2 Statement and geometry of the problem

As portrayed in figure 8.1, the present problem is designed in such a way that a guiding PV layer of finite thickness $h_0$ is ad-joint to a PP half-space, and this joint or bonding is characterized with the help of a buffer layer which is of width $h$. As it exists at the joining of PV layer and PP half-space, the buffer layer is composed of a material which is a mixture of PV and PP, and hence called a functionally graded material. The upper free surface of the PV layer is coated with a thin, sensitive film, which introduces mass loading sensitivity in the model. In terms of Cartesian coordinates, the origin is taken at the interface of PV and buffer layer, the $x$-axis lies along the interface, and the positive direction of the $y$-axis is along the PP half-space.

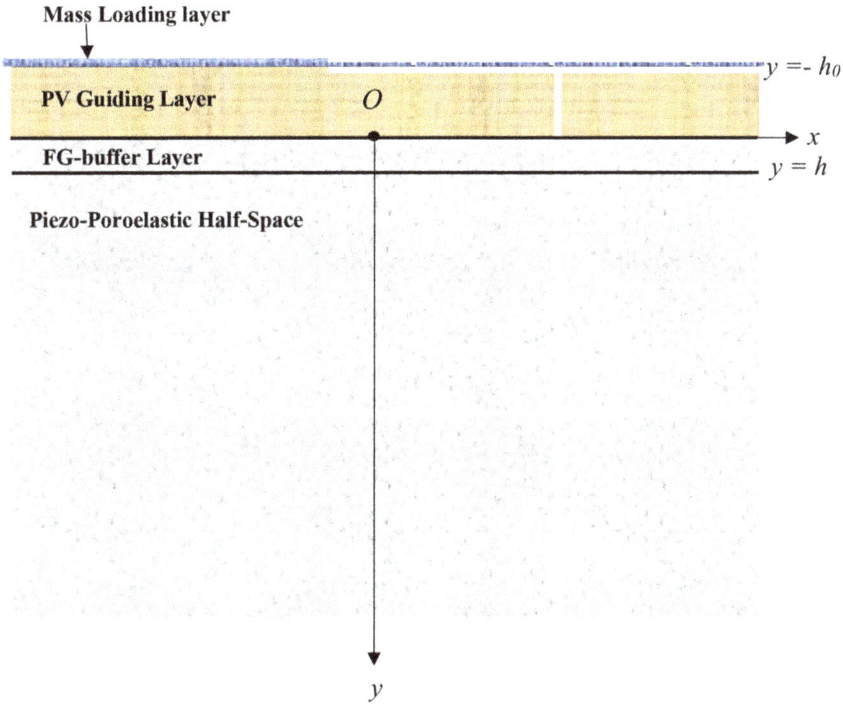


**Figure 8.1.** Schematic representation of the problem.

## 8.3 Constitutive and governing equations

### 8.3.1 For PV layer

The constitutive relations for PV material are

$$\begin{pmatrix}\sigma_{xx}^G\\ \sigma_{yy}^G\\ \sigma_{zz}^G\\ \sigma_{yz}^G\\ \sigma_{xz}^G\\ \sigma_{xy}^G\end{pmatrix}=\begin{pmatrix}c_{11}^G & c_{12}^G & c_{13}^G & 0 & 0 & 0\\ c_{12}^G & c_{22}^G & c_{23}^G & 0 & 0 & 0\\ c_{13}^G & c_{23}^G & c_{33}^G & 0 & 0 & 0\\ 0 & 0 & 0 & c_{44}^G & 0 & 0\\ 0 & 0 & 0 & 0 & c_{55}^G & 0\\ 0 & 0 & 0 & 0 & 0 & c_{66}^G\end{pmatrix}\begin{pmatrix}s_{xx}^G\\ s_{yy}^G\\ s_{zz}^G\\ 2s_{yz}^G\\ 2s_{xz}^G\\ 2s_{xy}^G\end{pmatrix}-\begin{pmatrix}0 & 0 & e_{31}^G\\ 0 & 0 & e_{32}^G\\ 0 & 0 & e_{33}^G\\ 0 & e_{24}^G & 0\\ e_{15}^G & 0 & 0\\ 0 & 0 & 0\end{pmatrix}\begin{pmatrix}E_x^G\\ E_y^G\\ E_z^G\end{pmatrix},$$

$$\begin{pmatrix}D_x^G\\ D_y^G\\ D_z^G\end{pmatrix}=\begin{pmatrix}0 & 0 & 0 & 0 & e_{15}^G & 0\\ 0 & 0 & 0 & e_{24}^G & 0 & 0\\ e_{31}^G & e_{32}^G & e_{33}^G & 0 & 0 & 0\end{pmatrix}\begin{pmatrix}s_{xx}^G\\ s_{yy}^G\\ s_{zz}^G\\ 2s_{yz}^G\\ 2s_{xz}^G\\ 2s_{xy}^G\end{pmatrix}+\begin{pmatrix}\xi_{11}^G & 0 & 0\\ 0 & \xi_{22}^G & 0\\ 0 & 0 & \xi_{33}^G\end{pmatrix}\begin{pmatrix}E_x^G\\ E_y^G\\ E_z^G\end{pmatrix}, \tag{8.1}$$

where the symbols $c_{ij}^G$, $e_{ij}^G$, and $\xi_{ij}^G$ represent mechanical stiffness, piezoelectric constants, and dielectric constants, respectively, and $G$ as superscript stands for the guiding layer. For incorporating viscoelasticity, the model of Kelvin–Voigt [39, 40] is taken up, which suggests that the material properties of PV must be defined as

$$c_{ij}^G = c_{ij}^R + i\omega c_{ij}^I,\ e_{ij}^G = e_{ij}^R + i\omega e_{ij}^I,\ \xi_{ij}^G = \xi_{ij}^R + i\omega \xi_{ij}^I,$$

where $\rho$ is the mass density; the terms with superscripts $R$ and $I$ represent the real and imaginary parts of the respective properties, and $\omega$ is the angular frequency.

Ignoring the body forces and electric forces for mathematical simplicity, the governing equations of the PV composite are

$$\sigma_{ij,j}^G = \rho^G \ddot{u}_i^G, \tag{8.2}$$

$$D_{i,i}^G = 0, \tag{8.3}$$

where $u_i^G$ is the component of mechanical displacement.

### 8.3.2 For PP half-space

The mechanical stresses of solid and fluid phase of PP material are [41]

$$\begin{aligned}\sigma_{ij}^H &= c_{ijkl}^H \epsilon_{kl}^H + m_{ij}^H \epsilon^{*H} - e_{kij}^H E_k^H - \zeta_{kij}^H E_k^{*H},\\ \sigma^{*H} &= m_{ij}^H \epsilon_{ij}^H + R^H \epsilon^{*H} - \zeta_k^H E_k^H - e_k^{*H} E_k^{*H},\end{aligned} \tag{8.4}$$

respectively, and the corresponding expression of electric field displacements are

$$
\begin{aligned}
D_i^H &= e_{ikl}^H \epsilon_{kl}^H + \zeta_i^H \epsilon^{*H} + \xi_{il}^H E_l^H + A_{il}^H E_l^{*H}, \\
D_i^{*H} &= \zeta_{ikl}^H \epsilon_{kl}^H + e_i^{*H} \epsilon^{*H} + A_{il}^H E_l^H + \xi_{il}^{*H} E_l^{*H}.
\end{aligned} \tag{8.5}
$$

Here, the stress tensors, strain tensors, electric displacement, and electric field vectors are $\sigma_{ij}^H$, $\sigma^{*H}$, $\epsilon_{ij}^H$, $\epsilon^{*H}$, $D_i^H$, $D_i^{*H}$, and $E_i^H$, $E_i^{*H}$. The material constants are $c_{ijkl}^H$, $m_{ij}^H$, $R^H$, piezoelectric constants are $e_{ijkl}^H$, $e^{*H}$, $\zeta_{ijkl}^H$, $\zeta_i^H$, and dielectric constants are $\xi_{ij}^H$, $\xi_{ij}^{*H}$, $A_{ij}^H$; $H$ as superscript stands for PP half-space. Throughout this work, * as superscript denotes the fluid phase.

Furthermore, we have the following relations relating strain with displacement and electric field with potential:

$$
\begin{aligned}
&\epsilon_{ij}^H = \frac{1}{2}\left(u_{i,j}^H + u_{j,i}^H\right), \quad \epsilon^{*H} = U_{i,i}^{*H}, \\
&E_i^H = -\Phi_{,i}^H, \quad E_i^{*H} = -\Phi_{,i}^{*H}, \quad i, j = 1, 2, 3.
\end{aligned} \tag{8.6}
$$

The equations of the PP material free from body forces and electric forces are [41]

$$
\sigma_{ij,j}^H = (\rho_{11})_{ij}^H \ddot{u}_i^H + (\rho_{12})_{ij}^H \ddot{U}_i^H, \tag{8.7}
$$

$$
\sigma_{,i}^{*H} = (\rho_{12})_{ij}^H \ddot{u}_i^H + (\rho_{22})_{ij}^H \ddot{U}_i^H, \; D_{i,i}^H = 0, \tag{8.8}
$$

$$
D_{,i}^{*H} = 0, \tag{8.9}
$$

where $(\rho_{11})_{ij}^H$, $(\rho_{12})_{ij}^H$, $(\rho_{22})_{ij}^H$ are the dynamical mass coefficients and $i, j = 1, 2, 3$.

### 8.3.3 For the buffer layer and air medium

As the interface of PV and PP solids is considered to be a buffer, it constitutes a functionally graded material which is characterized by the material constants and volume fraction of PV and PP. All the material parameters of this buffer are functions of depth, i.e., $y$. As there is no porosity in the PV layer, the porosity of the buffer at the interface of buffer and PV layer is taken to be zero. PP material is a material of a more generalized nature in comparison to PV. Thus, the mechanics of the buffer is defined similarly as that of PP using a superscript '*B*'.

To keep the study realistic, the medium above the PV layer is not a vacuum, but it is considered an air medium whose mechanics is governed by the Laplace equation. If $\phi_0$ is the potential of the air, then for $y \leqslant -h_0$, the Laplace equation is

$$
\frac{\partial^2 \phi_0}{\partial x^2} + \frac{\partial^2 \phi_0}{\partial y^2} = 0. \tag{8.10}
$$

### 8.3.4 For the coated film

To study the mass loading sensitivity of the model, the Rayleigh hypothesis is considered, which suggests that the coated film must be infinitesimally thin and have no elasticity. This means that the material properties of the coated layer do not affect the propagation of the anti-plane wave in the model, and only the width and mass density will make an impact [35]. If $H'$ is the width and $\rho'$ is the mass density of the coated film, then the mass loading sensitivity ($Sc$) in terms of change in phase velocity due to the coated layer is defined as:

$$Sc = \frac{1}{c_0} \lim_{\triangle m \to 0} \frac{c - c_0}{\triangle m},$$

where $c$ and $c_0$ are the phase velocities with and without coating, respectively, and $m$ is the mass deposited on the surface per unit area. Since the surface of the guiding layer was initially free from any deposition of mass, therefore the change in mass deposited must be $\triangle m = m(=\rho' H')$. Following this, the above condition reduces to

$$Sc = \frac{c - c_0}{c_0\, \rho' H'}.$$

Now, as per the geometry of the problem, the conditions for anti-plane wave propagation are

$$u_1^G = 0,\ u_2^G = 0,\ u_3^G = \bar{W}_1,\ \Phi^G = \Phi_1, \tag{8.11}$$

$$u_1^B = 0,\ u_2^B = 0,\ u_3^B = \bar{W}_2,\ \Phi^B = \Phi_2,\ U_1^B = 0,\ U_2^B = 0,\ U_3^B = W_2^*,\ \Phi^{*B} = \Phi_2^*, \tag{8.12}$$

$$u_1^H = 0,\ u_2^H = 0,\ u_3^H = \bar{W}_3,\ \Phi^H = \Phi_3,\ U_1^H = 0,\ U_2^H = 0,\ U_3^H = W_3^*,\ \Phi^{*H} = \Phi_3^*. \tag{8.13}$$

Here, the mechanical displacements ($\bar{W}_1$, $\bar{W}_2$, $\bar{W}_3$, $W_2^*$, $W_3^*$) and the electric potentials ($\Phi_1$, $\Phi_2$, $\Phi_3$, $\Phi_2^*$, $\Phi_3^*$) are chosen for $u_3^G$, $u_3^B$, $u_3^H$, $U_3^B$, $U_3^H$, $\Phi^G$, $\Phi^B$, $\Phi^H$, $\Phi^{*B}$ and $\Phi^{*H}$, respectively, which are functions of $x$, $y$, and $t$.

## 8.4 Boundary conditions

1. As the coated film's elastic characteristics are negligible, the conventional stress-free boundary condition at $y = -h_0$ is [37]

$$\tau_{13}^{(1)} = -\rho' H' \ddot{w}_1(-h_0). \tag{8.14}$$

2. At $y = -h_0$, there are two electric conditions which are electric short and open conditions denoted by

$$\Phi_1(x, -h_0) = 0, \tag{8.15}$$

and

$$\Phi_1(x, -h_0) = \phi_0(x, -h_0), \tag{8.16}$$

$$D_y^G(x, -h_0) = D_{y_0}(x, -h_0), \tag{8.17}$$

respectively.

3. At $y = 0$, the continuity conditions are

$$\bar{W}_1(x, 0) = \bar{W}_2(x, 0), \tag{8.18}$$

$$\sigma_{yz}^G(x, 0) = \sigma_{yz}^B(x, 0), \tag{8.19}$$

$$\Phi_1(x, 0) = \Phi_2(x, 0) + \Phi_2^*(x, 0), \tag{8.20}$$

$$D_y^G(x, 0) = D_y^B(x, 0) + D_y^{*B}(x, 0). \tag{8.21}$$

4. At $y = h$, the continuity conditions are

$$\bar{W}_2(x, h) = \bar{W}_3(x, h), \tag{8.22}$$

$$\sigma_{yz}^B(x, h) = \sigma_{yz}^H(x, h), \tag{8.23}$$

$$\Phi_2(x, h) = \Phi_3(x, h), \tag{8.24}$$

$$\Phi_2^*(x, h) = \Phi_3^*(x, h), \tag{8.25}$$

$$D_y^B(x, h) = D_y^H(x, h), \tag{8.26}$$

$$D_y^{*B}(x, h) = D_y^{*H}(x, h). \tag{8.27}$$

## 8.5 Solution procedure involved

### 8.5.1 For PV layer

Keeping equation (8.11) in mind, equations (8.2)–(8.3) reduce to two coupled equations

$$c_{44}^G\left(\frac{\partial^2 \bar{W}_1}{\partial y^2} + \frac{\partial^2 \bar{W}_1}{\partial x^2}\right) + e_{15}^G\left(\frac{\partial^2 \Phi_1}{\partial y^2} + \frac{\partial^2 \Phi_1}{\partial x^2}\right) = \rho^G \frac{\partial^2 \bar{W}_1}{\partial t^2}, \tag{8.28}$$

$$e_{15}^G\left(\frac{\partial^2 \bar{W}_1}{\partial y^2} + \frac{\partial^2 \bar{W}_1}{\partial x^2}\right) - \xi_{11}^G\left(\frac{\partial^2 \Phi_1}{\partial y^2} + \frac{\partial^2 \Phi_1}{\partial x^2}\right) = 0, \tag{8.29}$$

whose solution can be assumed as follows:

$$\{\bar{W}_1,\ \Phi_1\}(x, y, t) = \{w_1(y),\ \phi_1(y)\}e^{ik(x-ct)}.$$

With the help of this solution and using equation (8.29) in equation (8.28), we obtain

$$\frac{d^2 w_1}{dy^2} + k^2 a_G^2 w_1 = 0, \tag{8.30}$$

where $a_G^2 = \left(\frac{c}{\beta_1}\right)^2 - 1$, $\beta_1 = \sqrt{\frac{\bar{c}_{44}}{\rho^G}}$, and $\bar{c}_{44} = c_{44}^G + \frac{(e_{15}^G)^2}{\xi_{11}^G}$.

The expressions for mechanical displacement and electrical potential function derived from these equations are

$$\begin{aligned} \bar{W}_1 &= [A_1 \cos ka_G y + A_2 \sin ka_G y]e^{ik(x-ct)}, \\ \Phi_1 &= \left[A_3 e^{-ky} + A_4 e^{ky} + \frac{e_{15}^G}{\xi_{11}^G}(A_1 \cos ka_G y + A_2 \sin ka_G y)\right]e^{ik(x-ct)}, \end{aligned} \tag{8.31}$$

where $A_i$, $i = 1, 2, 3, 4$ are arbitrary constants.

### 8.5.2 For the buffer layer

For the buffer layer, in view of equation (8.12), the governing equations reduce to

$$\begin{aligned} \frac{\partial}{\partial y}\left(c_{44}^B \frac{\partial \bar{W}_2}{\partial y}\right) + \frac{\partial}{\partial x}\left(c_{44}^B \frac{\partial \bar{W}_2}{\partial x}\right) + \frac{\partial}{\partial y}\left(e_{15}^B \frac{\partial \Phi_2}{\partial y}\right) + \frac{\partial}{\partial x}\left(e_{15}^B \frac{\partial \Phi_2}{\partial x}\right) = (\rho_{11})_{33}^B \\ \frac{d^2 \bar{W}_2}{dt^2} + (\rho_{12})_{33}^B \frac{d^2 \bar{W}_2^*}{dt^2}, \end{aligned} \tag{8.32}$$

$$(\rho_{12})_{33}^B \frac{d^2 \bar{W}_2}{dt^2} + (\rho_{22})_{33}^B \frac{d^2 \bar{W}_2^*}{dt^2} = 0, \tag{8.33}$$

$$\begin{aligned} \frac{\partial}{\partial y}\left(e_{15}^B \frac{\partial \bar{W}_2}{\partial y}\right) + \frac{\partial}{\partial x}\left(e_{15}^B \frac{\partial \bar{W}_2}{\partial x}\right) - \frac{\partial}{\partial y}\left(\xi_{11}^B \frac{\partial \Phi_2}{\partial y}\right) - \frac{\partial}{\partial x}\left(\xi_{11}^B \frac{\partial \Phi_2}{\partial x}\right) \\ - \frac{\partial}{\partial y}\left(A_{11}^B \frac{\partial \Phi_2^*}{\partial y}\right) - \frac{\partial}{\partial x}\left(A_{11}^B \frac{\partial \Phi_2^*}{\partial x}\right) = 0, \end{aligned} \tag{8.34}$$

$$\frac{\partial}{\partial y}\left(A_{11}^B \frac{\partial \Phi_2}{\partial y}\right) + \frac{\partial}{\partial x}\left(A_{11}^B \frac{\partial \Phi_2}{\partial x}\right) + \frac{\partial}{\partial y}\left(\xi_{11}^{*B} \frac{\partial \Phi_2^*}{\partial y}\right) + \frac{\partial}{\partial x}\left(\xi_{11}^{*B} \frac{\partial \Phi_2^*}{\partial x}\right) = 0, \tag{8.35}$$

whose solution can be assumed as:

$$\{\bar{W}_2,\ \Phi_2,\ \Phi_2^*\}(x, y, t) = \{w_2(y),\ \phi_2(y),\ \phi_2^*(y)\}e^{ik(x-ct)}.$$

This buffer layer is a functionally graded material composed of PV and PP material. Now, the volume fractions of the PV layer and PP half-space are defined with the help of gradient parameter $p$ as

$$V_1(y) = 1 - \frac{1 - e^{py/h}}{1 - e^p} \quad \text{and} \quad V_2(y) = \frac{1 - e^{py/h}}{1 - e^p},$$

and the expressions of the buffer layer's material parameters are then defined as

$$f^B(y) = f^G V_1(y) + f^H V_2(y),$$

where $f^B(y)$ is any material parameter of buffer layer.

It can be observed from these two definitions that the buffer layer at the interface with PV layer and PP half-space is identical with them, respectively.

Using the assumed solutions, the governing equations become

$$c_{44}^B\left[\frac{d^2w_2}{dy^2} - k^2w_2\right] + (c_{44}^B)'\frac{dw}{dy} + e_{15}^B\left[\frac{d^2\phi_2}{dy^2} - k^2\phi_2\right] + (e_{15}^B)'\frac{d\phi}{dy} + (\rho^p)^B c^2k^2w_2 = 0, \quad (8.36)$$

$$e_{15}^B\left[\frac{d^2w_2}{dy^2} - k^2w_2\right] + (e_{15}^B)'\frac{dw_2}{dy} - \xi_{11}^B\left[\frac{d^2\phi_2}{dy^2} - k^2\phi_2\right] - (\xi_{11}^B)'\frac{d\phi_2}{dy} \quad (8.37)$$

$$-A_{11}^B\left[\frac{d^2\phi_2^*}{dy^2} - k^2\phi_2^*\right] - (A_{11}^B)'\frac{d\phi_2^*}{dy} = 0, \quad (8.38)$$

$$A_{11}^B\left[\frac{d^2\phi_2}{dy^2} - k^2\phi_2\right] + (A_{11}^B)'\frac{d\phi_2}{dy} + \xi_{11}^{*B}\left[\frac{d^2\phi_2^*}{dy^2} - k^2\phi_2^*\right] + (\xi_{11}^{*B})'\frac{d\phi_2^*}{dy} = 0, \quad (8.39)$$

where $(\rho^p)^B = (\rho_{11})_{33}^B - \frac{((\rho_{12})_{33}^B)^2}{(\rho_{22})_{33}^B}$.

These are ODEs with variable coefficients, and to obtain their solution, the power series technique may be employed. Under this series solution method, the material parameters are taken in the form of a polynomial function of infinite order, such as [32]:

$$\begin{aligned} c_{44}^B &= \sum_{n=0}^{\infty} c_n^{(1)}\left(\frac{y}{h}\right)^n, \quad e_{15}^B = \sum_{n=0}^{\infty} c_n^{(2)}\left(\frac{y}{h}\right)^n, \quad \xi_{11}^B = \sum_{n=0}^{\infty} c_n^{(3)}\left(\frac{y}{h}\right)^n, \\ A_{11}^B &= \sum_{n=0}^{\infty} c_n^{(4)}\left(\frac{y}{h}\right)^n, \quad \xi_{11}^{*B} = \sum_{n=0}^{\infty} c_n^{(5)}\left(\frac{y}{h}\right)^n, \quad (\rho^p)^B = \sum_{n=0}^{\infty} c_n^{(6)}\left(\frac{y}{h}\right)^n, \end{aligned} \quad (8.40)$$

where $c_n^{(i)}$ are arbitrary constants which are determined from the relations between the material parameters and the corresponding Taylor expansions.

Complying with the power series technique, the expressions of mechanical displacement and potential functions appearing in equations (8.36)–(8.38) are assumed as polynomial functions in the following form

$$w_2 = \sum_{n=0}^{\infty} q_n\left(\frac{y}{h}\right)^n, \quad \phi_2 = \sum_{n=0}^{\infty} r_n\left(\frac{y}{h}\right)^n, \quad \phi_2^* = \sum_{n=0}^{\infty} s_n\left(\frac{y}{h}\right)^n, \quad (8.41)$$

where $q_n$, $r_n$ and $s_n$ can be obtained using the recursion relation. Implementing the polynomial forms of material constants, mechanical displacement, and potential functions, the governing equations (8.36)–(8.38) yield

$$\begin{aligned}&\left(\sum_{n=0}^{\infty} c_n^{(1)}\left(\frac{y}{h}\right)^n\right)\left(\sum_{n=0}^{\infty} q_{n+2}^{(1)}(n+1)(n+2)\left(\frac{y}{h}\right)^n - (kh)^2\sum_{n=0}^{\infty} q_n^{(1)}\left(\frac{y}{h}\right)^n\right)\\&\quad+\left(\sum_{n=0}^{\infty} c_{n+1}^{(1)}(n+1)\left(\frac{y}{h}\right)^n\right)\left(\sum_{n=0}^{\infty} q_{n+1}^{(1)}(n+1)\left(\frac{y}{h}\right)^n\right)+\left(\sum_{n=0}^{\infty} c_n^{(2)}\left(\frac{y}{h}\right)^n\right)\\&\left(\sum_{n=0}^{\infty} r_{n+2}^{(1)}(n+1)(n+2)\left(\frac{y}{h}\right)^n - (kh)^2\sum_{n=0}^{\infty} r_n^{(1)}\left(\frac{y}{h}\right)^n\right)+\left(\sum_{n=0}^{\infty} c_{n+1}^{(2)}(n+1)\left(\frac{y}{h}\right)^n\right)\\&\left(\sum_{n=0}^{\infty} r_{n+1}^{(1)}(n+1)\left(\frac{y}{h}\right)^n\right)+c^2(kh)^2\sum_{(n=0)}^{(\infty)} c_n^6\left(\frac{y}{h}\right)^n\sum_{(n=0)}^{(\infty)} q_n\left(\frac{y}{h}\right)^n = 0,\end{aligned}\tag{8.42}$$

$$\begin{aligned}&\left(\sum_{n=0}^{\infty} c_n^{(2)}\left(\frac{y}{h}\right)^n\right)\left(\sum_{n=0}^{\infty} q_{n+2}^{(1)}(n+1)(n+2)\left(\frac{y}{h}\right)^n - (kh)^2\sum_{n=0}^{\infty} q_n^{(1)}\left(\frac{y}{h}\right)^n\right)\\&\quad+\left(\sum_{n=0}^{\infty} c_{n+1}^{(2)}(n+1)\left(\frac{y}{h}\right)^n\right)\left(\sum_{n=0}^{\infty} q_{n+1}^{(1)}(n+1)\left(\frac{y}{h}\right)^n\right)-\left(\sum_{n=0}^{\infty} c_n^{(3)}\left(\frac{y}{h}\right)^n\right)\\&\left(\sum_{n=0}^{\infty} r_{n+2}^{(1)}(n+1)(n+2)\left(\frac{y}{h}\right)^n - (kh)^2\sum_{n=0}^{\infty} r_n^{(1)}\left(\frac{y}{h}\right)^n\right)-\left(\sum_{n=0}^{\infty} c_{n+1}^{(3)}(n+1)\left(\frac{y}{h}\right)^n\right)\\&\left(\sum_{n=0}^{\infty} r_{n+1}^{(1)}(n+1)\left(\frac{y}{h}\right)^n\right)-\left(\sum_{n=0}^{\infty} c_n^{(4)}\left(\frac{y}{h}\right)^n\right)\left(\sum_{n=0}^{\infty} s_{n+2}^{(1)}(n+1)(n+2)\left(\frac{y}{h}\right)^n\right)-(kh)^2\\&\sum_{n=0}^{\infty} s_n^{(1)}\left(\frac{y}{h}\right)^n-\left(\sum_{n=0}^{\infty} c_{n+1}^{(4)}(n+1)\left(\frac{y}{h}\right)^n\right)\left(\sum_{n=0}^{\infty} s_{n+1}^{(1)}(n+1)\left(\frac{y}{h}\right)^n\right)=0,\end{aligned}\tag{8.43}$$

$$\begin{aligned}&\left(\sum_{n=0}^{\infty} c_n^{(4)}\left(\frac{y}{h}\right)^n\right)\left(\sum_{n=0}^{\infty} r_{n+2}^{(1)}(n+1)(n+2)\left(\frac{y}{h}\right)^n - (kh)^2\sum_{n=0}^{\infty} r_n^{(1)}\left(\frac{y}{h}\right)^n\right)\\&\quad+\left(\sum_{n=0}^{\infty} c_{n+1}^{(4)}(n+1)\left(\frac{y}{h}\right)^n\right)\left(\sum_{n=0}^{\infty} r_{n+1}^{(1)}(n+1)\left(\frac{y}{h}\right)^n\right)+\left(\sum_{n=0}^{\infty} c_n^{(5)}\left(\frac{y}{h}\right)^n\right)\\&\quad\left(\sum_{n=0}^{\infty} s_{n+2}^{(1)}(n+1)(n+2)\left(\frac{y}{h}\right)^n - (kh)^2\sum_{n=0}^{\infty} s_n^{(1)}\left(\frac{y}{h}\right)^n\right)\\&\quad+\left(\sum_{n=0}^{\infty} c_{n+1}^{(5)}(n+1)\left(\frac{y}{h}\right)^n\right)\left(\sum_{n=0}^{\infty} s_{n+1}^{(1)}(n+1)\left(\frac{y}{h}\right)^n\right)=0.\end{aligned}\tag{8.44}$$

Now, equating the coefficient of $\left(\frac{y}{h}\right)^n$ to zero in equations (8.42)–(8.44) yields the recursive relations for $q_n$, $r_n$ and $s_n$

$$\sum_{i=0}^{n} c_{n-i}^{(1)}[q_{i+2}(i+2)(i+1)-(kh)^2 q_i]+\sum_{i=0}^{n} c_{n-i+1}^{(1)}(n-i+1)q_{i+1}(i+1)+$$
$$\sum_{i=0}^{n} c_{n-i}^{(2)}[r_{i+2}(i+2)(i+1)-(kh)^2 r_i]+\sum_{i=0}^{n} c_{n-i+1}^{(2)}(n-i+1)r_{i+1}(i+1) \quad (8.45)$$
$$+(ckh)^2\sum_{i=0}^{n} c_{n-i}^{(6)}q_i=0,$$

$$\sum_{i=0}^{n} c_{n-i}^{(2)}[q_{i+2}(i+2)(i+1)-(kh)^2 q_i]+\sum_{i=0}^{n} c_{n-i+1}^{(2)}(n-i+1)q_{i+1}(i+1) \quad -$$
$$\sum_{i=0}^{n} c_{n-i}^{(3)}[r_{i+2}(i+2)(i+1)-(kh)^2 r_i]-\sum_{i=0}^{n} c_{n-i+1}^{(3)}(n-i+1)r_{i+1}(i+1) \quad - \quad (8.46)$$
$$\sum_{i=0}^{n} c_{n-i}^{(4)}[s_{i+2}(i+2)(i+1)-(kh)^2 s_i]-\sum_{i=0}^{n} c_{n-i+1}^{(4)}(n-i+1)s_{i+1}(i+1)=0,$$

$$\sum_{i=0}^{n} c_{n-i}^{(4)}[r_{i+2}(i+2)(i+1)-(kh)^2 r_i]+\sum_{i=0}^{n} c_{n-i+1}^{(4)}(n-i+1)r_{i+1}(i+1)+$$
$$\sum_{i=0}^{n} c_{n-i}^{(5)}[s_{i+2}(i+2)(i+1)-(kh)^2 s_i]+\sum_{i=0}^{n} c_{n-i+1}^{(5)}(n-i+1)s_{i+1}(i+1)=0. \quad (8.47)$$

Equations (8.45)–(8.47) provide linear recursive relations in $q_n$, $r_n$ and $s_n$ ($n = 0, \ldots, \infty$). Furthermore, for $i \geqslant 2$, $q_i$, $r_i$ and $s_i$ can be expressed as linear functions of $q_0$, $q_1$, $r_0$, $r_1$, $s_0$, and $s_1$.

A $6 \times 6$ identity matrix $I$ is used to decouple $q_0$, $q_1$, $r_0$, $r_1$, $s_0$, and $s_1$, using the following relation [32]

$$\left[q_{0j}, q_{1j}, r_{0j}, r_{1j}, s_{0j}, s_{1j}\right] = I; \quad j = 1-6. \quad (8.48)$$

With the help of this relation, the expression of displacement and potential functions can be rewritten as

$$w_2=\sum_{j=1}^{6}\bar{A}_j\sum_{n=0}^{\infty}q_{nj}\left(\frac{y}{h}\right)^n,\quad \phi_2=\sum_{j=1}^{6}\bar{A}_j\sum_{n=0}^{\infty}r_{nj}\left(\frac{y}{h}\right)^n, \times \phi_2^*=\sum_{j=1}^{6}\bar{A}_j\sum_{n=0}^{\infty}s_{nj}\left(\frac{y}{h}\right)^n, \quad (8.49)$$

where $\bar{A}_j$ are arbitrary constants to be determined. The terms $q_{nj}$, $r_{nj}$ and $s_{nj}$ in equation (8.49) may be determined with the aid of equation (8.48). In view of equation (8.49), the final expressions of displacement and potential functions of the buffer layer are

$$\bar{W}_2 = \left[\sum_{j=1}^{6}\bar{A}_j\sum_{n=0}^{\infty}q_{nj}\left(\frac{y}{h}\right)^n\right]e^{ik(x-ct)}, \quad \Phi_2 = \left[\sum_{j=1}^{6}\bar{A}_j\sum_{n=0}^{\infty}r_{nj}\left(\frac{y}{h}\right)^n\right]e^{ik(x-ct)}, \times \Phi_2^* = \left[\sum_{j=1}^{6}\bar{A}_j\sum_{n=0}^{\infty}s_{nj}\left(\frac{y}{h}\right)^n\right]e^{ik(x-ct)}. \tag{8.50}$$

For the other values of $n$; $q_{nj}$, $r_{nj}$ and $s_{nj}$ can be determined by solving equations (8.45)–(8.47), while in these recursive equations, $q_{nj}$, $r_{nj}$ and $s_{nj}$ are replaced by $q_n$, $r_n$ and $s_n$.

### 8.5.3 For PP half-space

Following the conditions for propagation mentioned in equation (8.13), the governing equations of PP (equations (8.7)–(8.9)) become

$$\frac{\partial}{\partial y}\left(c_{44}^H\frac{\partial\bar{W}_3}{\partial y}\right) + \frac{\partial}{\partial x}\left(c_{44}^H\frac{\partial\bar{W}_3}{\partial x}\right) + \frac{\partial}{\partial y}\left(e_{15}^H\frac{\partial\Phi_3}{\partial y}\right) + \frac{\partial}{\partial x}\left(e_{15}^H\frac{\partial\Phi_3}{\partial x}\right) = (\rho_{11})_{33}^H\frac{d^2\bar{W}_3}{dt^2} + (\rho_{12})_{33}^H\frac{d^2\bar{W}_3^*}{dt^2}, \tag{8.51}$$

$$(\rho_{12})_{33}^H\frac{d^2\bar{W}_3}{dt^2} + (\rho_{22})_{33}^H\frac{d^2\bar{W}_3^*}{dt^2} = 0, \tag{8.52}$$

$$\frac{\partial}{\partial y}\left(e_{15}^H\frac{\partial\bar{W}_3}{\partial y}\right) + \frac{\partial}{\partial x}\left(e_{15}^H\frac{\partial\bar{W}_3}{\partial x}\right) - \frac{\partial}{\partial y}\left(\xi_{11}^H\frac{\partial\Phi_3}{\partial y}\right) - \frac{\partial}{\partial x}\left(\xi_{11}^H\frac{\partial\Phi_3}{\partial x}\right) - \frac{\partial}{\partial y}\left(A_{11}^H\frac{\partial\Phi_3^*}{\partial y}\right) - \frac{\partial}{\partial x}\left(A_{11}^H\frac{\partial\Phi_3^*}{\partial x}\right) = 0, \tag{8.53}$$

$$\frac{\partial}{\partial y}\left(A_{11}^H\frac{\partial\Phi_3}{\partial y}\right) + \frac{\partial}{\partial x}\left(A_{11}^H\frac{\partial\Phi_3}{\partial x}\right) + \frac{\partial}{\partial y}\left(\xi_{11}^{*H}\frac{\partial\Phi_3^*}{\partial y}\right) + \frac{\partial}{\partial x}\left(\xi_{11}^{*H}\frac{\partial\Phi_3^*}{\partial x}\right) = 0, \tag{8.54}$$

whose solutions may be assumed as

$$\left\{\bar{W}_3,\ \Phi_3,\ \Phi_3^*\right\}(x, y, t) = \left\{w_3(y),\ \phi_3(y),\ \phi_3^*(y)\right\}e^{ik(x-ct)}.$$

Using this solution and applying the same procedure used in section 8.5.1 we have

$$c_{44}^H\left[\frac{d^2w_3}{dy^2} - k^2w_3\right] + e_{15}^H\left[\frac{d^2\phi_3}{dy^2} - k^2\phi_3\right] + (\rho^p)^H c^2k^2w_3 = 0, \tag{8.55}$$

$$e_{15}^H\left[\frac{d^2w_3}{dy^2} - k^2w_3\right] - (\xi_{11}^p)^H\left[\frac{d^2\phi_3}{dy^2} - k^2\phi_3\right] = 0, \tag{8.56}$$

where $(\rho^p)^H = (\rho_{11})_{33}^H - \frac{((\rho_{12})_{33}^H)^2}{(\rho_{22})_{33}^H}$ and $(\xi_{11}^p)^H = \xi_{11}^H - \frac{(A_{11}^H)^2}{\xi_{11}^{*H}}$.

Since the anti-plane wave in half-space dies out as $y \to 0$, thus, the displacement and potential functions follow $\bar{W}_3 \to 0, \Phi_3 \to 0$ and $\Phi_3^* \to 0$ as $y \to \infty$, therefore, the solutions of equations (8.55) and (8.56) are

$$\bar{W}_3 = A_{10}e^{-ka_H y}e^{ik(x-ct)}, \quad \Phi_3 = \left[A_{11}e^{-ky} + \frac{e_{15}^H}{(\xi_{11}^p)^H}A_{10}e^{-ka_H y}\right]e^{ik(x-ct)},$$
$$\Phi_3^* = \left[A_{12}e^{-ky} - A_{11}\frac{A_{11}^H}{\xi_{11}^{*H}}e^{-ky} - A_{10}\frac{e_{15}^H A_{11}^H}{\xi_{11}^{*H}\xi_{11}^H}e^{-ka_H y}\right]e^{ik(x-ct)}, \tag{8.57}$$

where $a_H^2 = 1 - \left(\frac{c}{\beta_2}\right)^2$, $\beta_2 = \sqrt{\frac{\bar{\bar{c}}_{44}}{(\rho^p)^H}}$ and $\bar{\bar{c}}_{44} = c_{44}^H + \frac{(e_{15}^H)^2}{(\xi_{11}^p)^H}$.

### 8.5.4 For air medium

Since the air medium is considered in the half-space $y \leqslant -h_0$, the condition for the potential of the anti-plane wave is $\phi_0 \to 0$ as $y \to -\infty$. Therefore, the solution of equation (8.10) is

$$\phi_0 = A_{13}e^{ky}e^{ik(x-ct)}. \tag{8.58}$$

## 8.6 Dispersion relation

To derive the dispersion relation of propagating anti-plane wave in the considered layered structure, the solution of the expression of displacement and potential functions are substituted in the respective boundary conditions.

Particularly, for the electrically open (EO) case using equations (8.31), (8.50), (8.57), and (8.58) in the boundary conditions equations (8.14), (8.16)–(8.27), and eliminating the arbitrary constants, the following determinant is obtained

$$\left|D_{ij}^O\right| = 0, \quad i, j = 1 - 13. \tag{8.59}$$

The above determinant, on equating to zero, provides the dispersion relation in the EO case.

Similarly, for the electrically short (ES) case using equations (8.31), (8.50), (8.57), and (8.58) in the boundary equations (8.14), (8.15), (8.18)–(8.27), and eliminating the arbitrary constants, the following determinant is obtained

$$\left|D_{ij}^S\right| = 0, \quad i, j = 1 - 12. \tag{8.60}$$

The above determinant, on equating to zero, provides the dispersion relation in ES case.

The existence of viscoelasticity accounts for attenuation of anti-plane waves in the model. In order to analyze the attenuation we have taken the velocity of the propagating wave in complex form. Thus, these dispersion relations contain real and imaginary parts, and separating them we can have relations for dispersion and attenuation of the anti-plane wave.

The entries $D_{ij}^{O}$ and $D_{ij}^{S}$ are provided in appendix A and the superscripts $O$ and $S$ stand for the EO and ES cases, respectively.

## 8.7 Special cases pertaining to this study

In this section some pre-existing works are considered and the present work is reduced to a similar structure and then the dispersion equation is validated with them.

### 8.7.1 Case 1—validation with the work of [32]

In their work, [32] considered a piezoelectric layer bonded with a piezoelectric half-space having a functionally graded buffer layer at their interface, and studied the propagation of SH wave in this model. In the present work, if it is assumed that there is no coating at the free surface (i.e., $h' \to 0$), the viscoelastic counterpart of each material parameter tends to zero, and the terms associated with fluid phase and the coupling parameters of solid and fluid phase tend to zero. In these conditions, the present model theoretically becomes similar to the work of [32]. When these changes are mathematically implemented to the dispersion relation in EO and ES cases, similar dispersion relations as those of [32] are obtained. This validates the present study.

### 8.7.2 Case 2—validation with the work of [42]

In their work, [42] studied the propagation of Love wave in a structure comprised of an isotropic layer bonded with an isotropic substrate. In the present work, if it is assumed that there is no coating at the free surface (i.e., $h' \to 0$), the viscoelastic counterpart of each material parameter tends to zero, the piezoelectric and dielectric coefficients of the PV layer and PP half-space tends to zero, and the terms associated with fluid phase, the coupling parameters tends to zero. Also, no buffer layer exists at the interface (i.e., $h \to 0$). Then in this situation the present model theoretically becomes an isotropic layer attached with an isotropic half-space. On implementing these changes (i.e., $\{(c_{44}^{I})^{G}, (e_{15})^{G}, (\xi_{11})^{G}\} \to 0, \quad \{(e_{15})^{H}, (\xi_{11})^{H}, \xi_{11}^{*H}, A_{11}^{H}, (\rho_{12})_{33}^{H}, (\rho_{22})_{33}^{H}\} \to 0$) the dispersion relation of the present work (in both EO and ES cases) reduces to

$$\tan\left[kH\sqrt{\frac{c^2}{(\beta_1^0)^2} - 1}\right] = \frac{\mu_2\sqrt{1 - \dfrac{c^2}{(\beta_2^0)^2}}}{\mu_1\sqrt{\dfrac{c^2}{(\beta_1^0)^2} - 1}}, \tag{8.61}$$

where, $\beta_1^0 = \sqrt{\mu_1/\rho_1}$ and $\beta_2^0 = \sqrt{\mu_2/\rho_2}$.

Here, equation (8.61) is similar to the dispersion relation obtained by [42].

## 8.8 Numerical discussion

In this section the propagation of the anti-plane wave is analyzed numerically and graphically for the material data of Epoxy-BNKLBT as PV layer and PZT−1 as PP half-space (table 8.1). The coating of thin film is considered to be of $SiO_2$.

**Table 8.1.** Material parameters.

| Material | Epoxy-BNKLBT [43, 44] | | PZT−1[45] Half-space |
|---|---|---|---|
| Elastic constant (N m$^{-2}$) | $C_{44}^R = 7.511 \times 10^9$ | $C_{44}^I = 9.725\,2 \times 10^9$ | $C_{44}^H = 22.2 \times 10^9$ |
| Piezoelectric constant (C m$^{-2}$) | $e_{15}^R = -0.091\,9$ | $e_{15}^I = -0.000\,285$ | $e_{15}^H = 9.8$ |
| Dielectric constant (F m$^{-1}$) | $\xi_{11}^R =- 0.304\,1 \times 10^{-9}$ | $\xi_{11}^I = -0.02 \times 10^{-12}$ | $\xi_{11}^H = 8.76 \times 10^{-9}$<br>$\xi_{11}^{*H} = 10.9 \times 10^{-9}$ |
| Mass density (kg m$^{-3}$) | $\rho = 3949$ | | $(\rho_{11})_{33}^H = 4950$<br>$(\rho_{12})_{33}^H = -1125$<br>$(\rho_{22})_{33}^H = 4800$ |

An equivalent or averaged thickness of guiding (PV) layer to describe the wave propagation features is defined as follows:

$$H = h_0 + \int_0^h V_1(y)dy. \tag{8.62}$$

Equation (8.62) shows that the width of guiding and buffer layer are associated with each other in such a way that a thicker guiding layer implies a thinner buffer layer.

In order to maintain a slow variance in the material properties of the buffer layer, the material properties of PV layer and PP half-space must satisfy

$$|(f^G - f^H)/f^G| \leqslant 1.$$

Keeping this condition in mind all the numerical results satisfy the convergence condition

$$\frac{|c_{k=N} - c_{k>N}|}{c_{k=N}} < \epsilon,$$

of the power series, with $c_k$ being the phase velocity derived by using the first $k$ terms of the power series and $\epsilon \approx 0.001$.

In the upcoming subsections, the figures are plotted against the dimensionless wave number ($kH$) showing variation of dimensionless phase velocity ($c/\beta_1$) in figures 8.2 and 8.3; EMC parameter ($K^2$) in figure 8.4 and mass loading sensitivity ($|Sc|$) in figure 8.5.

### 8.8.1 Effect of guiding layer width

Figures 8.2(a) and (b) represent the impact of the width ratio ($h_0/H$) of the guiding layer on the phase velocity of the anti-plane wave in electrically open and short cases, respectively. The curves 1, 2 and 3 represent the fundamental mode and 4, 5 and 6 represent the 2nd mode of vibration. It is seen in both figures that the impact

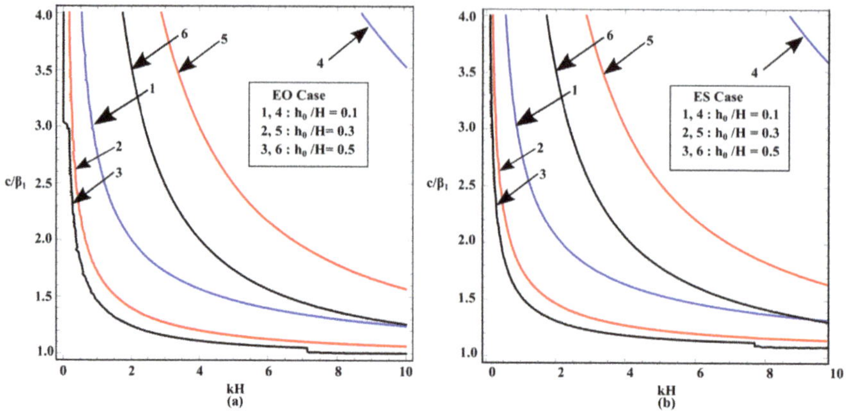


**Figure 8.2.** Impact of width ratio of guiding layer ($h_0/H$) on phase velocity (dimensionless) of the anti-plane wave.

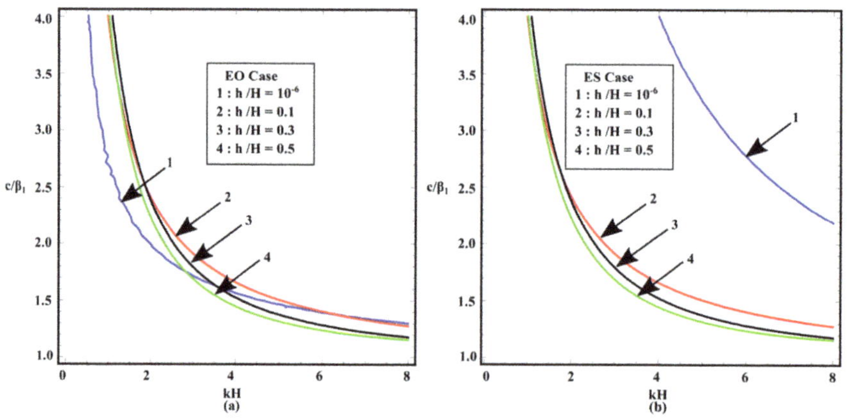


**Figure 8.3.** Impact of width ratio of buffer layer ($h/H$) on phase velocity (dimensionless) of the anti-plane wave.

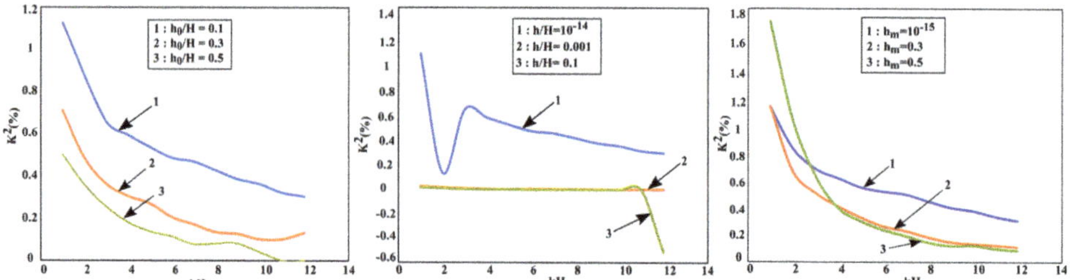


**Figure 8.4.** Variation of EMC parameter for distinct value *s* of (a) guiding layer width, (b) buffer layer width, (c) coated film width.

of increasing $h_0/H$ is decreasing on $c/\beta_1$ in both cases and in both modes of vibration. Additionally, for both the EO and ES cases considering $h_0/H = 0.1$, the 2nd vibrational mode is observed at a high wave number compared to other values of width ratio.

### 8.8.2 Effect of sandwiched FG-buffer layer

Figures 8.3(a) and (b) represent the impact of the width ratio ($h/H$) of the buffer layer on the phase velocity of the anti-plane wave in electrically open and short cases, respectively. It is to be noted that $h/H = 10^{-6}$ indicates the case when the buffer layer ceases to exist. Considering the EO case, it can be seen that for $h/H = 10^{-6}$, initially phase velocity is lowest among all considered values of $h/H$. However, after a certain wave number phase velocity becomes highest. Except for some initial values of wave number, as $h/H$ keeps taking larger values, the phase velocity keeps decreasing. For the ES case it can be seen from figure 8.3(b) that for $h/H = 10^{-6}$ the phase velocity is quite high compared to other values of $h/H$. Overall, the impact of increasing $h/H$ is decreasing on phase velocity.

### 8.8.3 Electromechanical coupling parameter ($K^2$)

Figures 8.4(a)–(c) represent the impact of width ratio ($h_0/H$) of the guiding layer, width ratio ($h/H$) of buffer layer, and width of coated sensitive layer on EMC, respectively, on the EMC parameter ($K^2$). It can be seen from figure 8.4(a) that initially for low wave numbers the magnitude of EMC is high, which decreases with increasing $kH$. For $h_0/H = 0.5$, it tends to 0 after $kH = 10.5$. The impact of increasing $h_0/H$ is decreasing on EMC.

In the case of figure 8.4(b), when $h/H = 10^{-14}$ which means that the buffer layer ceases to exist, EMC initially decreases till $kH = 2$, following which it increases till $kH = 4$, and thereafter it keeps declining. For $h/H = 0.001$, 0.1 EMC is close to 0 for almost all values of $kH$. Impact of increasing $h/H$ is decreasing on EMC.

From figure 8.4(c), the nature of EMC against $kH$ is observed to be decreasing for the increasing widths of coated sensitive layer past $kH = 4$. Similar to figures 8.4(a) and (b), in this case also, EMC is high for low wave numbers and keeps on decreasing with increasing $kH$.

### 8.8.4 Effect of mass loading sensitivity

The mass loading sensitivity $|Sc|$ of the present model is derived and plotted for a coating of $SiO_2$ having mass density $\rho' = 2650$ kg m$^{-3}$ and thickness $H' = 0.05$ m. $|Sc|$ as observed from figure 8.5 is high for low wave numbers and it reduces as the wave number increases for both EO and ES cases. Figure 8.5 also depicts that $|Sc|$ is high for the ES case as compared to the EO case till $kH = 4$, following which the sensitivity becomes high for the EO case and lowers for the ES case.

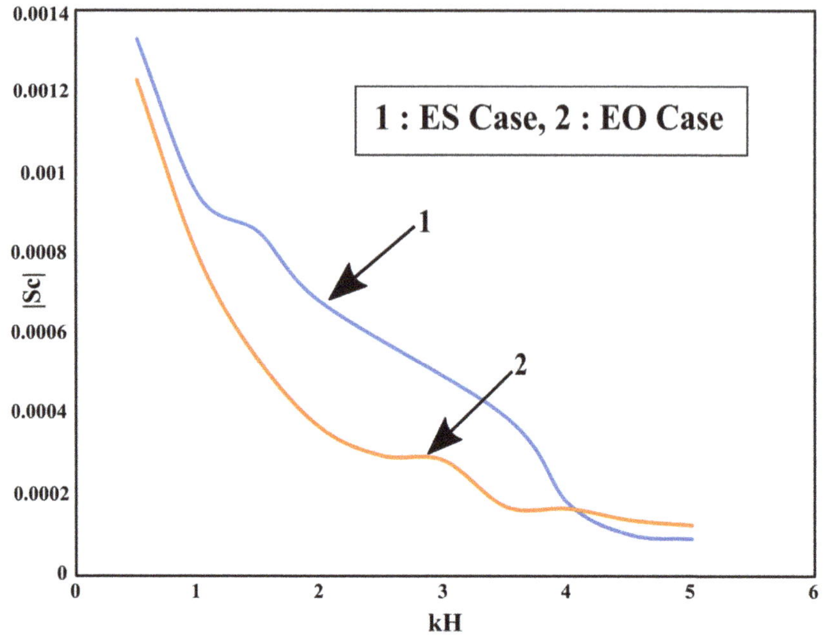


**Figure 8.5.** Mass loading sensitivity.

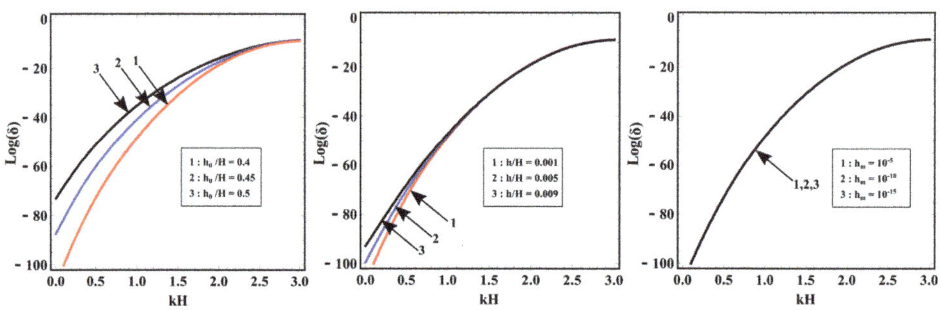


**Figure 8.6.** Impact of (a) width ratio of guiding layer ($h_0/H$), (b) width ratio of buffer layer ($h/H$), and (c) width of coated layer ($h_m$) on attenuation coefficient log($\delta$) of the anti-plane wave.

### 8.8.5 Attenuation of anti-plane wave

From figures 8.6(a) and (b), it is clear that impact of increasing $h_0/H$ and $h/H$ is increasing on log($\delta$). However, from figure 8.6(c) it is found that impact of $h_m$ is negligible on log($\delta$) as all the three curves are overlapped.

## 8.9 Conclusions

In this chapter, the propagation of anti-plane wave in a mass-loaded tri-layered structure (PV layer/functionally graded buffer layer/PP half-space) is investigated in detail. The material parameters and displacement and potential function associated with the functionally graded buffer layer are considered as a polynomial function of depth of infinite order. The governing equations of the buffer layer are ODEs with variable coefficients. Owing to its high precision and extensive applicability, the power series approach is used to solve the governing equations of the functionally graded buffer layer. At the top surface of the PV layer, two electric conditions (electrically open and short) are assumed. Dispersion relation bearing the influence of gradient parameter and mass loading layer widths is derived in all cases. The dispersion and attenuation of the anti-plane wave are derived by separating the real and imaginary parts of the complex dispersion equation. The dispersion and attenuation are portrayed, demonstrating the impact of layer widths and gradient parameter. The EMC and mass loading sensitivity of the model are also discussed.

The following outcomes are drawn from this study:

- For increasing wave number the phase velocity decreases whereas the attenuation increases.
- A thicker guiding layer is favorable for low phase velocity in both open and short electric cases.
- A thicker buffer layer is suitable for low phase velocity after some initial values of wave number for both open and short electric cases.
- The electromechanical coupling parameter ($K^2$) is high for thin guiding layer. A similar impact of buffer layer and coated film width is observed on $K^2$.
- Mass loading sensitivity is high for the short electric case.
- A thin guiding and buffer layer is suitable for low attenuation.
- Thickness of coated film has an insignificant impact on attenuation of the anti-plane wave.

## Appendix A

$$D_{11}^{o} = \bar{c}_{44}ka_G \sin(ka_G h_0) + \rho' H' \cos(ka_G h_0), \quad D_{12}^{o} = \bar{c}_{44}ka_G \cos(ka_G h_0) + \rho' H' \sin(ka_G h_0),$$

$$D_{14}^{o} = - e_{15}^{G} k e^{-kh_0}, \quad D_{1j}^{o} = 0, \quad j = 5 - 13, \quad D_{2j}^{o} = 0 \quad \text{for } j = 1, 2, \quad D_{23}^{o} = k\xi_{11}^{G} e^{kh_0},$$

$$D_{24}^{o} = - k\xi_{11}^{G} e^{-kh_0}, \quad D_{2j}^{o} = 0 \quad \text{for } j = 5 - 12, \quad D_{2-13}^{o} = k\xi_{11}^{A} e^{-kh_0}, \quad D_{31}^{o} = \frac{e_{15}^{G}}{\xi_{11}^{G}} \cos(ka_G h_0),$$

$$D_{32}^{o} = - \frac{e_{15}^{G}}{\xi_{11}^{G}} \sin(ka_G h_0)\ D_{33}^{o} = e^{kh_0}, \quad D_{34}^{o} = e^{-kh_0}, \quad D_{3-13}^{o} = -e^{-kh_0}, \quad D_{3j} = 0$$

$$\text{for } j = 5 - 12, \quad D_{41} = 1, \quad D_{41} = 1, \quad D_{4j} = 0 \quad \text{for } j = 2 - 13 \quad \text{and} \quad j \neq 5,$$

$$D_{51} = 0,$$

$$D_{52} = \bar{c}_{44}ka_G, \quad D_{53} = -ke_{15}^{G}, \quad D_{54} = ke_{15}^{G}, \quad D_{55} = 0, \quad D_{56} = -c_{44}^{0}/h, \quad D_{57} = 0,$$

$$D_{58} = -e_{15}^{0}/h,$$

$$D_{5j} = 0 \quad \text{for } j = 9 - 13, \quad D_{61} = e_{15}/\xi_{11}, \quad D_{62} = 0, \quad D_{63} = 1, \quad D_{64} = 1, \quad D_{65} = 0,$$

$$D_{66} = 0,$$

$$D_{67} = - 1, \quad D_{6j}^{o} = 0 \quad \text{for } j = 8 - 13, \quad D_{71}^{o} = 0, \quad D_{72}^{o} = 0, \quad D_{73}^{o} = k\xi_{11}^{G}, \quad D_{74}^{o} = -k\xi_{11}^{G},$$

$$D_{75}^{o} = 0, \quad D_{76}^{o} = -e_{15}^{0}/h, \quad D_{78}^{o} = (A_{11}^{0} + \xi_{11}^{0})/h, \quad D_{79}^{o} = (A_{11}^{0} + \xi_{11}^{*0})/h, \quad D_{7j}^{o} = 0$$

$$\text{for } j = 10 - 13,$$

$$D_{8j}^{o} = 0 \quad \text{for } j = 1 - 4, \quad D_{85}^{o} = 1, \quad D_{86}^{o} = 1, \quad D_{8-10}^{o} = -e^{-ka_H h},$$

$$D_{8j}^{o} = 0 \quad \text{for } j = 7 - 13 \quad \text{and} \quad j \neq 10,$$

$$D_{96}^{o} = c_{44}^{h}/h, \quad D_{98}^{o} = e_{15}^{h}/h, \quad D_{9-10}^{o} = \bar{\bar{c}}_{44}ka_H e^{-ka_H h}, \quad D_{9-10}^{o} = ke_{15}^{H} e^{-kh},$$

$$D_{9j}^{o} = 0 \quad \text{for } j = 1 - 13 \quad \text{and}$$

$$j \neq 6, 8, 10, 11, \quad D_{10-j}^{o} = 0 \quad \text{for } j = 1 - 6, \quad D_{10-7}^{o} = 1, \quad D_{10-8}^{o} = 1, \quad D_{10-9}^{o} = 0,$$

$$D_{10-10}^{o} = - e_{15}^{H}/\xi_{11}^{p}, \quad D_{10-11}^{o} = -e^{-kh}, \quad D_{10-12}^{o} = 0, \quad D_{10-13}^{o} = 0, \quad D_{1j}^{o} = 0 \quad \text{for } j = 1 - 8,$$

$$D_{11-9}^{o} = 1, \quad D_{11-10}^{o} = \frac{e_{15}^{H} A_{11}^{H}}{\xi_{11}^{H} \xi_{11}} e^{-ka_H h}, \quad D_{11-11}^{o} = A_{11}^{H}/\xi_{11}^{*H} e^{-kh}, \quad D_{11-12}^{o} = -e^{-kh},$$

$$D_{11-13}^{o} = 0, D_{12-j}^{o} = 0 \quad \text{for } j = 1 - 5, D_{12-6}^{o} = e_{15}^{h}/h, D_{12-8}^{o} = -\xi_{11}^{h}/h, D_{12-9}^{o} = -A_{11}^{h}/h,$$

$$D_{12-11}^{o} = - k\xi_{11}^{H} e^{-kh}, \quad D_{12-12}^{o} = -kA_{11}^{H} e^{-kh}, \quad D_{12-13}^{o} = 0 \quad D_{13-j}^{o} = 0 \quad \text{for } j = 1 - 7,$$

$$D_{13-8}^{o} = - A_{11}^{h}/h, \quad D_{13-9}^{o} = -\xi_{11}^{*h}/h, \quad D_{13-10}^{o} = 0, \quad D_{13-11}^{o} = 0, \quad D_{13-12}^{o} = -k\xi_{11}^{*H} e^{-kh},$$

$$D_{13-13}^{o} = 0, \quad D_{1j}^{s} = D_{1j}^{o} \quad \text{for } j = 1 - 12, \quad D_{2j}^{s} = D_{3j}^{o} \quad \text{for } j = 1 - 12,$$

$$D_{3j}^{s} = D_{4j}^{o} \quad \text{for } j = 1 - 12,$$

$$D_{4j}^{s} = D_{5j}^{o} \quad \text{for } j = 1 - 12, \quad D_{5j}^{s} = D_{6j}^{o} \quad \text{for } j = 1 - 12, \quad D_{6j}^{s} = D_{7j}^{o} \quad \text{for } j = 1 - 12,$$

$$D_{7j}^{s} = D_{8j}^{o} \quad \text{for } j = 1 - 12, \quad D_{8j}^{s} = D_{9j}^{o} \quad \text{for } j = 1 - 12, \quad D_{9j}^{s} = D_{10-j}^{o} \quad \text{for } j = 1 - 12,$$

$$D_{10-j}^{s} = D_{11-j}^{o}$$

$$\text{for } j = 1 - 12, \quad D_{11-j}^{s} = D_{12-j}^{o} \quad \text{for } j = 1 - 12, \quad D_{12-j}^{s} = D_{13-j}^{o} \quad \text{for } j = 1 - 12,$$

the terms with superscripts 0 and $h$ correspond to the data associated with PV and PP materials, respectively.

# References


[1] Auld B A 1990 Acoustic resonators *Acoustic Fields Waves in Solids* **2** 250–9

[2] Harvey A P, Craine R E and Syngellakis S 1992 Propagation of nonlinear surface acoustic waves on elastic and piezoelectric solids *J. Mech. Phys. Solids* **40** 1529–42

[3] Sosa H and Khutoryansky N 1996 New developments concerning piezoelectric materials with defects *Int. J. Solids Struct.* **33** 3399–414

[4] Walia V, Sharma J N and Sharma P K 2009 Propagation characteristics of thermoelastic waves in piezoelectric(6 mm class) rotating plate *Eur. J. Mech.* A **28** 569–81

[5] Pankaj K K, Sahu S A and Kumari S 2020 Surface wave transference in a piezoelectric cylinder coated withreinforced material. *Appl. Math. Mech.* **41** 123–38

[6] Guha S, Singh A K and Das A 2021 Analysis on different types of imperfect interfaces between twodissimilar piezothermoelastic half-spaces on reflection and refraction phenomenon of plane waves *Waves Random Complex Media* **31** 660–89

[7] Chen C, Li S, Dai L and Qian C 2014 Buckling and stability analysis of a piezoelectric viscoelasticnanobeam subjected to van der waals forces *Commun. Nonlinear Sci. Numer. Simul.* **19** 1626–37

[8] Zenkour A M and Sobhy M 2018 Nonlocal piezo-hygrothermal analysis for vibration characteristics of a piezoelectric Kelvin-Voigt viscoelastic nanoplate embedded in a viscoelastic medium *Acta Mech.* **229** 3–19

[9] Chunyu L and Weng G J 2002 Antiplane crack problem in functionally graded piezoelectric materials *J. Appl. Mech.* **69** 481–8

[10] Li X Y, Wang Z K and Huang S H 2004 Love waves in functionally graded piezoelectric materials *Int. J. Solids Struct.* **41** 7309–28

[11] Singh S and Singh A K 2021 Anti-plane surface and interfacial waves influenced by layer reinforcement in piezo-electro-magnetic structures with surface energy *Eur. Phys. J. Plus* **136** 312

[12] Singh S, Singh A K and Guha S 2021 Impact of interfacial imperfections on the reflection andtransmission phenomenon of plane waves in a porous-piezoelectric model *Appl. Math. Model.* **100** 656–75

[13] Singh A K, Rajput P, Guha S and Singh S 2022 Propagation characteristics of love-type wave at theelectro-mechanical imperfect interface of a piezoelectric fiber-reinforced composite layer overlying a piezoelectric half-space *Eur. J. Mech.* A **93** 104527

[14] Singh A K and Guha S 2020 Reflection of plane waves from the surface of a piezothermoelastic fiber-reinforced composite half-space *Mech. Adv. Mater. Struct.* **28** 2370–82

[15] Singh A K, Mahto S and Guha S 2021 Analysis of plane wave reflection phenomenon from the surface of amicro-mechanically modeled piezomagnetic fiber-reinforced compositehalf-space *Waves Random Complex Media* https://doi.org/10.1080/17455030.2021.2006822

[16] Sayantan G and Singh Abhishek Kumar 2020 Effects of initial stresses on reflection phenomenon of plane waves at the free surface of a rotating piezothermoelastic fiber-reinforced composite half-space *Int. J. Mech. Sci.* **181** 105766

[17] Singh A K, Mahto S and Guha S 2021 Analysis of plane wave reflection and transmission phenomenon at the interface of two distinct micro-mechanically modeled rotating initially stressed piezomagnetic fiber-reinforced half-spaces *Mech. Adv. Mater. Struct.* https://doi.org/10.1080/15376494.2021.2003490

[18] Guha S and Singh A K 2021 Analysis of plane wave reflection and transmission phenomenon at the interface of two distinct micro-mechanically modeled rotating initially stressed piezomagnetic fiber-reinforced half-spaces *Mech. Adv. Mater. Struct.* **88** 104242

[19] Guha S and Singh A K 2021 Influence of varying fiber volume fractions on plane waves reflecting from the stress-free/rigid surface of a piezoelectric fiber-reinforced composite half-space *Mech. Adv. Mater. Struct.* https://doi.org/10.1080/15376494.2021.1964046

[20] Singh A K and Guha S 2022 Mathematical study of reflection and transmission phenomenon of plane waves at the interface of two dissimilar initially stressed rotating micro-mechanically modeled piezoelectric fiber-reinforced composite half-spaces *Wave Dynamics* (Singapore: World Scientific) pp 131–62

[21] Singh A K, Singh S, Kumari R and Ray R 2020 Impact of point source and mass loading sensitivity on the propagation of an SH wave in an imperfectly bonded fgppm layered structure *Acta Mech.* **231** 2603–27

[22] Singh A K, Rajput P and Chaki M S 2022 Analytical study of love wave propagation in functionally graded piezo-poroelastic media with electroded boundary and abruptly thickened imperfect interface *Waves Random Complex Media* **32** 463–87

[23] Qian Z, Jin F, Wang Z and Kishimoto K 2007 Transverse surface waves on a piezoelectric material carrying a functionally graded layer of finite thickness *Int. J. Eng. Sci.* **45** 455–66

[24] Liu J, Cao X S and Wang Z K 2006 Propagation of love waves in a smart functionally gradedpiezoelectric composite structure *Smart Mater. Struct.* **16** 13

[25] Vlasie V and Rousseau M 2004 Guided modes in a plane elastic layer with gradually continuous acoustic properties *NDT E Int.* **37** 633–44

[26] Collet B, Destrade M and Maugin Gérard A 2006 Bleustein-Gulyaev waves in some functionally graded materials *Eur. J. Mech.* A **25** 695–706

[27] Lefebvre J E, Zhang V, Gazalet J, Gryba T and Sadaune V 2001 Acoustic wave propagation in continuous functionally graded plates: an extension of the Legendre polynomial approach *IEEE Trans. Ultrason. Ferroelectr. Freq. Control* **48** 1332–40

[28] Daneshjou K, Bakhtiari M and Tarkashvand A 2017 Wave propagation and transient response of a fluid-filled FGM cylinder with rigid core using the inverse laplace transform *Eur. J. Mech.* A **61** 420–32

[29] Movchan A B 2006 Multi-structures: asymptotic analysis and singular perturbation problems. *Eur. J. Mech.* A **25** 677–94

[30] Abdelmoula R and Leger A 2008 Singular perturbation analysis of the buckling of circular cylindrical shells *Eur. J. Mech.* A **27** 706–29

[31] Narisetti R K, Leamy M J and Ruzzene M 2010 A perturbation approach for predicting wave propagation in one-dimensional nonlinear periodic structures *J Vib. Acoust.* **132** 031001

[32] Cao X, Jin F, Jeon I and Lu T J 2009 Propagation of love waves in a functionally graded piezoelectric material (FGPM) layered composite system *Int. J. Solids Struct.* **46** 4123–132

[33] Cao X-S, Jin F, Mo X-Y and Shi J-P 2011 Love waves in piezoelestric layered structure with functionally graded materail half space *Symp. on Piezoelectricity, Acoustic Waves and Device Applications (SPAWDA)* (Piscataway, NJ: IEEE) pp 240–4

[34] Ren D, Shen X, Li C and Cao X 2018 The fractional Kelvin-Voigt model for Rayleigh surface waves in viscoelastic FGM infinite half space *Mech. Res. Commun.* **87** 53–8

[35] Wang Z, Jen C K and Cheeke J D N 1994 Mass sensitivity of two-layer shear horizontal plate wave sensors *Ultrasonics* **32** 209–15

[36] Zimmermann C, Mazein P, Rebiere D, Dejous C, Josse F and Pistre J 2004 A theoretical study of love wave sensors mass loading and viscoelastic sensitivity in gas and liquid environments *IEEE Ultrasonics Symp.* vol 2 (Piscataway, NJ: IEEE) pp 813–6

[37] Li P and Jin F 2015 Excitation and propagation of shear horizontal waves in a piezoelectric layer imperfectly bonded to a metal or elastic substrate. *Acta Mech.* **226** 267–84
[38] Singh S, Singh A K and Guha S 2022 Shear waves in a piezo-fiber-reinforced-poroelastic composite structure with sandwiched functionally graded buffer layer: power series approach *Eur. J. Mech.* A **92** 104470
[39] Voigt W 1892 Ueber innere reibung fester körper, insbesondere der metalle *Ann. Phys., Lpz.* **283** 671–93
[40] Thomson W 1865 IV. On the elasticity and viscosity of metals *Proc. R. Soc.* **14** 289–97
[41] Sharma M D 2010 Piezoelectric effect on the velocities of waves in an anisotropic piezo-poroelastic medium *Proc. R. Soc.* A **466** 1977–92
[42] Maurice Ewing W, Jardetzky W S, Press F and Beiser A 1957 Elastic waves in layered media *Phys. Today* **10** 27
[43] Li J and Dunn M L 2001 Viscoelectroelastic behavior of heterogeneous piezoelectric solids *J. Appl. Phys.* **89** 2893–903
[44] Kwok K W, Lee T, Choy S H and Chan H L W 2010 Lead-free piezoelectric transducers for microelectronic wirebonding applications *Piezoelectric Ceramics* **3** 145–64
[45] Vashishth A K, Dahiya A and Gupta V 2015 Generalized Bleustein–Gulyaev type waves in layered porous piezoceramic structure *Appl. Math. Mech.* **36** 1223–42

# Chapter 9

# Vibration analysis of single-link robotic manipulator by polynomial based Galerkin method in uncertain environment

**Priya Rao, S Chakraverty and Debanik Roy**

In the field of robotics, a single-link manipulator has many applications. The exact solution of the equations associated with such single-link manipulator may not always exist. To deal with the above problem, this chapter focuses on solving the governing differential equations by the Galerkin method, where the solution has been approximated as a polynomial function. First four natural frequencies of the single-link manipulator are computed and presented in corresponding tables. The results are compared with the previous studies, which show good agreement. Further, solutions have been again obtained by the polynomial based Galerkin method to investigate the problem under an uncertain (fuzzy) environment. In a similar fashion, the multi-link flexible manipulator may also be investigated (in an uncertain environment) using various polynomial approximation methods such as the Galerkin method, Collocation method, Rayleigh–Ritz method, etc.

## 9.1 Introduction

In recent years, robotics has become an emerging field of research in both engineering and science. It is never easy to control real-time robots while coping with *in-situ* vibrations. While designing a high-performance robot, many problems can arise, such as nonlinear strain displacement relations, non-inertial frame effects, and time-varying boundary conditions. Engineers and mathematicians can collude to solve problems by developing new and efficient models.

The equation of motion for a planar two-link flexible manipulator was derived by Cannon [1]. The mathematical modelling of lightweight flexible manipulators has been discussed by Usoro *et al* [2], using the finite element/Lagrangian technique. Clamped-mass boundary condition in model links has been studied by Luca and Siciliano [3] using the Euler–Bernoulli beam theory. The assumed mode approach

doi:10.1088/2053-2563/ac9580CH009

has been used by them to obtain a finite-dimensional model. For flexible manipulator arms with rotating joints [4] derived dynamic equations of motion. Based on the Timoshenko beam assumption and Hamilton's principle, dynamic modelling of a flexible hub–beam system is examined in the study [5], with the influence of shear deformation. The planar motion of a manipulator made up of two flexible links and two rotating joints have been studied using a Lagrangian dynamics approach [6]. In addition, a strategy based on classical principles is used to construct a multivariable control system. A system of ordinary differential equations whose order is connected to the required order of approximation for certain robots is discussed by Tomei and Tornambe [7]. A single-link manipulator with a payload at one end has made significant contributions [8, 9]. In [10], the dynamics of a single-link flexible robot with a hard mass attached to the tip have been discussed. Nonetheless, Chakraverty [11] has investigated the solution of the eigenvalue problem of a beam, which is identical to the eigenvalue problem associated with the vibration of flexible link.

It could be wise to point out that a dynamic model of a robot that switches states will closely resemble real-world instances of an unorganised and chaotic robotic workspace. In light of this viewpoint, our study focuses on incorporating 'uncertainty'. According to Roy [12–14], a recently developed formulation for the single-link flexible manipulator with uncertainty may also be expanded for multi-linked flexible robots with grippers. However, a functional prototype of a single-link, compact-volume, flexible robot with a small gripper will be effective for real-world applications.

An analytical solution of the differential equation of the flexible-link manipulator is always not possible. To get around this, various numerical techniques like Galerkin, collocation, etc are available. Additionally, polynomials are simple to manage and require simpler calculations. Various studies using polynomial approximations have been done. The approximate solution of linear and nonlinear second-order ordinary differential equations has been demonstrated in [15] using a 'Sinc–Galerkin' method. Solutions of tenth-order boundary value problem have been discussed in [16] using the finite element method incorporating the Galerkin method and quintic B-splines as basis functions. Several problems have been solved using the Galerkin method [17–21]. The Chebyshev Wavelet Collocation Method has been used in the study [22], which investigates free vibration problems of non-uniform Euler–Bernoulli beams under various supporting conditions. Galerkin's method based on Legendre and Laguerre polynomials has been used to investigate Euler beam equations by Bassuony *et al* [23].

According to the studies mentioned above, the scaling parameters for either of the models are precise or deterministic. In practice, however, it is not always possible to gather the parameters in a precise or predictable manner due to human error, environmental factors, wear and tear, and flawed tests. Because vibration-related parameters are unreliable in nature, the use of interval or fuzzy uncertainties is required. The foundational theoretical idea of fuzzy theory was introduced by Zadeh [24], which sparked additional research and provided solutions to a range of real-world issues in other domains which have also been studied [25–31]. The uncertain beam equation has been solved by Tapaswini and Chakraverty [32].

The goal of this research is to solve the differential equation of the single-link robot using the polynomial approximation method as it is not always possible to find the exact solution. To achieve this, the Galerkin method has been used which is in good agreement with the exact solution. Another objective is to develop a novel fuzzy model to describe the uncertain frequencies obtained from solving the systems equations using the Galerkin method of oscillation in a realistic fuzzy environment. This model will be able to handle the uncertainty brought on by the structural characteristics of a single-link flexible manipulator. Gaussian fuzzy numbers (GFNs) have been employed in parameters to deal with uncertainty. Raksha's model [10] has been utilised in this situation as a benchmark, using parameters $E$ (Young's modulus) and $\rho$ (density) as GFNs for three different cases, including: (1) when $E$ is GFN, (2) when $\rho$ is GFN, and (3) when $E$ and $\rho$ both are considered as GFNs.

## 9.2 Preliminaries

### 9.2.1 Fuzzy number

The fuzzy number is defined to be as a normalised convex fuzzy set $X$ of real line $R$ such that

$$\{\mu_{\widetilde{X}}(x)\colon R \to [0,1],\ \forall\ x \in \mathbb{R}\}$$

where $\mu_{\widetilde{X}}$ is the membership function of the fuzzy set which is piecewise continuous.

### 9.2.2 Gaussian fuzzy number

*9.2.2.1 Definition [33]*

A GFN $\widetilde{X} = (\delta, \sigma_l, \sigma_r)$ is a fuzzy number where the membership function $\mu_{\widetilde{X}}$ is defined as:

$$\mu_{\widetilde{X}}(x) = \begin{cases} \exp\left[-\frac{(x-\delta)^2}{2\sigma_l{}^2}\right], & x \leqslant \delta \\ \exp\left[-\frac{(x-\delta)^2}{2\sigma_r{}^2}\right], & x \geqslant \delta \end{cases} \quad \forall\ x \in \mathbb{R}$$

where $\delta$ is the modal value and $\sigma_l$, $\sigma_r$ is the spreading to the Gaussian distribution, for symmetric GFN $\sigma_l = \sigma_r = \sigma$.

*9.2.2.2 α-Cut approach of GFN [33]*

The GFN $\widetilde{X} = (\delta, \sigma_l, \sigma_r)$ can be mirrored by an ordered pair of functions using the $\alpha$-cut approach as $(x,x)$ where,

$$\underline{x} = \delta - \sigma\sqrt{-2\log\alpha}$$

$$\overline{x} = \delta + \sigma\sqrt{-2\log\alpha}$$

where $\alpha \in [0,1]$.

*9.2.2.3 Double Parametric Form of GFN [33]*

From the above definition, we have $\widetilde{X} = [\underline{x}, \overline{x}]$ rewriting the double parametric form as

$$X(\alpha, \beta) = \beta(x(\alpha) - x(\alpha)) + x(\alpha)$$

where $\alpha$ and $\beta \in [0,1]$.

### 9.2.3 Fuzzy Arithmetic

For any two fuzzy numbers $\tilde{u} = [\underline{u}(\alpha), \bar{u}(\alpha)]$, $\tilde{v} = [\underline{v}(\alpha), \bar{v}(\alpha)]$, and scalar κ, fuzzy arithmetic is defined as:

I. $u = v$ iff $u(\alpha) = v(\alpha)$ and $u(\alpha) = v(\alpha)$.

II. $u + v = (u(\alpha) + v(\alpha), u(\alpha) + v(\alpha))$.

III. $u \times v = [\min(W), \max(W)]$, where $W = \{u(\alpha) \times v(\alpha), u(\alpha) \times v(\alpha),$

$$u(\alpha) \times v(\alpha), u(\alpha) \times v(\alpha)\}.$$

IV. $$\kappa\tilde{u} = \begin{cases} [\kappa\bar{u}(\alpha), \kappa\underline{u}(\alpha)], & \kappa < 0 \\ [\kappa\underline{u}(\alpha), \kappa\bar{u}(\alpha)], & \kappa \geqslant 0 \end{cases}.$$

V. $$\frac{\tilde{u}}{\tilde{v}} = [\underline{u}(\alpha), \bar{u}(\alpha)] \times \left[\frac{1}{\bar{v}(\alpha)}, \frac{1}{\underline{v}(\alpha)}\right], \text{ where } 0 \notin v.$$

## 9.3 Mathematical modelling of single-link manipulator

The conceptual (typical) representation of a typical single-link flexible manipulator is shown in figure 9.1 with a gripper fitted at the free end. The model of a single-link manipulator has been taken using the Newton–Euler approach.

Figure 9.2 represents the single-link robot with link length $l$ with a rectangular cross-sectional area. A payload at the end of the link with mass $m$ is attached.

Based on Euler's theory, the equation of motion of rotating flexible link is [34]

$$BY^{IV}(z, t) + \rho[\ddot{Y} + (b + z)\ddot{\Theta}] = 0 \tag{9.1}$$

with boundary conditions as

$$\begin{gathered} Y(0, t) = Y'(0, t) = 0 \\ -BY'''(l, t) + m[\ddot{Y}(l, t) + (b + l)\ddot{\Theta}] + mc[\ddot{Y}'(l, t) + \ddot{\Theta}] = 0 \\ BY''(l, t) + mc[\ddot{Y}(l, t) + (b + l)\ddot{\Theta}] + I[\ddot{Y}'(l, t) + \ddot{\Theta}] = 0 \end{gathered} \tag{9.2}$$

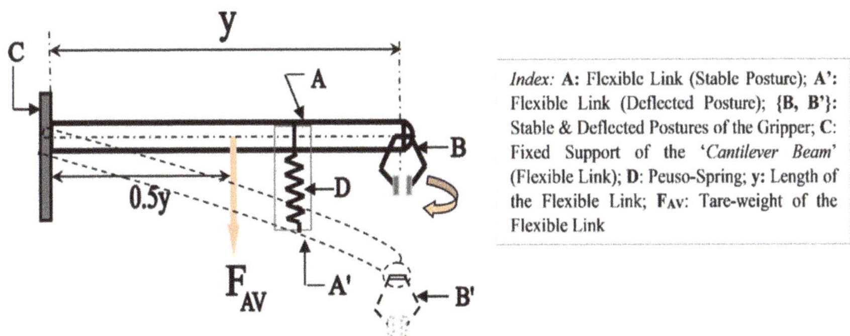


**Figure 9.1.** Conceptual disposition of a single-link flexible manipulator with gripper.

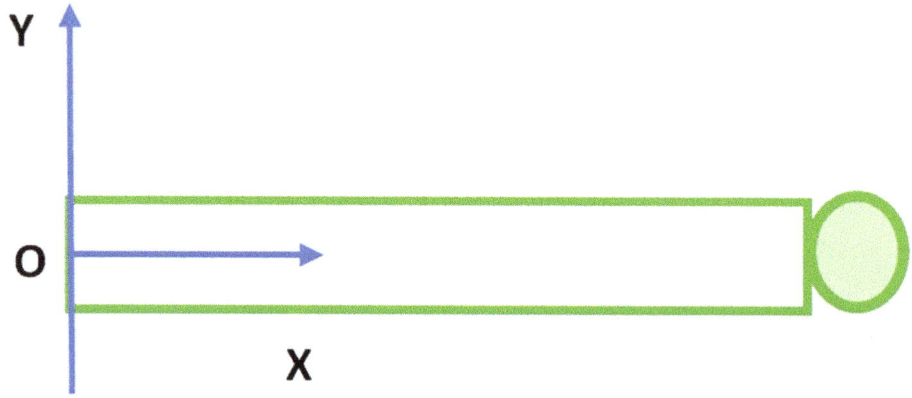


**Figure 9.2.** Single-link manipulator with mass attached.

where $B = EI$ is the constant flexural rigidity of the beam, with $E$ as Young's modulus, $I$ is the moment of inertia, $b$ is the radius of the circular body, $c$ is the distance between link end to the centre of tip mass and θ is the angular displacement. When there are no external effects (i.e., $\Theta = 0$) on the system, natural restricted motion occurs in the system, and the dynamics given by [10]

$$BY^{IV}(z, t) + \rho[\ddot{Y}(z, t)] = 0 \tag{9.3}$$

with boundary conditions as [10]

$$\begin{aligned} Y(0, t) &= Y'(0, t) = 0 \\ -BY'''(l, t) + m[\ddot{Y}(l, t)] + mc[\ddot{Y}'(l, t)] &= 0 \\ BY''(l, t) + mc[\ddot{Y}(l, t)] + I[\ddot{Y}'(l, t)] &= 0. \end{aligned} \tag{9.4}$$

Let us assume a variable separation of the form

$$Y(z, t) = y(z)\cos\omega t \tag{9.5}$$

where $\omega$ is the natural frequency and $y(z)$ is the mode shape. Hence

$$\ddot{Y}(z, t) = -\omega^2 Y(z, t). \tag{9.6}$$

Substituting equation (9.6) into equations (9.3) and (9.4) we get the differential equation of the system [10]

$$By''''(z) - \rho\omega^2 y(z) = 0 \tag{9.7}$$

with boundary conditions [10]

$$y(0) = 0 \tag{9.8a}$$

$$y'(0) = 0 \tag{9.8b}$$

$$-By'''(l) - m\omega^2 y(l) - m\omega^2 cy'(l) = 0 \tag{9.8c}$$

$$By''(l) - mc\omega^2 y(l) - I\omega^2 y'(l) = 0. \tag{9.8d}$$

The solution of $y(z)$ in equation (9.7) is

$$y(z) = A_1\cosh(rz) + A_2\sinh rz + A_3\cos rz + A_4\sin rz \tag{9.9}$$

where

$$r = \frac{\sqrt[4]{\rho}\,\sqrt{\omega}}{\sqrt[4]{B}}.$$

## 9.4 Application of the Galerkin method in the present model

In the Galerkin method, we assume a polynomial approximation as

$$y(z) = \sum_{k=0}^{n} c_k z^k \tag{9.10}$$

where $n$ is the degree of polynomial assumed.

The residue could be found by putting equation (9.10) into equation (9.7)

$$R(z) = B\left(\sum_{k=4}^{n} k(k-1)(k-2)(k-3)c_k z^{k-4}\right) - \rho\omega^2 \sum_{k=0}^{n} c_k z^k. \tag{9.11}$$

Using the boundary conditions (9.8a)–(9.8d), $C_0$, $C_1$, $C_2$, and $C_3$ can be eliminated from equation (9.10) and corresponding weight functions can be obtained.

Setting the integral over $[0,l]$ of weighted residual to zero, we have

$$\int_0^l R(z)w_i(z)\mathrm{d}z = 0 \text{ for } i = 4,5, \ldots, n. \tag{9.12}$$

Now from equation (9.12), we get a homogenous system with constant coefficients $C_4$, $C_5$, …, $C_n$. In order to find the value of ω, the homogenous system obtained is solved for the non-trivial case.

For $n = 5$ we have,

$$y(z) = C_0 + C_1 z + C_2 z^2 + C_3 z^3 + C_4 z^4 + C_5 z^5. \tag{9.13}$$

The corresponding residue will be,

$$R(z) = B(24C_4 + 120C_5 z) - \rho\omega^2(C_0 + C_1 z + C_2 z^2 + C_3 z^3 + C_4 z^4 + C_5 z^5).$$

Using the boundary conditions (9.8a)–(9.8d), $C_0$, $C_1$, $C_2$, and $C_3$ can be eliminated from $y(z)$ in equation (9.10) as; $C_0 = C_1 = 0$ and from boundary conditions i.e., equations (9.8c) and (9.8d) we have,

$$\begin{aligned} &C_2(-m\omega^2 l^2 - 2mc\omega^2 l) + C_3(-6B - m\omega^2 l^3 - 3mc\omega^2 l^2) \\ &+ C_4(-24Bl - m\omega^2 l^4 - 4mc\omega^2 l^3) + C_5(-60Bl^2 - m\omega^2 l^5 - 5mc\omega^2 l^4) = 0 \end{aligned} \tag{9.14}$$

$$C_2(2B - mc\omega^2 l^2 - 2I\omega^2 l) + C_3(6Bl - mc\omega^2 l^3 - 3I\omega^2 l^2) \\ + C_4(12Bl^2 - mc\omega^2 l^4 - 4I\omega^2 l^3) + C_5(20Bl^3 - mc\omega^2 l^5 - 5I\omega^2 l^4) = 0. \tag{9.15}$$

Eliminating $C_2$ and $C_3$ from equations (9.14) and (9.15), then proceeding as equation (9.12), we get a homogenous system of equations as

$$M(\omega)X = 0 \tag{9.16}$$

where $X = \{C_4, C_5\}$ is the non-zero Eigenvector and $M(\omega)$ is the coefficient matrix.

## 9.5 Proposed model in fuzzy environment using Gaussian fuzzy number

Unanticipated induced vibrations may develop in many application-centric real-world problems. As a result of errors in the material properties of flexible link can also occur, which could cause the system to become extremely unstable. The best way to handle these issues is by including fuzzy parameters in the model. In this study, Young's modulus ($E$) and material density ($\rho$) are considered as fuzzy parameters. For three distinct situations, the investigation is done viz. when (a) $E$ is GFN (b) $\rho$ is GFN, and (c) $E$ and $\rho$ both are GFNs.

$$\widetilde{E} = \left(E, \sigma_E, \sigma_E\right) \text{ and } \widetilde{X} = \left(\rho, \sigma_\rho, \sigma_\rho\right).$$

**Table 9.1.** Parameters for single-link manipulator.

| Parameter | Symbol | Units | Values |
|---|---|---|---|
| Length | $L$ | m | 1 |
| Breadth | $h$ | m | 0.030 |
| Width | $b$ | m | 0.003 |
| Modulus of elasticity | $E$ | N m$^{-2}$ | $71\times10^9$ |
| Density | $\rho$ | kg m$^{-3}$ | 2712.6 |
| Tip load mass | $m$ | kg | 0.61 |
| Tip load inertia | $J$ | kg m$^2$ | $1.5\times10^{-3}$ |

**Table 9.2.** Frequencies using Galerkin's method.

| $n$ | $\omega_1$ | $\omega_2$ | $\omega_3$ | $\omega_4$ |
|---|---|---|---|---|
| 5 | 0.235 100 351 98 | 1.518 462 0651 | — | — |
| 7 | 0.235 099 023 24 | 1.473 265 3913 | 4.129 249 3957 | 9.108 984 0052 |
| 9 | 0.235 099 023 16 | 1.473 205 5273 | 4.124 609 7287 | 8.109 659 7670 |
| 11 | 0.235 099 023 16 | 1.473 205 4923 | 4.124 587 3432 | 8.081 777 9294 |
| 13 | 0.235 099 023 16 | 1.473 205 4923 | 4.124 587 3301 | 8.081 505 4222 |
| 15 | 0.235 099 023 16 | 1.473 205 4923 | 4.124 587 3301 | 8.081 504 3889 |
| Exact frequency | 0.235 099 023 16 | 1.473 205 4923 | 4.124 587 3301 | 8.081 504 3869 |

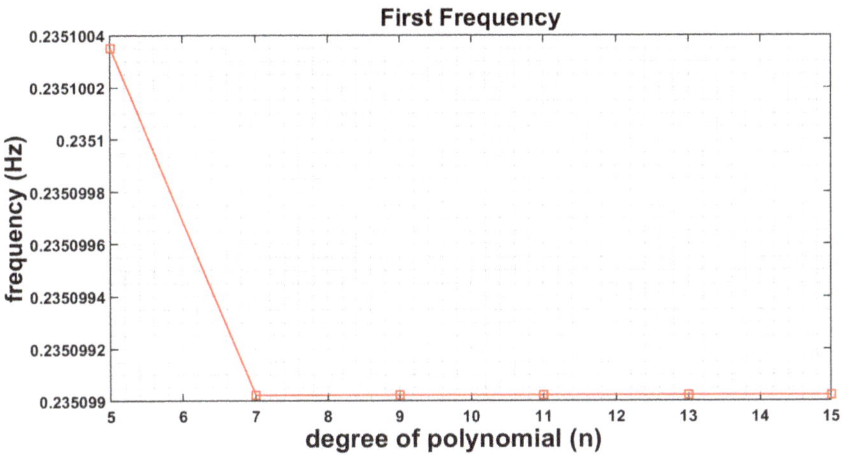


**Figure 9.3.** (a) Variation of the first frequency with respect to degree of the polynomials. (b) Variation of the second frequency with respect to degree of the polynomials. (c) Variation of the third frequency with respect to degree of the polynomials. (d) Variation of the fourth frequency with respect to degree of the polynomials.

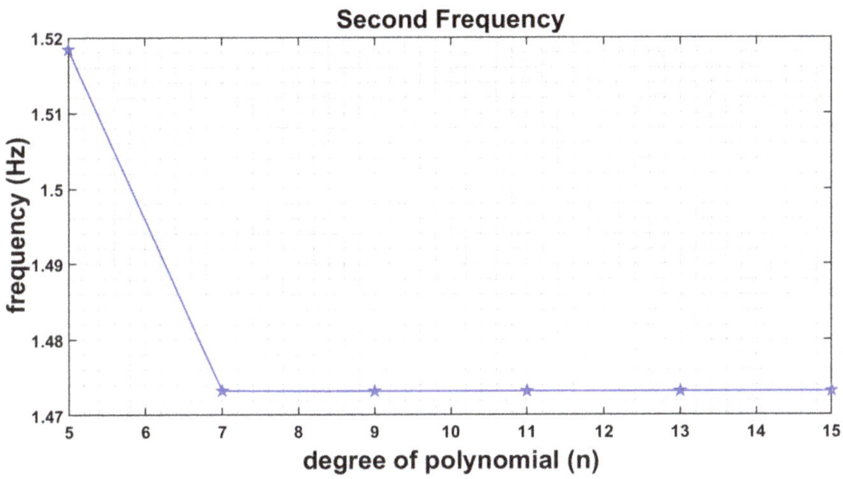


**Figure 9.3.** (Continued.)

The dynamic with fuzzy parameters for the single-link flexible manipulator using Euler beam theory is given by:

$$(E + (2\beta - 1)\sigma\sqrt{-2\log\alpha})Iy''''(z) - \left(\rho + \left(2\beta - 1\right)\sigma\sqrt{-2\log\alpha}\right)\omega^2 y(z) = 0 \quad (9.17)$$

with boundary conditions

$$y(0) = 0$$

$$y'(0) = 0$$

$$-(E + (2\beta - 1)\sigma\sqrt{-2\log\alpha})Iy'''(l) - m\omega^2 y(l) - m\omega^2 cy'(l) = 0 \quad (9.18)$$

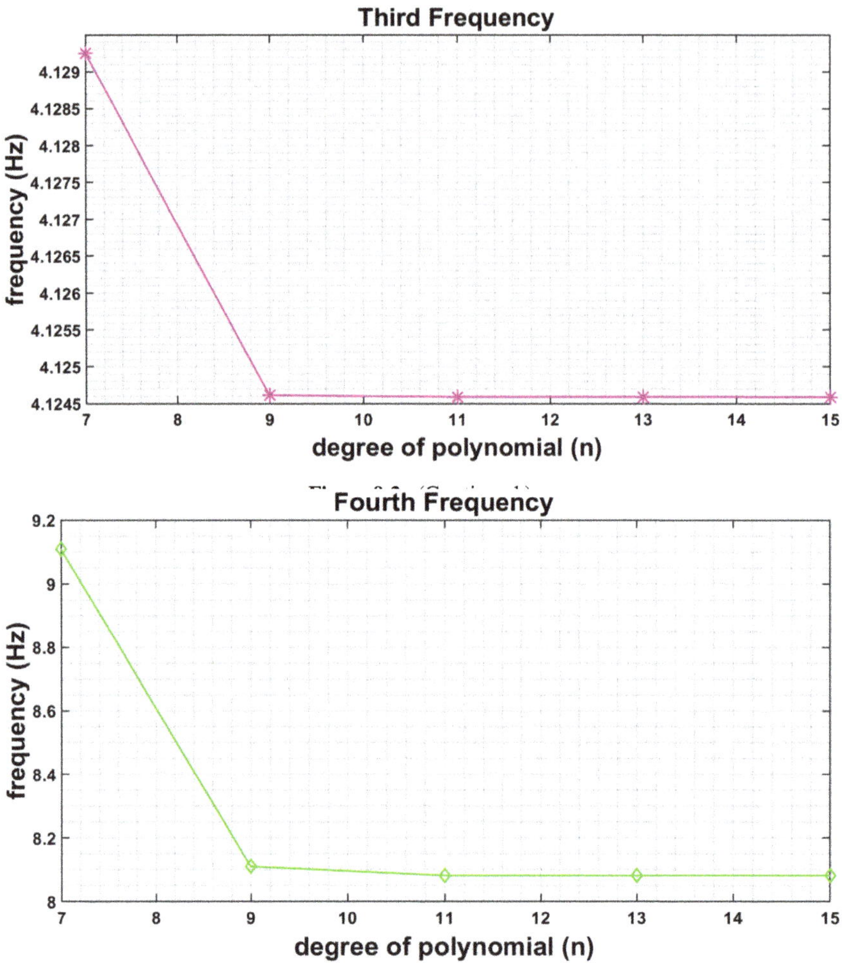


**Figure 9.3.** (Continued.)

$(E + (2\beta - 1)\sigma\sqrt{-2\log\alpha})Iy''(l) - mc\omega^2 y(l) - I\omega^2 y'(l) = 0$. The polynomial solution is approximated as

$$y(z) = \sum_{k=0}^{n} c_k z^k \tag{9.19}$$

where $n$ is the degree of polynomial assumed.

Residue with $E$ and ρ as GFNs,

$$\widetilde{R}(z) = \left(E + (2\beta - 1)\sigma\sqrt{-2\log\alpha}\right)I\left(\sum_{k=4}^{n} k(k-1)(k-2)(k-3)c_k z^{k-4}\right)$$

$$-(\rho + (2\beta - 1)\sigma\sqrt{-2\log\alpha})\omega^2 \sum_{k=0}^{n} c_k z^k. \tag{9.20}$$

**Table 9.3.** Error calculation with different degrees of polynomial.

| Degree of Polynomial | Frequency | Galerkin Method | Absolute Error |
|---|---|---|---|
| $n = 7$ | $\omega_1$ | 0.235 099 023 24 | $0.8\times10^{-11}$ |
| | $\omega_2$ | 1.473 265 391 30 | $0.59\times10^{-4}$ |
| | $\omega_3$ | 4.129 249 395 70 | $0.46\times10^{-2}$ |
| | $\omega_4$ | 9.108 984 005 20 | 1.02 |
| $n = 9$ | $\omega_1$ | 0.235 099 023 16 | 0 |
| | $\omega_2$ | 1.473 205 527 30 | $0.35\times10^{-7}$ |
| | $\omega_3$ | 4.124 609 728 70 | $0.22\times10^{-4}$ |
| | $\omega_4$ | 8.109 659 767 00 | 0.028 |
| $n = 13$ | $\omega_1$ | 0.235 099 023 16 | 0 |
| | $\omega_2$ | 1.473 205 492 30 | 0 |
| | $\omega_3$ | 4.124 587 330 10 | 0 |
| | $\omega_4$ | 8.081 505 422 20 | $0.1\times10^{-5}$ |

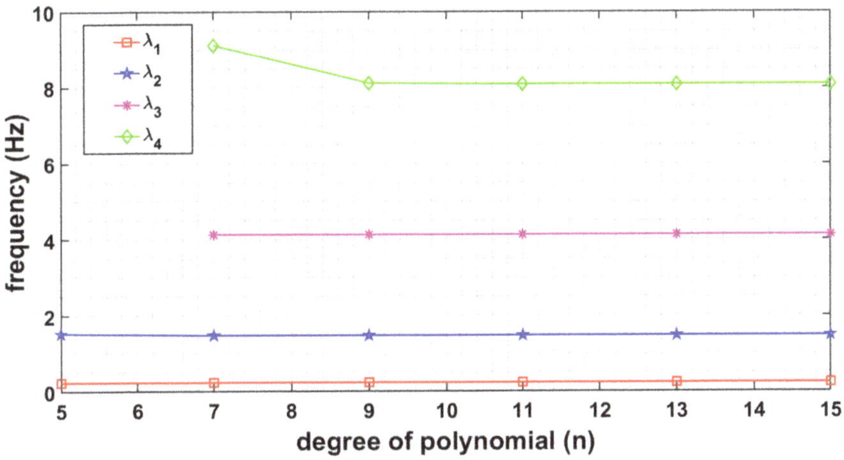


**Figure 9.4.** Comparison of modal frequencies with respect to the degree of the polynomial.

Eliminating $C_0$, $C_1$, $C_2$, and $C_3$ using equation (9.18) and setting the integral over $[0,l]$ of weighted residual to zero, we have

$$\int_0^l \widetilde{R}(z)\widetilde{w_i}(z)dz = 0 \text{ for } i = 4,5, \ldots, \quad n. \tag{9.21}$$

Now from equation (9.21), we get a homogenous system which results in an eigenvalue problem as

$$\widetilde{M}(\omega)X = 0 \tag{9.22}$$

where $X = \{C_4, C_5, \ldots\}$ and $M(\omega)$ is the coefficient matrix. Solving for the non-trivial case, $\omega$ is obtained. The computations are performed for three distinct situations, including when (a) $E$ is GFN (b) $\rho$ is GFN, and (c) $E$ and $\rho$ both are GFNs.

## 9.6 Results and discussions

In this section, frequencies have been computed by taking $y(z)$ as a polynomial function for different degrees, and those are compared with the frequency obtained by exact solution from equation (9.9). Parameters are taken as given in table 9.1 [10] with $c = 0.225$ m. Using GFNs as uncertain parameters computations have also been done for 15-degree polynomial.

First four exact frequencies using equation (9.9) and frequencies obtained by Galerkin's method for different degrees of polynomial are incorporated in table 9.2.

**Table 9.4.** First four frequencies when Young's modulus is GFN.

| $\alpha$ | $\lambda_1$ | $\lambda_2$ | $\lambda_3$ | $\lambda_4$ |
|---|---|---|---|---|
| $\beta = 0$ | | | | |
| 0.2 | 0.232 109 619 62 | 1.454 472 9359 | 4.072 141 1064 | 7.978 743 9542 |
| 0.4 | 0.232 846 967 00 | 1.459 093 3898 | 4.085 077 1605 | 8.004 090 1936 |
| 0.6 | 0.233 419 570 35 | 1.462 681 5051 | 4.095 122 9381 | 8.023 773 3739 |
| 0.8 | 0.233 990 372 16 | 1.466 258 3313 | 4.105 137 1094 | 8.043 394 6264 |
| 1.0 | 0.235 099 023 16 | 1.473 205 4923 | 4.124 587 3301 | 8.081 504 3889 |
| $\beta = 1$ | | | | |
| 0.2 | 0.238 050 889 22 | 1.491 702 8269 | 4.176 374 9947 | 8.182 974 4768 |
| 0.4 | 0.237 329 710 18 | 1.487 183 6889 | 4.163 722 5988 | 8.158 184 0226 |
| 0.6 | 0.236 766 563 43 | 1.483 654 8316 | 4.153 842 7280 | 8.138 825 9116 |
| 0.8 | 0.236 202 470 60 | 1.480 120 0460 | 4.143 946 2593 | 8.119 435 2796 |
| 1.0 | 0.235 099 023 16 | 1.473 205 4923 | 4.124 587 3301 | 8.081 504 3889 |

**Table 9.5.** First four frequencies when density is GFN.

| $\alpha$ | $\lambda_1$ | $\lambda_2$ | $\lambda_3$ | $\lambda_4$ |
|---|---|---|---|---|
| $\beta = 0$ | | | | |
| 0.2 | 0.235 879 624 86 | 1.478 096 0951 | 4.138 276 8672 | 8.108 320 0424 |
| 0.4 | 0.235 687 294 78 | 1.476 891 1151 | 4.134 903 9487 | 8.101 713 0311 |
| 0.6 | 0.235 537 841 24 | 1.475 954 7635 | 4.132 282 9597 | 8.096 578 9272 |
| 0.8 | 0.235 388 776 23 | 1.475 020 8457 | 4.129 668 7823 | 8.091 458 1638 |
| 1.0 | 0.235 099 023 16 | 1.473 205 4923 | 4.124 587 3301 | 8.081 504 3889 |
| $\beta = 1$ | | | | |
| 0.2 | 0.234 326 1203 | 1.468 363 1153 | 4.111 032 7567 | 8.054 953 0408 |
| 0.4 | 0.234 515 134 61 | 1.469 547 3251 | 4.114 347 5484 | 8.061 446 2199 |
| 0.6 | 0.234 662 648 59 | 1.470 471 5273 | 4.116 934 5362 | 8.066 513 7376 |
| 0.8 | 0.234 810 337 48 | 1.471 396 8250 | 4.119 524 5896 | 8.071 587 258 |
| 1.0 | 0.235 099 023 16 | 1.473 205 4923 | 4.124 587 3301 | 8.081 504 3889 |

**Table 9.6.** First four frequencies when both Young's modulus and density are GFNs.

| $\alpha$ | $\lambda_1$ | $\lambda_2$ | $\lambda_3$ | $\lambda_4$ |
|---|---|---|---|---|
| $\beta = 0$ | | | | |
| 0.2 | 0.232 880 295 57 | 1.459 301 3522 | 4.085 656 5742 | 8.005 218 6331 |
| 0.4 | 0.233 429 603 47 | 1.462 743 7075 | 4.095 294 9544 | 8.024 105 2536 |
| 0.6 | 0.233 855 2537 | 1.465 411 1367 | 4.102 763 5933 | 8.038 740 2258 |
| 0.8 | 0.234 278 758 85 | 1.468 065 1242 | 4.110 194 5992 | 8.053 301 4625 |
| 1.0 | 0.235 099 023 16 | 1.473 205 4923 | 4.124 587 3301 | 8.081 504 3889 |
| $\beta = 1$ | | | | |
| 0.2 | 0.237 268 281 92 | 1.486 799 6498 | 4.162 650 2322 | 8.156 089 7542 |
| 0.4 | 0.236 740 281 53 | 1.483 490 8120 | 4.153 385 6591 | 8.137 935 5360 |
| 0.6 | 0.236 327 093 68 | 1.480 901 4748 | 4.146 135 6534 | 8.123 728 9327 |
| 0.8 | 0.235 912 429 96 | 1.478 302 8897 | 4.138 859 7566 | 8.109 471 6022 |
| 1.0 | 0.235 099 023 16 | 1.473 205 4923 | 4.124 587 3301 | 8.081 504 3889 |

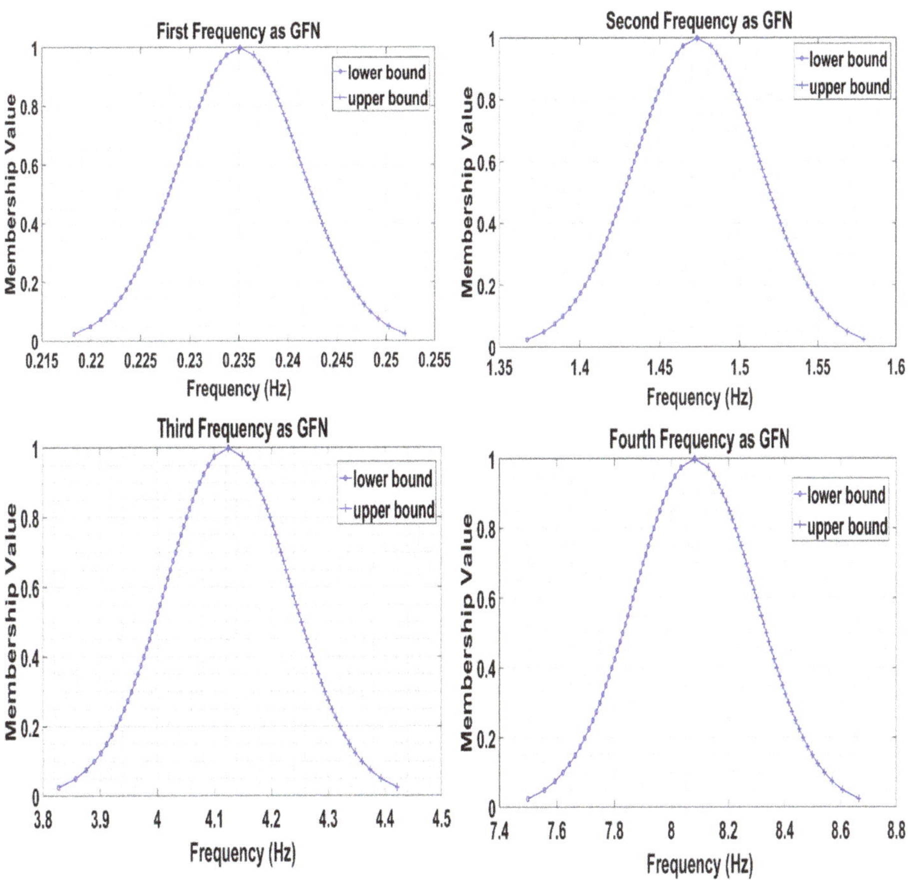


**Figure 9.5.** First four frequencies when $E$ is taken as an uncertain parameter as GFN.

From figures 9.3(a)–(d), we can see that all the four frequencies show higher deviation in lower degrees but as the degree of polynomial increases, it converges towards the exact frequency as the graph tends to be a straight line. Hence assuming the solution $y(z)$ of equation (9.10) to be a higher degree polynomial the frequency is nearer to the exact frequency which results in the polynomial being an excellent solution to differential equation (9.7). Table 9.3 shows the absolute error between the exact solution and the solution obtained by Galerkin for $n = 7$, 9 and 13.

The first four frequencies for different degrees of polynomial are displayed in figure 9.4. According to the figure, the fourth frequency exhibits a greater divergence than the initial frequencies. But for higher degree polynomials, the fourth frequency also showing good agreement. From figure 9.3, it is clear that a polynomial function is the best approximation to a differential equation (9.7) because all frequencies excellently converge at a 9th-degree polynomial. Tables 9.4–9.6 are showing the results for uncertain parameters in three cases for 15-degree polynomial.

Further, $E$ and $\rho$ are considered as GFNs to examine the model in an uncertain environment. The study is conducted for three distinct situations, when

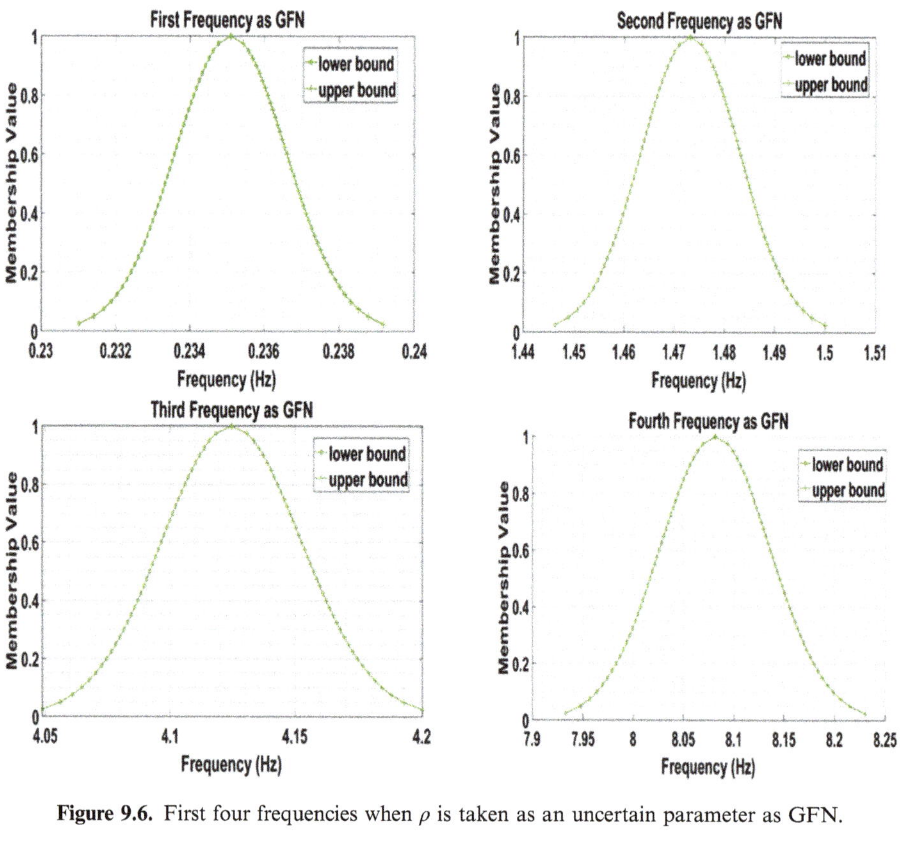


**Figure 9.6.** First four frequencies when $\rho$ is taken as an uncertain parameter as GFN.

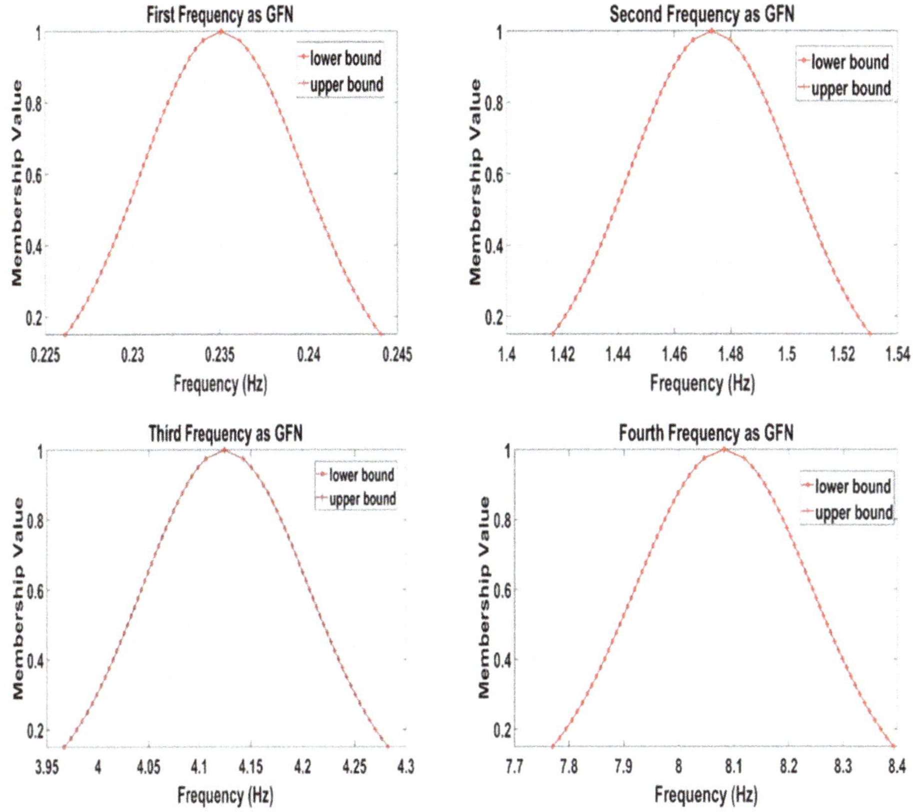


**Figure 9.7.** First four frequencies when $E$ and $\rho$ are taken as an uncertain parameter as GFNs.

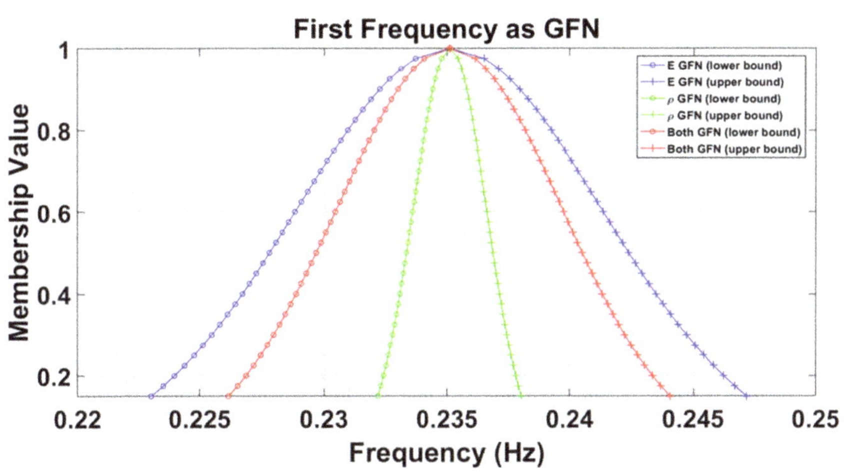


**Figure 9.8.** First frequency as GFN for all three cases.

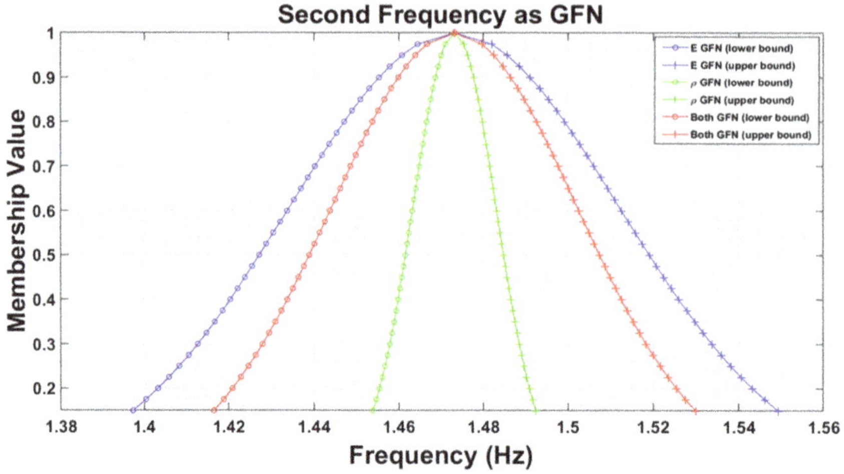


**Figure 9.9.** Second frequency as GFN for all three cases.

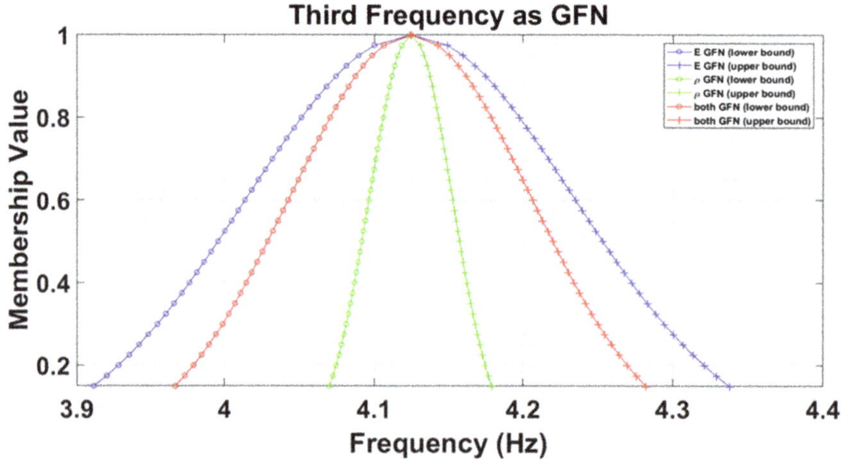


**Figure 9.10.** Third frequency as GFN for all three cases.

(a) $E$ is GFN (b) $\rho$ is GFN, and (c) $E$ and $\rho$ both are GFNs. Figures 9.5–9.7 show the variation for each mode of frequency for each case and figures 9.8–9.11 show all three cases for the first four frequencies. From the plots, it is clear that Young's modulus has a greater impact on the system because it exhibits a large deflection from actual frequency, whereas density is considered as GFN it exhibits a smaller deviation than $E$. When both density and $E$ are taken into account as GFN, the system exhibits a deflection that is intermediate between the first and second case. It is a terrific idea to use GFNs to study the behaviour in uncertain environment because each frequency can get an infinite variety of values for each mode.

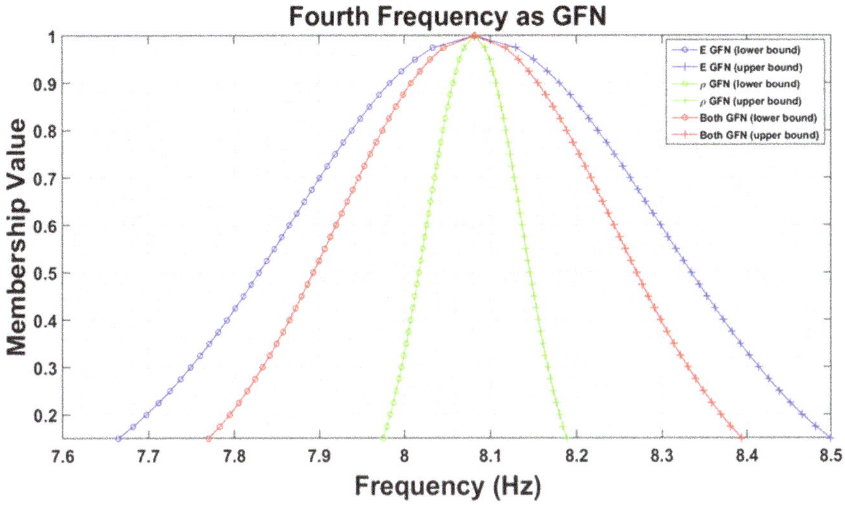


**Figure 9.11.** Fourth frequency as GFN for all three cases.

## 9.7 Conclusion

In this chapter, analysis of generalised single-link robot vibration has been carried out by approximating the solution of governing differential equation of single-link manipulator to a polynomial using Galerkin's method. The use of boundary conditions in the approximate solution of the governing differential equation yields excellent convergence with the actual frequencies. Frequencies obtained from the polynomial based Galerkin's and analytical methods are included in table 9.3 and comparisons are shown via figures 9.3 and 9.4. The results obtained show that the frequencies converge rapidly to exact frequencies for higher degree polynomial. Further, frequency fluctuations are investigated in uncertain environments using various cases of uncertainty in parameters. Three distinct situations are addressed that is when (a) $E$ is GFN (b) $\rho$ is GFN, and (c) $E$ and $\rho$ both are GFNs. It may be seen here that Young's modulus has a greater impact on the system and so large deviation may be seen from actual frequency when $E$ is GFN.

## References


[1] Oakley C M and Cannon R H 1989 Equations of motion for an experimental planar two-link flexible manipulator *Proc. ASME Winter Annual Meeting* pp 267–78

[2] Usoro P B, Nadira R and Mahil S S 1986 A finite element/Lagrange approach to modeling lightweight flexible manipulators *J. Dyn. Syst. Meas. Control* **108** 198–205

[3] De Luca A and Siciliano B 1991 Closed-form dynamic model of planar multilink lightweight robots *IEEE Trans. Syst. Man Cybern* **21** 826–39

[4] Book W J 1984 Recursive lagrangian dynamics of flexible manipulator arms *The Int. J. Rob. Res* **3** 87–101

[5] Chaolan Y, Jiazhen H and Guoping C 2006 Modeling study of a flexible hub–beam system with large motion and with considering the effect of shear deformation *J. Sound Vib.* **295** 282–93

[6] Ower J and De Vegte J 1987 Classical control design for a flexible manipulator: modeling and control system design *IEEE J. Robotics Automation* **3** 485–89

[7] Tomei P and Tornambe A 1988 Approximate modeling of robots having elastic links *IEEE Trans. Syst. Man Cybern.* **18** 831–40

[8] Bellezza F, Lanari L and Ulivi G 1990 Exact modeling of the flexible slewing link *Proc. IEEE Int. Conf. on Robotics and Automation* (Piscataway, NJ: IEEE) pp 734–9

[9] White M W D and Heppler G R 1995 A Timoshenko model of a flexible slewing link *Proc. of 1995 American Control Conf.-ACC'95* vol 4 (Piscataway, NJ: IEEE) pp 2815–9

[10] Rakhsha F and Goldenberg A 1985 Dynamics modelling of a single-link flexible robot *Proc. 1985 IEEE Int. Conf. on Robotics and Automation* vol 2 (Piscataway, NJ: IEEE) pp 984–9

[11] Chakraverty S 2008 *Vibration of Plates* (Boca Raton, FL: CRC Press)

[12] Roy D 2019 Control of inherent vibration of flexible robotic systems and associated dynamics *Lecture Notes in Mechanical Engineering: Recent Trends in Wave Mechanics and Vibrations* (Berlin: Springer)

[13] Roy D 2020 Design, modeling and indigenous firmware of patient assistance flexible robotic system-type I: beta version *Advances in Robotics and Mechanical Engineering* **2** 148–59

[14] Roy D 2019 Towards the control of inherent vibration of flexible robotic systems and associated dynamics: new proposition and model *International Journal of Robotics Research, Applications and Automation* **1** 6–17

[15] Stenger F 1979 A 'Sinc-Galerkin' method of solution of boundary value problems *Math. Comput.* **33** 85–109

[16] Viswanadham K K and Ballem S 2014 Numerical solution of tenth order boundary value problems by galerkin method with quintic B-splines *Int. J. Appl. Sci. Eng.* **2** 288–94

[17] Wang H and Zhang X 2015 A high-accuracy preserving spectral Galerkin method for the Dirichlet boundary-value problem of variable-coefficient conservative fractional diffusion equations *J. Comput. Phys.* **281** 67–81

[18] El-Gamel M, Cannon J and Zayed A 2004 Sinc-Galerkin method for solving linear sixth-order boundary-value problems *Math. Comput.* **73** 1325–43

[19] Rahman M M, Hossen M A, Islam M N and Ali M S 2012 Numerical solutions of second order boundary value problems by Galerkin method with Hermite polynomials *Annals of Pure and Applied Mathematics* **1** 138–48

[20] Lei Y, Friswell M I and Adhikari S 2006 A Galerkin method for distributed systems with non-local damping *Int. J. Solids Struct.* **43** 3381–400

[21] Liew K M, Lim H K, Tan M J and He X Q 2002 Analysis of laminated composite beams and plates with piezoelectric patches using the element-free Galerkin method *Comput. Mech.* **29** 486–97

[22] Çelik İ 2018 Free vibration of non-uniform Euler–Bernoulli beam under various supporting conditions using Chebyshev wavelet collocation method *Appl. Math. Modell.* **54** 268–80

[23] Bassuony M A, Abd-Elhameed W M, Doha E H and Youssri Y H 2018 A Legendre–Laguerre–Galerkin method for uniform Euler–Bernoulli beam equation *East Asian J. Appl. Math* **8** 280–95

[24] Zadeh L 1965 Fuzzy sets *Inf. Control* **8** 338–53

[25] Chang S L and Zadeh L A 1972 On fuzzy mapping and control *IEEE Trans. Syst. Man Cybern.* **2** 30–4

[26] Dubois D and Prade H 1982 Towards fuzzy differential calculus part 3: differentiation *Fuzzy Sets Syst.* **8** 225–33

[27] Kaleva O 1990 The Cauchy problem for fuzzy differential equations *Fuzzy Sets Syst.* **35** 389–96

[28] Khastan A, Nieto J J and Rodrıguez-López R 2011 Variation of constantformula for first order fuzzy differential equations *Fuzzy Sets Syst* **177** 20–33

[29] Mikaeilvand N and Khakrangin S 2012 Solving fuzzy partial differential equations by fuzzy two dimensional differential transform method *Neural Comput Appl* **21** 307–12

[30] Khastan A, Nieto J J and Rodrıguez-López R 2013 Periodic boundary value problems for first-order linear differential equations with uncertainty under generalized differentiability *Info. Sci.* **222** 544–58

[31] Chakraverty S and Perera S (ed) 2018 *Recent Advances in Applications of Computational and Fuzzy Mathematics* (Berlin: Springer)

[32] Tapaswini S and Chakraverty S 2014 Dynamic response of imprecisely defined beam subject to various loads using adomian decomposition method *Appl. Soft Comput.* **24** 249–63

[33] Chakraverty S, Tapaswini S and Behera D 2016 *Fuzzy Differential Equations and Applications for Engineers and Scientists* (Boca Raton, FL: CRC Press)

[34] Thomson W T 1981 *Theory of Vibration with Applications* (Englewood Cliffs, NJ: Prentice-Hall)

**IOP** Publishing

Polynomial Paradigms
Trends and applications in science and engineering
**S Chakraverty**

# Chapter 10

## Solving Type-2 Fuzzy Differential Equations Using Collocation Method with Type-2 Fuzzy Polynomials

**Dhabaleswar Mohapatra and S Chakraverty**

Several practical problems are inextricably linked with differential equations. So, it is of great interest of investigation to find the solution of a system's governing differential equation. However, it may not be possible to obtain the exact solution of each governing equation. As a result, various numerical methods for approximating the solution have been devised based on polynomials. Furthermore, as in a real-world problem, ambiguity may not always be minimal. To deal with such uncertainty, a probability/interval/fuzzy theory approach may be beneficial. However, the application of probability theory in all uncertain cases may be ineffective because it may generate less reliable results for small data sets. Accordingly, interval and fuzzy theory approach may be used to handle the uncertainties. Moreover, because of the uncertain membership grade, the use of type-2 fuzzy numbers over type-1 fuzzy numbers in uncertain problems is more realistic. As such, a numerical approach viz. polynomials based collocation method is developed to solve type-2 fuzzy differential equations. In this regard, the polynomials in the assumed solution to use in the collocation method are type-2 fuzzy polynomials. This study includes a few numerical examples and one application problem to demonstrate the validity of the proposed technique. Finally, the obtained type-2 fuzzy solutions are compared with previously obtained results in various special cases, such as type-1 fuzzy, interval, and crisp.

doi:10.1088/978-0-7503-5067-9ch10  

**List of Abbreviation**

| Full form | Abbreviation |
|---|---|
| Fuzzy differential equation | FDE |
| Type-2 fuzzy set | T2FS |
| Type-2 fuzzy number | T2FN |
| Triangular perfect quasi type-2 fuzzy number | TPQT2FN |
| Footprint of uncertainty | FOU |
| Lower-membership function | LMF |
| Upper-membership function | UMF |
| Type-2 fuzzy differential equation | T2FDE |
| Type-2 fuzzy boundary value problem | T2FBVP |

## 10.1 Introduction

Differential equations are used in various fields, including physics, engineering, biology, etc. In almost all problems involving differential equations, the parameters involved are assumed to be crisp for computational convenience. However, in the actual scenario, there may be some uncertainty in the values of the parameters, which could be due to an observational error or a computational error, and these uncertainties can be handled by fuzzy theory using fuzzy numbers. As a result, the use of fuzzy parameters in differential equations has increased significantly, which leads to the concept of FDEs. The use of FDEs in many practical cases has attracted researchers to investigate and propose various efficient methodologies analytically as well as numerically to solve the FDEs. Further, many situations may arise where the use of type-1 fuzzy numbers is practically not reliable to handle the uncertainty in the application problem. In such cases, type-2 fuzzy sets/numbers may be helpful.

Polynomial functions are vital for determining the solution of different application problems governed by differential and integral equations. In this regard, several research works have been done on polynomial based numerical methods. Non-linear Burgers' equation has been solved by the polynomial based differential quadrature method [1]. Different types of wavelet methods, viz. Legendre wavelet, Haar wavelet, cosine and sine (CAS) wavelet have been formulated based on different special types of polynomials [2–5]. Other approaches have been made using polynomials for solving various mathematical problems [6–8]. The concept of fuzzy theory was pioneered by Zadeh [9]. Following this concept of fuzzy theory, different fuzzy numbers, viz. TFN, TrFN, and GFN were defined. Also, depending upon the nature of the membership grade Zadeh has defined different types of fuzzy sets, viz. type-1, type-2, ..., type-*n* fuzzy sets. Further, several concepts related to the arithmetic and calculus of fuzzy numbers have been proposed by researchers. Chang and Zadeh [10] have developed the concept of fuzzy derivatives. Dubois and Prade [11] have used the concept of extension principle to define the fuzzy derivatives in their approach. Researchers have discussed several other approaches to define the concept of fuzzy derivatives [12–15, 17]. The existence and uniqueness of the solution for FDEs under H-derivative and Seikkala derivative have been investigated by Seikala [18] and Kaleva [19].

Different analytical and numerically efficient methods have been proposed to solve the FDEs, from which a few are mentioned here to provide a better understanding for the readers. Buckley and Feuring [20, 21] have implemented two analytical methods for solving *n*th order linear differential equations with fuzzy initial conditions. FDEs have been solved using the concept of generalized H-differentiability in the lateral sense [22]. Numerical approach based on the classical Euler method has been developed to solve the FDE [23]. The solutions to Cauchy's problem in a fuzzy environment have been solved in [24] using two approaches, viz. differential inclusions and Zadeh's extension principle. Also, first-order fuzzy initial and boundary value problems have been studied [25]. Along with these, different other studies have been conducted related to FDE using the concepts of Hukuhara and Seikkala derivatives [26–35]. Further, based on the concept of differential inclusions, several works have been carried out to solve FDE [36–38]. Different researchers have used the concept of standard generalized Hukuhara derivative to define FDE and proposed several effective techniques for the solution [39–43]. FDEs have been studied under the concept of generalized Hukuhara derivative [44–47].

From the above literature survey, one can mark that in most of the studies related to the FDEs, researchers have used type-1 fuzzy sets(/numbers). In type-1 fuzzy sets (/numbers), the membership grade is crisp and lies in [0, 1]. But, there may be a possibility that the value of this membership function is not always crisp, i.e. impreciseness in the membership value. To handle such circumstances, fuzzy membership grades may be helpful. Depending upon the nature of the membership function fuzzy sets are of different types. In a type-$n$ fuzzy set, the associated membership is a type-$(n - 1)$ fuzzy set. So, if we consider a type-2 fuzzy set, then the associated membership is a type-1 fuzzy set, i.e. fuzzy set. The concept of the type-2 fuzzy set was introduced by Zadeh [48]. Different arithmetic operations and set operations related to type-2 fuzzy sets have been developed [49–52]. Researchers have solved different application problems in type-2 fuzzy uncertain environment [53, 54]. Also, research works have been conducted related to type-2 fuzzy differential equations [55–58].

This chapter is dedicated to investigating the T2FBVP numerically using the proposed technique based on the collocation method. In this regard, the uncertainty is considered both in trial polynomial functions and in unknown coefficients represented by type-2 fuzzy numbers, viz. TPQT2FNs [56]. The T2FBVP is first paramaetrized with two parameters $r_1$ and $r_2$. Then, using the definition of lower and upper bounds of the type-2 fuzzy parameters and following the simplification process, unknown type-2 fuzzy coefficients are calculated, which will be further utilized to approximate the required type-2 fuzzy solution.

The remainder of the chapter is categorized into several sections. Section 10.2 discusses some fundamental concepts of fuzzy set theory and fuzzy calculus. The proposed methodology is discussed in section 10.3. Section 10.4 illustrates some numerical examples to check the validation and effectiveness of the proposed methodology. In section 10.5, the obtained numerical results are discussed in tables and plots. Finally, the concluding remarks are discussed in the last section.

## 10.2 Preliminaries

### 10.2.1 Type-1 fuzzy numbers

A fuzzy set that is convex and normalized is known as 'type-1 fuzzy number' or simply 'fuzzy number'. A fuzzy set $\tilde{A}$ is an ordered pair [16, 17], given by

$$\tilde{A} = \{(x, \nu_{\tilde{A}}(x))|x \in X \subset \mathbf{R}\}. \quad (10.1)$$

Here $\nu_{\tilde{A}}(x)$ is the membership function of $x$.

### 10.2.2 Parametric form of fuzzy number

Fuzzy numbers can be expressed as pair of functions using the $\alpha$ – cut, where $\alpha \in [0, 1]$, as [16]

$$\tilde{A}(\alpha) = [\underline{A}(\alpha), \bar{A}(\alpha)]. \quad (10.2)$$

Which is a standard closed interval, where $\underline{A}(\alpha)$ and $\bar{A}(\alpha)$ are lower and upper bounds, with the following criteria

1. $\underline{A}(\alpha)$ is a bounded left continuous non-decreasing function over [0, 1].
2. $\bar{A}(\alpha)$ is a bounded right continuous non-increasing function over [0, 1].
3. $\underline{A}(\alpha) \leqslant \bar{A}(\alpha)$, $0 \leqslant \alpha \leqslant 1$.

### 10.2.3 Type-2 fuzzy set

A type-2 fuzzy set $\tilde{\tilde{W}}$ [52] is defined as,

$$\tilde{\tilde{W}} = \{((x, v), \nu_{\tilde{\tilde{W}}}(x, v))|x \in X, v \in V_x \subseteq [0, 1]\} \quad (10.3)$$

where $v$ is the primary membership of $x$ and $\nu_{\tilde{\tilde{W}}}(x, v)$ is the secondary membership function of $v$ such that $0 \leqslant \nu_{\tilde{\tilde{W}}}(x, u) \leqslant 1$. Here $x$ is the primary variable, $X$ is the domain of the fuzzy set and $V_x$ is the domain of the secondary membership function at $x$.

### 10.2.4 Vertical slice of type-2 fuzzy set

Vertical slice of a type-2 fuzzy set $\tilde{\tilde{W}}$ [52] is a type-1 fuzzy set, obtained by fixing the primary variable. It is given by,

$$\nu_{\tilde{\tilde{W}}}(x = x^*, v) = \nu_{\tilde{\tilde{W}}}(x^*) = \int \frac{f_{x^*}(v)}{v}\, dv$$

where $v \in V_{x^*} \subseteq [0, 1]$ and $0 \leqslant f_{x^*}(v) \leqslant 1$.

### 10.2.5 $r_1$-plane of type-2 fuzzy set

Let $r_1 \in [0, 1]$ be any real number, the union of all secondary domains of the type-2 fuzzy set $\tilde{A}$ [52] with secondary grade greater or equal to $r_1$ is called the $r_1$ – plane of $\tilde{A}$.

### 10.2.6 FOU of a type-2 fuzzy set

FOU of a type-2 fuzzy set [52] is obtained from $\tilde{r}_1$–plane of the type-2 fuzzy set by substituting $r_1 = 0$.

### 10.2.7 LMF and UMF of a type-2 fuzzy set

The lower and upper bounds of a type-2 fuzzy set at any level $r_1$ [52], are known as lower-membership function and upper-membership function respectively. i.e. $\tilde{A}_{r_1} = (\underline{A}_{r_1}, \bar{A}_{r_1})$.

### 10.2.8 Principle set of $\tilde{A}$

Principle set is a type-1 fuzzy set, which is achieved for $r_1 = 1$ in the $r_1$ – plane form of the type-2 fuzzy set.

### 10.2.9 $r_2$-cut of $r_1$-plane

$r_2$ – cut of a type-2 fuzzy set $\tilde{A}$ at level $r_1$ is given by $\tilde{A}(r_1, r_2) = (\underline{A}(r_1, r_2), \bar{A}(r_1, r_2))$, i.e. $r_2$ – cut of LMF and UMF of $\tilde{A}_{r_1}$.

### 10.2.10 Triangular perfect quasi type-2 fuzzy numbers

A T2FS $\tilde{A}$ [52] is said to be a perfect T2FN, if it holds the conditions below;

1. LMF and UMF of footprint of uncertainty of $\tilde{A}$ are type-1 fuzzy numbers.
2. LMF and UMF of principle set of $\tilde{A}$ are also type-1 fuzzy numbers.

Perfect quasi type-2 fuzzy number [52] is a kind of perfect type-2 fuzzy number, where all vertical slices are type-1 fuzzy numbers and piece-wise functions of the same kind. Triangular perfect QT2FNs was introduced in [56], as

$$\tilde{t} = (\bar{L}_{t_0}, L_{t_1}, \underline{L}_{t_0}, p, \underline{R}_{t_0}, R_{t_1}, \bar{R}_{t_0}) \tag{10.4}$$

satisfying,

$$\bar{L}_{t_0} \leqslant L_{t_1} \leqslant \underline{L}_{t_0} \leqslant p \leqslant \underline{R}_{t_0} \leqslant R_{t_1} \leqslant \bar{R}_{t_0}.$$

A schematic diagram of the triangular perfect QT2FN $\tilde{t}$ is given in figure 10.1. Now, the $r_1$, $r_2$ form is [56]

$$\tilde{t}(r_1, r_2) = (\underline{t}(r_1, r_2), \bar{t}(r_1, r_2))$$

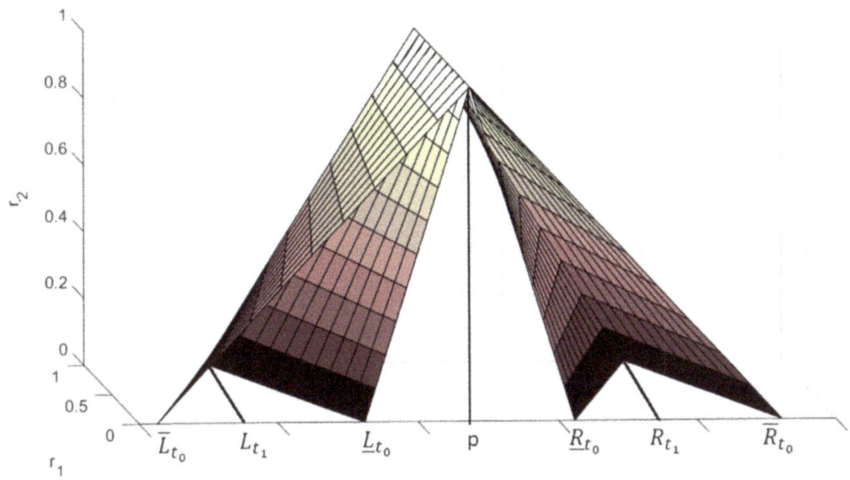


**Figure 10.1.** Schematic diagram of triangular perfect QT2FN $\tilde{t}$.

where

$$\underline{t}(r_1, r_2) = (\underline{L}_t(r_1, r_2), \underline{R}_t(r_1, r_2))$$

$$\underline{L}_t(r_1, r_2) = L_{t_1}(r_2) - (1 - r_1)(L_{t_1}(r_2) - \underline{L}_{t_0}(r_2))$$

$$\underline{R}_t(r_1, r_2) = R_{t_1}(r_2) - (1 - r_1)(R_{t_1}(r_2) - \underline{R}_{t_0}(r_2))$$

and

$$\bar{t}(r_1, r_2) = (\bar{L}_t(r_1, r_2), \bar{R}_t(r_1, r_2))$$

$$\bar{L}_t(r_1, r_2) = L_{t_1}(r_2) - (1 - r_1)(L_{t_1}(r_2) - \bar{L}_{t_0}(r_2))$$

$$\bar{R}_t(r_1, r_2) = R_{t_1}(r_2) - (1 - r_1)(R_{t_1}(r_2) - \bar{R}_{t_0}(r_2)).$$

Further,

$$\bar{L}_{t_0}(r_2) = p - (1 - r_2)(p - \bar{L}_{t_0})$$

$$L_{t_1}(r_2) = p - (1 - r_2)(p - L_{t_1})$$

$$\underline{L}_{t_0}(r_2) = p - (1 - r_2)(p - \underline{L}_{t_0})$$

$$\bar{R}_{t_0}(r_2) = p - (1 - r_2)(p - \bar{R}_{t_0})$$

$$R_{t_1}(r_2) = p - (1 - r_2)(p - R_{t_1})$$

$$\underline{R}_{t_0}(r_2) = p - (1 - r_2)(p - \underline{R}_{t_0}).$$

### 10.2.11 Type-2 fuzzy functions

A type-2 fuzzy function $\tilde{f}$ may be defined as,

$$\tilde{f} :[a, b] \in \mathbf{R} \to E_2 \tag{10.5}$$

where $E_2$ is the set consists of type-2 fuzzy numbers. Similarly for TPQT2FN, the type-2 fuzzy function $\tilde{f}$ may be represented by,

$$\tilde{f}(t) = \left(\bar{L}_{f_0}(t),\, L_{f_1}(t),\, \underline{L}_{f_0}(t), f(p),\, \underline{R}_{f_0}(t),\, R_{f_1}(t),\, \bar{R}_{f_0}(t)\right). \tag{10.6}$$

### 10.2.12 Hukuhara differential of type-2 fuzzy numbers ($H_2$ differential)

Let $E_2$ be the collection of perfect type-2 fuzzy numbers over **R**. Suppose $\tilde{g}$ be a type-2 fuzzy function over $(a, b)$, then it is called $H_2$-differentiable at $x_0 \in (a, b)$, if for $h > 0$ and very very close to zero, there exist $H_2$-differences $\tilde{g}(x_0 + h)\ominus_{H_2}\tilde{g}(x_o)$, $\tilde{g}(x_0)\ominus_{H_2}\tilde{g}(x_0 - h)$ and an element $\tilde{g}'(x_0)$ such that,

$$\lim_{h\to o}\frac{g(x_0 + h)\ominus_{H_2}g(x_o)}{h} = \lim_{h\to o}\frac{g(x_0)\ominus_{H_2}g(x_0 - h)}{h} = g'(x_0). \tag{10.7}$$

## 10.3 Proposed method

In this section, we introduce a method based on the collocation method to solve the T2FDE. In this regard, let us consider a T2FDE of the form

$$\tilde{y}^n(x; r_1, r_2) + a_1(x)\tilde{y}^{n-1}(x; r_1, r_2) + \cdots + a_n(x)\tilde{y}(x; r_1, r_2) = \tilde{f}(x; r_1, r_2), \tag{10.8}$$

where, $a_i(x)$, $1 \leqslant i \leqslant n$ are continuous functions on some interval $[a, b]$ and subjected to the type-2 fuzzy boundary conditions,

$$\begin{aligned}\tilde{y}(a; r_1, r_2) &= (\underline{A}_1(r_1, r_2), \bar{A}_1(r_1, r_2)),\\ \tilde{y}(b; r_1, r_2) &= (\underline{B}_1(r_1, r_2), \bar{B}_1(r_1, r_2)).\end{aligned} \tag{10.9}$$

These boundary conditions may vary depending upon the problem. Now, approximating the solution of the given T2FBVP using the type-2 fuzzy polynomial trial functions $\tilde{\psi}_i$, $1 \leqslant i \leqslant n$ as,

$$\tilde{y}(x; r_1, r_2) = \tilde{\psi}_0(x; r_1, r_2) + \sum_{i=1}^{n}\tilde{c}_i(r_1, r_2)\tilde{\psi}_i(x; r_1, r_2), \tag{10.10}$$

where,

- $\tilde{c}_i(r_1, r_2)$ are type-2 fuzzy coefficients.
- $\tilde{\psi}_0(x; r_1, r_2) = (M\underline{A}_1(r_1, r_2) + N\underline{B}_1(r_1, r_2),\, M\bar{A}_1(r_1, r_2) + N\bar{B}_1(r_1, r_2))$ with $M = \frac{b-x}{b-a}$, $N = \frac{x-a}{b-a}$.
- $\tilde{\psi}_i(x; r_1, r_2) = \tilde{a}x^{i-1}(x - a)(b - x)$.
- $\tilde{a}$ is a TPQT2FN. In this chapter, we have assumed $\tilde{a} = (1 - \frac{2}{k}, 1 - \frac{1}{k}, 1 - \frac{1}{2k}, 1, 1 + \frac{1}{2k}, 1 + \frac{1}{k}, 1 + \frac{2}{k})$, where $k$ is a positive integer. Which controls the degree of uncertainty of the uncertain parameters.

Now, substituting equation (10.10) into equation (10.8), the residual may be written as,

$$
\begin{aligned}
&\tilde{R}(x; r_1, r_2, \tilde{c}_1, \ldots, \tilde{c}_n) \\
&\quad = \left(\tilde{\psi}_0(x; r_1, r_2) + \sum_{i=1}^{n} \tilde{c}_i(r_1, r_2)\tilde{\psi}_i(x; r_1, r_2)\right)^n + a_1(x)\left(\tilde{\psi}_0(x; r_1, r_2)\right. \\
&+ \left.\sum_{i=1}^{n} \tilde{c}_i(r_1, r_2)\tilde{\psi}_i(x; r_1, r_2)\right)^{n-1} \\
&\quad +\cdots + \left(\tilde{\psi}_0(x; r_1, r_2) + \sum_{i=1}^{n} \tilde{c}_i(r_1, r_2)\tilde{\psi}_i(x; r_1, r_2)\right) - \tilde{f}\left(x; r_1, r_2\right).
\end{aligned} \tag{10.11}
$$

Now, forcing the residual to zero at the collocation points in the given interval $[a, b]$. Suppose, $x_i$, $1 \leqslant n$ be $n$ collocation points, then,

$$\tilde{R}(x_i; r_1, r_2, \tilde{c}_1, \ldots, \tilde{c}_n) = 0. \tag{10.12}$$

Now, the system of linear equations (10.9) can be solved to obtain the values of the type-2 fuzzy constants $\tilde{c}_i$, $1 \leqslant n$ in terms of the parameters $r_1$ and $r_2$. Further, putting the above values in equation (10.7), the approximate type-2 fuzzy solution of the T2FBVP (10.5), may be achieved.

## 10.4 Numerical examples

In this section three numerical examples and one application problem are given to show the validation of the proposed method.

**Example 10.1.** Let us consider a second order T2FBVP, given by

$$\tilde{y}''(x) + \tilde{y} + x = 0 \tag{10.13}$$

subject to the type-2 fuzzy boundary conditions

$$\tilde{y}(0) = \tilde{y}(1) = (-0.15, -0.1, -0.05, 0, 0.05, 0.1, 0.15). \tag{10.14}$$

Now, let us assume the solution $\tilde{y}(x; r_1, r_2)$ as a three-term type-2 fuzzy polynomial approximation given by,

$$\tilde{y}(x; r_1, r_2) = \tilde{\psi}_0(x; r_1, r_2) + \sum_{i=1}^{2} \tilde{c}_i(r_1, r_2)\tilde{\psi}_i(x; r_1, r_2), \tag{10.15}$$

where, $\tilde{\psi}_0(x; r_1, r_2) = (\underline{\psi}_0(x; r_1, r_2), \bar{\psi}_0(x; r_1, r_2))$, $\underline{\psi}_0(x; r_1, r_2) = [(1 - x)\underline{L}_{A_1} + x\underline{L}_{B_1}, (1 - x)\underline{R}_{A_1} + x\underline{R}_{B_1}]$ and $\bar{\psi}_0(x; r_1, r_2) = [(1 - x)\bar{L}_{A_1} + x\bar{L}_{B_1}, (1 - x)\bar{R}_{A_1} + x\bar{R}_{B_1}]$.

Now, suppose $\tilde{\psi}_1(x) = \tilde{a}x(1 - x)$ and $\tilde{\psi}_2(x) = \tilde{a}x^2(1 - x)$ and substituting these above polynomial approximations in equation (10.13), we have the residual as,

$$\begin{aligned} \tilde{R}(x; r_1, r_2, c_1, c2) = (\tilde{\psi}_0(x; r_1, r_2) + \sum_{i=1}^{2} \tilde{c}_i(r_1, r_2)\tilde{\psi}_i(x; r_1, r_2))'' \\ + \tilde{\psi}_0(x; r_1, r_2) + \sum_{i=1}^{2} \tilde{c}_i(r_1, r_2)\tilde{\psi}_i(x; r_1, r_2) + x. \end{aligned} \tag{10.16}$$

Now, forcing the residual to zero at the collocating points $x_1 = \frac{1}{3}$ and $x_2 = \frac{2}{3}$. Then, on further simplification one may obtain

$$\begin{aligned} \tilde{\psi}_0\left(\frac{1}{3}\right) - 2\tilde{a}\tilde{c}_1 + \frac{2}{9}\tilde{a}\tilde{c}_1 + \frac{2}{27}\tilde{a}\tilde{c}_2 + \frac{1}{3} = 0 \\ \tilde{\psi}_0\left(\frac{2}{3}\right) - 2\tilde{a}\tilde{c}_1 + \frac{2}{9}\tilde{a}\tilde{c}_1 - 2\tilde{a}\tilde{c}_2 + \frac{4}{27}\tilde{a}\tilde{c}_2 + \frac{2}{3} = 0. \end{aligned} \tag{10.17}$$

Now, using the double parametric representation of type-2 fuzzy numbers, the system of equations (10.17) is written as,

$$\begin{aligned} \tilde{\psi}_0\left(\tfrac{1}{3}; r_1, r_2\right) - 2\tilde{a}(r_1, r_2)\tilde{c}_1(r_1, r_2) + \frac{2}{9}\tilde{a}(r_1, r_2)\tilde{c}_1(r_1, r_2) + \frac{2}{27}\tilde{a}(r_1, r_2)\tilde{c}_2(r_1, r_2) \\ + \frac{1}{3} = 0 \tilde{\psi}_0\left(\tfrac{2}{3}; r_1, r_2\right) - 2\tilde{a}(r_1, r_2)\tilde{c}_1(r_1, r_2) + \frac{2}{9}\tilde{a}(r_1, r_2)\tilde{c}_1(r_1, r_2) \\ -2\tilde{a}(r_1, r_2)\tilde{c}_2(r_1, r_2) + \frac{4}{27}\tilde{a}(r_1, r_2)\tilde{c}_2(r_1, r_2) + \frac{2}{3} = 0. \end{aligned} \tag{10.18}$$

The above system further can be simplified by using definitions of lower and upper bounds of the involved type-2 fuzzy parameters and then solved for unknown constants, in terms of the parameters $r_1$ and $r_2$ to obtain the unknown constants $\tilde{c}_1$ and $\tilde{c}_2$. The values obtained after solving the system of equations are,

$$\begin{aligned} \tilde{c}_1 = \Bigg(\frac{1323k}{10\,400k + 20\,800}, \frac{1557k}{10\,400k + 10\,400}, \frac{1791k}{10\,400k + 5200}, 0.1947, \\ \frac{2259k}{10\,400k - 5200}, \frac{2493k}{10\,400k - 10\,400}, \frac{2727k}{10\,400k - 20\,800}\Bigg), \\ \tilde{c}_2 = \Bigg(\frac{9k}{52k + 104}, \frac{9k}{52k + 52}, \frac{9k}{52k + 26}, 0.173\,1, \\ \frac{9k}{52k - 26}, \frac{9k}{52k - 52s}, \frac{9k}{52k - 104}\Bigg). \end{aligned} \tag{10.19}$$

Further, the values of $\tilde{c}_1$ and $\tilde{c}_2$ are substituted in equation (10.15) to achieve the approximate solution of the given T2FBVP. In table 10.1, type-2 fuzzy solutions obtained by the proposed method are shown and also compared with the published type-1 fuzzy results in a special case.

**Example 10.2.** Now, let us consider the T2FBVP, given by

$$\tilde{y}''(x) + \tilde{y} = x \tag{10.20}$$

subject to the type-2 fuzzy boundary conditions

$$\begin{aligned} \tilde{y}(0) &= (2.5,\ 3,\ 3.5,\ 4,\ 4.5,\ 5,\ 5.5) \\ \tilde{y}(1) &= (-0.5,\ 0,\ 0.5,\ 1,\ 1.5,\ 2,\ 2.5). \end{aligned} \tag{10.21}$$

Let us consider the approximate solution as a four-term type-2 fuzzy polynomial function,

$$\tilde{y}(x; r_1, r_2) = \tilde{\psi}_0(x; r_1, r_2) + \sum_{i=1}^{3} \tilde{c}_i(r_1, r_2)\tilde{\psi}_i(x; r_1, r_2), \tag{10.22}$$

where, $\tilde{\psi}_0(x; r_1, r_2)$ and $\tilde{\psi}_i(x; r_1, r_2)$, $1 \leqslant i \leqslant 3$ can be assumed in a similar way as done in example 10.1. Now, substituting the values of $\tilde{y}$ in the given differential equation. The residual will be,

$$\begin{aligned} &\left(\tilde{\psi}_0(x; r_1, r_2) + \sum_{i=1}^{3} \tilde{c}_i(r_1, r_2)\tilde{\psi}_i(x; r_1, r_2)\right)'' \\ &+ \left(\tilde{\psi}_0(x; r_1, r_2) + \sum_{i=1}^{3} \tilde{c}_i(r_1, r_2)\tilde{\psi}_i(x; r_1, r_2)\right) - x. \end{aligned} \tag{10.23}$$

Suppose, the collocating points are $x_1 = 0.25$, $x_2 = 0.5$ and $x_3 = 0.75$, then forcing the residual to zero at these collocating points, we have

$$\begin{aligned} &\tilde{\psi}_0(0.25) - 2\tilde{a}\tilde{c}_1 + 0.187\,5\tilde{a}\tilde{c}_1 + 0.5\tilde{a}\tilde{c}_2 + 0.046\,875\tilde{a}\tilde{c}_2 + 0.75\tilde{a}\tilde{c}_3 \\ &\quad + 0.011\,718\,75\tilde{a}\tilde{c}_3 - 0.25 = 0,\ \tilde{\psi}_0(0.5) - 2\tilde{a}\tilde{c}_1 + 0.25\tilde{a}\tilde{c}_1 \\ &\quad -\tilde{a}\tilde{c}_2 + 0.125\tilde{a}\tilde{c}_2 + 0.062\,5\tilde{a}\tilde{c}_3 - 0.5 \\ &= 0,\ \tilde{\psi}_0(0.75) - 2\tilde{a}\tilde{c}_1 + 0.187\,5\tilde{a}\tilde{c}_1 - 2.5\tilde{a}\tilde{c}_2 + 0.140\,625\tilde{a}\tilde{c}_2 \\ &\quad -2.25\tilde{a}\tilde{c}_3 + 0.105\,468\,75\tilde{a}\tilde{c}_3 - 0.75 = 0. \end{aligned} \tag{10.24}$$

Further, using the double parametric representation in the above type-2 fuzzy system (10.24) yields,

$$\begin{aligned} &\tilde{\psi}_0(0.25; r_1, r_2) - 2\tilde{a}(r_1, r_2)\tilde{c}_1(r_1, r_2) + 0.187\,5\tilde{a}(r_1, r_2)\tilde{c}_1(r_1, r_2) \\ &\quad + 0.5\tilde{a}(r_1, r_2)\tilde{c}_2(r_1, r_2) + 0.046\,875\tilde{a}(r_1, r_2)\tilde{c}_2(r_1, r_2) \\ &\quad + 0.75\tilde{a}(r_1, r_2)\tilde{c}_3(r_1, r_2) + 0.011\,718\,75\tilde{a}(r_1, r_2)\tilde{c}_3(r_1, r_2) - 0.25 = 0, \\ &\tilde{\psi}_0(0.5; r_1, r_2) - 2\tilde{a}(r_1, r_2)\tilde{c}_1(r_1, r_2) + 0.25\tilde{a}(r_1, r_2)\tilde{c}_1(r_1, r_2) - \tilde{a}(r_1, r_2)\tilde{c}_2(r_1, r_2) \\ &\quad + 0.125\tilde{a}(r_1, r_2)\tilde{c}_2(r_1, r_2) + 0.062\,5\tilde{a}(r_1, r_2)\tilde{c}_3(r_1, r_2) - 0.5 = 0, \\ &\tilde{\psi}_0(0.75; r_1, r_2) - 2\tilde{a}(r_1, r_2)\tilde{c}_1(r_1, r_2) + 0.187\,5\tilde{a}(r_1, r_2)\tilde{c}_1(r_1, r_2) \\ &\quad - 2.5\tilde{a}(r_1, r_2)\tilde{c}_2(r_1, r_2) + 0.140\,625\tilde{a}(r_1, r_2)\tilde{c}_2(r_1, r_2) - 2.25\tilde{a}(r_1, r_2)\tilde{c}_3(r_1, r_2) \\ &\quad + 0.105\,468\,75\tilde{a}(r_1, r_2)\tilde{c}_3(r_1, r_2) - 0.75 = 0. \end{aligned} \tag{10.25}$$

Now, the above system can be solved for $\tilde{c}_1$, $\tilde{c}_2$ and $\tilde{c}_3$ by using the definition of lower and upper bounds of TPQT2FNs in terms of the parameters $r_1$ and $r_2$. These type-2 fuzzy coefficients are obtained as,

$$\begin{aligned}
\tilde{c}_1 = \bigg(&\frac{68\,921\,618k}{221\,527\,801k + 443\,055\,602}, \frac{205\,733\,288k}{443\,055\,602k + 443\,055\,602},\\
&\frac{273\,623\,340k}{443\,055\,602k + 221\,527\,801}, 0.7708, \frac{409\,403\,444k}{443\,055\,602k - 221\,527\,801},\\
&\frac{477\,293\,496k}{443\,055\,602k - 443\,055\,602}, \frac{272\,591\,774k}{221\,527\,801k - 443\,055\,602}\bigg),\\
\tilde{c}_2 = \bigg(&\frac{545\,978\,304k}{221\,527\,801k + 443\,055\,602}, \frac{1\,072\,800\,128k}{443\,055\,602k + 443\,055\,602},\\
&\frac{1\,053\,643\,648k}{443\,055\,602k + 221\,527\,801}, 2.334\,9, \frac{1\,015\,330\,688k}{443\,055\,602k - 221\,527\,801},\\
&\frac{996\,174\,208k}{443\,055\,602k - 443\,055\,602}, \frac{488\,508\,864k}{221\,527\,801k - 443\,055\,602}\bigg),\\
\tilde{c}_3 = \bigg(&\frac{-531\,369\,568k}{221\,527\,801k - 443\,055\,602}, \frac{-1\,115\,214\,208k}{443\,055\,602k - 443\,055\,602},\\
&\frac{-1\,167\,689\,280k}{443\,055\,602k - 221\,527\,801}, -2.754\,0, \frac{-1\,272\,639\,424k}{443\,055\,602k + 221\,527\,801},\\
&\frac{-1\,325\,114\,496k}{443\,055\,602k + 443\,055\,602}, \frac{-688\,794\,784k}{221\,527\,801k + 443\,055\,602}\bigg).
\end{aligned} \tag{10.26}$$

Then, these values can be substituted in equation (10.22) to approximate the type-2 fuzzy solution of the T2FBVP.

**Example 10.3.** Let us now consider a second order T2FBVP given by,

$$\tilde{y}'' - 3\tilde{y}' + 2\tilde{y} = 0, \tag{10.27}$$

subject to the type-2 fuzzy boundary conditions

$$\begin{aligned}
\tilde{y}(0) &= (1.7,\ 1.8,\ 1.9,\ 2,\ 2.1,\ 2.2,\ 2.3),\\
\tilde{y}(1) &= (2.7,\ 2.8,\ 2.9,\ 3,\ 3.1,\ 3.2,\ 3.3).
\end{aligned} \tag{10.28}$$

Let the approximate solution using a five-term polynomial function,

$$\tilde{y}(x;\, r_1,\, r_2) = \tilde{\psi}_0(x;\, r_1,\, r_2) + \sum_{i=1}^{4}\tilde{c}_i(r_1,\, r_2)\tilde{\psi}_i(x;\, r_1,\, r_2), \tag{10.29}$$

where the values of $\tilde{\psi}_0(x)$ and $\tilde{\psi}_i(x)$, $1 \leqslant i \leqslant 4$ can be chosen as discussed in the previous examples.

Now, suppose in this case the collocating points are $x_1 = 0.2$, $x_2 = 0.4$, $x_3 = 0.6$, $x_4 = 0.8$. Equating the residual $\tilde{R}(x;\, \tilde{c}_1,\, \tilde{c}_2,\, \tilde{c}_3,\, \tilde{c}_4)$ to zero at these collocating points.

Then, solving the system of type-2 fuzzy linear equations will result the type-2 fuzzy coefficients as,

$$\tilde{c}_1 = \left(\frac{29\,383\,115\,202\,528\,132k}{91\,479\,243\,845\,490\,304k + 182\,958\,487\,690\,980\,608}, \frac{34\,227\,489\,834\,565\,216k}{91\,479\,243\,845\,490\,304k + 91\,479\,243\,845\,490\,304}, \frac{39\,071\,864\,466\,602\,296k}{91\,479\,243\,845\,490\,304k + 45\,739\,621\,922\,745\,152}, 0.480\,1, \frac{48\,760\,613\,730\,676\,456k}{91\,479\,243\,845\,490\,304k - 45\,739\,621\,922\,745\,152}, \frac{53\,604\,988\,362\,713\,536k}{91\,479\,243\,845\,490\,304k - 91\,479\,243\,845\,490\,304}, \frac{58\,449\,362\,994\,750\,616k}{91\,479\,243\,845\,490\,304k - 182\,958\,487\,690\,980\,608}\right),$$

$$\tilde{c}_2 = \left(\frac{85\,155\,824\,520\,110\,700k}{182\,958\,487\,690\,980\,608k + 365\,916\,975\,381\,961\,216}, \frac{97\,921\,305\,087\,135\,425k}{182\,958\,487\,690\,980\,608k + 182\,958\,487\,690\,980\,608}, \frac{110\,686\,785\,654\,160\,150k}{182\,958\,487\,690\,980\,608k + 91\,479\,243\,845\,490\,304}, 0.674\,8, \frac{136\,217\,746\,788\,209\,600k}{182\,958\,487\,690\,980\,608k - 91\,479\,243\,845\,490\,304}, \frac{148\,983\,227\,355\,234\,325k}{182\,958\,487\,690\,980\,608k - 182\,958\,487\,690\,980\,608}, \frac{161\,748\,707\,922\,259\,050k}{182\,958\,487\,690\,980\,608k - 365\,916\,975\,381\,961\,216}\right),$$

$$\tilde{c}_3 = \left(\frac{1\,725\,625\,180\,679\,525k}{4\,691\,243\,274\,127\,708k + 9\,382\,486\,548\,255\,416}, \frac{1\,835\,046\,831\,475\,100k}{4\,691\,243\,274\,127\,708k + 4\,691\,243\,274\,127\,708}, \frac{1\,944\,468\,482\,270\,675k}{4\,691\,243\,274\,127\,708k + 2\,345\,621\,637\,063\,854}, 0.437\,8, \frac{2\,163\,311\,783\,861\,825k}{4\,691\,243\,274\,127\,708k - 2\,345\,621\,637\,063\,854}, \frac{2\,272\,733\,434\,657\,400k}{4\,691\,243\,274\,127\,708k - 4\,691\,243\,274\,127\,708}, \frac{2\,382\,155\,085\,452\,975k}{4\,691\,243\,274\,127\,708k - 9\,382\,486\,548\,255\,416}\right),$$

$$\tilde{c}_4 = \left( \frac{9\,707\,456\,590\,065\,625k}{60\,986\,162\,563\,660\,208k + 121\,972\,325\,127\,320\,416}, \right.$$
$$\frac{10\,462\,965\,104\,039\,375k}{60\,986\,162\,563\,660\,208k + 60\,986\,162\,563\,660\,208},$$
$$\frac{11\,218\,473\,618\,013\,125k}{60\,986\,162\,563\,660\,208k + 30\,493\,081\,281\,830\,104},$$
$$0.196\,3, \frac{12\,729\,490\,645\,960\,625k}{60\,986\,162\,563\,660\,208k - 30\,493\,081\,281\,830\,104}, \tag{10.30}$$
$$\frac{13\,484\,999\,159\,934\,375k}{60\,986\,162\,563\,660\,208k - 60\,986\,162\,563\,660\,208},$$
$$\left. \frac{14\,240\,507\,673\,908\,125k}{60\,986\,162\,563\,660\,208k - 121\,972\,325\,127\,320\,416} \right).$$

Then, substituting these values of type-2 fuzzy constants in equation (10.29), we obtain the approximate type-2 fuzzy solution.

**Example 10.4.** Let us consider an application problem related to a simply supported beam subject to concentrated moments at both the ends. The governing equation for the defection of the beam is given by [59],

$$EI\frac{d^2y}{dx^2} - M = 0, \quad x \in [0, L], \tag{10.31}$$

with the boundary conditions

$$y(0) = y(L) = 0, \tag{10.32}$$

where, $EI$ is the resistance to defection, $M$ is the applied moment and $L$ is the length of the beam. Assume that $M = 50$ Nm, $EI = 10$ N m$^{-1}$ and $L = 10$ m. Further, let us introduce the uncertainties in the measurement of the deflection and in the involved boundary conditions of the system. Assuming this impreciseness as type-2 fuzzy numbers (say TPQT2FNs), then the type-2 fuzzy governing equation is presented as,

$$EI\frac{d^2\tilde{y}}{dx^2} - M = 0, \quad x \in [0, L], \tag{10.33}$$

subject to the type-2 fuzzy boundary conditions,

$$\tilde{y}(0) = \tilde{y}(L) = (-0.15, -0.1, -0.05, 0, 0.05, 0.1, 0.15). \tag{10.34}$$

Assuming the solution $\tilde{y}(x; r_1, r_2)$ as type-2 fuzzy polynomial functions up to three terms,

$$\tilde{y}(x; r_1, r_2) = \tilde{\psi}_0(x; r_1, r_2) + \sum_{i=1}^{2} \tilde{c}_i(r_1, r_2)\tilde{\psi}_i(x; r_1, r_2), \tag{10.35}$$

where the trial functions satisfy the boundary conditions. Finally, following the proposed method the unknown type-2 fuzzy coefficients are obtained as,

$$\begin{aligned}\tilde{c}_1 &= \left(\frac{-5k}{2k-4}, \frac{-5k}{2k-2}, \frac{-5k}{2k-1}, -5/2, \frac{-5k}{2k+1}, \frac{-5k}{2k+2}, \frac{-5k}{2k-4}\right), \\ \tilde{c}_2 &= 0.\end{aligned} \tag{10.36}$$

Now, substituting the values of $\tilde{c}_1$ and $\tilde{c}_2$ in (10.35), the required type-2 fuzzy solution are achieved.

## 10.5 Results and discussions

The polynomial based collocation method is used in type-2 fuzzy environment to solve T2FBVPs. A few numerical examples are solved to show the efficacy of the method. In this regard, the obtained type-2 fuzzy solutions are shown in tables (tables 10.1–10.8) and three-dimensional plots (figures 10.9–10.12). Also, corresponding crisp and type-1 fuzzy solutions are compared with exact solutions

**Table 10.1.** Type-2 fuzzy solution of example 10.1 when $k = 5$ for the three-term polynomial approximation.

| $t$ | $\bar{L}$ | $l_1$ | $\underline{L}$ | $m$ | $\underline{R}$ | $r_1$ | $\bar{R}$ |
|---|---|---|---|---|---|---|---|
| 0.1000 | −0.1444 | −0.0900 | −0.0360 | 0.0191 | 0.0758 | 0.1347 | 0.2087 |
| 0.2000 | −0.1389 | −0.0803 | −0.0229 | 0.0367 | 0.0992 | 0.1658 | 0.2608 |
| 0.3000 | −0.1339 | −0.0718 | −0.0115 | 0.0518 | 0.1191 | 0.1919 | 0.3039 |
| 0.4000 | −0.1298 | −0.0650 | −0.0026 | 0.0633 | 0.1340 | 0.2112 | 0.3356 |
| 0.5000 | −0.1271 | −0.0606 | 0.0029 | 0.0703 | 0.1428 | 0.2223 | 0.3534 |
| 0.6000 | −0.1262 | −0.0594 | 0.0042 | 0.0717 | 0.1442 | 0.2237 | 0.3550 |
| 0.7000 | −0.1276 | −0.0621 | 0.0004 | 0.0663 | 0.1368 | 0.2137 | 0.3378 |
| 0.8000 | −0.1318 | −0.0693 | −0.0093 | 0.0533 | 0.1196 | 0.1908 | 0.2996 |
| 0.9000 | −0.1391 | −0.0817 | −0.0258 | 0.0315 | 0.0910 | 0.1534 | 0.2378 |

**Table 10.2.** Type-2 fuzzy solution of example 10.1 when $k = 10$ for the three-term polynomial approximation.

| $t$ | $\bar{L}$ | $l_1$ | $\underline{L}$ | $m$ | $\underline{R}$ | $r_1$ | $\bar{R}$ |
|---|---|---|---|---|---|---|---|
| 0.1000 | −0.1413 | −0.0877 | −0.0346 | 0.0191 | 0.0733 | 0.1283 | 0.1877 |
| 0.2000 | −0.1327 | −0.0759 | −0.0201 | 0.0367 | 0.0945 | 0.1536 | 0.2212 |
| 0.3000 | −0.1249 | −0.0654 | −0.0074 | 0.0518 | 0.1125 | 0.1749 | 0.2490 |
| 0.4000 | −0.1186 | −0.0570 | 0.0024 | 0.0633 | 0.1260 | 0.1906 | 0.2693 |
| 0.5000 | −0.1144 | −0.0517 | 0.0085 | 0.0703 | 0.1339 | 0.1997 | 0.2808 |
| 0.6000 | −0.1130 | −0.0502 | 0.0099 | 0.0717 | 0.1352 | 0.2008 | 0.2818 |
| 0.7000 | −0.1152 | −0.0535 | 0.0057 | 0.0663 | 0.1285 | 0.1926 | 0.2708 |
| 0.8000 | −0.1217 | −0.0623 | −0.0050 | 0.0533 | 0.1129 | 0.1740 | 0.2462 |
| 0.9000 | −0.1330 | −0.0775 | −0.0233 | 0.0315 | 0.0871 | 0.1435 | 0.2064 |

**Table 10.3.** Type-2 fuzzy solution of example 10.1 when $k = 20$ for the three-term polynomial approximation.

| $t$ | $\bar{L}$ | $l_1$ | $\underline{L}$ | $m$ | $\underline{R}$ | $r_1$ | $\bar{R}$ |
|---|---|---|---|---|---|---|---|
| 0.1000 | −0.1394 | −0.0864 | −0.0338 | 0.0191 | 0.0722 | 0.1256 | 0.1807 |
| 0.2000 | −0.1288 | −0.0733 | −0.0185 | 0.0367 | 0.0924 | 0.1485 | 0.2080 |
| 0.3000 | −0.1192 | −0.0617 | −0.0052 | 0.0518 | 0.1094 | 0.1677 | 0.2306 |
| 0.4000 | −0.1114 | −0.0525 | 0.0051 | 0.0633 | 0.1223 | 0.1820 | 0.2472 |
| 0.5000 | −0.1063 | −0.0466 | 0.0115 | 0.0703 | 0.1298 | 0.1901 | 0.2566 |
| 0.6000 | −0.1046 | −0.0449 | 0.0130 | 0.0717 | 0.1310 | 0.1911 | 0.2574 |
| 0.7000 | −0.1073 | −0.0485 | 0.0086 | 0.0663 | 0.1247 | 0.1838 | 0.2484 |
| 0.8000 | −0.1152 | −0.0583 | −0.0027 | 0.0533 | 0.1098 | 0.1669 | 0.2284 |
| 0.9000 | −0.1292 | −0.0751 | −0.0219 | 0.0315 | 0.0853 | 0.1393 | 0.1960 |

**Table 10.4.** Type-2 fuzzy solution of example 10.1 when $k = 20$ for the four-term polynomial approximation.

| $t$ | $\bar{L}$ | $l_1$ | $\underline{L}$ | $m$ | $\underline{R}$ | $r_1$ | $\bar{R}$ |
|---|---|---|---|---|---|---|---|
| 0.1000 | −0.1380 | −0.0855 | −0.0335 | 0.0187 | 0.0711 | 0.1237 | 0.1778 |
| 0.2000 | −0.1263 | −0.0716 | −0.0179 | 0.0362 | 0.0905 | 0.1453 | 0.2030 |
| 0.3000 | −0.1159 | −0.0594 | −0.0043 | 0.0512 | 0.1071 | 0.1636 | 0.2242 |
| 0.4000 | −0.1078 | −0.0500 | 0.0061 | 0.0627 | 0.1198 | 0.1775 | 0.2402 |
| 0.5000 | −0.1029 | −0.0443 | 0.0124 | 0.0696 | 0.1274 | 0.1858 | 0.2498 |
| 0.6000 | −0.1020 | −0.0432 | 0.0136 | 0.0709 | 0.1288 | 0.1874 | 0.2516 |
| 0.7000 | −0.1057 | −0.0476 | 0.0087 | 0.0655 | 0.1229 | 0.1808 | 0.2440 |
| 0.8000 | −0.1146 | −0.0581 | −0.0030 | 0.0526 | 0.1085 | 0.1650 | 0.2257 |
| 0.9000 | −0.1293 | −0.0753 | −0.0223 | 0.0310 | 0.0846 | 0.1385 | 0.1949 |

**Table 10.5.** Type-2 fuzzy solution of example 10.1 when $k = 20$ for the five-term polynomial approximation.

| $t$ | $\bar{L}$ | $l_1$ | $\underline{L}$ | $m$ | $\underline{R}$ | $r_1$ | $\bar{R}$ |
|---|---|---|---|---|---|---|---|
| 0.1000 | −0.1397 | −0.0868 | −0.0342 | 0.0186 | 0.0717 | 0.1250 | 0.1801 |
| 0.2000 | −0.1291 | −0.0737 | −0.0190 | 0.0361 | 0.0917 | 0.1477 | 0.2071 |
| 0.3000 | −0.1195 | −0.0620 | −0.0057 | 0.0512 | 0.1087 | 0.1668 | 0.2295 |
| 0.4000 | −0.1116 | −0.0528 | 0.0047 | 0.0628 | 0.1216 | 0.1811 | 0.2461 |
| 0.5000 | −0.1065 | −0.0469 | 0.0111 | 0.0697 | 0.1291 | 0.1893 | 0.2555 |
| 0.6000 | −0.1050 | −0.0454 | 0.0125 | 0.0710 | 0.1303 | 0.1903 | 0.2563 |
| 0.7000 | −0.1078 | −0.0491 | 0.0079 | 0.0656 | 0.1239 | 0.1828 | 0.2473 |
| 0.8000 | −0.1158 | −0.0590 | −0.0035 | 0.0525 | 0.1090 | 0.1659 | 0.2273 |
| 0.9000 | −0.1297 | −0.0757 | −0.0225 | 0.0309 | 0.0846 | 0.1386 | 0.1952 |

**Table 10.6.** Type-2 fuzzy solution of example 10.2 when $k = 20$ for the four-term polynomial approximation.

| $t$ | $\bar{L}$ | $l_1$ | $\underline{L}$ | $m$ | $\underline{R}$ | $r_1$ | $\bar{R}$ |
|---|---|---|---|---|---|---|---|
| 0.1000 | 2.2389 | 2.7553 | 3.2709 | 3.7879 | 4.3065 | 4.8268 | 5.3568 |
| 0.2000 | 1.9899 | 2.5212 | 3.0495 | 3.5804 | 4.1141 | 4.6509 | 5.2070 |
| 0.3000 | 1.7352 | 2.2787 | 2.8161 | 3.3569 | 3.9015 | 4.4501 | 5.0269 |
| 0.4000 | 1.4626 | 2.0145 | 2.5569 | 3.1034 | 3.6541 | 4.2095 | 4.8003 |
| 0.5000 | 1.1658 | 1.7216 | 2.2649 | 2.8124 | 3.3645 | 3.9214 | 4.5186 |
| 0.6000 | 0.8442 | 1.3988 | 1.9390 | 2.4833 | 3.0319 | 3.5852 | 4.1804 |
| 0.7000 | 0.5031 | 1.0513 | 1.5847 | 2.1217 | 2.6626 | 3.2076 | 3.7921 |
| 0.8000 | 0.1532 | 0.6899 | 1.2137 | 1.7402 | 2.2697 | 2.8023 | 3.3672 |
| 0.9000 | −0.1886 | 0.3318 | 0.8440 | 1.3577 | 1.8732 | 2.3904 | 2.9270 |

**Table 10.7.** Type-2 fuzzy solution of example 10.3 when $k = 20$ for the five-term polynomial approximation.

| $t$ | $\bar{L}$ | $l_1$ | $\underline{L}$ | $m$ | $\underline{R}$ | $r_1$ | $\bar{R}$ |
|---|---|---|---|---|---|---|---|
| 0.1000 | 0.5742 | 1.0960 | 1.6215 | 2.1497 | 2.6807 | 3.2148 | 3.7600 |
| 0.2000 | 0.6496 | 1.1934 | 1.7448 | 2.3015 | 2.8638 | 3.4323 | 4.0233 |
| 0.3000 | 0.7282 | 1.2932 | 1.8690 | 2.4527 | 3.0448 | 3.6459 | 4.2805 |
| 0.4000 | 0.8120 | 1.3954 | 1.9926 | 2.5998 | 3.2178 | 3.8474 | 4.5199 |
| 0.5000 | 0.9025 | 1.4998 | 2.1130 | 2.7379 | 3.3752 | 4.0260 | 4.7272 |
| 0.6000 | 1.0011 | 1.6057 | 2.2268 | 2.8604 | 3.5073 | 4.1687 | 4.8842 |
| 0.7000 | 1.1093 | 1.7113 | 2.3292 | 2.9592 | 3.6022 | 4.2591 | 4.9691 |
| 0.8000 | 1.2280 | 1.8145 | 2.4142 | 3.0241 | 3.6450 | 4.2777 | 4.9554 |
| 0.9000 | 1.3580 | 1.9121 | 2.4742 | 3.0427 | 3.6179 | 4.2006 | 4.8114 |

using two-dimensional plots (figures 10.2–10.8) for different degrees of polynomial approximations.

Tables show approximate type-2 fuzzy results for three-term polynomial approximation of example 10.1 for different values of $x$ and $k$. Tables 10.4 and 10.5 show the obtained type-2 fuzzy results of example 10.1 for four and five-term polynomial approximations for $k = 20$ and different values of $x$. Further, figures show the convergence of the approximate solutions towards the exact solution. From these plots, it can be visualized that both the solutions are approximately equal. Figures compare the obtained type-1 fuzzy solution with the published type-1 fuzzy solution for examples 10.1–10.3. From these graphical results, one can see that the solutions by the proposed method agree well with the previously published solutions [30]. Lastly, the type-2 fuzzy solutions are shown using three-dimensional plots for fixed $x$, i.e. $x = 0.1$ for examples 10.1–10.3 and $x = 0.5$ for example 10.4.

**Table 10.8.** Type-2 fuzzy solution of example 10.4 when $k = 20$.

| $t$ | $\overline{L}$ | $l_1$ | $\underline{L}$ | $m$ | $\underline{R}$ | $r_1$ | $\overline{R}$ |
|---|---|---|---|---|---|---|---|
| 0.5000 | −12.0250 | −11.9750 | −11.9250 | −11.8750 | −11.8250 | −11.7750 | −11.7250 |
| 1.0000 | −22.6500 | −22.6000 | −22.5500 | −22.5000 | −22.4500 | −22.4000 | −22.3500 |
| 1.5000 | −32.0250 | −31.9750 | −31.9250 | −31.8750 | −31.8250 | −31.7750 | −31.7250 |
| 2.0000 | −40.1500 | −40.1000 | −40.0500 | −40.0000 | −39.9500 | −39.9000 | −39.8500 |
| 2.5000 | −47.0250 | −46.9750 | −46.9250 | −46.8750 | −46.8250 | −46.7750 | −46.7250 |
| 3.0000 | −52.6500 | −52.6000 | −52.5500 | −52.5000 | −52.4500 | −52.4000 | −52.3500 |
| 3.5000 | −57.0250 | −56.9750 | −56.9250 | −56.8750 | −56.8250 | −56.7750 | −56.7250 |
| 4.0000 | −60.1500 | −60.1000 | −60.0500 | −60.0000 | −59.9500 | −59.9000 | −59.8500 |
| 4.5000 | −62.0250 | −61.9750 | −61.9250 | −61.8750 | −61.8250 | −61.7750 | −61.7250 |
| 5.0000 | −62.6500 | −62.6000 | −62.5500 | −62.5000 | −62.4500 | −62.4000 | −62.3500 |
| 5.5000 | −62.0250 | −61.9750 | −61.9250 | −61.8750 | −61.8250 | −61.7750 | −61.7250 |
| 6.0000 | −60.1500 | −60.1000 | −60.0500 | −60.0000 | −59.9500 | −59.9000 | −59.8500 |
| 6.5000 | −57.0250 | −56.9750 | −56.9250 | −56.8750 | −56.8250 | −56.7750 | −56.7250 |
| 7.0000 | −52.6500 | −52.6000 | −52.5500 | −52.5000 | −52.4500 | −52.4000 | −52.3500 |
| 7.5000 | −47.0250 | −46.9750 | −46.9250 | −46.8750 | −46.8250 | −46.7750 | −46.7250 |
| 8.0000 | −40.1500 | −40.1000 | −40.0500 | −40.0000 | −39.9500 | −39.9000 | −39.8500 |
| 8.5000 | −32.0250 | −31.9750 | −31.9250 | −31.8750 | −31.8250 | −31.7750 | −31.7250 |
| 9.0000 | −22.6500 | −22.6000 | −22.5500 | −22.5000 | −22.4500 | −22.4000 | −22.3500 |
| 9.5000 | −12.0250 | −11.9750 | −11.9250 | −11.8750 | −11.8250 | −11.7750 | −11.7250 |

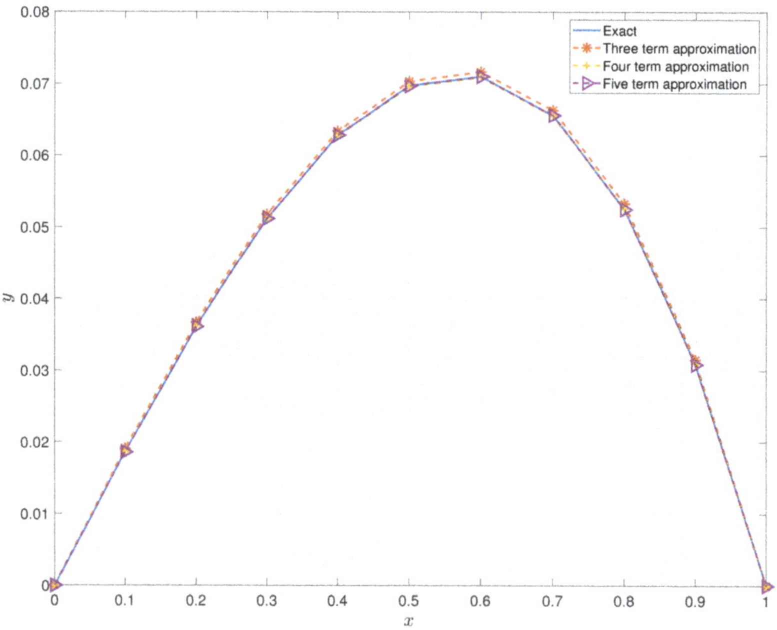


**Figure 10.2.** Comparison of crisp solution by the polynomial based collocation method with exact solution for example 10.1.

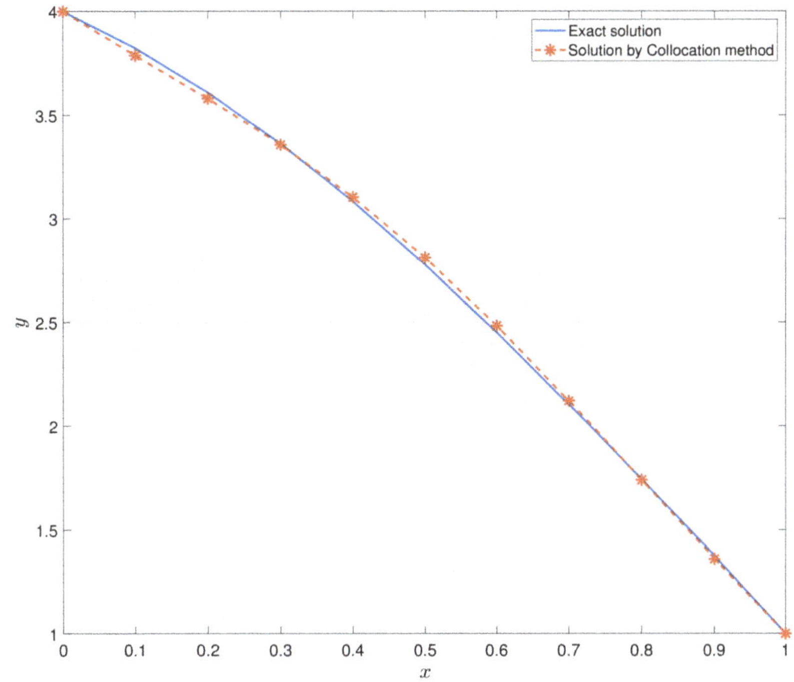


**Figure 10.3.** Comparison of crisp solution by the polynomial based collocation method with exact solution for example 10.2.

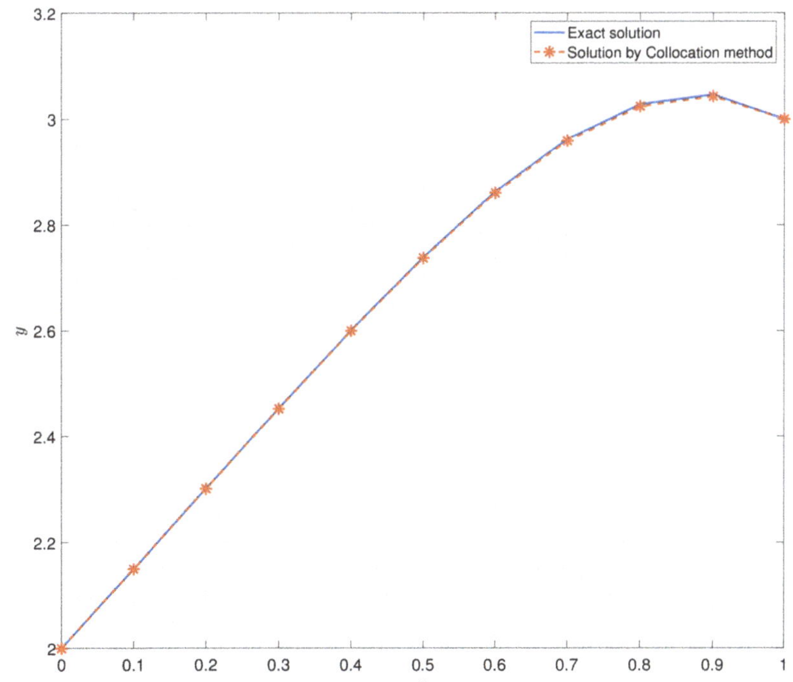


**Figure 10.4.** Comparison of crisp solution by the polynomial based collocation method with exact solution for example 10.3.

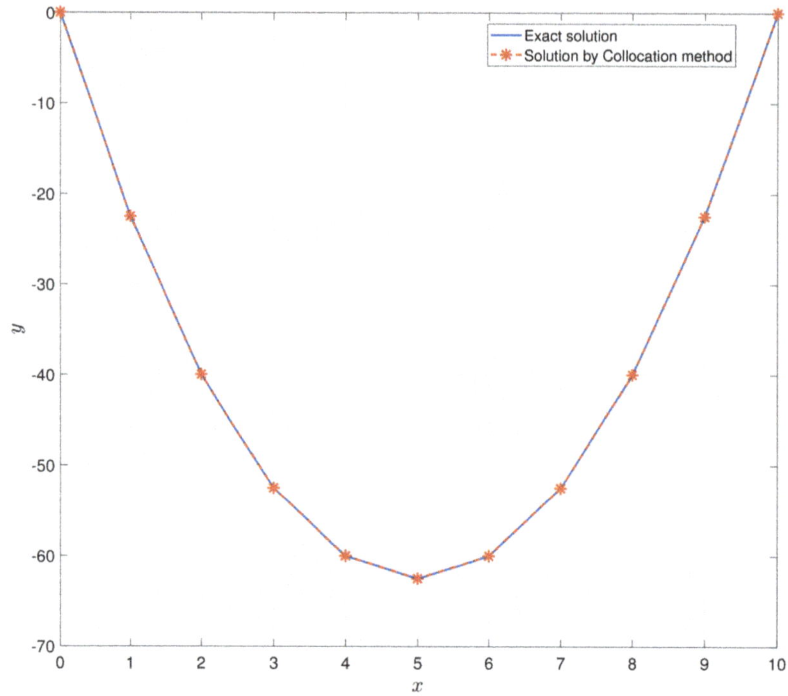


**Figure 10.5.** Comparison of crisp solution of simply supported beam by the polynomial based collocation method with exact solution for $k = 20$.

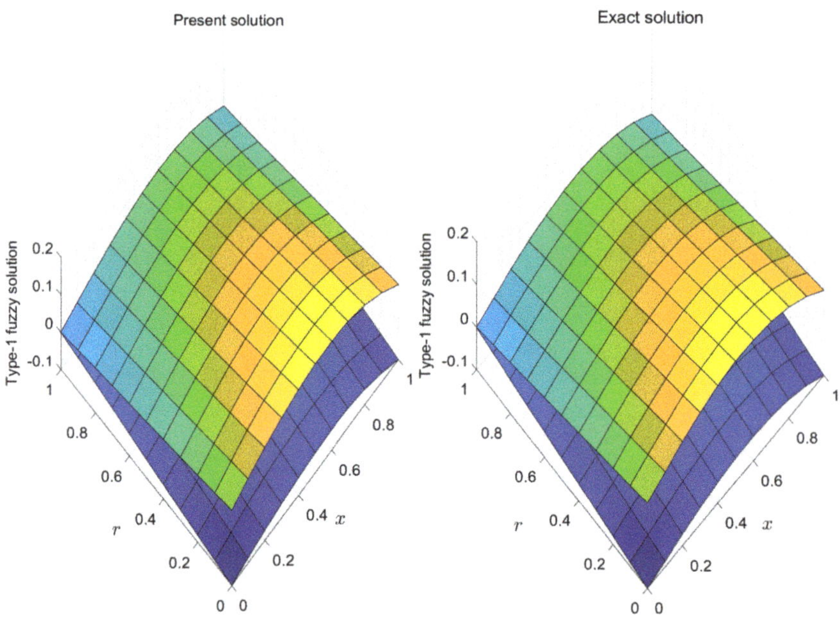


**Figure 10.6.** Comparison of present type-1 fuzzy solution and exact solution [30] for example 10.1.

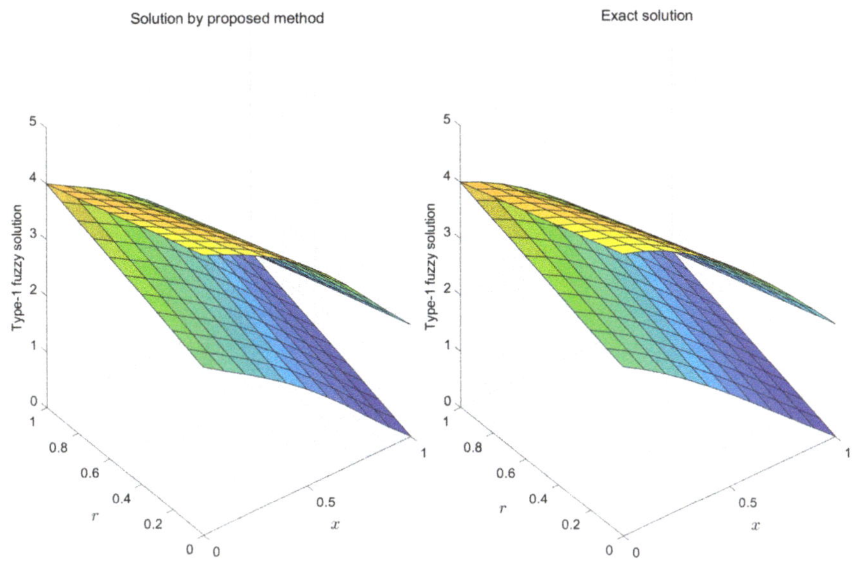


**Figure 10.7.** Comparison of present type-1 fuzzy solution and exact solution [30] for example 10.2.

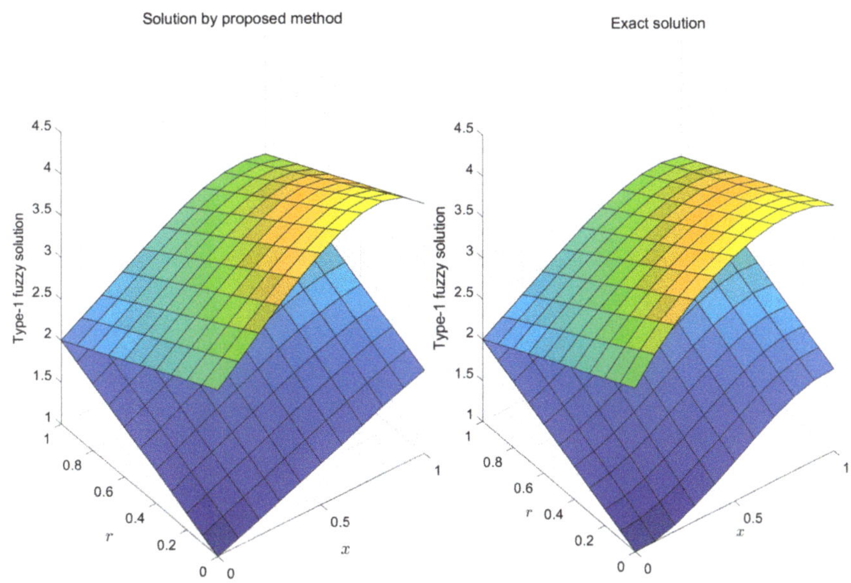


**Figure 10.8.** Comparison of present type-1 fuzzy solution and exact solution [30] for example 10.3.

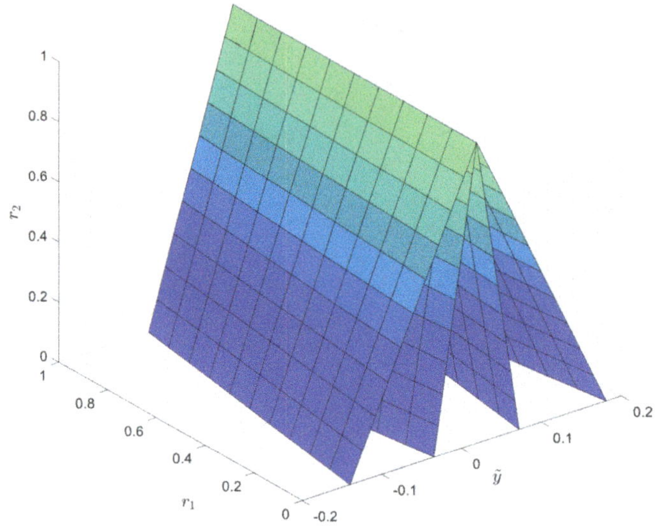


**Figure 10.9.** Type-2 fuzzy solution of example 10.1 at $x = 0.1$ for $k = 20$.

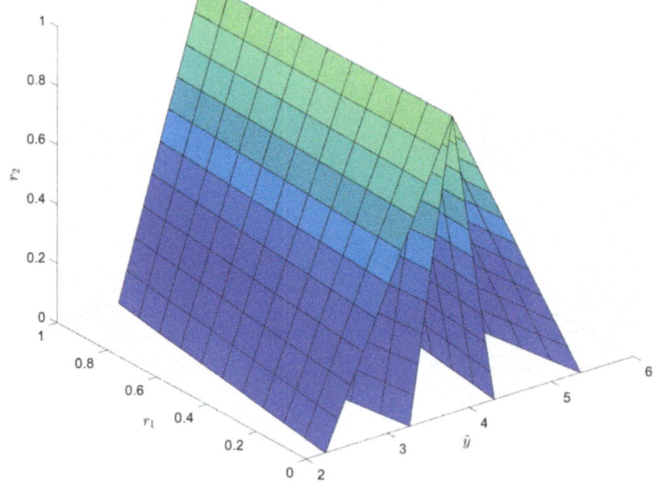


**Figure 10.10.** Type-2 fuzzy solution of example 10.2 at $x = 0.1$ for $k = 20$.

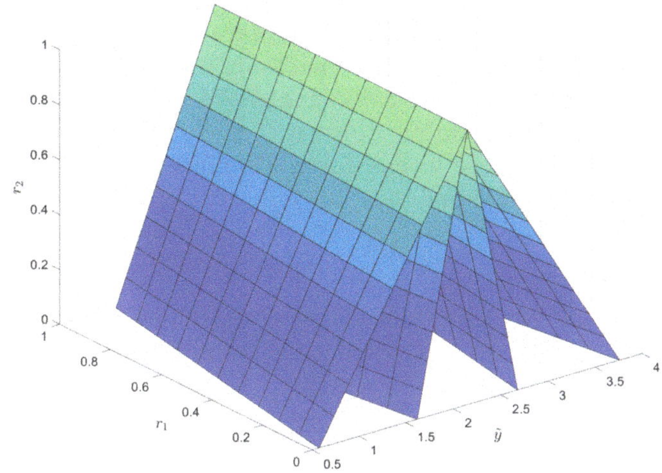


**Figure 10.11.** Type-2 fuzzy solution of example 10.3 at $x = 0.1$ for $k = 20$.

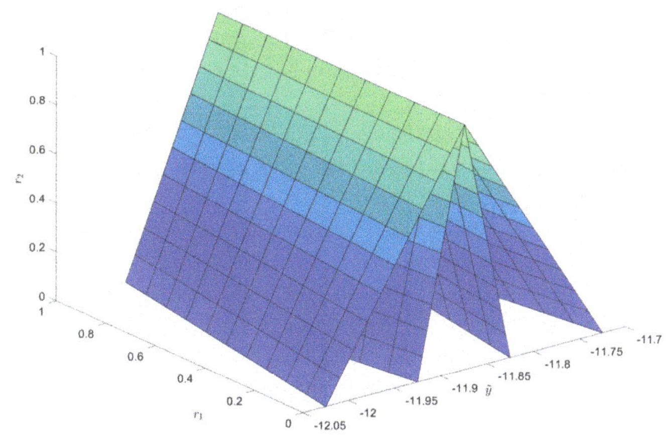


**Figure 10.12.** Type-2 fuzzy solution of simply supported at $x = 0.5$ for $k = 20$.

## 10.6 Conclusions

In this chapter, T2FBVPs have been successfully solved using the proposed methodology. The proposed methodology is based on the polynomial collocation method. This study approximated the type-2 fuzzy solution as a linear combination of trial polynomial functions satisfying the given boundary conditions. Both unknown coefficients and trial functions are taken in the type-2 fuzzy environment. Three numerical examples and one application problem are solved to check the efficacy of the proposed method. Also, solutions obtained by the present method are compared with the exact solutions in the crisp and type-1 fuzzy case under a special case using two-dimensional plots. The convergence of the polynomial collocation method has been successfully studied by increasing the degree of the approximated polynomial for example 10.1, which can also be shown for other numerical examples. Further, numerical tabulations of type-2 fuzzy solutions obtained for different values of $k$ and $x$ are given. Lastly, obtained type-2 fuzzy solutions are illustrated using three-dimensional plots.

## Acknowledgments

The first author would like to thank University Grant Commission(UGC), New Delhi for the financial support to carry out the research work.

## References

[1] Korkmaz A and Dag I 2011 Polynomial based differential quadrature method for numerical solution of nonlinear Burgers' equation *J. Franklin Inst.* **348** 2863–75

[2] Cohen A 2000 Wavelet methods in numerical analysis. *Handbook of Numerical Analysis* **vol 7** pp 417–711

[3] Mohammadi F, Hosseini M M and Mohyud-Din S T 2011 Legendre wavelet Galerkin method for solving ordinary differential equations with non-analytic solution *J. Franklin Inst.* **42** 579–85

[4] Lepik U 2007 Numerical solution of evolution equations by the Haar wavelet method *Appl. Math. Comput.* **185** 695–704

[5] Saeedi H, Moghadam M M, Mollahasani N and Chuev G N 2011 A CAS wavelet method for solving nonlinear Fredholm integro-differential equations of fractional order *Commun. Nonlinear Sci. Numer. Simul.* **16** 1154–63

[6] Brisebarre N and Joldeş M 2010 *Chebyshev interpolation polynomial-based tools for rigorous computing Proc. 2010 Int. Symp. on Symbolic and Algebraic Computation* pp 147–154

[7] Khan M A *et al* 2006 A numerical method based on polynomial sextic spline functions for the solution of special fifth-order boundary-value problems *Appl. Math. Comput.* **181** 356–61

[8] Chakraverty S, Mahato N, Karunakar P and Rao T D 2019 *Advanced Numerical and Semi-Analytical Methods for Differential Equations* (New York: Wiley)

[9] Zadeh L A 1978 Fuzzy sets as a basis for a theory of possibility *Fuzzy Sets Syst.* **1** 3–28

[10] Klir G J and Yuan B (ed) 1996 *Fuzzy Sets, Fuzzy Logic, and Fuzzy Systems: Selected Papers by Lotfi A Zadeh* vol 6 (Singapore: World Scientific)

[11] Dubois D and Prade H 1982 Towards fuzzy differential calculus part 3: differentiation *Fuzzy Sets Syst.* **8** 225–33

[12] Mizumoto M and Tanaka K 1976 The four operations of arithmetic on fuzzy numbers *Syst. Comput. Controls* **7** 73–81

[13] Nahmias S 1978 Fuzzy variables *Fuzzy Sets Syst.* **1** 97–110

[14] Goetschel R Jr and Voxman W 1986 Elementary fuzzy calculus *Fuzzy Sets Syst.* **18** 31–43

[15] Bede B and Gal S G 2005 Generalizations of the differentiability of fuzzy-number-valued functions with applications to fuzzy differential equations *Fuzzy Sets Syst.* **151** 581–99

[16] Zimmermann H J 2001 Fuzzy control *Fuzzy Set Theory and Its Applications* (Dordrecht: Springer) pp 223–64

[17] Chakraverty S and Perera S 2018 *Recent Advances in Applications of Computational and Fuzzy Mathematics* (Berlin: Springer)
[18] Seikkala S 1987 On the fuzzy initial value problem *Fuzzy Sets Syst.* **24** 319–30
[19] Kaleva O 1987 Fuzzy differential equations *Fuzzy Sets Syst.* **24** 301–17
[20] Buckley J J and Feuring T 2000 Fuzzy differential equations *Fuzzy Sets Syst.* **110** 43–54
[21] Buckley J J and Feuring T 2001 Fuzzy initial value problem for Nth-order linear differential equations *Fuzzy Sets Syst.* **121** 247–55
[22] Chalco-Cano Y and Roman-Flores H 2008 On new solutions of fuzzy differential equations *Chaos Solitons Fractals* **38** 112–9
[23] Ma M, Friedman M and Kandel A 1999 Numerical solutions of fuzzy differential equations *Fuzzy Sets Syst.* **105** 133–8
[24] Mizukoshi M T, d Barros L C, Chalco-Cano Y, Román-Flores H and Bassanezi R C 2007 Fuzzy differential equations and the extension principle *Inf. Sci.* **177** 3627–35
[25] O'Regan D, Lakshmikantham V and Nieto J J 2003 Initial and boundary value problems for fuzzy differential equations *Nonlinear Anal. Theory Methods Appl.* **54** 405–15
[26] Fard O S and Ghal-Eh N 2011 Numerical solutions for linear system of first-order fuzzy differential equations with fuzzy constant coefficients *Inf. Sci.* **181** 4765–79
[27] Nieto J J and Rodriguez-Lopez R 2013 Some results on boundary value problems for fuzzy differential equations with functional dependence *Fuzzy Sets Syst.* **230** 92–118
[28] Ramli A and Majid Z A 2016 Fourth order diagonally implicit multistep block method for solving fuzzy differential equations *J. Pure Appl. Math.* **107** 635–60
[29] Xu J, Liao Z and Hu Z 2007 A class of linear differential dynamical systems with fuzzy initial condition *Fuzzy Sets Syst.* **158** 2339–58
[30] Bede B 2008 Note on 'Numerical solutions of fuzzy differential equations by predictor-corrector method' *Inf. Sci.* **178** 1917–22
[31] Tapaswini S and Chakraverty S 2013 Approximate solution of fuzzy quadratic Riccati differential equation *Coupled Syst. Mech.* **23** 255–69
[32] Tapaswini S and Chakraverty S 2014 Euler-based new solution method for fuzzy initial value problems *J. Artif. IntellJ. Artif. Intell.* **4** 58–79
[33] Tapaswini S and Chakraverty S 2014 New analytical method for solving nth order fuzzy differential equations *Ann. Fuzzy Math. Inform.* **8** 231–44
[34] Tapaswini S, Chakraverty S and Nieto J J 2017 Numerical solution of fuzzy boundary value problems using Galerkin method *Sādhanā* **42** 45–61
[35] Chakraverty S, Tapaswini S and Behera D 2016 *Fuzzy Differential Equations and Applications for Engineers and Scientists* (Boca Raton, FL: CRC Press)
[36] Van Hung N, Tam V M and O'Regan D 2020 Existence of solutions for a new class of fuzzy differential inclusions with resolvent operators in Banach spaces *Comput. Appl. Math.* **39** 1–23
[37] Khastan A and Rodríguez-López R 2020 On linear fuzzy differential equations by differential inclusions' approach *Fuzzy Sets Syst.* **387** 49–67
[38] Mehmood N, Al-Rawashdeh A and Azam A 2019 On Darboux-type differential inclusions with uncertainty *Complexity* **2019** 2161230
[39] Gumah G, Naser N F M, Al-Smadi M, Al-Omari S K Q and Baleanu D 2020 Numerical solutions of hybrid fuzzy differential equations in a Hilbert space *Appl. Numer. Math.* **151** 402–12
[40] Attili B S 2019 The shooting method for solving second order fuzzy two-point boundary value problems *J. Appl. Math.* **32** 663

[41] Khastan A, Nieto J J and Rodríguez-López R 2014 Fuzzy delay differential equations under generalized differentiability *Inf. Sci.* **275** 145–67
[42] Behzadi S S and Allahviranloo T 2016 Solving fuzzy differential equations by using picard method *Iran. J. Fuzzy Syst.* **13** 71–81
[43] Shen Y and Chen W 2017 Laplace transform method for the Ulam stability of linear fractional differential equations with constant coefficients *Mediterr. J. Math.* **14** 1–17
[44] Mondal S P and Roy T K 2017 Solution of second order linear fuzzy ordinary differential equation by Lagrange multiplier method with application in mechanics *Opsearch* **54** 766–98
[45] Radi D, Sorini L and Stefanini L 2020 On the numerical solution of ordinary, interval and fuzzy differential equations by use of F-transform *Axioms* **9** 15
[46] Chalco-Cano Y, Rodríguez-López R and Jiménez-Gamero M 2016 Characterizations of generalized differentiable fuzzy functions *Fuzzy Sets Syst.* **295** 37–56
[47] Chehlabi M and Allahviranloo T 2018 Positive or negative solutions to first-order fully fuzzy linear differential equations under generalized differentiability *Appl. Soft Comput.* **70** 359–70
[48] Zadeh L A 1975 The concept of a linguistic variable and its application to approximate reasoning–I *Inf. Sci.* **8** 199–249
[49] Mizumoto M and Tanaka K 1976 Some properties of fuzzy sets of type 2 *Inf. Sci.* **31** 312–40
[50] Mizumoto M and Tanaka K 1981 Fuzzy sets and type 2 under algebraic product and algebraic sum *Fuzzy Sets Syst.* **5** 277–90
[51] Karnik N N and Mendel J M 1998 Introduction to type-2 fuzzy logic systems *IEEE Int. Conf. on Fuzzy Systems Proc. IEEE World Congress on Computational Intelligence (Cat. No. 98CH36228)* **vol 2** pp 915–20
[52] Mendel J M and John R B 2002 Type-2 fuzzy sets made simple *IEEE Trans. Fuzzy Syst.* **10** 117–27
[53] Mohapatra D and Chakraverty S 2022 Type-2 fuzzy linear eigenvalue problems with application in dynamic structures *Soft Computing in Interdisciplinary Sciences* pp 93–108
[54] Mohapatra D and Chakraverty S 2021 Non-linear dynamic problems with uncertainty in type-2 fuzzy environment *Modeling and Computation in Vibration Problems* **2** 10-1–10-21
[55] Najariyan M and Qiu L 2021 Interval type-2 fuzzy differential equations and stability *IEEE Trans. Fuzzy Syst.* **30** 2915–29
[56] Mazandarani M and Najariyan M 2014 Differentiability of type-2 fuzzy number-valued functions *Commun. Nonlinear Sci. Numer. Simul.* **19** 710–25
[57] Tudu S, Mondal S P, Ahmadian A, Mahmood A K, Salahshour S and Ferrara M 2021 Solution of generalised type-2 fuzzy boundary value problem *Alex. Eng. J.* **60** 2725–39
[58] Singh M, Srivastava S, Gupta J R P and Hanmandlu M 2007 A new algorithm-based type-2 fuzzy controller for diabetic patient *Int. J. Biomed. Eng. Technol.* **1** 18–40
[59] Salih A 2016 *Weighted Residual Methods* (Thiruvananthapuram: Department of Aerospace Engineering, Indian Institute of Space Science and Technology)

# Chapter 11

# Shannon entropy determination for the elastic Euler–Bernoulli beam via random polynomials and stochastic finite difference method

**Marcin Kamiński**

The main aim of this work is to investigate an applicability of Shannon probabilistic entropy in uncertainty analysis for some linear elasticity problems. The methodology invented for this purpose uses the finite difference method implemented for Euler–Bernoulli beam resting on the Winkler foundation with random compliance. Response polynomials of structural behavior like beam deflection are numerically recovered thanks to the successive solutions of difference equations with subsoil compliance varying in-between the given upper and lower bounds. These random polynomials serve further in triple stochastic analyses, where Monte-Carlo simulation, the generalized iterative stochastic perturbation technique as well as the semi-analytical probabilistic approach are employed to calculate the first four probabilistic coefficients of the structural responses. Finally, the Monte-Carlo simulation is engaged to calculate Shannon probabilistic entropy fluctuations for various input uncertainty levels and to contrast it with the resulting coefficient of variation for the extreme deflections. This comparison shows that Shannon entropy may be used to study uncertainty propagation in elastic problems solved via the stochastic finite difference method. Application of random polynomials guarantees relatively high quality of the results with minimum computational time effort.

## 11.1 Introduction

It is well known that probabilistic entropy of the given random variables represents the average level of information or uncertainty related to its possible realizations. Probabilistic entropy available in information theory [5], combinatorics or machine learning is some extension of thermodynamical entropy and both quantities differ by the Boltzmann constant only (except physical interpretation). This analogy can be explained in the following manner—numerical values of the given random variable represent the energies of microstates, and then the Gibbs formula is formerly identical to the Shannon definition.

doi:10.1088/978-0-7503-5067-9ch11

Probabilistic entropy in computational science and engineering [3] is most frequently determined according to the classical Shannon definition [10] and represents some universal uncertainty measure similar to the coefficient of variation (CoV) known from modern computational stochastic mechanics; some other definitions proposed by Renyi or Tsallis have been also considered. It is noticeable that probabilistic entropy was accidentally used in practical engineering case studies until now (most frequently in the context of maximum entropy principle), to study stochastic structural dynamics response of the beams with uncertain cross-sectional area [2]. A presentation as well as discussion of its detailed numerical values and their parametric variabilities was postponed as less important and useless while studying uncertainty propagation.

Therefore, this work is focused on the Monte-Carlo simulation (MCS) [6, 9] of probabilistic entropy for some fundamental engineering problem of the Euler–Bernoulli beam resting upon a single-parameter elastic Winkler foundation. This problem has been numerically solved with the use of three various implementations of the stochastic finite difference method (SFDM) (stochastic perturbation technique, semi-analytical approach close to the MCS estimation), where the local response functions relating random compliance of the subsoil with deflections of this beam are adopted as the polynomial basis of this compliance. Let us note that the second-order second-moment perturbation technique was used in conjunction with the FDM to study some fundamental problems in engineering mechanics [12]. Probabilistic Shannon entropy is calculated here and contrasted with the CoV for the extreme deflection fundamental for the reliability assessment and it is done while assuming Gaussian distribution of the input compliance due to its wide popularity in different areas of engineering studies. Numerical determination of the given structural response is a result of some optimization process based upon the least squares method (LSM) [4] having also statistical character, where optimal order of polynomial basis minimizes approximation error and maximizes the fitting variance. Common application of both FDM and LSM techniques in probabilistic engineering mechanics has been presented before, at least in [19]. The entire numerical analysis including the FDM solver, statistical simulation, probabilistic integration, complex visualization and data post-processing has been programmed in the computer algebra system MAPLE 2019; most of the algorithms applied in this study have been displayed in [8]. The idea of usage of the symbolic analysis in determination of probabilistic entropy and its propagation analysis is available in the literature and has been presented at least in [20]. This software would serve in a future for numerical error analysis of probabilistic entropy determination for various orders of the structural responses polynomials and would be extended towards other FDM solutions of engineering problems.

## 11.2 Problem statement

Find $H_S\big(u_i(x_j,\ t)\big)$ for $u_i(x_j,\ t)$ with $i, j = 1, 2, 3$ belonging to the kinematically admissible space of displacements fulfilling the following partial differential equations system [13]:

$$\frac{\partial \sigma_{ij}}{\partial x_j} + \rho f_i = \rho \ddot{u}_i, \qquad t \in [t_0,\ \infty) \tag{11.1}$$

$$\sigma_{ij} = C_{ijkl} \varepsilon_{kl} \tag{11.2}$$

$$\varepsilon_{ij} = \frac{1}{2}\left(\frac{\partial u_i}{\partial x_j} + \frac{\partial u_j}{\partial x_i}\right) \tag{11.3}$$

with the following boundary conditions:

$$\sigma_{ij}n_j = \hat{t}_i, \qquad x_i \in \partial\Omega_\sigma, \tag{11.4}$$

$$u_i = \hat{u}_i, \qquad x_i \in \partial\Omega_u, \qquad \partial\Omega_\sigma \bigcap \partial\Omega_u = \{\emptyset\} \tag{11.5}$$

and preserving also the following initial conditions:

$$u_i = 0; t = t_0 \tag{11.6}$$

$$\dot{u}_i = 0; t = t_0. \tag{11.7}$$

This problem is primarily solved for the stress tensor $\sigma_{ij}(x_k)$, strain tensor $\varepsilon_{ij}(x_k)$, and also for the displacement vector $u_i(x_k)$, where $C_{ijkl}$ stands for the constitutive tensor and $\rho f_i$ denotes mass forces vector. Let us note that equations system (11.1)–(11.7) are solved with the use of the FDM, so that partial differential equation (11.1) is replaced with its difference counterpart as follows:

$$\frac{\Delta\sigma_{ij}}{\Delta x_j} + \rho f_i = \rho\ddot{u}_i, \qquad t \in [t_0, \infty). \tag{11.8}$$

However, most frequently initial static or dynamic equilibrium equations system is formulated in displacements, which can be rewritten as

$$(\lambda + \mu)\,\frac{\partial^2 u_j}{\partial x_j \partial x_i} + \mu\frac{\partial^2 u_i}{\partial x_j^2} + \rho f_i = \rho\ddot{u}_i, \;\; i, j = 1,2,3 \tag{11.9}$$

where $\lambda$ and $\mu$ denote here material Lame constants. The differential equation (11.9) needs more extended difference formulation than its stresses counterpart (11.8), which may demand the second or even the fourth-order differencing scheme. Let us assume for simplicity that it is provided for a regular rectangular grid $\Delta x_1 \times \Delta x_2$ of $M \times N$ discrete points. It enables for the following representation of the second-order derivatives in equation (11.9):

$$\begin{gathered} \left(\frac{\Delta^2 u_j}{\Delta x_i}\right)_{m,\,n} = \frac{(u_j)_{m+1,\,n} - 2(u_j)_{m,\,n} + (u_j)_{m-1,\,n}}{\Delta x_i^2}, \\ \left(\frac{\Delta^2 u_j}{\Delta x_i \Delta x_j}\right)_{m,\,n} = \frac{(u_j)_{m+1,\,n+1} - (u_j)_{m+1,\,n-1} - (u_j)_{m-1,\,n+1} + (u_j)_{m-1,\,n-1}}{4\Delta x_i \Delta x_j}, \\ i, j = 1,2; \quad m = 1,\ldots, M; \quad n = 1,\ldots, N. \end{gathered} \tag{11.10}$$

These relations used in a conjunction with equation (11.9) form the algebraic equations system relevant to the static equilibrium when the RHS inertia component vanishes; the dynamic equilibrium solution needs an additional time differencing.

$H_S(u_i(x_j, t))$ denotes in this problem time fluctuations of the Shannon probabilistic entropy related to the probability density function $p_{u_i}$ of the displacements being a solution to equations (11.1)–(11.7) and they are calculated for any $\tilde{t} \in [t_0, \infty)$ as [17]

$$H_S(u_i(x_j, \tilde{t})) = \int_{-\infty}^{+\infty} p_{u_i}(x, \tilde{t}) \log(p_{u_i}(x, \tilde{t})) \, dx. \tag{11.11}$$

## 11.3 Probabilistic response with polynomial bases

Let us consider a continuous random variable $b$ with its probability density function $g_b(x)$ discretized by a set of subintervals of constant length $\delta$. Applying the mean value theorem, the following representation of $p_b(x)$ can be made in each $i$th subinterval:

$$p_b(x_i) = \frac{1}{\delta} \int_{i\delta}^{(i+1)\delta} p_b(x) \, dx. \tag{11.12}$$

Let us introduce further the quantized random variable $\hat{x}$, defined as

$$\hat{x} = x_i \text{ for } i\delta \leqslant \hat{x} \leqslant (i+1)\delta \tag{11.13}$$

with probability equal to

$$p_i = \int_{i\delta}^{(i+1)\delta} p_b(x) \, dx = \delta \; p_b(x_i). \tag{11.14}$$

Then, probabilistic Shannon entropy corresponding to a partition of $p_b(x)$ into $n$ equal subsets can be represented as [1, 18]

$$\begin{aligned} H(\hat{x}) &= -\sum_{i=1}^{n} p_i \log(p_i) = -\sum_{i=1}^{n} \delta p_b(x_i) \log(\delta p_b(x_i)) \\ &= -\sum_{i=1}^{n} \delta p_b(x_i) \log(p_b(x_i)) - \log(\delta). \end{aligned} \tag{11.15}$$

The symbol $H$ refers here to the $H$-theorem proposed by Boltzmann and it is traditionally used to denote probabilistic entropy. The total number of subintervals in a given probability distribution may affect the entropy value.

It is assumed that the input uncertainty source and it is distributed according to the Gaussian probability density function and that this parameter is uniquely defined by its expectation $E[b]$ and some small variability interval for its CoV, i.e. $\alpha(b) \in [0.00, 0.20]$. Shannon entropy is obviously time-independent and is defined for this variable as

$$H_S(b) = \int_{-\infty}^{+\infty} p_b(x) \log(p_b(x)) \, dx. \tag{11.16}$$

This entropy is engaged further to study uncertainty propagation in this system via structural response function $u_i(x_j, b)$, whose analytical form is proposed in the form of random polynomials family

$$u_i(x_j, b) \cong \sum_{r=1}^{R} A_{\mathrm{ir}}(x_j)b^r, \qquad i, j = 1,2,3. \tag{11.17}$$

These polynomials are sought via the weighted least squares method (WLSM) approximation, and finally, statistical optimization of numerical error allows to detect optimal order of the polynomial basis.

## 11.4 Computational implementation

Let us consider the following ordinary fourth-order differential equation for the linear elastic isotropic and statistically homogeneous beam exposed to the transversally distributed load q($x$) (N m$^{-1}$):

$$\frac{d^2}{dx^2}\left(e(x)J(x)\frac{d^2u(x)}{dx^2}\right) = -ku(x) + \mathrm{q}(x) \tag{11.18}$$

fulfilling the typical boundary conditions applicable in engineering theories of the elastic beams, where $E(x)$ (Pa) denoted here the Young modulus of the beam material, $J(x)$ (m$^4$) is the inertia moment of this structure, $k$ (Pa) means the compliance coefficient of the subsoil and $u(x)$ (m) denotes the deflection function of the beam (figure 11.1).

Let us also note that we further neglect the influence of the longitudinal and transversal forces as well as we reduce the analysis to the small deformations, which apparently simplifies the final form of the differential equilibrium equation; the entire domain of the beam length $l$ (m) is partitioned here into $n$ equidistant subdomains. The following approximations of spatial derivatives of the function $u(x)$ are employed according to the central differencing scheme [14, 15]:

(i) the first derivative

$$\left(\frac{du}{dx}\right)_i \cong \left(\frac{\Delta u}{\Delta x}\right)_i = \frac{u_{i+1} - u_{i-1}}{2\Delta x} \tag{11.19}$$

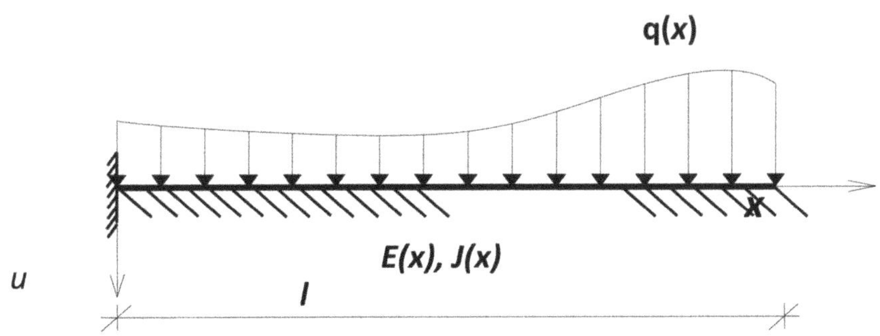


**Figure 11.1.** Elastic beam resting upon the Winkler foundation.

(ii) the second derivative

$$\left(\frac{d^2u}{dx^2}\right)_i \cong \left(\frac{\Delta^2 u}{\Delta x^2}\right)_i = \frac{u_{i+1} - 2u_i + u_{i-1}}{\Delta x^2} \tag{11.20}$$

(iii) third derivative

$$\left(\frac{d^3u}{dx^3}\right)_i \cong \left(\frac{\Delta^3 u}{\Delta x^3}\right)_i = \frac{-u_{i+2} + 2u_{i+1} - 2u_{i-1} + u_{i-2}}{2\Delta x^3} \tag{11.21}$$

(iv) fourth derivative

$$\left(\frac{d^4u}{dx^4}\right)_i \cong \left(\frac{\Delta^4 u}{\Delta x^4}\right)_i = \frac{u_{i+2} - 4u_{i+1} + 6u_i - 4u_{i-1} + u_{i-2}}{\Delta x^4}. \tag{11.22}$$

It is seen that further both deterministic and stochastic solutions are highly affected by a choice of the independent spatial variable $\Delta x$ and relatively good numerical convergence and stability is obtained with $\Delta x = [0.01\ l, \ldots, 0.05\ l]$. Let us note also that when the bending stiffness varies along the beam, the full equilibrium equation may be rewritten in a difference form as

$$\begin{aligned} &E_{i-1}J_{i-1}u_{i-2} - 2(E_{i-1}J_{i-1} + E_iJ_i)u_{i-1} + (E_{i-1}J_{i-1} + 4E_iJ_i + E_{i+1}J_{i+1} + k\Delta x^4)u_i + \\ &-2(E_iJ_i + E_{i+1}J_{i+1})u_{i+1} + E_{i+1}J_{i+1}u_{i+2} = q_i\Delta x^4. \end{aligned} \tag{11.23}$$

A particular case of constant bending stiffness $E(x)J(x) = \text{const.} = EJ$ enables to transform this formula into a decisively simpler form

$$u_{i-2} - 4u_{i-1} + \left(6 + \frac{k\Delta x^4}{EJ}\right)u_i - 4u_{i+1} + u_{i+2} = \frac{q_i\Delta x^4}{EJ}. \tag{11.24}$$

The parameter $k$ has been identified as the uncertainty source in this equations system, whose probability distribution has been defined as Gaussian. Therefore, the following statistical and probability definitions are used to calculate in turn expectations, standard deviations, skewness, and also kurtosis of the structural responses [7]:

$$E[u(x_i)] \cong \frac{1}{M}\sum_{l=1}^{M} u^{(l)}(x_i)E[u(x_i)] = \int_0^{+\infty} u(x_i, k)\ p_k(t)dt \tag{11.25}$$

$$\begin{aligned} \sigma(u(x_i)) &\cong \sqrt{\frac{1}{M-1}\sum_{l=1}^{M}(u^{(l)}(x_i) - E[u(x_i)])^2}, \\ \sigma(u(x_i)) &= \sqrt{\int_0^{+\infty} (u(x_i, k) - E[u(x_i)])^2\ p_k(t)dt}, \end{aligned} \tag{11.26}$$

$$\beta(u(x_i)) = \frac{\mu_3(u(x_i))}{\sigma^3(u(x_i))}, \qquad \kappa(u(x_i)) = \frac{\mu_4(u(x_i))}{\sigma^4(u(x_i))} - 3. \tag{11.27}$$

They are computed for each point of the computational FDM grid separately, for the few different total numbers of random trials in Monte-Carlo statistical approach denoted here by $M$, while $\mu_3(u(x_i))$, $\mu_4(u(x_i))$ stand here for the third and the fourth central probabilistic moments of the beam deflection at the given node of the grid. Substitution of the polynomial representation given in equation (11.17) to the aforementioned relations (11.25)–(11.27) is the basis of semi-analytical methods and Monte-Carlo simulation. As is well known, the stochastic perturbation method demands the additional Taylor series expansion with random coefficients [11], which means that

$$u(x_i;\, b) \cong u^0\left(x_i;\, b^0\right) + \frac{\varepsilon^l}{l!}\sum_{l=1}^{n}\frac{\partial^l u(x_i;\, b)}{\partial b^l}\Bigg|_{b=b^0}(b - b^0)^l. \tag{11.28}$$

This formula includes $n$ as the order of Taylor expansion, $u^0(x_i;\, b^0)$ is the mean value of the deflection calculated for the mean value $b^0$ of the input random parameter $b$. Traditionally, $\varepsilon$ represents the perturbation parameter here, where unit value is most frequently recommended.

## 11.5 Numerical experiments

This case study is devoted to the cantilever I-beam made of stainless steel $E = 210.0$ GPa having constant cross-sectional area, inertia moment $J = 1.71 \times 10^{-6}$ m$^4$ (IPE 100), and resting on the elastic foundation. The coefficient of compliance is quite naturally adopted here as the most influential structural parameter and treated as Gaussian input random variables; its expected value has been assumed as equal to $k = 3.5 \times 10^7$ N m$^{-2}$. The external distributed load throughout this beam equals $q = 10$ kN m$^{-1}$, whose length has been adopted as equal $l = 5.0$ m. The solution has been achieved with the FDM program implemented in the computer algebra software MAPLE 2019, where the least squares method solver has been employed. A deformation line of this structure has been presented in figure 11.2 below for the few exemplary integer compliance coefficients, i.e.

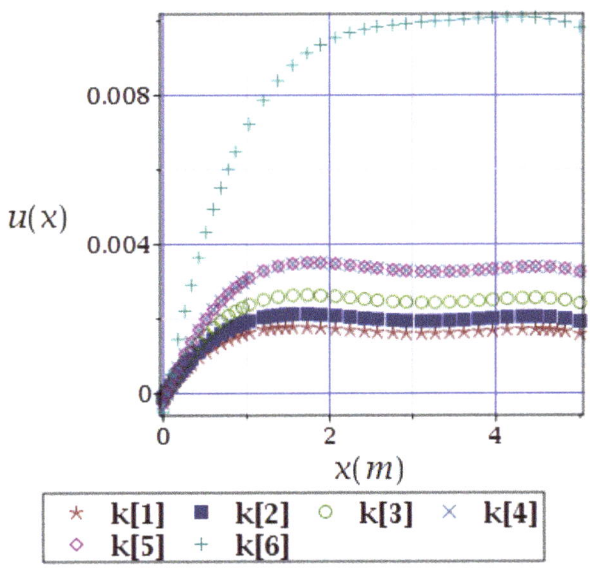


**Figure 11.2.** Deterministic solutions with varying subsoil compliance.

$k = \{1.0–6.0\} \times 10^7\ \mathrm{N\ m^{-2}}$. It is consistent with engineering intuition that the difference in the neighboring deflection functions increase together with the mean value of this coefficient, so that this effect is especially remarkable for the maximum and the second maximum values of the parameter $k$. The deflection line corresponding to the compliance expected value is further taken as the zeroth order solution in the perturbation-based stochastic analysis.

Further stochastic analysis of this beam deflection has been delivered using three different numerical methods, namely statistical crude Monte-Carlo simulation approach (a population includes $M = 10^6$ random trials), the iterative generalized stochastic perturbation technique of the tenth order and also—with the use of semi-analytical approach based upon symbolic integration inherent in probability theory. All these three concurrent probabilistic techniques have been based upon the same response polynomial of the fourth order recovered from 11 tests computed with the set of trial compliances equal in turn $k = \{1.0, 1.5,\ldots,6.0\} \times 10^7\ \mathrm{N\ m^{-2}}$. Polynomial recovery has been made with the use of the LSM and determination of the final order of polynomial basis has been carried out via statistical optimization (common minimization of fitting variance and maximization of the correlation coefficient). The entire computational implementation for the FDM solution, LSM approximation, three probabilistic methods solutions as well as probabilistic Shannon entropy determination have been delivered in the computer algebra system MAPLE 2019. This introductory stochastic analysis would be necessary to recognize a precision and convergence of all the methods engaged for precise calculation and estimation of the resulting expectations and CoV. Particularly, one needs to verify when alternative faster methods like stochastic perturbation technique start to be inefficient and Monte-Carlo simulation must replace it in the first two moment analysis.

Probabilistic methods employed in this work enabled for determination of the expected values for the extreme deflection (figure 11.3), its CoV (figure 11.4) as well as kurtosis (figure 11.5) and kurtosis (figure 11.6). This is all done for some interval of input uncertainty, which means $\alpha(b) \in [0.00,0.20]$. The results obtained allow to conclude that a coincidence of the first two probabilistic characteristics is perfect for the given uncertainty interval. Then, one can notice from this evidence that uncertainty propagation in elastic systems may be provided with the use of higher order stochastic perturbation technique within the given bounds on the input CoV. Further, the expected values decrease a little bit while increasing input statistical scattering of the subsoil compliance, while CoV of the extreme deflection is even larger than the input one. Therefore, the reliability index for structural deformation during the ultimate limit state verification should include for sure the subsoil deformations uncertainty.

The skewness and kurtosis demonstrate some divergence for $\alpha(b) \in [0.00,0.15]$ and keep rather close to 0, so that the resulting probability distribution does not seem to be very distant from the Gaussian one; some specific statistical tests and relative probabilistic entropy determination could confirm this observation.

A coincidence of the given three probabilistic numerical methods indicates that the iterative generalized stochastic perturbation technique could be reliably used in four-moments structural responses determination when the input uncertainty level does not exceed the level of $\alpha(b) \cong 0.15$. The SFDM implemented with this

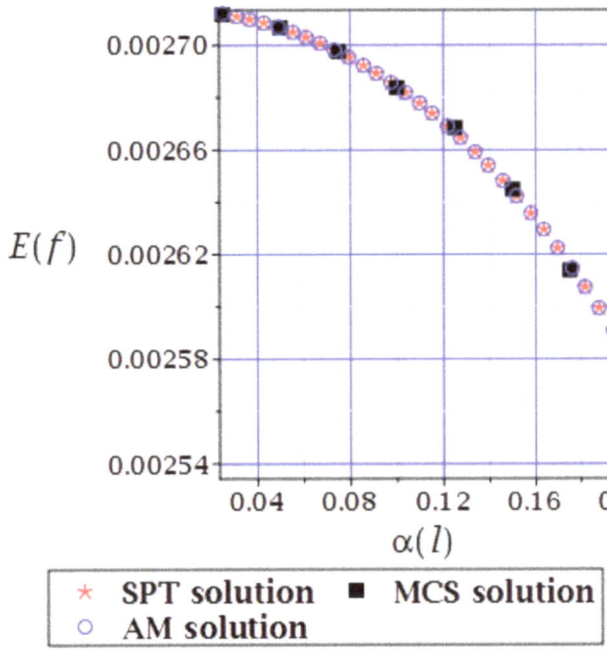


**Figure 11.3.** Expected values fluctuations for the extreme displacements.

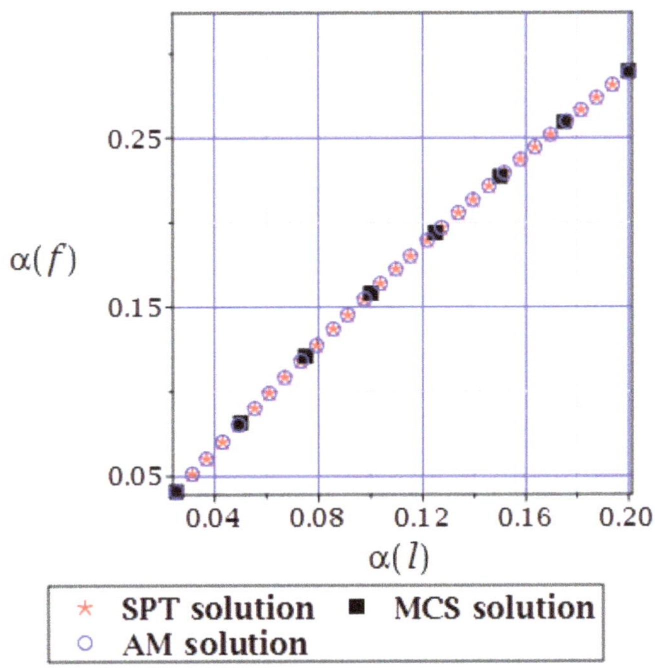


**Figure 11.4.** Coefficients of variation fluctuations for the extreme displacements.

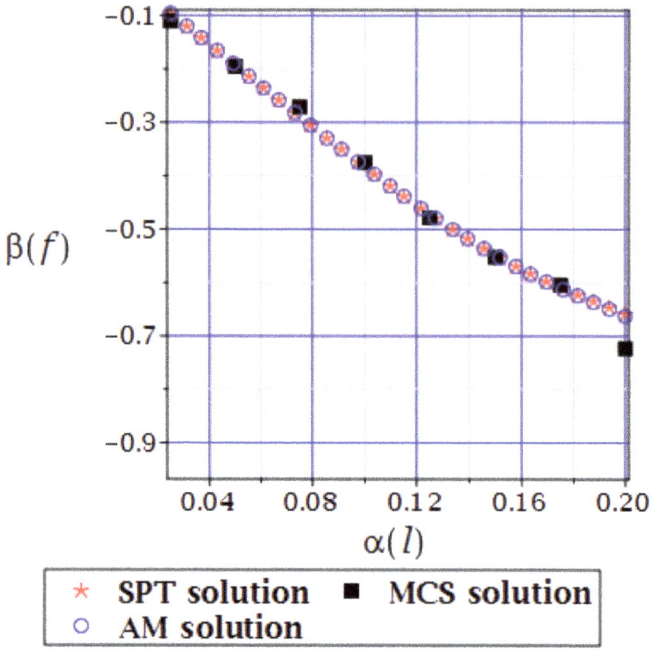


**Figure 11.5.** Skewness fluctuations for the extreme displacements.

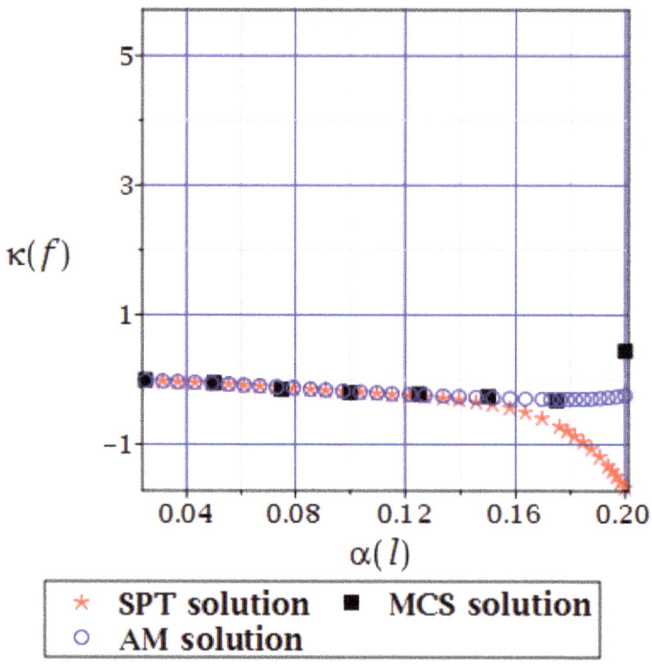


**Figure 11.6.** Kurtosis fluctuations for the extreme displacements.

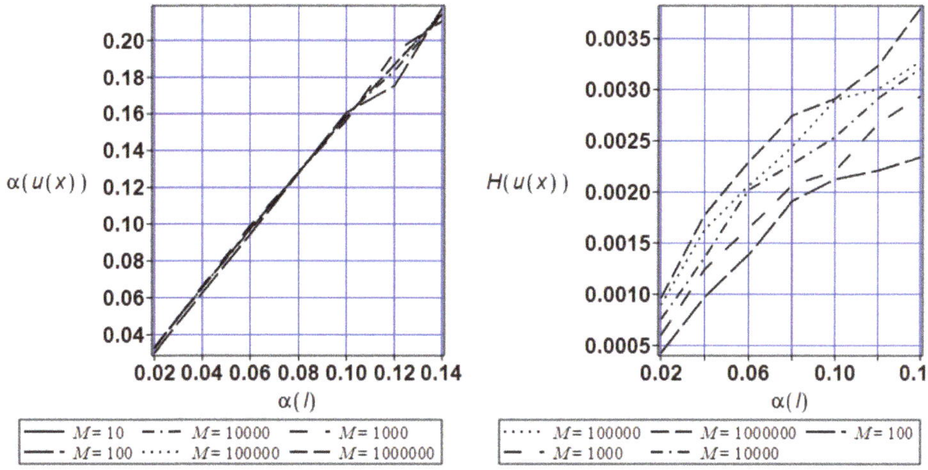


**Figure 11.7.** Coefficient of variation (left) versus Shannon entropy (right) for the extreme deformation of the beam.

technique would be the fastest numerical techniques here (time effort close to deterministic solution), which could be additionally available without any computer algebra system (unlike the semi-analytical method using symbolic integration).

Finally, a comparison of the CoV for the resulting extreme deformation of the beam (figure 11.7) had been provided together with Shannon entropy of the same physical variable. This has been done in some small input uncertainty interval, $\alpha(b) \in [0.00, 0.14]$, for the few mostly explored values of the Monte-Carlo random trials, $M = [10^p]$, $p = 2,3,4,5,6,7$. A single simulation has been delivered for seven different input CoVs, with ten subintervals for the resulting PDF partition while computing the entropy as well as with an arithmetic precision equal to 10 digits.

There is no doubt that probabilistic entropy, at least that proposed by Shannon, is much more sensitive to the numerical simulation process than the CoV. The resulting CoV shows clear linear dependence upon the input uncertainty, while entropy varies in a complex and nonlinear mode. The series of CoVs are almost constant with negligible variations at higher input uncertainty, while Shannon entropies asymptotically and rather slowly converge together with an increasing of the parameter $M$. This is due to the fact that the entropy is calculated via post-processing of the resulting discrete probability distribution (and its histogram), while CoV follows directly the finite set of random trials. Further, one notices that probabilistic entropy is noticeably more sensitive to the simulation parameters and less stable while increasing a total number of the Monte-Carlo simulation random trials. Remarkable benefit of the usage of Shannon entropy to study uncertainty propagation instead of the CoV is the decisive shortening of numerical experiments time effort (when this CoV follows Monte-Carlo method). These time savings have been contrasted in table 11.1 and it seems that this time effort is almost two times smaller while analyzing probabilistic entropy. A fundamental reason for this disproportion is that the CoV needs prior estimate of expectation, while entropy

**Table 11.1.** Computational time effort for the resulting CoVs and Shannon entropies.

| $M$ | $H$ cost (s) | $\alpha$ cost (s) | $H$ cost (MB) | $\alpha$ cost (MB) |
|---|---|---|---|---|
| $10^2$ | 3.95 | 0.65 | 36.00 | 36.00 |
| $10^3$ | 4.51 | 0.82 | 36.00 | 36.00 |
| $10^4$ | 4.67 | 2.15 | 36.00 | 56.24 |
| $10^5$ | 7.76 | 17.22 | 80.22 | 102.51 |
| $10^6$ | 50.39 | 211.20 | 410.72 | 486.17 |
| $10^7$ | 1112.76 | 3719.18 | 3447.34 | 4803.34 |

needs interval probabilities and their logarithms summation, so that relatively simple operations on random samples only. A separate methodological question is whether the PDF-based entropy is more accurate and more representative that statistics-based CoV in uncertainty propagation analysis in physical and mechanical systems. Additionally, reliable computation of the resulting CoV in some engineering case studies may be connected with larger numerical error, worse stability and convergence (when i.e. $\alpha(b)>1.0$) whereas probabilistic entropy determination accuracy remains constant in this context.

## 11.6 Concluding remarks

(1) It has been documented that Shannon entropy can be efficiently determined in elastic Euler–Bernoulli beams, whose deterministic solution can be achieved with the use of the FDM. It is a very promising result considering further applications in simple beams, rectangular or circular plates (with various thickness) modeling as well as some boundary value problems discretized using irregular grids [14]. This entropy has been found as a relatively efficient measure of uncertainty propagation in such structures, whose some geometrical (or materials) design parameters can be treated as random variables. Computational time and power consumption is remarkably lower while computing Shannon entropy especially for higher input statistical dispersion, so that this way of uncertainty propagation analysis could be recommended in solid mechanics would be advised for nonlinear and transient problems. Moreover, in turn, probabilistic entropy could be useful in further reliability analysis of such structures, even due to Eurocode 0 demands without any expensive engineering software. There is no doubt that further study concerning applicability of Shannon as well as other probabilistic entropies with the FDM, and also with other modern numerical techniques [13] deserves further attention and some effort of the brand new computational implementations.

(2) The key majority of the software invented for this study and presented in this paper is common computation of both engineering problem deterministic solution, polynomial structural response recovery, application of three independent stochastic methods to get up to the 4th probabilistic characteristics of this response and final determination of Shannon entropy. A modification of such a computer program

towards deformations or the eigen-vibrations analysis [16] of other elastic beams, rectangular and/or circular plates of different thickness would be relatively easy. Some fundamental unsteady problems with time and space differences, even with micro-heterogeneity, could be programmed too. An implementation of the reliability assessment is quite straightforward and may be provided with the use of simple first-order reliability method by adding just a definition of the reliability index or, alternatively, by using the relative probabilistic entropy either (according to Bhattacharyya model, for instance).

## Acknowledgment

This paper has been written in the framework of the research grant OPUS no 2021/41/B/ST8/02432 entitled 'Probabilistic entropy in engineering computations' and sponsored by The National Science Center in Poland.

## References

[1] Al-Omari A I 2014 Estimation of entropy using random sampling *J. Comput. Appl. Math.* **261** 95–102

[2] Batou A, Ritto T G and Sampaio R 2014 Entropy propagation analysis in stochastic structural dynamics: application to a beam with uncertain cross sectional area *Comput. Mech.* **54** 591–601

[3] Beirlant J, Dudewicz E J, Gyoerfi L and van der Meulen E C 1997 Non-parametric entropy estimation: an overview *Int. J. Math. Stat. Sci.* **6** 17–39

[4] Björck A 1996 *Numerical Methods for Least Squares Problems* (Philadelphia, PA: SIAM)

[5] Cover T M and Thomas J A 2006 *Elements of Information Theory* 2nd edn (New York: Wiley-Interscience)

[6] Cruz M E and Patera. A T 1995 A parallel Monte-Carlo finite element procedure for the analysis of multicomponent media *Int. J. Numer. Methods Eng.* **38** 1087–121

[7] Feller W 1965 *An Introduction to Probability Theory and Applications* (New York: Wiley)

[8] Forsythe G E, Malcolm M A and Moler C B 1977 *Computer Methods for Mathematical Computations* (Englewood Cliffs, NJ: Prentice-Hall)

[9] Hurtado J E and Barbat A H 1998 Monte-Carlo techniques in computational stochastic mechanics *Archives of Comp. Methods Eng.* **5** 3–30

[10] Kamiński M 2021 On Shannon entropy computations in selected plasticity problems *Int. J. Numer. Methods Eng.* **122** 5128–43

[11] Kamiński M 2013 *The Stochastic Perturbation Method for Computational Mechanics* (Chichester: Wiley)

[12] Kamiński M and Spanos P D 2001 Stochastic second order perturbation approach to the finite difference based vibration analysis *Int. J. Appl. Mech. Eng.* **6** 369–93

[13] Kleiber M 1998 *Handbook of Computational Mechanics: Survey and Comparison of Contemporary Methods* (Berlin: Springer)

[14] Liszka T and Orkisz J 1980 The finite difference method at arbitrary irregular grids and its applications in applied mechanics *Computers & Structures* **11** 83–95

[15] Pietrzak J, Rakowski G and Wrześniowski K 1986 *Matrix Analysis of Structures* (Warsaw: Polish Scientific Publishers)

[16] Shu C, Wu W X, Ding H and Wang C M 2007 Free vibration analysis of plates using least-square-based finite difference method *Comp. Methods in Appl. Mech. Eng.* **196** 1330–43
[17] Shannon C E 1948 A mathematical theory of communication *Bell Syst. Tech. J.* **27** 623–56
[18] Shao Y and Hahn M G 1995 Limit theorems for the logarithm of sample spacings *Statistics & Probability Lett.* **24** 121–32
[19] To C W S and Kiu M L 1994 Random responses of discretized beams and plates by the stochastic central difference method with time co-ordinate transformation *Computers & Structures* **53** 727–38
[20] Xu Z, Shi Y, Zhao Q, Li W and Liu K 2019 Extreme interval entropy based on symbolic analysis and a self-adaptive method *Entropy* **21** 238–56

# Chapter 12

# Polynomials in hybrid artificial intelligence

**Saba Sajadi and Majid Amirfakhrian**

Artificial intelligence (AI) is a vast field whose goal is creating intelligent machines, something that has been achieved many times depending on how we define intelligence. Today, AI is everywhere—in factories we work, the school we study, banks, hospitals, and even in our phones.

We believe that it requires a more comprehensive, holistic approach to sustainably realize the full benefits of AI solutions. So, we introduce hybrid AI. Hybrid AI is the combination of symbolic AI and non-symbolic AI. Symbolic AI and non-symbolic AI are two types of AI that combine with their advantages. In other words, hybrid AI must be more than a combination of the approaches of symbolic and non-symbolic AI. There are attractive applications of hybrid AI in the real world like operations management, design engineering, etc.

There are some polynomials in hybrid AI that help us to optimize, forecast, predict some events such as ARMA, linear regression, multiple linear regression, polynomial linear regression, genetic algorithm, etc. It's so important to estimate a polynomial that is convergent and has the minimum error.

## 12.1 Artificial intelligence

Artificial intelligence (AI) is a new science that is very old and whose foundations derive from science such as philosophy, mathematics, psychology, and computer science. And it is associated with them and can be found in the social sciences, computer science, engineering, and life sciences. Progress and evolution occur rapidly in AI fields. The science of AI could fundamentally change our lives on Earth in the not too distant future. The existence of self-driving cars on the roads, the evolution of robots with different capabilities instead of hiring human resources, running stores and supermarkets with intelligent robots, rapid identification and diagnosis of diseases such as cancer masses, and performing various surgeries with high accuracy are examples of change in the not too distant future. In other words, by serving a cup of coffee with robots, we can taste AI.

doi:10.1088/978-0-7503-5067-9ch12  

### 12.1.1 Introduction

AI is a vast intersection of old and new sciences, and technologies. The main ideas of AI can be found in philosophy, linguistics, mathematics, psychology, neurology, and physiology. Its various branches, subdivisions, and applications are found in computer science, engineering, biology and medicine, communication science, and many other fields.

An AI system isn't artificial or intelligent. It's a goal-oriented device that solves the problem artificially. Their function depends on knowledge, experience, and human reasoning patterns. In general, AI refers to anything that makes tools or devices intelligent. Intelligence means thinking, searching, extracting information, and making decisions in unexperienced situations. Thus, the intelligent being is an entity that can think, search, analyze, and make decisions in unexperienced circumstances. Intelligent systems are systems that can behave and react like humans [20].

The concept of AI was first used by John McCarthy, known as the 'father of intelligent machine science.' He was the pioneer in AI science and explained the earliest definition in 1955: 'The goal of AI is to develop machines that behave as though they were intelligent.'

After McCarthy, scientists and philosophists presented other definitions.

1. The exciting new effort to make computers think. Machines with full minds in the literal sense (1985, Haugeland).

   The automation of activities that we associate with human thinking, activities such as decisions, making problem-solving, and learning (1978, Bellman) [2].
2. The study of mental faculties through the use of computational models (1985, Charniak and McDerrot). The study of computations that make it possible to perceive, reason, and act (1992, Winston).

   The act of creating machines that perform functions that require intelligence when performed by people (1990, Kurzweil).
3. The act of creating machines that perform functions that require intelligence when performed by people (1990, Kurzweil).

   The study of how to make computers do things at which, at the moment, people are better (1991, Rich and Knight).
4. Field of study that explains and emulates intelligent behavior in terms of computational processes (1990, Schalkoff).

   The branch of computer science is concerned with the automation of intelligent behavior (1993, Luger and Stubblefield).

If we concentrate on the definitions provided by different scientists at different times, we recognize that we can divide intelligent systems into four different categories [4].

1. Systems that think like a human.
2. Systems that think rationally.
3. Systems that act like human.
4. Systems that act rationally.

The purpose of the 'acting like a human' approach is to design and build systems that behave like humans so they can make decisions and act accordingly. The Turing test described earlier defines intelligent behavior as the ability to reach the level of human presence in all perceptual tasks so that it can even deceive another human being. For example, the user says a sentence, and the system reacts appropriately to it, laughs at a humorous sentence, gets upset at an insulting sentence, and so on. This approach depends on understanding human introversion, feelings, and perception, so studying how emotions are processed is essential. People behave based on their thoughts and perceptions, which means that machines should be able to recognize and implement human nature.

Awareness and understanding of the function of the human brain contribute to designing a machine or program that 'thinks like a human'. Psychological testing, introspection, and brain imaging are three methods used to understand how the human brain functions.

By knowing the actual function of the brain, we can adapt and determine the input–output and how the machine works with the human brain.

Aristotle, the Greek scientist, was one of the first people who try to methodize the right thinking. His famous analogy states that the correct data and information will lead to the right results. Socrates, for example, is a human being. Man is mortal, so Socrates is mortal. At the time, the belief was the mind operated based on laws. As a result, this assumption gave rise to the science of logic.

Until 1965, some programs could find a solution to a problem if they had enough time, memory, and problem-solving logic and if the problem had solution. However, there were two troubles in this method: first, expressing the problem in logical language and the form of acronyms, and second, there is a fundamental difference between solving a problem based on theory and approach on paper and solving it in practice.

'Rational action' is an action that achieves a goal. An intelligent agent is just an entity that observes and acts. In the law of thought method, all emphasis is on correct inferences. Correct inferences are sometimes part of being an intelligent agent.

One way to perform an intelligent action is to make a logical argument that leads to a conclusion and does it to achieve the goal. Sometimes, even though we have to act, there is no correct or reliable action to take. There are rational actions that do not require any inference or reasoning. Pulling your hand away from a hot object is a reaction from which we do not need to reason and deduce. Designing intelligent systems as intelligent agents has two strengths. First, this method is more common than the laws of thought. The correct inference is only applicable to achieving rationality, not every necessary action or reaction. Second, this method is more flexible than human-based methods.

The goal of AI is generally to build a machine that can 'think.' While, to classify and define thinking machines, one must define 'intelligence.' We also need to define 'awareness' and 'perception', as well as a standard for measuring the intelligence of machines. How can one measure intelligence?

AI was defined for the first time by John McCarthy. More than half a century ago, when silicon chips had not yet created, Alan Turing asked one of history's most challenging philosophical questions. 'Can the car think?' He asked.

He designed a test that he called the 'imitation game'. He described the imitation test like this: a questioner—a human being—talks to two people simultaneously. Both individuals are in separate rooms, and the questioner cannot see either of them. Humans are one, while machines, such as computers, are the other. Figure 12.1 shows the schematic diagram of the Turing Test. The questioner should start a conversation with these two people and try to understand which of the two is human and which is a machine. The machine is intelligent if it can answer in a way that the questioner cannot distinguish a human from a machine. Besides simplifying the test and avoiding complications, Turing limited it to textual and on-paper dialogues, so we didn't have to deal with deviant issues such as translating the text into spoken speech and adjusting accent and tone of voice.

Experts have criticized the AI methodology as a weakness, while some experts see it as a natural evolution related to the short history of computer science. AI can solve well-defined problems, such as many common computation problems, like physics calculations and payroll calculations that have an algorithm to solve, so we don't have to search for solutions.

In general, Turing's theory itself has serious opponents. Some of these critics do not accept machine intelligence at all, while others merely question the effectiveness of the Turing test to prove intelligence. Simulating intelligence is one of the most important topics in this field. It is often asked, does the mere fact that a machine can simulate the way humans speak mean that it is intelligent? For example, you may have heard about Chatter Bots. These robots use imitation methods and are, in a sense, a modern and Internet example of the Turing test.

There are other criticisms of the Turing test. While it may be an intelligent machine, it cannot communicate like a human. Second, the Turing test assumes that a human being who answers the examiner's questions and sits in a closed area is not intelligent enough.

Based on Turing's argument, we can conclude that many individuals, like children and illiterates, fail the test, not because of the machine's intelligence but because they do not possess enough communication skills through correspondence [2].

### 12.1.2 Abilities

As stated in section 12.1.1, AI is a technology that is somehow capable of thinking. This way of thinking is very different from what we know as human thinking, while it tries to mimic it. We define AI based on three principles:

1. Learning
2. Argument
3. Understand.

AI programming focuses on three cognitive skills: learning, reasoning and self-correction. Learning processes. This aspect of AI programming focuses on acquiring data and creating rules for how to turn the data into actionable information.

Some applications of AI:

Consumer marketing

Identification technology

Inturisum detection

Machine translation.

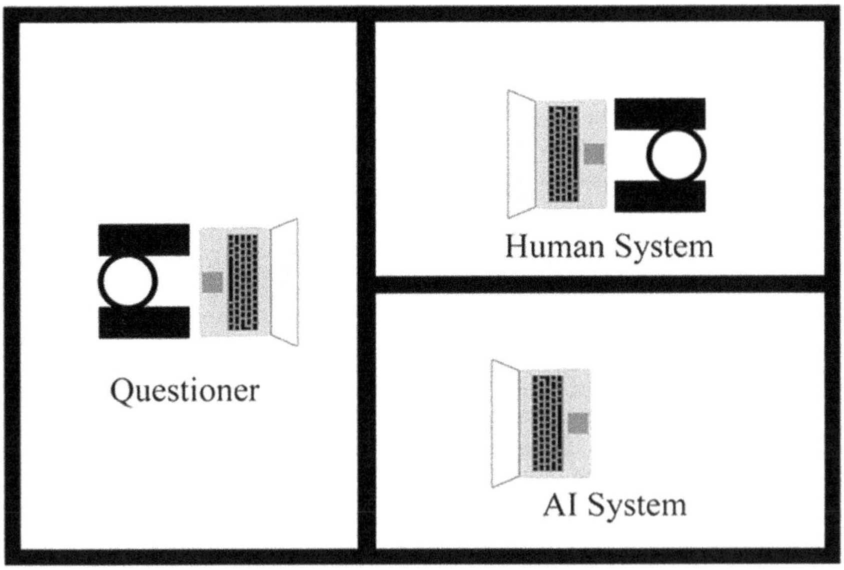


**Figure 12.1.** The Turing test.

### 12.1.3 Advantages and disadvantages

Even though AI is close to reality today, there are numerous underlying reasons why the evolved form of intelligence that Turing envisioned has yet to emerge. A philosopher named John Searle is one of the most famous critics of this field. He argues that the debate over the intelligence of non-biological machines is fundamentally irrelevant. He used the Chinese room debate as an example in theoretical discussions of AI to demonstrate his point. Initially, Searle critiqued machine intelligence in 1980 and later expanded it to a complete article in 1990.

The story of the Chinese room is as follows: suppose a person is sitting in a room and has a book of rules of Chinese language symbols. Chinese phrases and symbols are written on the paper and sent in from under the door in the room. He must refer to the book of rules to prepare the appropriate answer and send it back on paper. Assuming the reference book is complete enough, this person can answer the questions without even understanding the meaning of a symbol in the Chinese language. Is it possible to conclude that the respondent is intelligent?

The main argument of John Searle and other critics of simulation is that it is possible to build a machine that translates phrases and terms (for example, dictionary software). It means that a system receives the words and symbols as input and produces the symbols and words as output; without the system itself understanding the meaning of these symbols. Therefore, even if the Turing test is successful, it cannot prove that a machine is intelligent.

If machines can interact with the surrounding environment, then they can think. It means that computers, like us, have a sense of sight, hearing, touch, and other sensations. In this case, the simultaneous combination of 'imitative responses' with 'appropriate response to the environment' is the same as intelligence. John Searle has similar thoughts; the difference is that he considers the ideal form of action and reaction required to be the same as biological interaction.

Other important aspects of AI discussion are format and content. Human intelligence includes paying attention to content format and the content itself. In other words, consider the words 'homograph'. For example, when we say novel, the word alone has different meanings. However, when the same word is in a sentence, it has only one correct meaning. The human mind can recognize the meanings of words by following how they relate to the sentence.

Similar intelligence is present in all five human senses. For instance, scientists showed that when someone is paying attention to another person in a crowded environment, their ears can complete words and phrases they don't hear. Also, the human eye can recognize blurry parts of an image with a previous visual experience. With prior visual experience, the human eye can also distinguish blurry parts of images. Therefore, experts believe that the background knowledge or 'mental archive' of an intelligent being plays a crucial role in his intelligence.

According to comments and criticisms, AI has disadvantages from another perspective. The development of AI in medicine, industry, aerospace, voice recognition and speech recognition, etc, will cause AI to replace humans. Human jobs will be severely affected, and the unemployment rate will rise unless humans can eliminate unemployment. In the future, human resources will have no initiative, and creativity will rely on programmers alone. Dependence on technology is increasing sharply. Immoral and incorrect use of these machines will cause irreparable damage. AI makes people lazy by running programs that automate much of the work.

AI can replace human resources with robots, so human intervention is reduced, which can pose a significant problem in exploitation standards. Every organization seeks to exchange the least qualified people with AI robots that may perform similar tasks more efficiently. Machines can only perform tasks that have been designed or programmed for them, and they do so effectively. But they cannot replace the human connection that builds teamwork. Machines can't communicate with humans, when communication is a crucial aspect of the team working and management.

Despite the challenges and disadvantages of AI, the advantages outweigh the disadvantages. Proper managing and planning can overcome the obstacles and lead them in the right direction. One of the most obvious differences between humans and AI machines is the need for food and rest. AI machines can be programmed for long periods without becoming tired, unlike humans who need sleep and rest.

Intelligent machines can perform difficult and complex tasks that are difficult or impossible for humans to perform. They are more likely to be able to perform tasks faster than humans. Humans make mistakes from time to time, thus coining the term human error. However, computers do not make these mistakes if they are appropriately programmed. AI makes choices based on previously collected information, using a specific set of algorithms. Thus the errors are reduced, and therefore it is possible to achieve accuracy with a greater degree of accuracy.

Another valuable benefit of AI is judgment. The only thing that should influence our decisions is the evidence, not emotions. Even with our best efforts, it is a well-known fact that our emotions always negatively affect on our decisions.

As a result of AI and other technologies, machines can make decisions faster than humans and take action faster. However, humans analyze many factors both emotionally and practically before making decisions. AI devices work on what is planned and deliver results more quickly.

### 12.1.4 Symbolic and non-symbolic artificial intelligence

The creation and development of techniques and essential methods for complexity management should be a core component of all future scientific and research efforts in computer science, and in particular AI. In order to solve problems that could not be solved by mathematical methods or functional programming, AI techniques have been developed. Currently, AI can be divided into two categories:

1. Symbolic AI (good old fashion)
2. Non-symbolic AI (connection-oriented intelligence).

For most of the past six decades, symbolic AI has dominated the field, also known as **'classical AI'**, **'rule-based AI'**, and **'old-fashioned AI'**.

Symbolic AI is an approach that trains AI the same way the human brain learns. It learns to understand the world by forming internal symbolic representations of its world.

Symbols represent people and objects. The use of symbols is essential for human thinking and reasoning. If someone tells you about a trip to the North Pole, you imagine snow, ice, cold, polar bears, and Inuits in your mind. Snow, ice, and polar bears are symbols that define objects (snow, ice, airplane, etc) and people (Inuit, traveler, tour leader, etc). Additionally, symbols can represent abstract concepts (emotion, thinking, metaphors, etc) or objects that do not physically exist (web pages, social media, email, etc). They can also describe actions (running, drinking, talking, etc) or states (sadness, happiness, amazement, etc). Symbols can be organized into hierarchies (a classroom has doors, windows, boards, benches, etc). They can use it to describe other symbols (a small genius student, an eerie giant smart home, etc).

One of the essential aspects of intelligence is the ability to communicate between symbols. Symbols have, therefore, also played a crucial role in the creation of AI [5].

According to symbolic AI theory, the world can be understood as structured representations. In this theory, symbols that represent objects in the world form the basis of cognition. Based on symbolic AI, rules or operations are applied for symbolic processing on the set of symbols, and it uses for understanding encoding. This set of rules contains a large base of if/then statements. Human experts develop the knowledge base by adding data and symbols and providing the knowledge base with updated information. After that, the knowledge base determines the appropriate rules based on the symbols. By doing this, the user can draw conclusions based on queries from the knowledge base and reasoning. Semantic networks are examples of symbolic AI.

For example, when we Google the phrase Coronavirus, the information will display in the upper box below our query. Essentially, these systems are a series of nested if-then statements that draw conclusions about the relationships between entities.

In the early decades, AI researchers tried to model existing knowledge and rules of human behavior and implement them in computers. The research was somewhat successful, while in recent years, symbolic AI has almost disappeared due to the existence of non-symbolic or connectionist AI capabilities. Neural networks are an example of connectionist AI.

AI researchers tried to design systems that model the thinking and attitude of humans. One of the general goals of machine learning is to develop autonomous systems that think and act like humans. The system should mimic the learning process of humans. To simulate the human learning process, scientists must develop models of how humans represent the world and frameworks for logic and reasoning. Two main paradigms of symbolic AI and connectionism AI emerged from these studies.

The connectionist theory states that intelligent decision making can be done through an interconnected system of neurons of the unit size. The neuron is the smallest information processing unit that forms the basis of communication networks. The degree of activation of each neuron determines by its connection to other neurons.

Every system train by using input data. By training the system, the activation level of each neuron changes. The weight matrix encodes the weight contribution of the activation value of a neuron and sends it as an input signal to another neuron. The input function determines how the input signals are combined to set the state of the receiving neuron. The most common input function is the dot product of input activation vectors. Then, the transfer function applies a transformation to the combined input signals to calculate the activation state of a neuron. The learning rule determines how the weights of a network should change in response to new data.

Learning rules include backpropagation, perceptron, Hopfield, etc. In addition, the way in which training data is coded in the model environment is explained, usually as inputs and outputs pair. The network must be able to interpret the model environment. According to the mentioned materials, communication models have seven main features, which are:

1. a set of units
2. activation states
3. weight matrices
4. an input function
5. a transfer function
6. a learning rule
7. a model environment.

After training the data, this system can learn deep models of data generation distribution and can make intelligent decisions based on it, such as regression and classification [9].

## 12.2 Hybrid artificial intelligence

AI is a vast field in computer science whose goal is to create intelligent machines. The intelligence concept depends on how we can define it. If you compare a five year old child and intelligent machines, find out computers are not as great as the ability of a five year old child. In comparison, a five year old child can act intelligently while walking, picking up objects, recognizing people's emotions, and communicating about various topics, while intelligence machines are trained to do specific tasks and can solve problems that are trained to solve, but not much else.

Hybrid AI is one of the best methods to solve the problems that intelligent computers and AI are still struggling with. Hybrid AI is a combination of symbolic and non-symbolic AI.

### 12.2.1 Introduction

Hybrid AI is currently a controversial topic in the fields of AI. 'Hybrid' refers to the combination of two types or species.

AI is still limited in the domain of activity, but humans generally are not. It performs well-defined tasks based on a specific data type and in a controlled environment. Unlike humans, who can only learn a few examples and cannot work with specialized data types such as soft data, AI requires a large amount of educational data. It is where humans have a unique competitive advantage, and it is momentous to remember that.

Because the brain and AI use different basic algorithms. As a result, each succeeds while the other fails. Machine learning algorithms are better at detecting complex and subtle patterns in large human datasets. In spite of noise and ambiguity in the input, the brain is capable of processing information efficiently. Or even when situations change unexpectedly. That is why humans and AI must work together as a combined intelligence and join forces. According to research, this is exactly what future managers envision. According to 67% of them, AI enables humans and robots to work together to master their related skills.

Using hybrid AI plays a role in optimizing or generalizing AI models based on machine learning, deep learning, and neural networks with specialized knowledge. And it contributes to increased accuracy and increased probability of predicting an event. To overcome the problem, scientists investigated the brain's biological neural networks and found that the brain works by understanding different situations and perceptions and communicating with external stimuli. Then they tried to improve the performance of artificial neural networks.

For example, consider cutting your hand with a knife and the resulting pain. When you cut your hand for the first time, the brain is affected by pain, and information from external stimuli is created in the brain. The brain records information about the sharpness of the knife and the pain caused by it. And in this way, a person learns not to use a knife from a young age. Through the same neural network, we can perform many general learnings such as 'pizza is very delicious and tastes very yummy, there are always clouds before the rain, or all birds can fly'. Some of these statements are not right because some pizzas are tasteless, or some birds cannot fly, like penguins. The results can be refined through experience and create the conditions for adaptive learning.

Scientists designed artificial neural networks with the inspiration of the human brain. The function of the artificial neural network is exactly like the function of the human brain. An artificial neural network attempts to find a pattern or algorithm between the data that enters the system to solve the problem. In response to new data entering the system, the network adjusts and optimizes its response. This process continues until the optimal answer is obtained after the system has been trained fully. A trained neural network can solve different problems in the same way as it learned from the previous problem.

Symbolic AI, as described in 12.1.4, tries to represent facts and events using symbols. Then link them together using logical rules to make a meaningful system. And in contrast, non-symbolic AI uses large amounts of training data to draw conclusions and make optimal decisions. Non-symbolic AI includes models in machine learning, deep learning, and neural networks.

According to the classical view, symbolic AI is a 'supplier' of non-symbolic AI, which sends coded data as training data, and the learning process begins there. A hybrid AI approach goes beyond combining symbolic with non-symbolic AI approaches. In hybrid AI, non-symbolic AI such as neural networks, machine learning, and deep learning systems are combined with symbolic AI or embedding human intelligence.

Supervised learning is one of the basic hybrid AI because, in supervised learning, the work starts by importing a dataset containing training attributes and target attributes. A supervised learning algorithm determines the relationship between training data and specific target variables. It uses the learned relationship to classify completely new inputs (without targets).

### 12.2.2 Why we need hybrid AI?

In logic and computing, symbolic AI shows primary progress. Rule-based programs are easy to visualize, communicate, and troubleshoot.

Symbolic AI is useful for settings with a clear set of rules. It allows us to gather input and convert it into symbols or codes easily. Many computer programs, including those that use deep learning, are still based on rule-based systems.

Consider computer vision, which enables computers to understand images and videos. Suppose the police force has a picture of a burglar, and they want to create a program that detects images containing him. A rule-based program compares the pixels in updated images to those in the original image and responds to whether the burglar is present [6].

For this to work, you must provide your program with an exact copy of the original image. The burglar won't be identified if the picture is slightly different. Taking a picture from a different angle will cause the program to fail. A solution would be to take photos of the burglar from different angles and create new rules to compare each input with those images. No matter how many pictures you take of the burglar, you will not be able to account for all possibilities. Lighting conditions or the image background can cause the pixel value to change and cause the program to fail. You'll need millions of other pictures and rules for those.

AI programs based on the logic of symbolic AI have struggled with these kinds of problems for a long time. Defining rules for messy data in the real world is impossible. In such a case, we cannot define the rules for a self-driving car to detect all types of pedestrians it might face.

Numerous attempts have been made to create complicated symbolic AI systems that can solve complex problems with large amounts of data and multiple rules.

Generalizing these systems was impossible, so they were only applied in specific fields, such as medicine and economics. A weakness of this complicated AI system was its inability to deal with a lot of data. Data and new rules increase tremendously when a problem becomes generalized in specific fields, and more professional experts will be required. As the number of data increases, more rules are added. The number of rules in a system increases encoded knowledge, but more rules cannot override previous knowledge. Additionally, symbolic AI has the flaw of not comprehending the meaning of symbols. So the main challenge is how to distinguish the symbols, or relate them to other forms to allow computers to map raw emotions into symbols and then reason about them.

It took almost 40 years to complete complex projects with symbolic AI. Due to symbolic AI to act like human intelligence, become general, and solve complicated problems successfully, it is necessary to apply implicit knowledge or procedures that humans have in a formalized way using symbols and rules.

It should be noted that applying procedural or implicit knowledge that relies on sensory or movement processes is much more complicated in symbolic AI. Symbolic AI has had limited success, and neural networks are generally better suited for such applications.

A hybrid AI system is proposed in order to overcome these problems and shortcomings. Flowcharts 12.2 can depict the symbolic AI and non-symbolic AI clearly.

### 12.2.3 Abilities of hybrid AI

As mentioned in section 12.2.1, hybrid AI was proposed to eliminate the defect of symbolic AI. Hybrid AI, which is made from a combination of symbolic and non-symbolic AI, is capable of learning from training data and test data, based on which they can design a good decision-making model and make decisions in new experience. In this process, asking good questions is essential to guide hybrid AI and get the appropriate answer. Instead of learning everything automatically from the data, AI combines knowledge and data to solve the problem.

Implicating artificial neural networks and deep learning in non-symbolic AI add to the capabilities of hybrid AI. Hybrid AI can solve any issue related to general accuracy, uncertainty, ambiguity, and high dimensions.

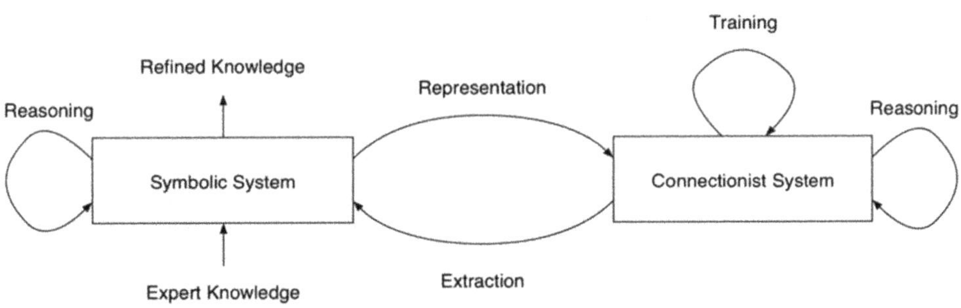


**Figure 12.2.** The scheme of symbolic and non-symbolic AI. Reproduced with permission from [27].

### 12.2.4 Applications of hybrid AI

Due to the capabilities of hybrid AI, its application can be seen in various fields of medicine, economics, industry, judgment, etc.

Websites that convert currencies, lengths, weights, or other units are examples processed by symbolic and non-symbolic AI. When a user selects 1 pound to transform into the ounce unit, the explore engine uses symbolic AI to detect the weight conversion problem and non-symbolic AI to convert web results to a pattern and output through a website or a widget presents.

Another area of AI is the performance of self-driving cars. Pedestrians, traffic lights, stop signs, speed control signs, and obstacles are marked on the map. Data collection is independent of non-symbolic AI, and all data is converted into symbols and code by symbolic AI. Automotive vehicles need to understand the basic rules of environmental signal processing to make real-time decisions. Understanding and processing the rules is performed by non-symbolic AI, and it contributes to making decisions with the minimum error if observing obstacles or other signs in the car direction.

Hybrid AI helps police and detectives to identify and arrest thieves from public centers such as banks and offices. Consider a bank that is monitored by a hybrid AI. First, the information of all staff and employees with details such as access level, check-in and check-out hours, bank working hours to provide customer service, and 24-hour security is stored. All the previous information is training data. Business days information and federal holidays information are also training data. Training data is converted into symbols and codes by symbolic AI.

In addition to the training data, the information of people who enter and leave the bank without receiving services and people who enter the bank and pay attention to more than one camera is stored as test data. Non-symbolic AI, based on training data and test data, identifies the information of people who enter the bank after the closure of the bank as a thief and gives an alarm to the guard and the relevant people. In the event of a robbery, the police will receive the suspect's list by asking questions. These questions can include the busiest people in the last ten days, the arrival and departure times of the mentioned people, and identifying the person or persons who most closely resemble the recorded images from the moment of the robbery. Asking good questions contributes to identifying a more accurate thief list.

If the police department is equipped with hybrid AI, the list of suspected people can be used as test data. Non-symbolic AI can check the training and test data to identify the robber or robbers. A police officer or detective can ask questions, such as who is not in prison or who has the bank robbery on his resume, etc?

Insurance companies also use hybrid AI to detect the amount of damage caused by accidents or burglaries, etc. Insurance experts take pictures of the customer from the accident scene, and the images and data are transformed into symbols and codes by symbolic AI. They use non-symbolic AI (deep learning) to check whether the airbag is open or not and what parts of the car are damaged.

In some cases where severe damage occurs or technical problems happen, such as burning the electronic control unit, appropriate data and images are not directly available. So a computer model or in-depth learning to generate data is used. In other words, simulating and receiving data allow us to use the same symbols as if someone had collected the data manually.

Practical implementation of these examples requires more precise conditions and constraints so that the output has the least amount of error [11].

## 12.3 Polynomials in hybrid artificial intelligence

Polynomials are one of the practical concepts which they applied in various fields such as mathematics, mechanics, economy, pharmacology, AI, etc. For example, astrophysicists use polynomials to calculate the speed and distance of a star from another object in space. Civil engineers use polynomials to design structures such as bridges, dams, etc. In addition, nursing assistants and medical assistants use polynomials to determine patient progress records and schedules. Economists use polynomials to model economic growth patterns, etc.

Polynomials are algebraic expressions with terms separated using the operators '+' and '−' in which the exponents of variables are always nonnegative integers. For example, $x^3 + x + 1$, $y^2 + 1$, and $2x^3 - 4y + 2$ are some polynomial.

The use of polynomials in hybrid AI is one of the fascinations of this concept.

### 12.3.1 Introduction of some polynomials such as regression, Rss, etc

Most data acquired from observing a phenomenon over time is significant, reviewing, and citing. For example, in business and economics, the amount of profit and benefit per week, stock prices in the stock market, and monthly indicators are noticeable. The lowest and highest daily temperatures, annual rainfall, drought indicators, and wind speed are considerable in meteorology. In agriculture, annual items related to harvest and livestock production, soil erosion, and export sales are perceptible. Electrical activity of the heart at intervals of one-thousandth of a second is considerable in biological sciences. The mentioned examples are areas that analyze and review data contribute to prediction, judgment, and decision making.

As mentioned in the previous subsection, the data obtained from symbolic AI are solved based on rule-based algorithms with mathematical models. However, many problems require data science and analysis instead of mathematical modeling. Since our knowledge of the data is incomplete, we have to compensate for this deficiency by finding patterns and using them to solve problems.

It is crucial to find an accurate model that can be used to predict, judge, and make decisions. Sometimes it is not simple to find a pattern in the data, so we must create authentic patterns. There are three essential steps in creating a pattern, some of which may be used multiple times [25]:

1. Pattern recognition or identification
2. Pattern fit
3. Pattern accuracy detection.

Each pattern contains one or more parameters whose values must estimate from the observed dataset. Pattern fitting finding the best possible estimates involves the unknown parameters of a given pattern. Measures such as least squares and maximum likelihood are considered for estimation. Checking the accuracy of the pattern is related to analyzing the quality of the pattern that we identified and estimated.

Francis Galton used regression for the first time in 1877. Regression is an equation that finds and calculates imprecise relationships between dependent variables (response) and a set of independent variables (predictor). In other words, the regression equation shows the best prediction of a dependent variable from several independent variables. In most cases, the dependent variable is continuous, while the independent variables can be discrete or continuous. Different regression models can be used depending on the type of variables and the purpose of the equation, such as linear regression, logistic regression, exponential regression, etc.

One of the simplest regression models is linear, that includes simple linear and multiple linear models. The linear regression method is used when there is a linear relationship between the response variable and the regression variables. Logistic regression is appropriate when the dependent variable has two states of presence or absence. If the dependent variable has stated such as improving or not improving, buying or not buying, going or not going, etc, we also use logistic regression. In addition to linear and logistic regression, there are other types of non-linear regression, such as power, inverse, logarithmic, parabolic, exponential, compound, growth, S curve, etc.

**Linear regression** is a supervised machine learning algorithm that predicts continuous values. Linear regression assumes that there is a linear relationship present between dependent and independent variables. In other words, it finds the best fitting line/plane that describes two or more variables.

Simple linear regression is used when only one independent variable is used to predict the dependent variable. Equation (12.1) shows simple linear regression.

$$Y = \beta_1 x + \beta_0 + \epsilon. \tag{12.1}$$

Equation (12.1) is the linear relationship between the dependent variables $\beta_0$ and $\beta_1$ with $Y$ to which the $\epsilon$ term is added. $\beta_1$ is the slope of the line, and $\beta_0$ is the $y$ intercept of equation (12.1). $\beta_0$ and $\beta_1$ are unknown parameters, and $\epsilon$ is the error term. The slope of the regression line shows the sensitivity of the dependent variable to the independent variable. The intercept elevation of the regression line represents the value of the dependent variable, which is calculated for the value of the independent variable equal to zero.

Since the relationship between dependent and independent variables doesn't predict accurately, therefore we consider the error term as an independent variable whose changes are independent of X. Thus, the error value is independent of the independent variable [24].

To derive the best-fitted line, first, we assign random values to $\beta_0$ and $\beta_1$ and calculate the corresponding value of $Y$ for a given $x$. The $Y$ value is the output value. The error can be defined in different ways and minimized in different ways. The least squares method is used in regression analysis to solve systems of equations whose number of equations exceeds the number of unknowns. The most important application of the least squares method is in data curve fitting. The model fitted to the data is the model in which the quantity $\chi^2$ is minimal

$$\chi^2 = \Sigma_{i=1}^{N} \frac{(y_i - f(x_i, \beta))^2}{\sigma_{y_i}^2}. \tag{12.2}$$

The 'mean squared' error (MSE) method calculates the residual error, and the function is called the 'loss function'.

$$L = \frac{1}{n}\Sigma(y - \hat{y})^2 \tag{12.3}$$

where $n$ is the number of observations.

To achieve the best-fitted line, we have to minimize the value of the loss function. Using the 'gradient descent' technique contributes to minimizing the loss function.

The 'classical least squares' method takes values as estimates $\beta_0$ and $\beta_1$ to minimize the following relationship.

$$Q(\beta_0, \beta_1) = \Sigma_{t=1}^{n}(Z_t - \beta_0 - \beta_1 t)^2. \tag{12.4}$$

To get the answer, we calculate its relative derivatives with respect to $\beta_0$ and $\beta_1$ and set them equal to zero. And we solve the resulting linear equations with respect to $\beta_0$ and $\beta_1$. If we show the answers in terms of $b_0$ and $b_1$, we will get

$$b_1 = \frac{\Sigma_{t=1}^{n}(Z_t - \bar{Z})(t - \bar{t})}{\Sigma_{t=1}^{n}(t - \bar{t})} \tag{12.5}$$

$$b_0 = \hat{Z} - b_t\bar{t} \tag{12.6}$$

where $\hat{t} = \frac{(n+1)}{2}$ and $1, 2, \dots, n$.

**Multiple linear regression** (MLR) is a statistical technique that uses several explanatory variables to predict the outcome of a response variable. Multiple linear regression aims to model the linear relationship between the explanatory (independent) variables and response (dependent) variables. MLR is used extensively in econometrics and financial inference.

The multiple linear regression equation is as follows:

$$y_i = \beta_0 + \beta_1 x_{i1} + \beta_2 x_{i2} + \cdots + \beta_p x_{ip} + \epsilon \tag{12.7}$$

where $i = n$ observations, $y_i$ is dependent variable, $x_i$ is explanatory variables, $\beta_0$ is $y$-intercept (constant term), $\beta_p$ is slope coefficients for each explanatory variable, $\epsilon$ is the model's error term (also known as the residuals).

Simple linear regression is a function that allows us to make predictions about one variable based on the information of another variable. In contrast, the multiple regression model extends to several explanatory variables. The relationship between independent and dependent variables is expressed as a line equation in simple linear regression. In multiple regression, if two independent variables are in a linear relationship with a dependent variable, the shape of this relationship will be a plane. If more than two independent variables are used in the linear regression model, the model appears as a hyperplane.

There is not much correlation between the independent variables in MLR. $Y_i$ observations are selected independently and randomly from the population. The coefficient of determination ($R$-squared) is a statistical metric used to measure how much variation in outcome can be explained by the variation in the independent variables. $R^2$ always increases as more predictors are added to the MLR model, even though the predictors may not be related to the outcome variable [14].

$R^2$ by itself can't thus be used to identify which predictors should be included in a model and which should be excluded. $R^2$ can only be between 0 and 1, where 0 indicates that any of the independent variables cannot predict the outcome and 1 indicates that the outcome can be predicted without error from the independent variables.

When interpreting the results of multiple regression, $\beta$ coefficients are valid while holding all other variables constant ('all else equal').

**Logistic regression** is used to classify elements of a set into two groups (binary classification) based on their probabilities. Determining a probability threshold is the initial step in logistic regression. If the probability of a particular element is higher than the probability threshold, then we classify that element in one group or vice versa.

First, we determine the best-fitted line by the linear regression to calculate the binary separation. The regression line obtained from linear regression is highly susceptible to outliers data. Due to this, it will not be able to classify two classes accurately. As a result, the predicted value is converted into probability using the sigmoid function. The transfer function has received an input value in the range of $-\infty$ to $+\infty$ and generated output value between 0 and 1. This function is expressed by the following general equation:

$$a = f_s(n) = \text{sigmoid}(n) = \frac{1}{1 + e^{-cn}}, \quad c > 0. \tag{12.8}$$

The sigmoid function is plotted in figure 12.3 for $c = 1$.

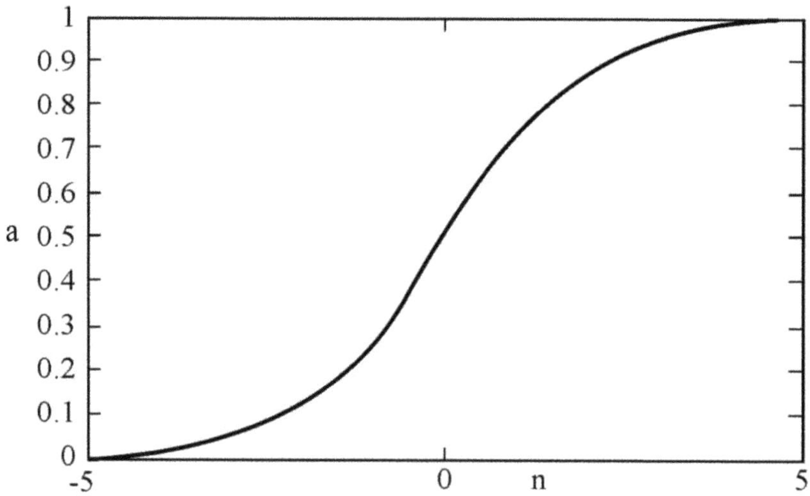


**Figure 12.3.** Sigmoid function.

The value of $c$ is determined linearity of the function.

Therefore, if we import the output value $\hat{y}$ to the sigmoid function, the possible value will be between 0 and 1. Based on the threshold value, the output value of the sigmoid function is converted to 0 or 1 (discrete values). Binary classification can be obtained by considering a threshold value of 0.5.

There are similarities between linear regression and logistic regression. Both of them are supervised machine learning algorithms. Both are parametric and use linear equations for prediction.

However, functionality-wise these two are different. Regression problems are handled using linear regression, while classification problems are handled using logistic regression. Linear regression provides a continuous output, while logistic regression provides discreet output data. The purpose of linear regression is to find the best-fitted line, while logistic regression is one step ahead and fitting the line values to the sigmoid curve. The method for calculating loss function in linear regression is the mean squared error, whereas the maximum likelihood estimation method is proper for logistic regression.

### 12.3.2 Approximation of polynomials

'Optimization' refers to the process of changing and manipulating inputs, mathematical operations, or characteristics of a device to reach an optimal output or solution. In the optimization process, the input data is the variables of the problem, which is supposed to be solved by one of the algorithms of 'numerical optimization' methods and its optimal solutions are determined. The output of the optimization process is called 'fitness'; the process or function that is supposed to be processed by the optimization algorithms is called the objective function. A significant subfield of evolutionary computation is the development of numerical optimization algorithms based on natural processes. Numerical optimization algorithms are used to solve various problems and optimize them numerically. One of the most prominent examples of such algorithms is the genetic algorithm (GA) (and other algorithms based on this computing method), which is inspired by mechanisms of biological evolution (such as 'mutation', 'recombination', 'like production'). Reproduction, natural selection, and survival of the fittest are used to generate candidate solutions to solve an optimization problem. These mechanisms are also called evolutionary operators. Candidate answers play the role of samples in the population. 'cost function' also specifies the fitness of the environment in which the candidate solution is located (a candidate solution for the optimization problem). Algorithms such as genetic algorithm, ant colony algorithm, firefly algorithm, cuckoo optimization algorithm, and others are among the most widely used evolutionary algorithms. Numerical optimization techniques and tools based on biologically-inspired algorithms have become effective increasingly in recent years. In particular, they have appeared very powerful in solving problems with NP-hard complexity, such as the traveling salesman problem, and converge to the globally optimal solution in a reasonable amount of time. Meanwhile, algorithms based on 'swarm intelligence' and 'multi-agent meta-heuristic algorithms' have a special place in the numerical optimization field. In addition, algorithms are considered as one of the hottest research subjects in this field. Algorithms such as particle swarm optimization (PSO), which is a subset of multi-agent meta-heuristic algorithms based on crowd intelligence, are among the most advanced algorithms developed in optimization and other similar applications.

The PSO algorithm was invented in 1995 by two scientists named Kennedy and Eberhart. The PSO algorithm is inspired by the 'swarm behaviors' of certain species of living organisms, such as the social behavior of birds flying to find food or group movement in the same direction as fish. As mentioned before, such behavior in living organisms, which inspired the development of some well-known optimization algorithms, is called swarm intelligence.

Although the PSO algorithm has many similarities with the GA, structurally, it is much simpler than this algorithm; because in the PSO algorithm, 'evolution operators' such as 'crossover' and 'mutation' are not used. In this algorithm, evolutionary mechanisms such as 'randomness' of real numbers and 'global communications' between swarm particles doesn't use to achieve evolution and reach the optimal global solution.

On the other hand, since all the representations generated from the population, the corresponding 'encodings' of the 'population' entities and the evolutionary mechanisms used to converge to optimal solutions (in the particle optimization algorithm) are based on operations that are mainly on 'real numbers' are done, its implementation will be easier than the GA.

**Standard particle swarm optimization algorithm**

In the PSO algorithm, each 'agent' in the population is called particle. In the PSO algorithm, each of the factors corresponding to piece paths are formed pseudo-randomly by positional vectors. By adjusting the paths of each of the factors in the population (particles), the objective functions are searched to find the optimal solution or the region of the optimal solution.

The movement of particles in the search space depends on two fundamental components: a stochastic component and a deterministic component. Each particle in the population simultaneously is tending to move randomly in the problem space and is attracted to the point where the global optimum or parameter $g\star$ is located. In other words, each of the particles in the population moves to the location of the best point in the problem space where the particles have been so far or are attracted to the $x_i^\star$ parameter. The best point is measured based on the history of particle movement in the environment.

When one of the particles in the population (particle $i$) finds a place in the search space that is better than the other searched places in this space, the newly found place is the best point (in the problem space) that particle $i$ has been to so far. The new location in the problem space will be known as the best possible solution for all $n$ particles in the population.

The ultimate goal of the PSO algorithm is to determine the globally optimal solution from the set of best possible solutions found (by the particles). The process of finding the best possible solution (by particles) continues until there is no significant improvement in the best possible solution after a certain number of iterations. The best solution in the set of found optimal solutions is defined as the optimal global solution of the problem.

The following relationship shows the movement of particles in the problem space, where the best point (answer) found by particle $i$ is denoted by $x_i^\star$.

$$g^\star \approx \min \quad \text{or} \quad \max f(x_i) \quad (i = 1, 2, \ldots, n). \tag{12.9}$$

Equation (12.9) is used to determine the globally optimal solution (the best solution in the set of optimal solutions found by the particles). The new velocity vector is obtained through equation (12.10), where $x_i$ and $\mathrm{v}_i$ are the location and velocity vectors of particle $i$, respectively.

$$v_i^{t+1} = v_i^t + \alpha\epsilon_1.\ (g^\star - x_i^t) + \beta\epsilon_2.\ (x_i^\star - x_i^t) \tag{12.10}$$

where $\epsilon_1$ and $\epsilon_2$ are two random vectors, and each element of these vectors takes a random value between 0 and 1. 'Hadamard product' of two $u.\ v$ matrices is defined in the form of 'entry-wise product' and displayed as follows:

$$|(u.\ v)|_{ij} = u_{ij}v_{ij} \tag{12.11}$$

$\alpha$ and $\beta$ parameters are called learning parameters or acceleration constants. The two parameters usually set as $\alpha \approx \beta \approx 2$. The initial values of the parameter $X_i^{t=0}$ are determined using the upper and lower limits $a = \min(x_i)$, $b = \max(x_j)$. Also, the initial value of the parameter $V_i^{t=0}$ is considered equal to zero $V_i^{t=0} = 0$. According to the previous relationships and parameter values, the new location of particle $i$ in the search space determines through the following relationship:

$$X_i^{t+1} = X_i^t + V_i^{t+1}. \tag{12.12}$$

Although the parameter $V_i$ can take any value, the value of this parameter is usually chose from the range $[0,\ V_{\max}]$.

The most significant improvement that has been made in different versions of the particle swarm optimization algorithm and its components is using the 'inertia function' in the form of $\theta(t)$. In versions of the PSO algorithm that uses the inertial function, the term $\theta(t)V_i^t$ is used instead of the $V_i^t$ parameter; in this expression, the parameter $\theta$ takes values between 0 and 1.

In the simplest case, the inertial function can be considered a constant in the PSO algorithm, which takes values of $\theta \approx 0.5 \approx 0.9$. Adding the inertial function (in the form of a fixed parameter) can be thought of as a 'virtual mass' that stabilizes the movement of particles in the problem search space. As a result, it is expected that the PSO algorithm will converge more quickly to the globally optimal solution (or the particles in the population will converge faster towards the area containing the globally optimal solution).

**Fireflies Algorithm**

Fireflies emit flashing light because of a biological process called 'bioluminescence', which contributes to their glowing appearance. Researchers have discussed and studied the function of the flashing lights to date, but they have not reached a consensus on their actual functions. However, the two fundamental functions of the bioluminescence process and the resulting night glow are:

Attracting the opposite sex for mating and copulation (communication function, bioluminescence process)

Attracting potential prey to themselves.

Additionally, researchers have found that fireflies use flashing lights as a protective mechanism to send a warning to other fireflies in the environment. The rhythm or frequency of flashing light, the rate of light flashing, and the duration of light flashing by fireflies form different parts of the communication system between worms (to mate with the opposite sex or to warn other fireflies in the environment about possible dangers).

For example, during the mating of fireflies, the female worm reacts to the unique flashing pattern of the male worms (both male and female are the same species). Also, the female worms of some firefly species, including Photuris, can imitate the flashing patterns of other species for mating. By doing so, the female firefly can trap and eat male worms of other species that have moved towards the female to mate [15].

The light flashing pattern of fireflies worm is that the light intensity at a given distance, $r$, from the light source follows an inverse square law

$$I\alpha\frac{1}{r^2}. \tag{12.13}$$

In the relationship (12.13), $I$ is the light intensity, and $r$ is the distance, which has the opposite relationship. In other words, as the distance increases, the light intensity decreases. Furthermore, 'air' absorbs light, causing the intensity of the light to decrease as the distance increases.

The combination of these two significant factors causes fireflies to be visible only from a certain distance. Usually, in the darkness of night, the flashing light of fireflies can be seen from a distance of several hundred meters, which is enough to be observed by other worms and communicate with them.

To implement the optimization algorithm inspired by the firefly behavior, the characteristic features related to the behavior of the acceleration worm and the flashing light pattern produced by them will be formulated. The following rules simplify the formulation of the firefly algorithm:

1. All fireflies are unisex. In other words, fireflies, regardless of their gender, will be attracted to other fireflies in the problem space.
2. In the firefly algorithm (FA), the 'attractiveness' of a firefly will be proportional to the 'brightness' of that firefly. Whenever both fireflies flash the one with less light will be attracted to the one with more light. Therefore, the attractiveness of the firefly will be proportional to its brightness.
3. The brightness of a firefly is influenced by or determined by the characteristic features of the objective function. The brightness can be specified as proportional to the value of the fitness function in maximization problems.

The firefly algorithm is implemented based on these three rules. The basic steps of the firefly algorithm are shown in the pseudo-code below.

Objective function: $f(x)$, $x = (x_1, \ldots, x_d)^T$.

Production of the initial population of fireflies: $x_i$, $(i = 1, 2, 3, \ldots, n)$ Light intensity parameter $I_i$ In $x_i$, it is obtained by substituting the value of $x_i$ into the function $f(x_i)$.

Definition of 'Light Absorption Coefficient' or $\gamma$

While loop: while $t <$ Max Generation

For ring: for $i = 1$ to $n$, for all $n$ fireflies in the population

For loop: for $j = 1$ to $i$, for all $n$ fireflies in the population

if condition: if $I_j > I_i$, firefly $i$ moves to firefly $j$ (in the $d$-dimensional search space of the problem).

end if

The attractiveness parameter value is calculated and updated based on the distance $r$ and according to the relation $\exp(-\gamma r)$. New solutions (candidate answers) evaluate, and the light intensity parameter is updated.

End For

End For

Fireflies are evaluated, and the best candidate answer is determined in this generation.

End While

The final answers are postprocessed, and illustrations related to the final output.

To formulate the firefly algorithm, two parameters, 'variation of light intensity' and the mutual attraction of fireflies, should be taken into account. In the simplest case and for a maximization problem, the value of the brightness parameter $I$ of a firefly located at a specific location $X$ can be obtained through a relationship such as $I(X)\alpha f(x)$. However, the value of the attraction parameter $\beta$ of a firefly is relative and must be determined by other fireflies (in other words, the distance between fireflies will play a direct role in their attraction (attracting them to each other)). Therefore, the attractiveness parameter $\beta$, based on the distance $r_{ij}$ will change between firefly $i$ and firefly $j$.

In addition, as the distance from the light source increases, the light intensity decreases. Also, light is absorbed by it when it passes through 'mediums level' such as air. As a result, the firefly algorithm should be formulated in such a way that based on different 'degrees of absorption', the value of the attractiveness parameter will also change.

In the simplest possible case, the light intensity parameter $I(r)$ will change according to the 'inverse square law':

$$I(r) = \frac{I_s}{r^2} \tag{12.14}$$

where the $I_s$ parameter shows the light intensity in the source. Having a medium level (such as air) with a constant coefficient of light absorption, $\gamma$, the light intensity $I$ will change based on the distance $r$ and through the following relationship:

$$I = I_0 e^{-\gamma r}. \tag{12.15}$$

In this regard, $I_0$ shows the intensity of the original light. To avoid the occurrence of 'singularity' in the $\frac{I_s}{r^2}$ relationship and at the point, $r = 0$, the 'combined effect' of the inverse square law and the principle of light absorption by the air interface can approximate using the following function (which is in Gaussian form):

$$I(r) = I_0 e^{-\gamma r^2}. \tag{12.16}$$

Sometimes a function that can decrease uniformly and at a slower rate is needed. In such a case, the following 'approximation' can be used:

$$I(r) = \frac{I_0}{1 + \gamma r^2}. \tag{12.17}$$

In a short distance between two fireflies, two forms of the function $I(r)$ are almost equal to each other. Since the value of $r = 0$ series expansions of these two functions will be equivalent to each other, up to the order of $O(r^3)$:

$$\frac{1}{1 + \gamma r^2} \approx 1 - \gamma r^2 + \frac{1}{2}\gamma^2 r^4 + \cdots \tag{12.18}$$

$$e^{-}\gamma r^2 \approx 1 - \gamma r^2 + \frac{1}{2}\gamma^2 r^4 + \cdots. \tag{12.19}$$

Since the attractiveness of a firefly is proportional to the light intensity observed by neighboring fireflies, the attractiveness parameter $\beta$ of firefly can be defined by the following relation:

$$\beta(r) = \beta_0 e^{-\gamma r^2} \tag{12.20}$$

where $\beta_0$ is the attractiveness of the firefly at the distance $r = 0$. Since the calculation of the relation $\frac{1}{(1 + r^2)}$ is faster than calculating an exponential function, instead of the function (12.20), we can use the function (12.21) and calculate the value of the attractiveness parameter through this relationship:

$$\beta(r) = \frac{\beta_0}{1 + \gamma r^2}. \tag{12.21}$$

The relation (12.21) defines a parameter called the characteristic distance (length scale) as $\Gamma = \frac{1}{\sqrt{\gamma}}$, by which the attractiveness parameter changes significantly from the form $\beta_0$ to the form $\beta_0 e^{-1}$ [17]. When implementing the firefly algorithm, the function $\beta(r)$ can be represented in the form of monotonically decreasing functions, such as the 'generalized' form below: [18]

$$\beta(r) = \beta_0 e - \gamma r^m, \quad (m \geqslant 1). \tag{12.22}$$

Assume that the $\gamma$ parameter is a constant. When $m \to \infty$ the characteristic distance parameter or the length scale changes to the form $\Gamma = \gamma^{\frac{-1}{m}}$. On the other hand, by having the value of the characteristic distance or length scale parameter in an optimization problem, the $\gamma$ parameter can be used in the form of $\gamma = \frac{1}{\Gamma^m}$ and as the normal initial value. The distance between two fireflies $i$ and $j$, which are located in the spatial coordinates $x_i$ and $x_j$, respectively, is equal to the Cartesian distance between them in the following form:

$$r_{ij} = \left\|x_i - x_j\right\| = \sqrt{\Sigma_d^{k=1}(x_{i,k} - x_{j,k})^2} \tag{12.23}$$

where $x_{i,k}$ is equal to the $k$th component of the spatial coordinates associated with the $i$th firefly ($x_i$). In a two-dimensional search space, the distance between two fireflies $i$ and $j$ is calculated through the following equation:

$$r_{ij} = \left\|x_i - x_j\right\| = \sqrt{(x_i - x_j)^2 - (y_i - y_j)^2}. \tag{12.24}$$

The movement of a firefly (for example, firefly $i$) attracted by another more attractive (brighter) firefly is defined by the following relation:

$$x_i = x_i + \beta_0 e^{-\gamma rij^2}(x_j - x_i) + \alpha\left(\text{rand} - \frac{1}{2}\right). \tag{12.25}$$

Here, the second term on the right side of the relation shows the influence of the attractiveness parameter in calculating the movement value of firefly $i$. Also, the third term is used for randomization, where $\alpha$ plays the role of randomization parameter. The expression rand is also a random number generator that the random values produced by it follow the uniform distribution in the interval [0, 1].

It is worth noting that the calculation of the distance $r$ in the firefly algorithm is not limited to using Euclidean distance and can be used from any arbitrary equation to calculate the distances between fireflies. In other words, depending on the type of optimization problem that is supposed to be solved by the firefly algorithm, different equations can be used to calculate the distance $r$ in the $n$-dimensional space. Two essential limiting cases must be considered to prove the convergence of the firefly algorithm to the globally optimal solution:

When $\gamma \to 0$, the parameter $\gamma$ tends to zero.

When $\gamma \to \infty$, the parameter $\gamma$ tends to infinity.

In the case that $\gamma \to 0$, the parameter $\gamma$ tends to zero, the attractiveness parameter will be a constant value and equal to zero ($\beta = \beta_0$), and the scale parameter tends to infinity ($\Gamma \to \infty$). Therefore, the firefly algorithm converges to an 'optimal' point in the space of the search problem, and this optimum usually will be the globally optimum of the problem.

**Genetic Algorithm (GA)**

Many human inventions are inspired by nature. An example of such an innovation is artificial neural networks. Another such innovation is the development of the idea of a GA. By simulating the process of evolution in nature, GAs search in the candidate solution space intending to find the best possible solution to a problem. First, a set or a population of initial solutions is generated in the search process to find the optimal solution. Second, modified solution sets are produced in successive generations (in each generation of the GA, specific changes are made in the genes of the chromosomes that make up the population). The initial solutions are usually changed so that in each generation, the solutions population 'converges' towards the optimal solution.

This branch is inspired by the field of AI, based on the mechanism of evolution of living organisms and the production of more triumphant and graceful species in nature. In other words, the main idea of GAs is the survival of the fittest.

A chromosome is a long and complex string of 'Deoxyribonucleic Acid' or DNA. Hereditary factors that determine the characteristics or traits of an 'individual' are formed along these chromosomes. Human genes encode each trait by combining DNA. In the body of living organisms, there are usually four bases for producing chromosomes from DNA:

1. Base A or Adenine
2. Base C or Cytosine
3. Base G or Thymine
4. Base T or Guanine

Just as the alphabet defines the building blocks of a language, the meaningful combination of chromosomes (and their bases) produces specific instructions for cells. Changes in chromosomes occur during the reproduction process. Parents' chromosomes exchange randomly with each other through a process known as crossover. Therefore, the children inherit some characteristics or traits of the father and mother and display them.

A rare process called 'mutation' also causes changes in the characteristics of living organisms. Sometimes, an error may occur in the process of copying chromosomes, which is also called mitosis. Most mutations in living organisms result in their destruction. Throughout millions of years of evolution, mutations have created new and more graceful species.

The most important steps required to implement the GA and its different types are: generating an initial population of answers to a problem, specifying the objective function, the fitness function, and using genetic operators to create changes in the population of problem answers. The functional principles of the GA are:

Formulating the initial population consisting of the solutions to the problem
Initialization and randomization of the primary population consisting of problem solutions
Repeat loop.

Evaluation of the objective function of the problem

Find the appropriate fitness function

Performing operations on the population consisting of problem solutions using genetic operators

'Reproduction' operator

'Crossover' operator

'Mutation' operator

Until: the stop condition is satisfied [14].

The best performance of GAs usually occurs when a binary representation is used to encode the variables of the problem.

If the problem whose optimal solutions are to be determined has more than one variable, according to the number of variables of that problem, 'single-variable coding' corresponding to each of them is created and merged. In such a case, a multi-variable coding problem forms. One of the most prominent features of GAs is the simultaneous processing of several candidate solutions generated for the defined problem.

Therefore, in the first step of the implementation of the GA, a set consisting of $P$ entities or chromosomes is formed by pseudo random generators. Each entity with the chromosomes in this population represents a candidate and possible solution to the problem. In a solution space, also known as the initial solution, each entity represents a vector representation of the solution.

Since the search operation starts from a set of answers in the solution space (initial answers are scattered randomly in the answer space), an effective and unbiased search will be guaranteed in the GA. In the next step, all the initial solutions generated are evaluated to determine the value of the objective function.

At this step, an 'exterior penalty function' is usually used to transform the 'constrained optimization problem' into an 'unconstrained' optimization problem. Such a transformation will be different depending on different optimization problems.

In the third step, the objective function of the problem is mapped to a fitness function. Each member of the initial population is assigned a fitness value using the fitness function. After determining the fitness value of the candidate answers, GA operators are used to make changes to the candidate answers.

Suppose that the following optimization problem is given and the goal of the GA is 'maximizing' the following function:

$$\text{maximize } f(x),\ x_i^l \leqslant x_i \leqslant x_i^u, \quad i = 1, 2, \dots, N \tag{12.26}$$

where $x_i^l$ and $x_i^u$ are the upper and lower limits set for the values of the variable $x_i$. While the GA demonstrates using a maximization problem, can minimize the various problems.

For the GA to be able to maximize function (12.26), the variables $x_i$ must be encoded by a string structure. It is worth noting that coding the variables of the problem is not necessary. Coding based on binary strings is used to code problem variables.

In the GA, the length of the binary string is usually determined according to the desired and expected accuracy of the optimization problem. For example, if four bits are used to encode a two-variable optimization problem, the strings (0000 0000) and (1111 1111) correspond to two dots, respectively. $(x_1^l, x_2^l)^T$ and $(x_1^u, x_2^u)^T$ will be in the solution space of the problem; because the substrings (0000) and (1111) will display the upper and lower limits of a point in the search space of the problem [16].

A fixed mapping rule can map any eight-bit string to a point in the search space for problem variables. A linear mapping rule is usually used to map an eight-bit string to a point in the problem search space.

$$x_i = x_i^l + \frac{x_i^u - x_i^l}{2^\beta - 1} \Sigma_{j=0}^{\beta} \gamma_j^{2^j} \tag{12.27}$$

where, $x_i$ is encoded by substring $s_i$ with length $\beta$. The decoded value of substring $s_i$ is calculated through the expression $\Sigma_{j=0}^{\beta} \gamma_j^{2^j}$. Also, $s_i$ is represented by the values (0, 1) and the string $s$, in the format $(s_{\beta-1}, s_{\beta-2}, \dots, s_1, s_0)$. For example, using the relation $\Sigma_{j=0}^{\beta} \gamma_j^{2^j}$, the decoded value for the four-bit string (0111), will be calculated equal to the value 7.

Since each bit can only take the values 0 or 1, for a 4-bit string, only a 16 distinct substrings can be defined. The accuracy obtainable by four-bit coding is approximately one-sixteenth (1/16) of the problem's search space. While if the length of the string increases to 5, the accuracy obtainable by five-bit encoding will be approximately equal to one-thirtieth (1/32) of the problem's search space.

Substring lengths, therefore, depend directly on the desired and expected accuracy of the variables. The substrings length depends on the desired and expected accuracy of the algorithm results; the longer the strings, the more accurate the output. The relationship between the string length $\beta$ and the accuracy of the produced results $\alpha$ is obtained through the following equation:

$$(x_i^u - x_i^l)10^\alpha \leqslant (2\beta - 1). \tag{12.28}$$

After coding the variables of the problem, each of the points corresponding to the variables of the problem, $x = (x_1, x_2, \ldots, x_N)^T$, will be generated and placed in the problem search space. Once the points in the search space corresponding to the values of the problem variables, the function value at the point $x$ will be calculated by substituting $x$ into a given objective function, or $f(x)$.

A GA will be suitable for solving optimization problems (maximization or minimization). In general, a fitness function, or $F(i)$, is initially formulated using the objective function and is used in successive genetic operations in the generations of the GA.

Biological vision explains that fitness is a qualitative value that measures the efficiency of the reproduction of chromosomes. Fitness functions are used in GAs to estimate the probability of entities or chromosomes reproducing. In other words, it is used as a criterion to determine the goodness of chromosomes or candidate solutions to the problem.

In the GA, the chromosomes or entities that have the highest fitness will have a greater chance to combine and mutate (genetic operators) than other chromosomes in the population. The correct operation of some genetic operators depends on the definition of 'NonNegative' fitness functions; however, other genetic operators do not define the precondition of non-negativity of the fitness function to perform their operations. The fitness function can be equivalent to the objective function in maximization problems.

$$F(i) = O(i) \rightarrow \text{Objective function } (i) = \text{Fitness function } (i). \tag{12.29}$$

The objective function of the problem must be mapped to the fitness function when minimizing problems to produce nonnegative values in all cases and to inverse the proportionate fitness of strings. There are different methods to do such mapping.

$$F(x) = \frac{1}{1 + f(x)} \tag{12.30}$$

where $F(x)$ is the fitness function, and $f(x)$ is the objective function of the minimization problem. The transformation (12.30) does not change the location of the minimum solution in the problem–solution space. While it turns a minimization problem into an equivalent maximization problem. Another substitution function used to transform the objective function into the fitness value of chromosome $i$, or $F(i)$, is defined as follows.

$$F(i) = V - \frac{O(i)P}{\Sigma_{i=1}^{P} O(i)} \tag{12.31}$$

where $O(i)$ is the value of the objective function of the $i$th entity or chromosome, $P$ is the size of the population, and $V$ is a large value that guarantees the negation of fitness values. The value of $V$ is generally considered equal to the maximum value of the expression $\Sigma_{i=1}^{P} O(i)$ in this transformation. In such a case, the fitness value corresponding to the maximum value of the objective function becomes zero. This transformation also does not change the location of the minimum solution in the problem–solution space. Whereas it turns a minimization problem into an equivalent maximization problem. The value of the fitness function of a string is also called string fitness.

In GAs, the optimization process begins with the generation of an initial population of random strings. The strings are equivalent to chromosomes, entities, or candidate solutions to the problem. Random strings represent the design of the problem or the 'decision variables' associated with a problem. Then, the initial population is affected by the three operators in the genetic algorithm to produce a new population of points in the problem–solution space. The new population will consist of new chromosomes, entities, or solutions. The principal operators of the genetic algorithm are the reproduction operator, combination operator, and mutation operator.

As mentioned before, the GA can be considered a mechanism for maximizing the objective function. The mechanism is done by evaluating chromosomes or answer vectors. The purpose of operators in the GA is to select, combine and change the vectors corresponding to the solutions that are considered the best solution for the optimization problem in the current generation.

The newly generated population is further evaluated, and it continues until the termination of the operational processes in the GA. Until the stopping condition of the GA is satisfied, the population of chromosomes or answer vectors is manipulated and evaluated through reproduction, combination, and mutation operators. This procedure continues until the stopping criterion of the GA is satisfied.

'Generation' of the GA is the process of manipulating the population of chromosomes or answers through reproduction, combination, and mutation and their evaluation.

### 12.3.3 Homogenious and estimation of error

In hybrid AI, different algorithms of non-symbolic artificial intelligence are used to make decisions. Neural network, GA, particle swarm algorithm, and firefly algorithm are algorithms that can be combined to increase the capability of hybrid AI.

To obtain the most accurate results more than three various AI models propose. In many instances, there is no significant difference between one model's performance to another. The selection of input variables plays a crucial role in the model performance and the accuracy of the results. The pre-processing phase of the data should be carried out as efficiently as possible to avoid any noise in the data and to filter out the most reliable and accurate data to be used later as input.

Most of these parameters have no optimal value, and they are significantly different in each case. If the input data doesn't choose wisely, It leads to problems such as long computation times. The parameters are determined using a trial and error conventionally. Various optimization algorithms are categorized into two main categories, global and local optimization algorithms. Global optimization algorithms offer a much higher probability than the local algorithms discussed so far to the global or near-global optimum [7].

## 12.4 Polynomials' applications

Several advantages can be derived from using hybrid AI tools rather than statistical methods [26]

1. Adaptive learning: AI tools can learn how to perform their tasks based on the information they are given.
2. AI tools automatically organize and deliver the given information.
3. Real-time operators: the calculations of these methods can be designed and implemented in parallel. As a result, the AI tools provide optimal results.
4. Classification: these methods can categorize inputs to get the outputs.
5. Generalization: these methods enable the network to only obtain a general rule by dealing with a limited number of instances and generalize the results of this learning to previous data.
6. Flexibility: AI also maintains its captured information and can accept updated information without losing any previous data.

The mentioned advantages of AI contribute to making the best decision in inexperienced situations, forecasting, and predicting. Using different algorithms based on their ability and capabilities helps us to design a hybrid algorithm.

The main decision to design a framework to solve the feature selection problem with hybrid algorithms is the choice of solution representation and fitness function. The features selection problem is solved with the help of the invasive weeds optimization, cultural algorithm, PSO, and GA.

## References


[1] Armano G, Marchesi M and Murru A 2005 A hybrid genetic-neural architecture for stock indexes forecasting *Inf. Sci.* **170** 3–33

[2] Wolfgang E 2017 *Introduction to Artificial Intelligence* 4th edn (New York: Springer)

[3] Russell S J and Norvig P 2021 *Artificial Intelligence: A Modern Approach* 4th edn (Englewood Cliffs, NJ: Prentice-Hall)

[4] Corchado E and Abraham A 2009 *Artificial Intelligence Systems* (New York: Springer)

[5] Schroeder L F, Bazzan A L C, Valiati J, Engel P M and Ceroni S 2002 *A comparison between symbolic and non-symbolic machine learning techniques in automated annotation of the 'Keywords' field of SWISS–PROT, Brazil*

[6] Bou-Rabee M and Lodi K A 2020 *One-Month-Ahead Wind Speed Forecasting Using Hybrid AI Model for Coastal Locations* (NJ: IEEE)

[7] Ibrahim K S M K, Huang Y F, Ahmed A N, Koo C H and El-Shafie A 2022 A review of the hybrid artificial intelligence and optimization modelling of hydrological streamflow forecasting *Alex. Eng. J.* **61** 279–303
[8] Carlos A, Filho P and Filho R M 2010 Hybrid training approach for artificial neural networks using genetic algorithms for rate of reaction estimation: application to industrial methanol oxidation to formaldehyde on silver catalyst *Chem. Eng. J.* **157** 501–8
[9] Aibinu A M, Onumanyi A J, Adedigba A P, Ipinyomi M and Folorunso T A 2017 Development of hybrid artificial intelligent based handover decision algorithm *Int. J. Eng. Sci. Technol.* **20** 381–90
[10] Nayak J, Naik B and Behera H S 2015 A comprehensive survey on support vector machine in data mining tasks: applications and challenges *Int. J. Database Theory Appl.* **8** 169–86
[11] Tsai C-F, McGarry K and Tait J 2003 *Image Classification Using Hybrid Neural Networks, ACM, SIGIR'03, July 28–August 1 (Toronto, Canada)*
[12] Ghasemieh R, Moghdani R and Sana S S 2017 A hybrid artificial neural network with metaheuristic algorithms for predicting stock price *Cybern. Syst.* **48** 365–92
[13] Veit-Haibach P and Herrmann K 2022 *Machine Learning in Nuclear Medicine and Hybrid Imaging* (Berlin: Springer)
[14] Zakeri H, Nejad F M and Gandomi A H 2022 *Automation and Computational Intelligence for Road Maintenance and Management* (New York: Wiley)
[15] Dehghani M, Trojovský P, Hubálovský T, Alsenani T R and Singh J 2022 Three-periods optimization algorithm: a new method for solving various optimization problems *IETE J. Res.* https://doi.org/10.1080/03772063.2022.2052982
[16] Chakraborty I, Kumar V, Nair S B and Tiwari R 2003 Rolling element bearing design through genetic algorithms *Eng. Optim.* **35** 649–59
[17] Xin-She Y 2010 Firefly algorithm, Lévy flights and global optimization *Research and Development in Intelligent Systems XXVI*
[18] Yang X-S 2008 Swarm optimization *Introduction to Computational Mathematics* (Singapore: World Scientific)
[19] Rodas J, Azpeitia D, OchoaZezzatti A, Camarena R and Olivier T 2016 *Use of GVRP as a Model of Two Specific Real World Problems and Its Bioinspired Solution* (Hershey, PA: IGI Global) pp 452–69
[20] Zgurovsky M, Sineglazov V and Chumachenko E 2021 Artificial intelligence systems based on hybrid *Neural Networks: Theory and Applications* (Berlin: Springer)
[21] Bhattacharyya S, Snášel V, Pan I and De D 2020 *Hybrid Comutational Intelligence* (Boca Raton, FL: CRC Press)
[22] Hallam J and McKevitt P 1995 *Hybrid Problems, Hybrid Solutions (Frontiers in Artificial Intelligence and Applications)* (Washington, DC: IOS Press)
[23] Kandel A and Langholz G 1992 *Hybrid Architectures for Intelligent Systems* (Boca Raton, FL: CRC Press)
[24] Ismail H 2018 *Statistical Modeling, Linear Regression and ANOVA* (Morrisville, NC: Lulu)
[25] Illukkumbura A 2020 *Introduction to Regression Analysis* (Independently Published)
[26] Anand A and Suganthi L 2018 Hybrid GA-PSO optimization of artificial neural network for forecasting electricity demand *Energies* **11** 728
[27] Bader S and Hitzler P 2005 Dimensions of neural-symbolic integration - a structured survey *We Will Show Them: Essays in Honour of Dov Gabbay* vol 1 ed S Artemov, H Barringer, A S d'Avila Garcez, L C Lamb and J Woods (College Publications) pp 167–94

# Chapter 13

# Comparative study of Chebyshev and Legendre polynomial-based neural models for approximating multidimensional poverty for an Indian State

**Sandeep Kumar, Arup Kumar Sahoo and S Chakraverty**

The objective of this chapter is to analyse different neural models for estimating the condition of multidimensional poverty in an Indian state viz. Odisha. In this regard, two functional link neural networks (FLNN) and a single hidden layer artificial neural network (ANN) have been developed by using the open access data from NITI Aayog, India. Feed-forward neural network model with error back propagation (BP) has been used for modifying the network parameters and minimising the computed error function. Optimal hyper-parameter tuning has been done to increase the efficacy of models. The numerical efficiency and accuracy of the FLNN models are investigated by multidimensional poverty data. We have found that FLNNs are comparatively better approximators and computationally efficient as compared to ANN. The results obtained by these methods are compared with the existing ANN method and are found to be in good agreement.

## 13.1 Introduction

Odisha is a beautiful Indian state endowed with pure unadulterated natural beauty, vibrant culture, amazing wildlife, ancient temples, a long coastal line and rich natural resources. However, such resources have not been fully exploited for income-generating purposes. As a result, Odisha ranks 22 among the Indian states in terms of GSDP per capita income, and it has become one of the poorest states in the country [1]. The majority of residents of the state are multidimensionally impoverished. Therefore, it becomes vital to assess the condition of poverty and manage its alleviation for the people of Odisha using models that are both scientific and practical.

doi:10.1088/978-0-7503-5067-9ch13  

Poverty alleviation and eradication have a significant impact on the growth of the world economy [2]. This is the primary goal of the UN 2030 agenda for the sustainable development of any country [3]. Poverty is a multidimensional phenomenon with varied interpretations in different societies. For many years, poverty is often considered a deficiency of monetary resources, which is a one-dimensional approach. However, poverty is no longer explained through an income perspective only but exists in a multidimensional condition [4]. Monetary policy sometimes fails to address deprivations in other categories. But the multidimensional poverty measure analyses different categories such as health, income, living standards, education etc. As a result, it is necessary to determine the important factors that cause poverty so that governments can establish effective measures to reduce it.

Artificial Intelligence (AI) is touching new heights every day. ANN, robotics, natural language processing, and fuzzy logic are some of the AI techniques that have shown significant growth in the last few years [5]. Among all AI techniques, ANN has been shown to be a very effective tool for data analysis in different sectors. ANN is a massively parallel processor composed of processing units that mimic the training process of the human brain to predict patterns from given historical data [6, 7]. ANN is being used to study the performance improvement mechanism of targeted poverty alleviation by the creation of new algorithms [8]. It has also been used in multidimensional poverty research, such as the analysis of poverty attributes, poverty probability prediction, and poverty area identification [8]. The nonlinear characteristic of a multilayer neural network (MLNN) proved to be quite effective in revealing the fundamental link between geographic factors and poverty [9]. But the MLNN requires a large amount of computation cost, and the time taken by the CPU to perform learning and predict the output is high. To overcome these drawbacks, researchers came up with the idea of FLNN.

Given the difficulty of scaling up traditional data collection efforts and the accuracy of existing multidimensional measurements, a viable option for measuring poverty is to use neural methods in combining the property of **polynomials**. FLNN is a second-generation neural network with a single input layer, a single output layer and a functional expansion block. In 1995 for the first time the FLNN model was introduced by Pao and Phillips [10], which has fewer network parameters than in traditional multilayer ANN. FLNN is more computationally efficient than MLNN due to the absence of the hidden layer. In this way, the nonlinear relationship between the inputs and outputs is processed via a set of functional expansions block (i.e. polynomials), as shown in figure 13.1.

For the measurement of multidimensional poverty, various approaches have been developed by researchers. Among them, A–F double cutoffs [11], Foster–Greer–Thorbecke [12] and first-order dominance [13] dominated globally. Wang and Islam [14] proposed a BP neural network algorithm to develop a poverty alleviation fund audit model, which can assist auditors in precisely judging the basic condition of the audit object and determining the audit target scientifically and rationally. Pareek and Prema [15] proposed a methodology to classify poverty into two categories, i.e. below poverty line (BPL) and non-BPL, using MLNN. Azcarraga and Setiono [16] used the data extraction property of NN to gain insight into the characteristics of poverty. Alsharkawi *et al* did a case study in Jordan to predict the poverty status of

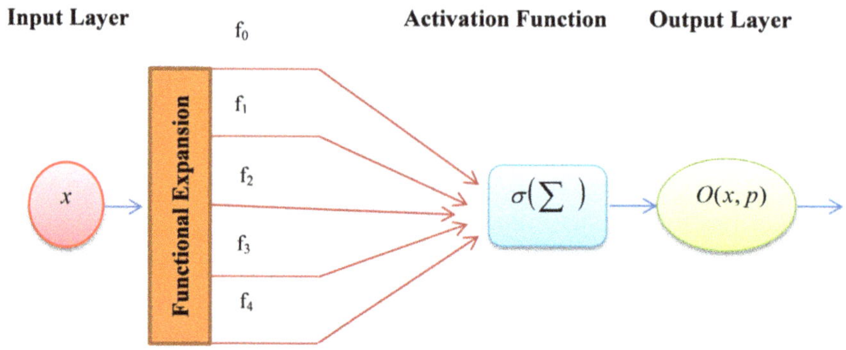


**Figure 13.1.** Schematic diagram of FLNN with single input and a single output.

Jordanian households [9]. In his investigation, multidimensional poverty classification was done by using a machine learning model.

In this investigation, two different FLNN models viz. Chebyshev neural network (ChNN) and Legendre neural network (LeNN) have been developed to identify and assist the poverty alleviating factors in the state of Odisha. The model has been trained with the data collected from the report of NITI Aayog, India. Different statistical measures have been calculated to show the efficacy and validity of the undertaken model. FLNN seems to be an effective alternative to the ANN models in the prediction of poverty alleviating factors.

The rest of this chapter is structured as follows. In section 13.2, we present preliminaries of the undertaken models. Section 13.3 discusses the methodological framework of the work. The results and discussions are given in section 13.4, and the conclusion of the chapter is discussed in section 13.5.

## 13.2 Preliminaries

The architecture of the FLNN models is made up of two parts: (i) the numerical transformation part and (ii) the learning part. In the numerical transformation part, each input data is expanded to several terms using given polynomials. The polynomials expansions block may contain random variables, generic basis, trigonometric functions and orthogonal polynomials as well. However, in other polynomial expansions, the function has a very small error near the initial point of the expansion, but the error increases rapidly as we employ more terms. But for the orthogonal polynomial, the same error criteria can be achieved with a much smaller magnitude of estimated parameters and fewer terms required than other functional expansion. Also, the nonlinear approximation capacity of the orthogonal polynomial is very powerful. For the recurrence property of the orthogonal polynomial, the neurons can be the same in structure. So combining the characteristics of the FLNN and orthogonal polynomial resulted in a new efficient FLNN. However, the robustness of FLNN using orthogonal polynomials in comparison to using other polynomials is high, and cost efficiency is low, as proved by researchers earlier [17]. Here the input data are expanded through Chebyshev and Legendre orthogonal polynomials to generate new input vectors.

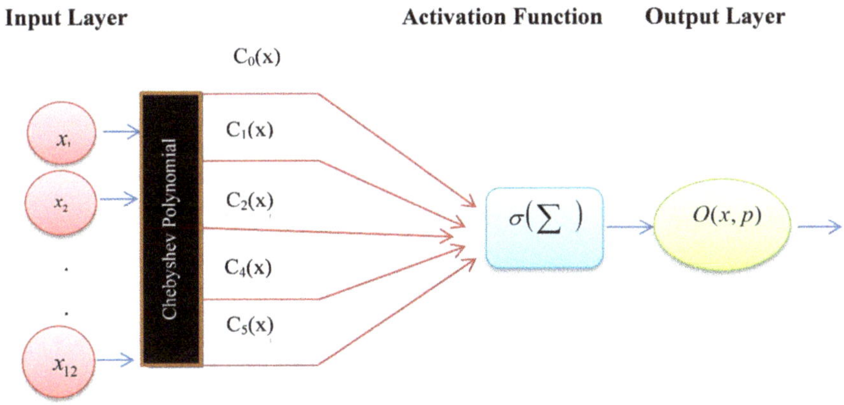


**Figure 13.2.** Schematic diagram of ChNN model with 12 input and single output nodes.

### 13.2.1 Chebyshev neural network (ChNN)

ChNN has been applied to various interdisciplinary problems, for instance, digital communication [18], function approximation [19], system identification [20], structural dynamics [21], etc. Figure 13.2 shows the structure of ChNN to predict the poverty pattern from the given data of NITI Aayog, India. The input datasets $x = (x_1, x_2, \ldots, x_{12})^T \in R^{12}$ are expanded through the $m$-dimensional Chebyshev polynomial to form a new set of input vectors of dimension $12m$.

The first few Chebyshev polynomials are written as [17]

$$\begin{aligned} C_0(x) &= 1, \\ C_1(x) &= x, \\ C_2(x) &= 2x^2 - 1. \end{aligned}$$

Other terms can be generated from the recursive formula

$$C_{r+1}(x) = 2x\, C_r(x) - C_{r-1}(x)\,.$$

### 13.2.2 Legendre neural network (LeNN)

The structure of LeNN is the same as ChNN, where Legendre polynomial replaces Chebyshev polynomial in the polynomial expansion block.

The first few terms of Legendre polynomial are written as [17]

$$\begin{aligned} L_0(x) &= 1, \\ L_1(x) &= x, \\ L_2(x) &= \frac{1}{2}(3x^2 - 1). \end{aligned}$$

Other terms can be generated from the recursive formula

$$L_{r+1}(x) = \frac{1}{n+1}[(2n+1)x\, L_r(x) - nL_{r-1}(x)]\,.$$

The LeNN model has also been widely applied in various areas like channel equalisation [22], prediction of machinery noise [23], ODE solver [24], stock value prediction [25] etc. Similar to the ChNN, the input dataset $x = (x_1, x_2, \ldots, x_{12})^T \in R^{12}$ are again expanded through the $m$-dimensional Legendre polynomial to form a new set of input vectors of dimension $12m$.

## 13.3 Methodology

### 13.3.1 Dataset construction

In this chapter, we have utilised data from the national multidimensional poverty index baseline report based on NFHS-4 (2015–16) available on the NITI Aayog, India website. Further, from this report, we have considered data for the state of Odisha, India. There are 12 input features, which are poverty indicators, and one output feature, which is headcount ratio. Indicators of poverty show the proportion of persons who are deficient in each individual indicator, while the headcount ratio shows the percentage of multidimensionally poor people.

### 13.3.2 Data preprocessing module

Machine learning estimators require the standardisation of a dataset because they may behave poorly if the individual features do not more or less resemble standard normally distributed data. For instance, several components of the L1 and L2 regularisers [26, 27] of linear models, which make up the objective function of a learning algorithm, assumed that all features be centred around 0 and have a variance that is distributed uniformly. If the variance of one variable is more than others, it may dominate the objective function and prevent the estimator from correctly drawing conclusions from other features as intended. Therefore, we have standardised features by subtracting the mean and scaling to unit variance before feeding data into the ANN or FLNN module. The standardisation of a feature $x$ is calculated as:

$$z = \frac{x - \mu}{\sigma},$$

where $\mu$ is the mean and $\sigma$ is the standard deviation of feature $x$.

### 13.3.3 Mechanism of MLNN model

The general architecture of MLNN [28] with one hidden layer having $n$ neuron and $K$ input training datasets $\boldsymbol{x}^k = (x_1^k, x_2^k, \ldots, x_m^k)^T$, $k = 1,2, \ldots, K$ with $m$ features and a single output $o^k$ is utilised for the training purpose (figure 13.3). So, the forward propagation for the described MLNN architecture can be written as follows:

$$x^k \to W^{[1]^T} x^k \to \sigma(W^{[1]^T} x^k) = a^{[1]} \to W^{[2]^T} a^{[1]} \to \sigma(W^{[2]^T} a^{[1]}) = a^{[2]} = o^k$$

where $W^{[1]}$ is of order $(m\times n)$ randomly weighted matrix for layer 1, $W^{[2]}$ is of order $(n\times 1)$ randomly weighted matrix for output layer, $\sigma$ is the activation function, $a^{[1]} = (a_1^{[1]}, a_2^{[1]}, \ldots, a_n^{[1]})^T$ is $(n \times 1)$ vector of first layer neuron output, and $a^{[2]}$ is the second layer neuron output.

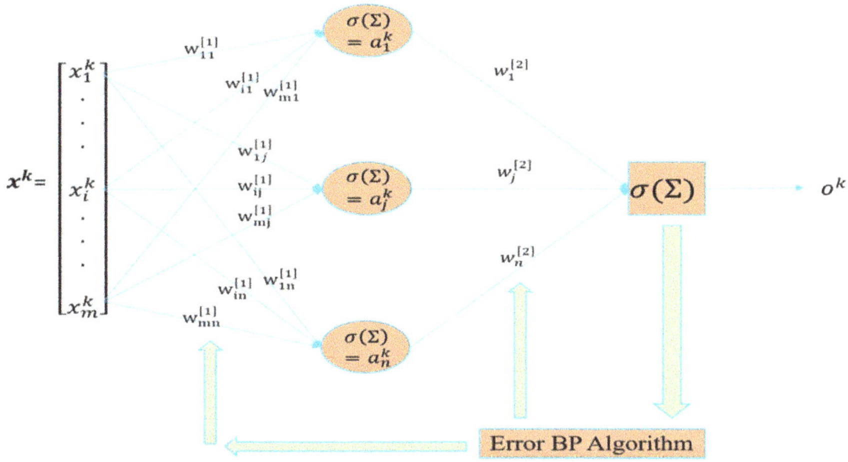


**Figure 13.3.** Architecture of ANN with one hidden layer with $m$ input and a single output.

Although, we give input with batch size $K$ at a time for the training, which is done by vectorisation. So, the forward propagation for batch training of the described ANN architecture is as follows:

$$X \rightarrow W^{[1]^T} X \rightarrow \sigma(W^{[1]^T} X) = A^{[1]} \rightarrow W^{[2]^T} A^{[1]} \rightarrow \sigma(W^{[2]^T} A^{[1]}) = A^{[2]} = o,$$

where $\boldsymbol{X} = \begin{bmatrix} x_1^1 & \cdots & x_1^k & \cdots & x_1^K \\ \vdots & \ddots & \vdots & & \vdots \\ x_i^1 & \cdots & x_i^k & \cdots & x_i^K \\ \vdots & & \vdots & \ddots & \vdots \\ x_m^1 & \cdots & x_m^k & \cdots & x_m^K \end{bmatrix}$ is $(m \times K)$ matrix of input neuron, $W^{[1]}$ is of order $(m \times n)$ randomly weighted matrix for layer 1,

$W^{[2]}$ is of order $(n \times 1)$ randomly weighted matrix for output layer,

$\sigma$ is the activation function,

$$A^{[1]} = \begin{bmatrix} a_1^{(1)[1]} & \cdots & a_1^{(k)[1]} & \cdots & a_1^{(K)[1]} \\ \vdots & \ddots & \vdots & & \vdots \\ a_j^{(1)[1]} & \cdots & a_j^{(k)[1]} & \cdots & a_j^{(K)[1]} \\ \vdots & & \vdots & \ddots & \vdots \\ a_n^{(1)[1]} & \cdots & a_n^{(k)[1]} & \cdots & a_n^{(K)[1]} \end{bmatrix}$$ is $(n \times K)$ matrix of first layer neuron output,

and

$A^{[2]} = (a^{(1)[2]}, a^{(2)[2]}, \ldots, a^{(K)[2]}) = (o^1, o^2, \ldots, o^K) = O$ is the $(1 \times K)$ vector of the second layer neuron output.

Followed by the forward propagation, we have utilised the BP algorithm [28] for learning the parameters. In addition, we used mean square error (MSE) for error computation and L2 regulariser for error regularisation in this process [27]. L2 regulariser is the square of the magnitude of coefficients which acts as a penalty to the error function and helps in keeping weights smaller. For the batch size of $K$ at a time, we compute cost function $J$ which is the average of all errors. Therefore, the cost function is given as [29]:

$$\mathcal{J} = \frac{1}{2K}\sum_{k=1}^{K}(d^k - o^k)^2 + \frac{\lambda}{2K}\|W^{[1]}\|^2 + \frac{\lambda}{2K}\|W^{[2]}\|^2$$

$$= \frac{1}{2K}\sum_{k=1}^{K}\left(d^k - \sigma(W^{[2]^T}.\, a^{[1]})\right)^2 + \frac{\lambda}{2K}\|W^{[1]}\|^2 + \frac{\lambda}{2K}\|W^{[2]}\|^2$$

$$= \frac{1}{2K}\sum_{k=1}^{K}\left(d^k - \sigma(W^{[2]^T}.\, \sigma(W^{[1]}\boldsymbol{x}^k))\right)^2 + \frac{\lambda}{2K}\|W^{[1]}\|^2 + \frac{\lambda}{2K}\|W^{[2]}\|^2$$

where $d = (d^1, d^2, \ldots, d^K)$ is the desired output, and $\lambda$ is a regularisation parameter.

After the computation of the cost function, we need to calculate the gradients of the cost function with respect to weights. Gradients are required for the weight update process with the help of the optimisation process.

Now, the gradient computation [28] can be done for layers one and two as:

$$\frac{\partial \mathcal{J}}{\partial W^{[1]}} = -\frac{1}{K}\sum_{k=1}^{K}(d^k - \sigma(W^{[2]T}.\, \sigma(W^{[1]T}.\, \boldsymbol{x}^k))).\, \sigma'(W^{[2]T}.\, \sigma(W^{[1]T}.\, \boldsymbol{x}^k)).\, W^{[2]T}.$$

$$\sigma'(W^{[1]T}.\, \boldsymbol{x}^k).\, \boldsymbol{x}^k + \frac{\lambda}{K}W^{[1]}$$

and

$$\frac{\partial \mathcal{J}}{\partial W^{[2]}} = -\frac{1}{K}\sum_{k=1}^{K}(d^k - \sigma(W^{[2]T}.\, \sigma(W^{[1]T}.\, \boldsymbol{x}^k))).\, \sigma'(W^{[2]T}.\, \sigma(W^{[1]T}.\, \boldsymbol{x}^k)).\, \sigma(W^{[1]T}.\, \boldsymbol{x}^k)$$

$$+ \frac{\lambda}{K}W^{[2]}.$$

Further, for the weight update process, we have utilised the Adam optimiser. This optimisation technique uses stochastic gradient descent and adaptive estimation of first and second-order moments. The Adam weight update rule is then, for $l = 1,2$ is written as follows [30]:

$$v_{dW^{[l]}} = \beta_1 v_{dW^{[l]}} + (1 - \beta_1)\ \frac{\partial \mathcal{J}}{\partial W^{[l]}}$$

$$v^{\text{corrected}}_{dW^{[l]}} = \frac{v_{dW^{[l]}}}{1-(\beta_1)^t}$$

$$s_{dW^{[l]}} = \beta_2 s_{dW^{[l]}} + (1 - \beta_2)\left(\frac{\partial \mathcal{J}}{\partial W^{[l]}}\right)^2$$

$$s^{\text{corrected}}_{dW^{[l]}} = \frac{s_{dW^{[l]}}}{1-(\beta_2)^t}$$

$$W^{[l]} = W^{[l]} - \alpha\frac{v^{\text{corrected}}_{dW^{[l]}}}{\sqrt{s^{\text{corrected}}_{dW^{[l]}}} + \varepsilon}$$

where $t$ counts the number of steps taken by Adam, $L$ is the number of layers, $\beta_1$ and $\beta_2$ are hyper parameters that control the two exponentially weighted averages, $\alpha$ is the learning rate, and $\varepsilon$ is a very small number to avoid dividing by zero.

### 13.3.4 Mechanism of FLNN model

The general architecture of FLNN model [10] has the functional expansion of $n$ nodes and $K$ input training datasets $\boldsymbol{x}^k = (x_1^k, x_2^k, \ldots, x_m^k)^T$, $k = 1,2, \ldots, K$ with $m$ features and a single output $o^k$ is considered for the training purpose (figure 13.4). So, the forward propagation for the described FLNN architecture is as follows:

$$x^k \rightarrow y(x^k) \rightarrow \sigma(W^T y(x^k)) = o^k$$

where $\sigma$ is the activation function, $W$ is $(n\times 1)$ weight matrix, and $y$ is the expansion function.

Although, we give input with batch size $K$ at a time for the training, which is done by vectorisation. So, the forward propagation for batch training of the described FLNN architecture is as follows:

$$X \rightarrow y(X) \rightarrow \sigma(W^T y(X)) = o$$

where $\boldsymbol{X} = \begin{bmatrix} x_1^1 & \cdots & x_1^k & \cdots & x_1^K \\ \vdots & \ddots & \vdots & & \vdots \\ x_i^1 & \cdots & x_i^k & \cdots & x_i^K \\ \vdots & & \vdots & \ddots & \vdots \\ x_m^1 & \cdots & x_m^k & \cdots & x_m^K \end{bmatrix}$ is $(m\times K)$ matrix of input neuron,

$W$ is $(n\times 1)$ weight matrix,

$y(\boldsymbol{X}) = \begin{bmatrix} y_1^1 & \cdots & y_1^k & \cdots & y_1^K \\ \vdots & \ddots & \vdots & & \vdots \\ y_j^1 & \cdots & y_j^k & \cdots & y_j^K \\ \vdots & & \vdots & \ddots & \vdots \\ y_n^1 & \cdots & y_n^k & \cdots & y_n^K \end{bmatrix}$ is $(m\times K)$ matrix of function expansion neurons,

and

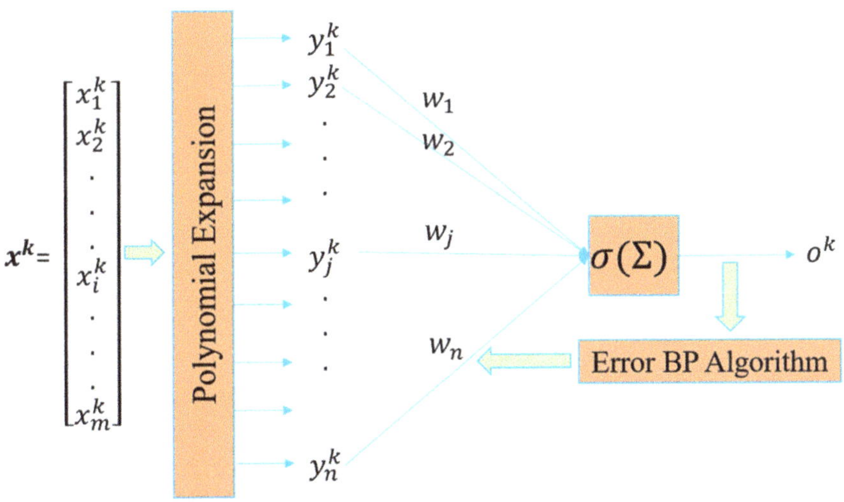


**Figure 13.4.** Architecture of FLNN with one functional expansion layer with $m$ input single output.

$O$ is (1×$K$) output matrix.

Followed by the forward propagation, we have utilised the BP algorithm for learning the parameters. In addition, we used MSE for error computation in this process. For the batch size of $K$ at a time, we computed cost function $J$, which is the average of all errors. Therefore, the cost function is given as [29]:

$$\mathcal{J} = \frac{1}{2K}\sum_{k=1}^{K}(d^k - o^k)^2$$
$$= \frac{1}{2K}\sum_{k=1}^{K}(d^k - \sigma(W^T y(\boldsymbol{x}^k)))^2$$

where $d = (d^1, d^2, \ldots, d^K)$ is the desired output.

After the computation of the cost function, we are required to calculate the gradients of the cost function with respect to weights. Gradients are required for the weight update process with the help of the optimisation process.

Now, the gradient computation can be done as follows [28]:

$$\frac{\partial \mathcal{J}}{\partial W} = -\frac{1}{K}\sum_{k=1}^{K}(d^k - \sigma(W^T y(\boldsymbol{X}))).\ \sigma'(W^T y(\boldsymbol{X})).\ y(\boldsymbol{X})$$

or

$$\frac{\partial \mathcal{J}}{\partial W_j} = -\frac{1}{K}\sum_{k=1}^{K}(d^k - \sigma(W^T(\boldsymbol{X}))).\ \sigma'(W^T y(\boldsymbol{X})).\ y_j(\boldsymbol{X})$$

where, $y_j(\boldsymbol{X}) = (y_j^1, y_j^2, \ldots, y_j^K)$ is the $K$ dimensional function expansion input for $j$th weight.

Further, for the weight update process, we have utilised the Adam optimiser [30].

## 13.4 Results and discussion

In this section, we have discussed the robustness of MLNN, ChNN and LeNN for the prediction of multidimensional poverty data. Moreover, this comparison is about whether functional link NNs are better approximators than traditional one hidden layer NNs. For this study, we have used ANN with one hidden layer with a BP algorithm and two different FLNNs, namely ChNN and LeNN, without hidden layer with a BP algorithm. In FLNN, the hidden layer is replaced by a functional expansion block for improving the input patterns using orthogonal polynomials like Chebyshev and Legendre. In all three models, we have utilised MSE for error computation. In the MLNN model, we have added L2 regularisation in error which is not used in FLNN models, so MLNN has an upper edge in the calculation of error. We have also utilised Adam optimiser for the weight update process in all three models.

For the comparison, we have taken MLNN with one hidden layer having 24 neurons in it, and for function expansion of FLNNs, we have taken second-order Chebyshev polynomial for ChNN and second-order Legendre polynomial for LeNN so that it expands 12 input parameters to twenty-four parameters. Now since MLNN has twenty-four neurons in the hidden layer and FLNNs also have twenty-four expanded inputs, we have trained the neural networks by forward propagation with Selu activation function and BP as defined in the methodology section for two-thousand iterations and studied the obtained results.

We found that ChNN and LeNN are computationally faster and better approximators than MLNN. For two-thousand iterations, ChNN took 47.49 s,

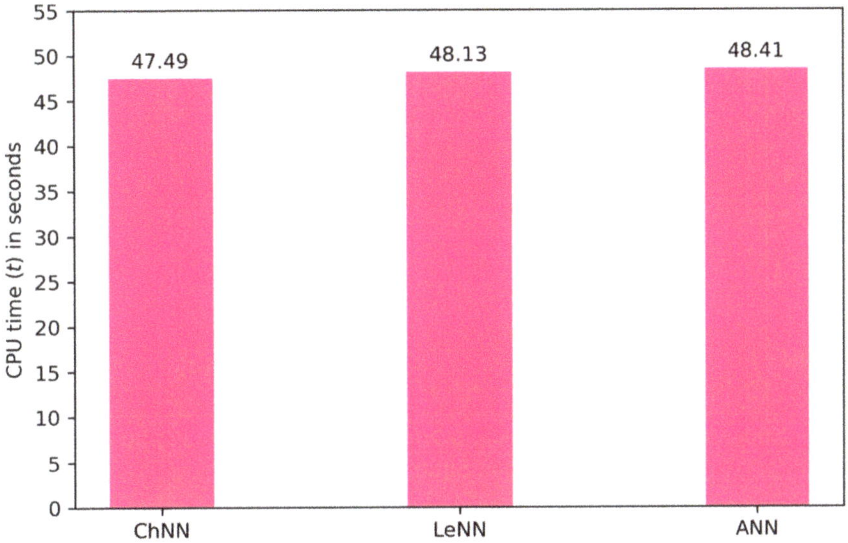


**Figure 13.5.** Computation time taken by neural network models.

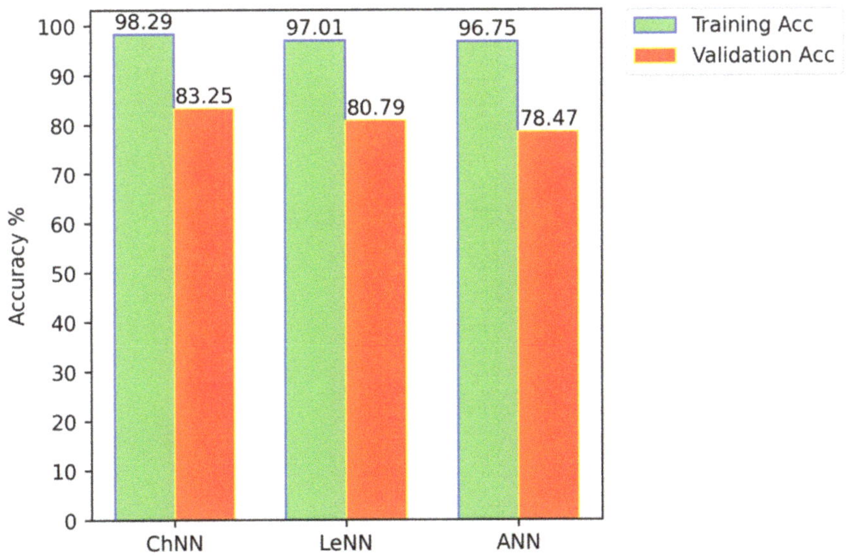


**Figure 13.6.** Training and validation accuracy of the neural network models.

followed by LeNN with 48.13 s and ANN with 48.41 s CPU time. The comparison of computation time for all three neural network models is shown graphically in figure 13.5. Also, in the training process, ChNN led with a training accuracy of 98.29% and validation accuracy of 83.25%, followed by LeNN with a training accuracy of 97.01% and validation accuracy of 80.79% and MLNN with a training

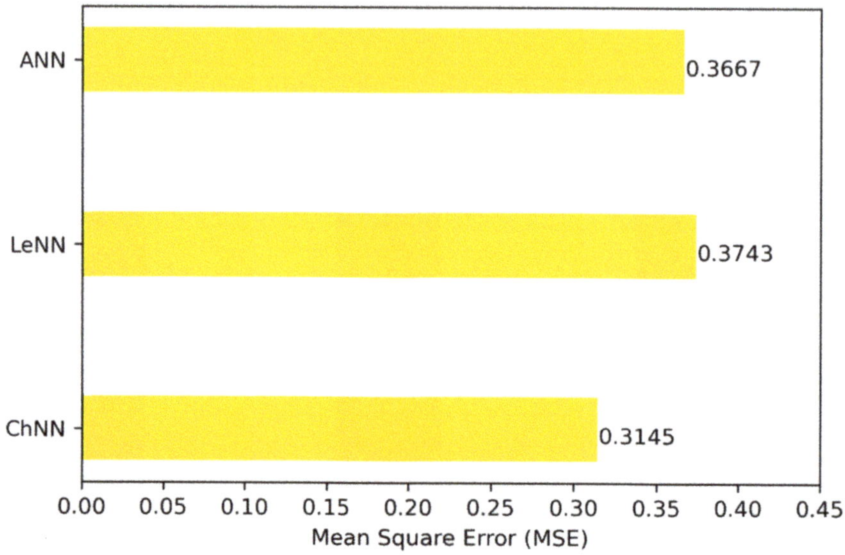


**Figure 13.7.** Mean squared error of different neural network models.

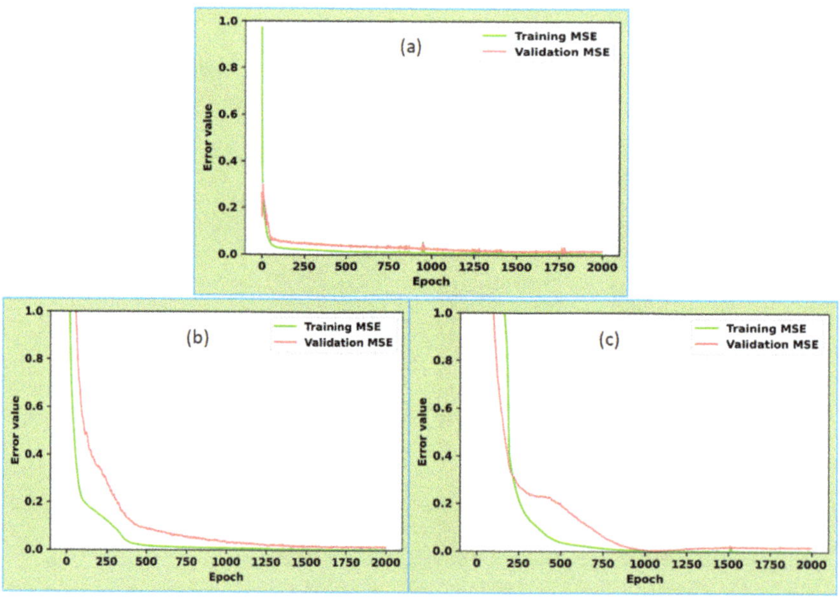


**Figure 13.8.** Training and validation loss plots of (a) MLNN (b) ChNN (c) LeNN.

accuracy of 96.75% and validation accuracy of 78.47%. These accuracy values can be viewed in figure 13.6. The MSE of ChNN was 0.3145, followed by 0.3667 of ANN and 0.3743 of LeNN. Bar graphs in figure 13.7 represent the MSE of different neural network models. From the error graphs (figure 13.8), we may conclude that error decreases when we increase the number of epochs. Also, table 13.1 shows the output of different neural networks with the original output.

**Table 13.1.** Output from different neural network models.

| Actual Output | ChNN Output | LeNN Output | MLNN Output |
|---|---|---|---|
| 24.57 | 24.55 | 24.35 | 24.58 |
| 27.49 | 27.42 | 27.28 | 27.63 |
| 24.42 | 24.44 | 24.31 | 24.12 |
| 24.9 | 24.79 | 24.90 | 24.84 |
| 33.03 | 33.11 | 33.35 | 33.20 |
| 28.43 | 28.44 | 28.55 | 27.98 |
| 14.97 | 14.20 | 14.61 | 14.72 |
| 37.1 | 37.20 | 37.01 | 37.11 |
| 30.08 | 30.19 | 30.13 | 30.39 |
| 38.76 | 38.55 | 39.01 | 39.09 |
| 21.88 | 21.93 | 21.76 | 21.81 |
| 11.83 | 12.62 | 13.03 | 12.57 |
| 20.75 | 20.85 | 20.84 | 20.67 |
| 18.62 | 18.81 | 18.71 | 18.74 |
| 47.28 | 47.24 | 47.31 | 47.33 |
| 44.75 | 44.80 | 44.55 | 45.16 |
| 21.67 | 21.44 | 21.27 | 21.18 |
| 41.78 | 41.94 | 42.09 | 42.03 |
| 15.49 | 15.46 | 15.82 | 15.68 |
| 51.14 | 51.19 | 51.12 | 51.25 |
| 58.71 | 58.69 | 58.68 | 58.81 |
| 44.9 | 44.88 | 44.98 | 44.93 |
| 59.32 | 59.24 | 59.24 | 59.31 |
| 20.49 | 20.50 | 20.04 | 20.51 |
| 37.98 | 37.96 | 38.06 | 37.91 |
| 11.64 | 11.26 | 11.72 | 12.30 |
| 48.14 | 49.65 | 49.61 | 48.33 |
| 24.53 | 27.99 | 27.16 | 26.59 |
| 28.05 | 27.45 | 26.78 | 26.70 |
| 24.75 | 24.91 | 25.27 | 26.72 |
| Accuracy | 98.62% | 98.41% | 98.39% |
| MSE | 0.314518139 | 0.374275597 | 0.366725575 |

## 13.5 Conclusion

In this study, we discovered that the FLNN approximators are superior to one hidden layer ANN models with the same number of hidden neurons as the polynomial expansion in FLNN in terms of accuracy and computation time. Between the two FLNN models, ChNN exhibits a better approximation than LeNN. As our dataset has been too small, thus the computing time difference is only a fraction of seconds, but in real-world scenarios with large datasets, the computation time difference may be much larger. Finally, it should be highlighted that though the main focus of the study is on estimating the condition of multidimensional poverty in Odisha, it may be applicable to estimating poverty in any other region.

## Acknowledgements

The first and third authors would like to acknowledge the Poverty Alleviation Research Center (PARC), National Institute of Technology, Rourkela and the second author thanks the Council of Scientific and Industrial Research (CSIR), New Delhi, India (File no: 09/983(0042)/2019-EMR-I), respectively, for the support and funding to pursue the present research work.

## References

[1] https://statisticstimes.com/economy/india/indian-states-gdp-per-capita.php
[2] Haushofer J and Fehr E 2014 On the psychology of poverty *Science* **344** 862–67
[3] Zhou Q, Chen N and Lin S 2022 A poverty measurement method incorporating spatial correlation: a case study in Yangtze River Economic Belt, China *ISPRS Int. J. Geo-Information* **11** 50
[4] Naraya Deepa R P and Schafft K 2000 *Voices of the Poor: Can Anyone Hear Us?* vol 1 (World Bank Publications)
[5] Chakraverty S and Jeswal S K 2021 *Applied Artificial Neural Network Methods for Engineers and Scientists* (Singapore: World Scientific)
[6] Sahoo A K and Chakraverty S 2021 Multilayer unsupervised symplectic artificial neural network model for solving Duffing and Van der Pol–Duffing oscillator equations arising in engineering problems *Modelling and Computation in Vibration Problems* 1st edn vol 2 ed S Chakraverty, F Tornabene and J N Reddy (Bristol: IOP Publishing) pp 1–13
[7] Sahoo A K and Chakraverty S 2022 Machine intelligence in dynamical systems: a state-of-art review *WIREs Data Mining and Knowledge Discovery* **12** e1461
[8] Liu H, Liu Y, Qin Z, Zhang R, Zhang Z and Mu L 2021 A novel DBSCAN clustering algorithm via edge computing-based deep neural network model for targeted poverty alleviation big data *Wireless Commun. Mobile Comput.* **2021** 5536579
[9] Alsharkawi A, Al-Fetyani M, Dawas M, Saadeh H and Alyaman M 2021 Poverty classification using machine learning: the case of Jordan *Sustainability (Switzerland)* **13** 1412
[10] Pao Y-H and Phillips S M 1995 The functional link net and learning optimal control *Neurocomputing* **9** 149–64
[11] Alkire S and Foster J 2011 Counting and multidimensional poverty measurement *J. Public Economics* **95** 476–87

[12] Foster J, Greer J and Thorbecke E 2010 The Foster–Greer–Thorbecke (FGT) poverty measures: 25 years later *J. Economic Inequality* **8** 491–524
[13] Arndt C, Mahrt K, Hussain M A and Tarp F 2018 A human rights-consistent approach to multidimensional welfare measurement applied to sub-Saharan Africa *World Development* **108** 181–96
[14] Wang H and Islam K 2019 An optimisation model for poverty alleviation fund audit mode based on BP neural network *J. Intell. Fuzzy Syst.* **37** 481–91
[15] Pareek P and v Prema K 2012 Classifying the population as BPL or non-BPL using multilayer neural network *Int. J. Sci. Research Publications* **2** 2250–3153
[16] Azcarraga A and Setiono R 2018 Neural network rule extraction for gaining insight into the characteristics of poverty *Neural Computing and Applications* **30** 2795–806
[17] Chakraverty S and Mall S 2017 *Artificial Neural Networks for Engineers and Scientists* (Boca Raton, FL: Taylor and Francis, CRC Press)
[18] Meher P K, Patra J C and Juhola M 2008 Intelligent sensors using computationally efficient Chebyshev neural networks *IET Sci. Measurement Technol.* **2** 68–75
[19] Lee T-T and Jeng J-T 1998 The Chebyshev-polynomials-based unified model neural networks for function approximation *IEEE Trans. on Systems, Man and Cybernetics, Part B (Cybernetics)* **28** 925–35
[20] Purwar S, Kar I N and Jha A N 2007 On-line system identification of complex systems using Chebyshev neural networks *Appl. Soft Comput.* **7** 364–72
[21] Sahoo D M and Chakraverty S 2018 Functional link neural network learning for response prediction of tall shear buildings with respect to earthquake data *IEEE Trans. on Systems, Man, and Cybernetics: Systems* **48** 1–10
[22] Patra J C, Meher P K and Chakraborty G 2009 Nonlinear channel equalisation for wireless communication systems using Legendre neural networks *Signal Process.* **89** 2251–262
[23] Nanda S K and Tripathy D P 2011 Application of functional link artificial neural network for prediction of machinery noise in opencast mines *Adv. Fuzzy Systems* **2011** 831261
[24] Mall S and Chakraverty S 2016 Application of Legendre neural network for solving ordinary differential equations *Appl. Soft Comput. J.* **43** 347–56
[25] Gupta A, Chaudhary D K and Choudhury T 2018 Stock prediction using functional link artificial neural network (FLANN) *Proc. 2017 Int. Conf. on Computational Intelligence and Networks, CINE 2017* pp 10–6
[26] Park M Y and Hastie T 2007 $L_1$-regularisation path algorithm for generalised linear models *J. Royal Stat. Soc. B* **69** 659–77
[27] Cortes C, Mohri M and Rostamizadeh A 2012 L2 regularization for learning kernels arXiv:1205.2653
[28] Zurada J M 1992 *Introduction to Artificial Neural Systems* (Eagan, MN: West Publishing Co)
[29] Haykin S 2009 *Neural Networks and Learning Machines* 3rd edn (Noida: Pearson Education)
[30] Kingma D P and Ba J 2014 Adam: A method for stochastic optimization arXiv:1412.6980

# Chapter 14

# Polynomial based model for solving unconstrained optimization problem with smoothing parameters

**Bhubaneswari Mishra, S Chakraverty and Rohtas Kumar**

The idea behind data smoothing is that it can identify simplified changes to help in predicting different trends and patterns. Smoothing techniques may be utilized extensively to solve complex mathematical problems. In this chapter, smoothing approaches are utilized to produce and solve support vector machine (SVM) problems reformulated as unconstrained optimization problems with fourth-order polynomial function and smoothing parameters. This technique is known as polynomial smooth support vector machine (PSSVM). PSSVM is solved using a rapid Broyden–Fletcher–Goldfarb–Shanno (BFGS) method that converges globally. After completing a number of experiments and analyzing the numerical data generated by these experiments, it has been seen that PSSVM is a more effective classifier than SSVM (smooth support vector machine). It handles the SVM problems more quickly and efficiently than previous algorithms with far more generalization power.

## 14.1 Introduction

Support vector machine (SVM) is a unique machine learning technique based on the structural risk minimization principle [1, 2] that can identify optimal global solutions to issues. Due to its generalization performance, it has attracted significant interest. In fact, SVM has overtaken neural network to become the most common statistically learning technique. As a potent tool for supervised learning, SVM has been effectively used in a wide range of real-world issues, such as mechanical flaws diagnosis, text classification, bioinformatics, and financial applications [3–7]. SVM has a global optimum solution that is unique and avoids the curse of dimensionality. There have been advancements in all classification problems [8–12]. The objective function of SVM cannot be differentiated at zero [7–11]. In the pioneering work of Li *et al* [4] the integral of the sigmoid function was used to obtain a smooth support vector (SSVM) model.

doi:10.1088/978-0-7503-5067-9ch14  

Different well-known techniques may be used to solve it, making it a highly significant and vital result for SVM. In 2005 Yuan *et al* [14] suggested two polynomial functions: a smooth quadratic polynomial function and a smooth fourth polynomial function. They obtained a fourth polynomial smooth support vector machine (FPSSVM) model and a quadratic polynomial smooth support vector machine (QPSSVM) model.

Recently, the QPSSVM was utilized to anticipate the direction of movement of financial time series [6]. In the study of Xiong *et al* [15], a significant recursive formula is derived and a class of smooth functions are obtained utilizing the interpolation functions method. In 2009 Purnami *et al* [16] introduced the multiple knot spline (MKS) function, which replaced the plus function. Afterwards, they acquired a multiple knot spline–smooth support vector machine (MKS–SSVM). In the study of Yuan *et al* [17], a three-order spline function is employed to smooth the objective function of SVM's unconstrained optimization problem, resulting in the three-order spline-smooth support vector machine (TSSVM) model. Yuan *et al* [17] developed a first-order continuously differentiable smooth spline function for approximating the plus function using interpolation theory in 2011 and obtained a corresponding smooth SVM [18].

This study focuses on the SVM method for data classification. The classification SVM model may be expressed as a nonsmooth unconstrained optimization problem. However, at 0, the objective function is not differentiable. In our method, we make a small modification to the model and employ smooth approaches with polynomials that have been widely used to solve significant mathematical programming issues. Li *et al* [4] have used a smooth technique to address the resultant optimization issue. Use of the integral sigmoid function is suggested,

$$p(x, k) = x + \frac{1}{k}\log(1 + \varepsilon^{kx}),\ k > 0. \tag{14.1}$$

They obtained a refined SSVM model. Here, $\varepsilon$ represents the natural logarithm and is referred to as the smooth parameter. Later, the same smooth function was used to SSVM regression [13].

Two polynomial functions are presented in [14] as following

$$q(x, k) = \begin{Bmatrix} x, & \text{if } x > \frac{1}{k} \\ \frac{k}{4}x^2 + \frac{1}{2}x + \frac{1}{4k}, & \text{if } -\frac{1}{k} \leqslant x \leqslant \frac{1}{k} \\ 0, & \text{if } x < -\frac{1}{k} \end{Bmatrix} \tag{14.2}$$

$$h(x, k) = \begin{Bmatrix} x, & \text{if } x > \frac{1}{k} \\ -\frac{k^3}{16}\left(x + \frac{1}{k}\right)^3\left(x - \frac{3}{k}\right), & \text{if } -\frac{1}{k} \leqslant x \leqslant \frac{1}{k} \\ 0, & \text{if } x < -\frac{1}{k} \end{Bmatrix}. \tag{14.3}$$

## 14.2 Preliminaries of support vector machine (SVM)

Figure 14.1 describes the schematic representation of the SVM. In order to facilitate comprehension, some notations that will be used in subsequent sections are presented

- **m**: the number of days for which financial time series are examined;
- **n**: the count of movement direction impactors;
- **A**: the database is represented by an $m \times n$ matrix;
- **D**: the movement direction matrix (diagonal matrix) whose diagonals are 1 or $-1$;
- **ω**: the decision vector used to calculate the hyperplane for movement direction forecasting;
- **γ**: forecasting parameter (mathematically, the distance between the decision hyperplane and the origin is primarily determined by a forecasting parameter);
- **e**: a vector whose elements are 1;
- **x**: the input vector for forecasting the movement direction.

### 14.2.1 Direction index matrix [6]

If the taken database is given by an $m \times n$ matrix $A$, according to the membership of each point $A_i$ in the classes 1 or $-1$ as specified by a given $m \times m$ diagonal matrix $D$ with 1 or $-1$ diagonals, called the direction index matrix [6].

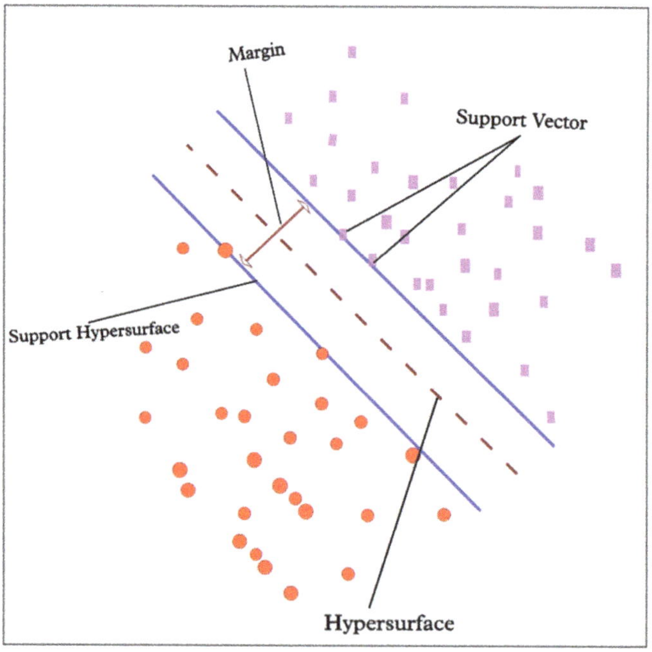


**Figure 14.1.** Schematic representation of SVM [20].

Let the linear separating hyper-surface is defined as follows which can be used to separate the $R^n$ in to two parts (left or right)

$$P = \{x | x \in R^n, x^T\omega = \gamma\} \tag{14.4}$$

The decision function can be defined as

$$f(x) = \begin{cases} +1, & x^T\omega \geqslant \gamma, \\ -1, & x^T\omega < \gamma. \end{cases} \tag{14.5}$$

The standard SVM can be written as [6]

$$\min_{(\omega,\gamma)\in R^{n1}} \frac{1}{2}\omega^T\omega, \ s.\ t.\ D(A\omega - e\gamma) \geqslant e. \tag{14.6}$$

The goal of SVM is to find the best solution $\omega$ and $\gamma$ for the given mathematical model. Adding a slack vectorand reformulating the standard SVM,

$$y \in R^n \quad \min_{(\omega,\gamma,y)\in R^{n+1+m}} ve^Ty + \frac{1}{2}\omega^T\omega, \tag{14.7}$$

s.t. $D(A\omega - e\gamma) + y \geqslant e$,

$y \geqslant 0$.

The first term in the objective function of (14.7) is the 1-norm of the slack variable $y$ with positive weight $v \in R$. The second term $\omega^T\omega$ is the square of the 2-norm of the vector $\omega$. Replacing the first term with the 2-norm vector $y$, the SVM problem (14.6) can be modified into the following form:

$$\begin{aligned} &\min_{(\omega,\gamma,y)\in R^{n+1+m}} \frac{v}{2} \parallel y \parallel_2^2 + \frac{1}{2}(\parallel \omega \parallel_2^2 + \gamma^2), \\ &\text{s.t. } D(A\omega - e\gamma) + y \geqslant e, \\ &y \geqslant 0. \end{aligned} \tag{14.8}$$

The solution of the problem (14.8), $y$ can be written as follows

$$y = (e - (D(A\omega - e\gamma)))_+. \tag{14.9}$$

Substituting $y$ in equation (14.8) yields,

$$\min_{(\omega,\gamma)\in R^{n+1}} \frac{v}{2} \|(e - (D(A\omega - e\gamma)))_+ \|_2^2 + \frac{1}{2}(\parallel \omega \parallel_2^2 + \gamma^2). \tag{14.10}$$

The above substitution converts the problem (14.8) into an equivalent unconstrained optimization problem. It is a strongly convex unconstrained minimization problem with unique solution.

In this work the proposed polynomial function referred from [14] can be stated as,

$$h(x, k) = \begin{cases} x, & \text{if } x > \frac{1}{\mathrm{k}}, \\ \frac{-k^3}{16}\left(x + \frac{1}{k}\right)^3\left(x - \frac{3}{k}\right), & \text{if } -\frac{1}{\mathrm{k}} \leqslant x \leqslant \frac{1}{k}, \\ 0, & \text{if } x < -\frac{1}{\mathrm{k}}. \end{cases} \tag{14.11}$$

The above mentioned is a fourth-order polynomial function.

The selection of polynomial function (14.11) is motivated by the fact that its objective function is not differentiable at zero. This prevents the application of several valuable optimization methods to solve the model (14.10). Essential requirement is that the objection function to be differentiable. To use algorithms to solve the model (14.10), we use the polynomial function (14.11) for approximation of the plus function in the model (14.10).

If the plus function in (14.10) is replaced by (14.11), a new SSVM is obtained called the PSSVM. It can be mathematically expressed as

$$\min_{(\omega,\gamma)\in R^{n+1}} F(\omega, \gamma, k) = \frac{v}{2} \| h\,(e - (D(A\omega - e\gamma)), k)\|_2^2 + \frac{1}{2}(\| \omega \|_2^2 + \gamma^2). \tag{14.12}$$

Objective function of (14.12) is convex and quadratic differential. Various efficient methods, such as the Newton method, the quasi-Newton method, and others, may be used to solve it. In this study, the BFGS algorithm approach is proposed.

### 14.2.2 Analysis of polynomial function

**Theorem 14.1** [6]

If the approximation function has formulation as (14.11), $h(x, k)$is continuous and quadratically differentiable for any given $k \in Z^+$.

Proof [6] Easily since,

$$\lim_{x \to +\frac{1}{k}} h\,(x, k) = \frac{1}{k}, \quad \lim_{x \to -\frac{1}{k}} h\,(x, k) = 0,$$

and

$$\left(\frac{d(h(x, k))}{dx}\right)\Big|\, x = -\frac{1}{k} = 0, \quad \left(\frac{d(h(x, k))}{dx}\right)\Big|\, x = \frac{1}{k} = 1.$$

Moreover, $h(x, k)$ satisfies

$$\left(\frac{d^2(h(x, k))}{dx^2}\right)\Big|\, x = -\frac{1}{k} = 0, \quad \left(\frac{d^2(h(x, k))}{dx^2}\right)\Big|\, x = \frac{1}{k} = 0.$$

These results demonstrate that it is continuous and quadratically differentiable. This concludes the proof [6].

**Theorem 14.2** [6]

For $x \in R$, the fourth polynomial function satisfies $h(x, k)^2 - x^2 \leqslant \frac{1}{19k^2}$. (14.14)

Proof [6]

From the definition the result $h(x, k) \geqslant x_+$ is obvious. In $\left(-\infty, -\frac{1}{k}\right)$ and $\left(\frac{1}{k}, \infty\right)$, the left hand side of (14.14) is equal to zero, so the inequality (14.14) holds.

For $x \in \left(0, \frac{1}{k}\right)$, we use the method of finding a maximized functional value and proof that the forth polynomial function satisfies

$$\max_{x \in \left(0, \frac{1}{k}\right)} \left(h(x, k)^2 - x_+^2\right) \leqslant \frac{1}{19k^2}$$

Theorem 14.2 shows that the difference between plus function and fourth-order polynomial function is very small. Hence, the fourth-order polynomial function can be used to smooth it.

### 14.2.3 Broyden–Fletcher–Goldfarb–Shanno (BFGS) algorithm for PSSVM [6]

BFGS technique [19] is well suited for unconstrained optimization with function and gradient value evaluation, and it may be stated that the BFGS approach is the most popular of the various quasi-Newton methods [6]. Figure 14.2 represents the detailed flow chart of BFGS algorithm to solve PSSVM.

Step 1: Given the control factor of algorithm accuracy $\epsilon$ and the initial smooth parameter $k$,
$H^0 = I$, $(\omega^0, \gamma^0) = p^0 \in R^{n+1}$, $\varepsilon = 10^{-8}$ and set $i = 0$;

Step 2: Evaluate $F^i = F(p^i, k)$ and $g^i = \nabla F(p^i, k)$;

Step 3: If $\| g^i \|_2^2 \leqslant \varepsilon$, then stop, and accept $p^i = (\omega^i \gamma^i)$ as the optimal solution of (14.11), else calculate $d^i = -H^i g^i$;

Step 4: Do a line search along direction $d^i$ to get a step length $\alpha^i > 0$; let
$p^{i+1} = p^i + \alpha^i d^i$, $s^i = p^{i+1} - p^i = -\alpha^i H^i g^i$, and evaluate
$F^{i+1}(k_j) = F\left(p^{i+1}, k_j\right)$, $g^{i+1} = \nabla F\left(p^{i+1}, k_j\right)$ and $y^i = g^{i+1} - g^i$;

Step 5: Update $H^i$ to get $H^{i+1}$: $H^{i+1} = \left(1 - \frac{s^i (y^i)^T}{(s^i)^T y^i}\right) H^i \left(1 - \frac{y^i (s^i)^T}{(s^i)^T y^i}\right) + \frac{s^i (s^i)^T}{(s^i)^T y^i}$;

Step 6: Set $i = i + 1$, go to step 3.

## 14.3 Result and discussion

For evaluation of performance of proposed BFGS algorithm, extensive simulation experiments are conducted and the margin, testing correctness (%), and objective function minimize values of PSSVM as well as the number of iterations required to get the super separating plane $P = \{x | x \in R^n, x^T \omega = \gamma\}$ are studied.

MATLAB R2021A is used to conduct tests on six randomly created databases with normal distribution on a PC with a 11th generation Intel Core i5 processor and 16 GB of RAM.

In the above mentioned results, '*m*' stands for number of points considered for classification, '*tn*' for the count of test points considered for SVM and '*n*' represents the dimension of data points or the number of features included in the database.

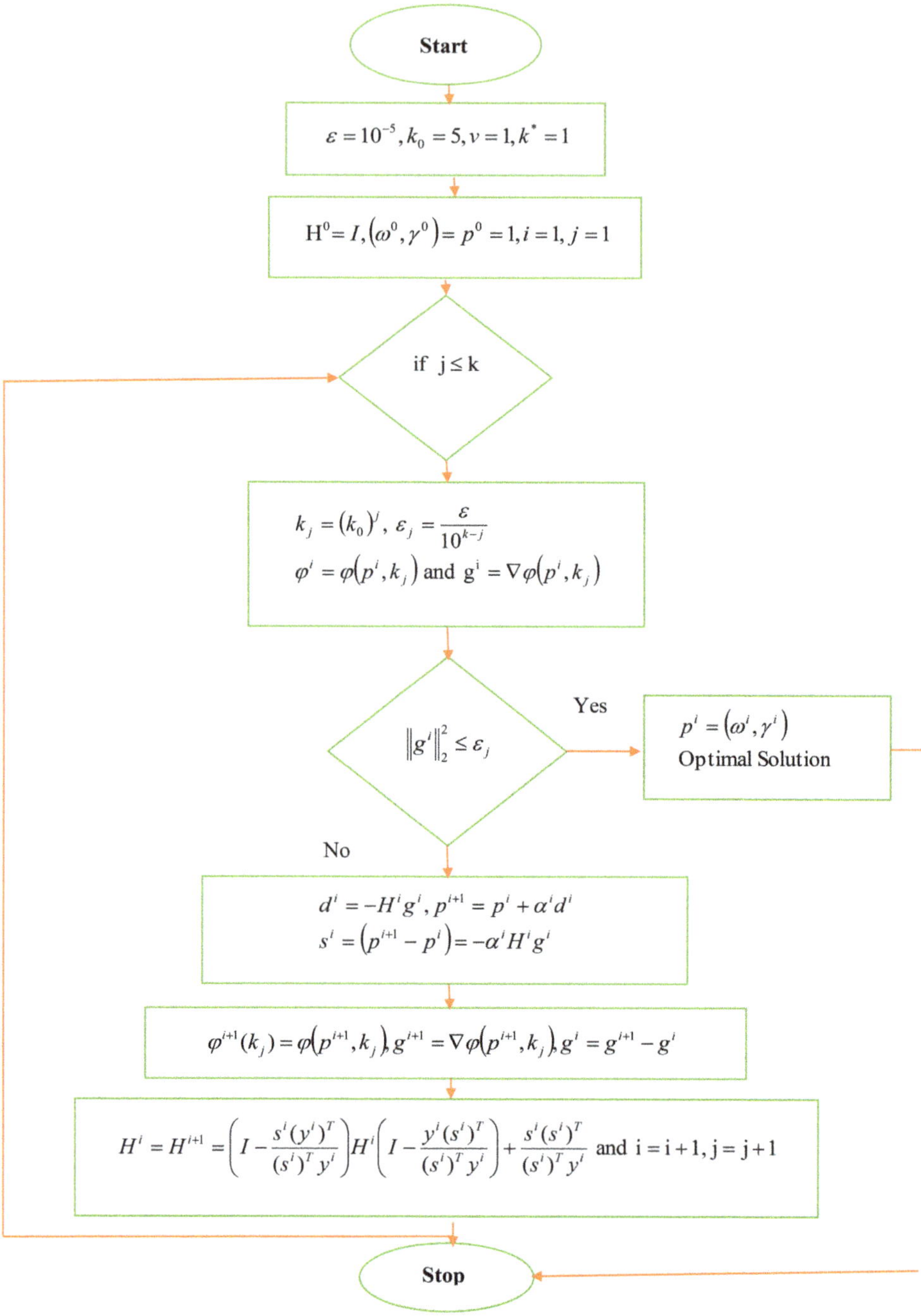


**Figure 14.2.** Flow chart of BFGS algorithm for solving PSSVM.

To illustrate the solving capability of PSSVM in larger problems, PSSVM is compared numerically with SSVM on randomly generated dataset. All used vectors are column vectors unless transposed. For vector $x$ in $R^n$, the plus function is

$(x_+)_i = \max\{0, x_i\}$, $i = 1,..., n$. A column vector of 1s with arbitrary dimension is denoted as $e$. All parameters used in the algorithm are chosen for optimal performance. In the experiment, initial values of the parameters are taken as $k_0 = 5$, $v = 1$, $\varepsilon = 10^{-8}$. QPSSVM represents quadratic polynomial smooth support vector machine, FPSSVM stands for fourth polynomial smooth support vector machine and SSVM is the smooth support vector machine.

Table 14.1 is the representation of results obtained from the experiments. Based on the data in table 14.1, the margins generated by the BFGS algorithm with

**Table 14.1.** Experimental results.

| Dataset size | | | | [$Z_1$] margin [$Z_2$] test correctness (%) [Z3] iteration number [Z4] the maximize value of objective | |
|---|---|---|---|---|---|
| | | | | BFGS method | |
| *m* | *n* | *tn* | | FPSSVM | SSVM |
| 100 | 20 | 1000 | $Z_1$ | 26.0953 | 21.5829 |
| | | | $Z_2$ | 92.68 | 92.30 |
| | | | $Z_3$ | 252 | 220 |
| | | | $Z_4$ | 0.0035 | 0.0058 |
| 500 | 20 | 1000 | $Z_1$ | 6.0852 | 5.6927 |
| | | | $Z_2$ | 98.60 | 98.70 |
| | | | $Z_3$ | 208 | 210 |
| | | | $Z_4$ | 0.0587 | 0.0749 |
| 1000 | 20 | 1000 | $Z_1$ | 2.5998 | 2.3589 |
| | | | $Z_2$ | 99.21 | 99.00 |
| | | | $Z_3$ | 227 | 242 |
| | | | $Z_4$ | 0.3351 | 0.38239 |
| 100 | 100 | 1000 | $Z_1$ | 92.1086 | 68.4456 |
| | | | $Z_2$ | 74.40 | 74.40 |
| | | | $Z_3$ | 701 | 382 |
| | | | $Z_4$ | $2.3744e^{-4}$ | $4.5879e^{-4}$ |
| 500 | 100 | 1000 | $Z_1$ | 25.1456 | 19.3682 |
| | | | $Z_2$ | 92.73 | 92.04 |
| | | | $Z_3$ | 1024 | 768 |
| | | | $Z_4$ | 0.0062 | 0.0074 |
| 1000 | 100 | 1000 | $Z_1$ | 12.9870 | 10.8427 |
| | | | $Z_2$ | 94.92 | 95 |
| | | | $Z_3$ | 939 | 759 |
| | | | $Z_4$ | 0.0121 | 0.0165 |

FPSSVM are greater than those obtained by SSVM. This circumstance cannot be altered by raising the scale of the data. On the other hand, the objective function minimize values obtained by the BFGS algorithm with FPSSVM are smaller than those obtained with SSVM. This circumstance cannot alter regardless of the amount of data collected.

## 14.4 Conclusion

This study proposes PSSVM, which is a traditional quadratic program associated with SVM expressed as a reformulation of smooth unconstrained optimization problem. The BFGS algorithm is used to solve the PSSVM. Several experiments are performed and from the numerical results it is concluded that PSSVM is a better classifier than SSVM. It is more effective and faster than the previous methods for solving SVM with better generalization ability.

## Acknowledgement

The first author would like to acknowledge CRISIL Ltd, Powai, Mumbai-400076, India for funding and support to carry out the present research work.

## References

[1] Burges C J and Crisp D 1999 Uniqueness of the SVM solution *Advances in Neural Information Processing Systems* **12** 223–9

[2] Zhang Y and Cai J 2013 A support vector classifier based on vague similarity measure *Math. Prob. Eng.* **2013** 928054

[3] Baccarini L M R, e Silva V V R, de Menezes B R and Caminhas W M 2011 SVM practical industrial application for mechanical faults diagnostic *Expert Syst. Appl.* **38** 6980–984

[4] Li K, Xie J, Sun X, Ma Y and Bai H 2011 Multi-class text categorization based on LDA and SVM *Procedia Engineering* **15** 1963–967

[5] Spinosa E J and Carvalho A C 2005 Support vector machines for novel class detection in bioinformatics *Genet Mol Res* **4** 608–15

[6] Yuan Y 2013 Forecasting the movement direction of exchange rate with polynomial smooth support vector machine *Math. Comput. Modell.* **57** 932–44

[7] Zheng J and Lu B L 2011 A support vector machine classifier with automatic confidence and its application to gender classification *Neurocomputing* **74** 1926–935

[8] Xin Z H O U, Ying W U and Wang D L 2011 Application of kernel methods in signals modulation classification *J. China Universities of Posts and Telecommunications* **18** 84–90

[9] Zhao Z C 2012 Combining SVM and CHMM classifiers for porno video recognition *J. China Universities of Posts and Telecommunications* **19** 100–06

[10] Lin H J and Yeh J P 2009 Optimal reduction of solutions for support vector machines *Appl. Math. Comput.* **214** 329–35

[11] Christmann A and Hable R 2012 Consistency of support vector machines using additive kernels for additive models *Comput. Stat. Data Anal.* **56** 854–73

[12] Shao Y H and Deng N Y 2012 A coordinate descent margin based-twin support vector machine for classification *Neural Netw.* **25** 114–21

[13] Lee Y J and Mangasarian O L 2001 SSVM: a smooth support vector machine for classification *Comput. Optim. Appl.* **20** 5–22

[14] Yuan Y B, Yan J and Xu C X 2005 Polynomial smooth support vector machine (PSSVM) *Jisuanji Xuebao (Chin. J. Comput.)* **28** 9–17

[15] Xiong J Z, Hu J L, Yuan H Q, Hu T M and Li G M 2007 Research on a new class of functions for smoothing support vector machines *Acta Electron. Sin.* **35** 366

[16] Purnami S W, Embong A, Zain J M and Rahayu S P 2009 A new smooth support vector machine and its applications in diabetes disease diagnosis *J. Computer Sci.* **5** 1003

[17] Yuan Y, Fan W and Pu D 2007 Spline function smooth support vector machine for classification *J. Industrial & Management Optimization* **3** 529

[18] Yuan B, Zhang W and Wu H 2010 New solution method to smoothing support vector machine with one control parameter smoothing function *2010s WRI Global Congress on Intelligent Systems* vol 1 (Piscataway, NJ: IEEE) pp 153–56

[19] Yuan Y X 1991 A modified BFGS algorithm for unconstrained optimization *IMA J. Numer. Anal.* **11** 325–32

[20] Suthaharan S 2016 Machine learning models and algorithms for big data classification. *Integr. Ser. Inf. Syst.* **36** 1–12

# Chapter 15

## Interval root finding and interval polynomials: methods and applications in science and engineering

**Hend Dawood and Yasser Dawood**

*Everything is vague to a degree you do not realize till you have tried to make it precise.*

—Bertrand Russell (1872–1970)

Polynomial systems are at the heart of a wide range of scientific fields. Multiple forms of polynomials are often involved in attempts to reduce currently open or hard problems to problems already solved. Because of their ever-increasing importance in many practical applications, polynomials span the entire spectrum of various disciplines of mathematics, science and engineering. In practice, a recurring problem is to determine the roots of polynomials under parametric uncertainty or imprecision. Interval mathematics is a subtle body of principles and methods for manipulating systems involving quantifiable uncertainties. Interval methods have proven successful and reliable for computing guaranteed enclosures of roots of polynomials under parametric uncertainty. This 'self-validating' feature of interval analysis makes it competitive and preferable to the ordinary approximation methods in many practical applications. In this connexion, by the pursuit of the quest of reliable knowledge amidst uncertainty, the *raison d'être* of the present work is a systematic investigation of *uncertain polynomials*. We begin by laying out a rigorous algebraic foundation for ordinary polynomials and some of their generalizations, and then, we go further to define two new generalizations of ordinary and trigonometric polynomials, namely, *generalized n-adic polynomials* and *n-adic* S*-polynomials*. These two generalizations are so framed as to provide a rigorous mathematical foundation for interval and set-valued polynomials. Next, we formalize the theories of interval algebra and interval polynomials and show how to obtain guaranteed interval enclosures of families of generalized real polynomials. Afterwards, we

doi:10.1088/978-0-7503-5067-9ch15

describe how to use infinite polynomials with Taylor models to compute finer interval enclosures. Moreover, we examine some of the interval methods for root finding, with special attention is paid to the interval branch-and-prune method. Finally, we put forward a more refined interval branch-and-prune algorithm. In order to illustrate the significance of our algorithm, we compute a finer interval enclosure for an example introduced by Eldon Hansen and G William Walster in 2003. We show that our interval enclosure is sharper than the optimal result given by Hansen and Walster. The algorithms of this work are coded using the software package, InCLosure. Many numerical examples are given, showing that all the results are guaranteed interval enclosures.

## 15.1 Introduction

Polynomial systems are most fundamental in a wide range of fields. Because of their ever-increasing importance in many practical applications, polynomials are heavily and actively researched across various disciplines of mathematics, science and engineering. Over the past decades, research into the study of different types of polynomials has played an indispensable role and reached its most advanced and sophisticated forms in several areas, such as numerical analysis, algebraic geometry, tropical geometry, statistics, dynamical systems, global optimization, control theory, artificial intelligence, computer-aided design, structural dynamics, and quantum mechanics, to mention a few (see, e.g., [1–8]).

In practice, a recurring problem is to determine the roots of polynomials under parametric uncertainty or imprecision. Interval mathematics is a subtle body of principles and methods for manipulating systems involving quantifiable uncertainties. Interval methods have proven successful and reliable for computing guaranteed enclosures of roots of polynomials under parametric uncertainty. This 'self-validating' feature of interval analysis makes it competitive and preferable to the ordinary approximation methods in many practical applications. It comes as no surprise that nowadays interval mathematics has many fruitful applications in diverse areas of science and engineering (see, e.g., [9–26]).

In this connection, and by the pursuit of the quest of reliable knowledge amidst uncertainty, the *raison d'être* of the present chapter is a systematic investigation of *uncertain polynomials*. The roadmap of this chapter is as follows. In section 15.2, we set the stage by formalizing some purely set-theoretical and algebraic ingredients we shall need. In section 15.3, we lay out a rigorous algebraic foundation for ordinary real polynomials. Section 15.4 presents an overview of the generalizations of polynomial systems. We explore three types of generalized and special polynomials, namely, *infinite polynomials*, *trigonometric and cylindrical polynomials*, and *polynomials over semirings*. After that, in section 15.5, we go further to define two new generalizations of ordinary and trigonometric polynomials, namely, *generalized n-adic polynomials* and *n-adic* S-*polynomials*. These two generalizations are so framed as to provide a rigorous mathematical foundation for interval and set-valued polynomials. In section 15.7, we formalize the theories of interval algebra and interval polynomials and show how to obtain guaranteed interval enclosures of families of generalized real polynomials. Afterwards, section 15.8 is devoted to describing how to use infinite polynomials with Taylor models to compute finer interval enclosures. Moreover, in section 15.9, we examine some of the interval methods for root finding, with special attention is paid to the interval branch-and-prune method. Furthermore, we put forward a more refined interval branch-and-prune algorithm. In order to illustrate the significance of our algorithm, we recompute a finer interval enclosure for an example introduced by Eldon Hansen and G. William Walster in [27]. We show that our interval enclosure is sharper than the optimal result given by Hansen and Walster. Finally, we conclude this work by stating briefly our major results and indicating directions for future research.

Some of the methods presented in this chapter are well known; some are believed to be new. Besides presenting the basic concepts and the mathematical underpinnings, we also discuss the computational algorithms, and explore the impact of systems of uncertain polynomials in science and engineering. Moreover, we provide an implementation of the algorithms, along with a number of numerical examples, illustrating the efficiency and significance of the methods and algorithms presented in the body of this chapter. The algorithms of this work are coded using the software package, InCLosure[1] [28]. The given numerical examples show that all the results are guaranteed interval enclosures.

## 15.2 On sets, relations, and structures: setting the stage

In order to make this chapter self-contained and reachable to a wide audience, this section introduces some basic set-theoretical and algebraic ingredients we shall need. Certain features of the concepts to be presented here are obvious and this section is then intended to serve as a propaedeutic for all the sections to follow. For further details about the notions introduced here, the reader may consult, e.g., [3, 16, 17, 19, 29, 30].

First, we rehearse some basic set-theoretical definitions. Let $\mathcal{U}^n$ be the $n$th Cartesian power of a set $\mathcal{U}$. A set $\mathfrak{R} \subseteq \mathcal{U}^n$ is an $n$-ary (finitary) relation on $\mathcal{U}$ iff $\mathfrak{R}$ is a binary relation from $\mathcal{U}^{n-1}$ to $\mathcal{U}$. Thus, for $\mathbf{v} = (x_1, \dots, x_{n-1}) \in \mathcal{U}^{n-1}$ and $y \in \mathcal{U}$, an $n$-ary relation $\mathfrak{R}$ is characterized to be $\mathfrak{R} \subseteq \mathcal{U}^n = \{(\mathbf{v}, y) | \mathbf{v} \in \mathcal{U}^{n-1} \wedge y \in \mathcal{U}\}$. In this sense, an $n$-ary relation is a binary relation and its *domain*, *range*, *field*, and *converse* are defined to be, respectively $\operatorname{dom}(\mathfrak{R}) = \{\mathbf{v} \in \mathcal{U}^{n-1} | (\exists\, y \in \mathcal{U})(\mathbf{v}\mathfrak{R}y)\}$, $\operatorname{ran}(\mathfrak{R}) = \{y \in \mathcal{U} | (\exists\, \mathbf{v} \in \mathcal{U}^{n-1})(\mathbf{v}\mathfrak{R}y)\}$, $\operatorname{fld}(\mathfrak{R}) = \operatorname{dom}(\mathfrak{R}) \cup \operatorname{ran}(\mathfrak{R})$, and $\widehat{\mathfrak{R}} = \{(y, \mathbf{v}) \in \mathcal{U}^n | \mathbf{v}\mathfrak{R}y\}$. Obviously $y\widehat{\mathfrak{R}}\mathbf{v} \Leftrightarrow \mathbf{v}\mathfrak{R}y$ and $\widehat{\widehat{\mathfrak{R}}} = \mathfrak{R}$ (see, [17, 19, 31]).

We then characterize some algebraic structures of particular importance to our purpose (see, [17, 19]).

**Definition 15.2.1 (Ringoid)** *A ringoid (or a ring-like structure) is a structure* $\mathfrak{R} = \langle \mathcal{R}; +_{\mathcal{R}}, \times_{\mathcal{R}} \rangle$ *with* $+_{\mathcal{R}}$ *and* $\times_{\mathcal{R}}$ *are total binary operations on the universe set* $\mathcal{R}$.

[1] InCLosure (Interval enCLosure) is an environment and a language for reliable scientific computing. The latest version of InCLosure is accessible for free download from CERN's archive via https://doi.org/10.5281/zenodo.2702404.

*The operations* $+_{\mathcal{R}}$ *and* $\times_{\mathcal{R}}$ *are called respectively the addition and multiplication operations of the ringoid* $\mathfrak{R}$.

**Definition 15.2.2 (S-Ringoid)** *An S-ringoid (or a subdistributive ringoid) is a ringoid that satisfies at least one of the following subdistributive criteria.*

(i) $(\forall x, y, z \in \mathcal{R})(x \times_{\mathcal{R}} (y +_{\mathcal{R}} z) \subseteq x \times_{\mathcal{R}} y +_{\mathcal{R}} x \times_{\mathcal{R}} z)$,
(ii) $(\forall x, y, z \in \mathcal{R})((y +_{\mathcal{R}} z) \times_{\mathcal{R}} x \subseteq y \times_{\mathcal{R}} x +_{\mathcal{R}} z \times_{\mathcal{R}} x)$.

Criteria (i) and (ii) in the above definition are called respectively left and right subdistributivity (or S-distributivity).

**Definition 15.2.3 (Semiring)** *Let* $\mathfrak{R} = \langle \mathcal{R}; +_{\mathcal{R}}, \times_{\mathcal{R}} \rangle$ *be a ringoid.* $\mathfrak{R}$ *is said to be a semiring iff*

(i) $\langle \mathcal{R}; +_{\mathcal{R}} \rangle$ *is a commutative monoid with identity element* $0_{\mathcal{R}}$,
(ii) $\langle \mathcal{R}; \times_{\mathcal{R}} \rangle$ *is a monoid with identity element* $1_{\mathcal{R}}$,
(iii) *Multiplication,* $\times_{\mathcal{R}}$, *left and right distributes over addition,* $+_{\mathcal{R}}$,
(iv) $0_{\mathcal{R}}$ *is an absorbing element for* $\times_{\mathcal{R}}$.

*A commutative semiring is one whose multiplication is commutative.*

**Definition 15.2.4 (S-Semiring)** *An* S-*semiring (or a subdistributive semiring) is an* S-*ringoid that satisfies criteria* (i), (ii), *and* (iv) *in definition* 15.2.3. *A commutative* S-*semiring is one whose multiplication is commutative.*

The definition of an S-semiring is so framed as to provide a rigorous algebraic basis for *set-valued algebras* (*uncertain structures*), including the algebras of intervals and interval polynomials. One might think of an S-semiring as a generalization of the concept of a *near*-semiring (for further details on near-semirings, the reader can refer to [32–34]).

## 15.3 The ring of polynomials and real polynomial arithmetic

In this section, we lay out a rigorous algebraic foundation for ordinary real polynomials. We begin, in section 15.3.1, by formalizing the fundamental notions of ordinary real polynomials. Thereupon, in section 15.3.2, we put on a systematic basis the algebra of real polynomials (for further background, the reader may consult, e.g., [1, 3–5, 8]).

### 15.3.1 Monomials, polynomials, and polynomial equations

In the following we will denote by $\mathbb{R}_{\langle x \rangle}$ the set of all univariate real-valued functions. The symbols $f$, $g$, and $h$ will be employed as variable symbols ranging over the elements of $\mathbb{R}_{\langle x \rangle}$.

It is natural to begin with the definition of a power function.

**Definition 15.3.1 (Power Function)** *A real power function is a single-term real function of the form* $f(x) = ax^r$, **where** $a \neq 0$ *and* $r$ *are real constants called respectively the coefficient and the exponent of* $f$, **and** $x$ *is the indeterminate.*

For example, $f(x) = 6$, $f(x) = 3x^{0.69}$, $f(x) = 1/x^3$, $f(x) = \sqrt[6]{x}$ are power functions, whereas $f(x) = 3^x$ is not.

**Definition 15.3.2 (Ordinary Real Polynomial)** *Let* $i \in \{0, 1, \ldots, n\}$ *be a nonnegative integer. An ordinary real polynomial (a univariate real polynomial, or in short, a real polynomial) of degree n is a real function of the form*

$$f_{a,n}(x) = \sum_{i=0}^{n} a_i x^i = a_0 + a_1 x + a_2 x^2 + \cdots + a_i x^i + \cdots + a_n x^n.$$

*The terms* $a_i x^i$ *and the constants* $a_i$ *are called respectively the monomials and the numerical coefficients of* $f_{a,n}$.

The name 'ordinary' in definition 15.3.2 serves to distinguish univariate real polynomials from the generalized polynomials to be discussed in sections 15.4 and 15.5. Saying that the polynomial $f_{a,n}(x)$ in the preceding definition is of degree $n$ implies that the coefficient $a_n$ is not zero. Terms with the same integral powers of $x$, e.g., $6x^3$ and $9x^3$, are known as *like* or *similar* terms. A one-term polynomial is called a *monomial*, a two-terms polynomial is called a *binomial*, and a three-terms polynomial is called a *trinomial*, and so forth. For example, 0, 1, and $9x^6$ are monomials, whereas $3x^6 + 6x^9 + 36$ is a trinomial.

Denote by $\mathsf{P}_{\mathbb{R}}$ the set of all univariate real-valued polynomials (polynomials over $\mathbb{R}$). Obviously $\mathsf{P}_{\mathbb{R}} \subset \mathbb{R}_{\langle x \rangle}$. Let $f_{a,n}$ denote a polynomial of degree $n$ and coefficients $a_i$, where $0 \leqslant i \leqslant n$. Analogously, in the sequel, the symbols $f_{a,n}$, $g_{b,m}$, and $h_{c,l}$ (with or without subscripts) will be deployed as variable symbols ranging over the set $\mathsf{P}_{\mathbb{R}}$ of real polynomials. That is, we will write

$$f_{a,n}(x) = \sum_{i=0}^{n} a_i x^i = a_0 + a_1 x + a_2 x^2 + \cdots + a_i x^i + \cdots + a_n x^n,$$

$$g_{b,m}(x) = \sum_{j=0}^{m} b_j x^j = b_0 + b_1 x + b_2 x^2 + \cdots + b_j x^j + \cdots + b_m x^m,$$

$$h_{c,l}(x) = \sum_{k=0}^{l} c_k x^k = c_0 + c_1 x + c_2 x^2 + \cdots + c_k x^k + \cdots + c_l x^l.$$

Also, in a manner analogous to real numbers, we will use the letters $a$, $b$, and $c$ to designate constant polynomials. In particular, we will use $0_{\mathsf{P}}$ and $1_{\mathsf{P}}$ to designate, respectively, the *zero* and *unit* polynomials.

Equality of univariate real-valued polynomials is characterized in the following definition.

**Definition 15.3.3 (Equality of Univariate Polynomials)** *Let* $f_{a,n}(x) = \sum_{i=0}^{n} a_i x^i$ *and* $g_{b,m}(x) = \sum_{j=0}^{m} b_j x^j$ *be in* $\mathsf{P}_{\mathbb{R}}$. *Then*

$$f_{a,n} = g_{b,m} \Leftrightarrow n = m \wedge (\forall i)(\forall j)\big(i = j \Rightarrow a_i = a_j\big).$$

Most polynomials occurring in science and engineering are real polynomials. However, in some practical situations polynomials may be complex. Definition 15.3.2 can be extended, via considering complex coefficients, to address *complex polynomials.*

A real polynomial equation and its solution set are characterized in the following definition.

**Definition 15.3.4 (Solution Set of a Polynomial Equation)** *Let* $f_{a,n}$ *be an ordinary real polynomial. The equation* $f_{a,n}(x) = 0$ *is called a real polynomial equation whose solution set is defined to be*

$$f_{a,\,n}^{-1}(x) = \{x \in \mathbb{R} | f_{a,n}(x) = 0\}.$$

For instance the solution set of the polynomial equation $4x^2 - 1 = 0$ is $\{1/2, -1/2\}$.

One may generalize definition 15.3.2 to involve more than one variable. This gives rise to multivariate polynomials.

**Definition 15.3.5 (Multivariate Real Polynomial)** *Let* $i \in \{1, \ldots, n\}$ *and* $j \in \{1, \ldots, \alpha\}$ *be nonnegative integers. A finitary multivariate real polynomial in n variables and* $\alpha$ *terms is a multivariate real function* $f_{n,\alpha}\colon \mathbb{R}^n \to \mathbb{R}$ *that can be written as the finite sum*

$$f_{n,\alpha} = \sum_{j=1}^{\alpha} t_j = t_1 + t_2 + \cdots + t_j + \cdots + t_\alpha.$$

*Each term* $t_j$ *is called a multivariate monomial of* $f_{n,\alpha}$ *and can be expressed as the product* $t_j = a_j \prod_{k=1}^{l \leqslant n} x_k^{\beta_k}$, *where* $a_j$ *are real constants,* $x_k$ *are real variables, and* $\beta_k$ *are integral exponents. The degree of each monomial* $t_j$ *is defined to be* $B_j = \beta_1 + \cdots + \beta_k + \cdots + \beta_l$, *with* $l \leqslant n$, *and the degree of the multivariate polynomial* $f_{n,\alpha}$ *is defined to be* $\delta_f = \max_{1 \leqslant j \leqslant \alpha}\{B_j\}$.

For instance, $f_{3,2} = 3x^3y^6z^9 + 6x^3y^6$ is a 3-variable polynomial with 2 terms and a degree $\delta_f = 18$. We will use the symbol $\mathsf{P}_{\mathbb{R}}^n$ to designate the set of all multivariate real-valued polynomials (polynomials over $\mathbb{R}^n$). In a way analogous to the one-variable case, the solution set of an $n$-variable polynomial equation $f(x_1, \ldots, x_n) = 0$ is defined to be

$$f^{-1}(x_1, \ldots, x_n) = \{(x_1, \ldots, x_n) \in \mathbb{R}^n | f(x_1, \ldots, x_n) = 0\}.$$

Two related notions are those of multinomials and rational functions. A *multinomial* is any algebraic combination of more than one term. Not every multinomial is a polynomial, since terms like $e^x$ and $\sqrt[3]{x}$ can be included in a multinomial. A *rational* real-valued function is a function obtained by means of a finite number of the basic real algebraic operations $\circ_{\mathbb{R}} \in \{+, \times\}$ and $\diamond_{\mathbb{R}} \in \{-, {}^{-1}\}$.

Let us illustrate by example how to evaluate real polynomials in InCLosure.

**Example 15.3.1 (InCLosure Evaluation of Real Polynomials)**
Consider the following real polynomials

$$f(x) = 3x^6 + 6x^9 + 36,$$
$$g(x, y) = 3x^3y^6 + 6x^6y^9.$$

To evaluate $f(x)$ at $x = 3$,

we write the following InCLosure command

```
rEval "3*x946+6*x949+36" "x=3"
```

which will result in 120321. And to evaluate $g(x, y)$ at $(x, y) = (3, 6)$, we write the following InCLosure command

```
rEval "3*x943*y946+6*x946*y949" "x=3" "y=6"
```

which will result in 4.408 362 144E10.

### 15.3.2 The algebra of real polynomials

On grounds of the notions characterized in section 15.3.1, we are now ready to formally construct the arithmetic of ordinary real polynomials. We begin by characterizing addition and multiplication on the set $\mathsf{P}_{\mathbb{R}}$ of real polynomials. Then, we provide some examples to illustrate how to perform real polynomial arithmetic in InCLosure.

**Theorem 15.3.1 (Polynomial Addition)** *Addition is a total operation on the set* $\mathsf{P}_{\mathbb{R}}$ *of real polynomials. That is*

$$\left(\forall f_{a,n} \in \mathsf{P}_{\mathbb{R}}\right)\left(\forall g_{b,m} \in \mathsf{P}_{\mathbb{R}}\right)\left(\exists h_{c,l} \in \mathsf{P}_{\mathbb{R}}\right)\left(h_{c,l} = f_{a,n} +_{\mathsf{P}} g_{b,m}\right),$$

*where* $c_i = a_i + b_i$ *and* $l = \max\{m, n\}$.

**Theorem 15.3.2 (Polynomial Multiplication)** *Multiplication is a total operation on the set* $\mathsf{P}_{\mathbb{R}}$ *of real polynomials. That is*

$$\left(\forall f_{a,n} \in \mathsf{P}_{\mathbb{R}}\right)\left(\forall g_{b,m} \in \mathsf{P}_{\mathbb{R}}\right)\left(\exists h_{c,l} \in \mathsf{P}_{\mathbb{R}}\right)\left(h_{c,l} = f_{a,n} \times_{\mathsf{P}} g_{b,m}\right),$$

*where* $c_i = a_0 b_i + a_1 b_{i-1} + \cdots + a_i b_0$ *and* $l = m + n$.

It can be easily proved that the polynomial structure $\langle \mathsf{P}_{\mathbb{R}}; +_{\mathsf{P}}, \times_{\mathsf{P}}; 0_{\mathsf{P}}, 1_{\mathsf{P}} \rangle$ is a ring.

The following example will illustrate how to perform polynomial arithmetic in InCLosure.

**Example 15.3.2 (Polynomial Arithmetic in InCLosure)** Consider the following three real polynomials

$$f(x) = x^2 + x^3 - 2,$$
$$g(x) = x^5 - x,$$
$$h(x) = -5x^4.$$

To compute the sum $f + g + h$, we write the following InCLosure command

```
PAdd "x^2+x^3-2" "x^5-x" "-5*x^4" "x"
```

which will result in

$$(-2.0) + (-1.0)*x + x\wedge 2 + x\wedge 3 + (-5.0)*x\wedge 4 + x\wedge 5.$$

And to compute the product $f \times g \times h$, we write the following InCLosure command

```
PMult "x^2+x^3-2" "x^5-x" "-5*x^4" "x"
```

which will result in

$$(-10.0)*x\wedge 5 + (5.0)*x\wedge 7 + (5.0)*x\wedge 8 + (10.0)*x\wedge 9 + (-5.0)*x\wedge 11 + (-5.0)*x\wedge 12.$$

## 15.4 Beyond ordinary polynomials: generalizations of polynomials

Despite the fact that real and complex polynomials are usually of most concern in scientific and engineering applications, there is no reason to restrict attention to polynomials over real or complex numbers. There are many ways to generalize a polynomial algebra and it is therefore possible to study polynomial algebras over other algebraic structures. For example, in areas like cryptography and computer algebra, where obtaining exact solutions is mandatory, polynomials over a finite field or the field of rational numbers are crucial. Also, in electromagnetism, kinematics and signal processing, trigonometric and cylindrical polynomials are of utmost importance. Moreover, there are many types of special polynomials that have worthwhile impact on both mathematics and practical applications. Among these, one may mention: Chebyshev polynomials, orthogonal polynomials, cyclotomic polynomials, Legendre polynomials, Laguerre polynomials, Hermite polynomials, Bessel polynomials, Bernoulli polynomials, Faulhaber polynomials, Euler polynomials, and Mittag-Leffler polynomials. There is an extensive literature on generalized and special polynomials and their applications. The interested reader may consult, e.g., [2, 5–8].

Here, we will not delve deep into these generalizations. This section is devoted to describing briefly three types of generalized and special polynomials, namely, *infinite polynomials*, *trigonometric and cylindrical polynomials*, and *polynomials over semirings*. In section 15.5, we will present two new generalizations of polynomials, namely, *generalized n-adic polynomials* and *n-adic* S-*polynomials*.

### 15.4.1 Formal power series: infinite polynomials

An important direction of generalizing an ordinary polynomial is to allow *infinitely many* nonzero terms. Such generalization results in the creation of the so-called *formal power series* or *infinite polynomials*. One major example of infinite polynomials is the case of the Taylor series. For example, the Taylor series of the exponential function $e^x$ may be written as the infinite sum

$$e^x = 1 + \frac{x^2}{2!} + \frac{x^3}{3!} + \cdots = \sum_{n=0}^{\infty} \frac{x^n}{n!}.$$

In section 15.8 (Taylor models), we will discuss how to compute sharper interval enclosures with infinite polynomials. For further details on infinite polynomials and related concepts, the reader can refer to [1, 3, 4, 6].

### 15.4.2 Trigonometric and cylindrical polynomials

Trigonometric polynomials play an indispensable role in several areas of pure and applied mathematics, and they arise in a wide range of fields: Fourier analysis, approximation theory, kinematics, geometry, and signal processing, to mention a few. In physics and engineering, for example, many problems are formulated using trigonometric polynomials (see, e.g., [2, 7, 8]).

A *real* trigonometric polynomial is a real-valued function of the form

$$f_{\mathrm{TR}}(x) = \sum_{\alpha=0}^{n} (a_\alpha \cos(\alpha x) + b_\alpha \sin(\alpha x)),$$

where $a_\alpha$ and $b_\alpha$ are real constants. By allowing the coefficients $a_\alpha$ and $b_\alpha$ to be *complex* numbers, the above function gives rise to a *complex trigonometric polynomial*. Another way of generalization is to allow *cylindrical functions*.

In section 15.5, we will present two new generalizations of ordinary and trigonometric polynomials, namely, *generalized n-adic polynomials* and *n-adic* S-*polynomials*.

### 15.4.3 Polynomials over semirings

One direction to generalize a polynomial ring is to allow the definition of polynomials over a semiring. Let $\mathfrak{R} = \langle \mathcal{R}; +_{\mathcal{R}}, \times_{\mathcal{R}} \rangle$ be a semiring. A polynomial over the semiring $\mathfrak{R}$ is a polynomial all of whose coefficients belong to the universe set $\mathcal{R}$ of the semiring. Such generalization results in the creation of the so-called a *polynomial rig*. Polynomials over semirings are crucial in areas like formal semantics, machine learning, and automata theory (see, e.g., [1, 4, 6]).

## 15.5 Two more generalizations of polynomials: *n*-adic polynomials and S-polynomials

As mentioned in section 15.4, there is no reason to restrict our attention to polynomials over real or complex numbers. Here we will go even further. Since the main business of this chapter is to compute guaranteed enclosures of roots of generalized polynomials under parametric uncertainty, it is then necessary to lay out a rigorous algebraic foundation for *uncertain polynomials* (*interval polynomials* in particular, and *set-valued polynomials* in general). Thus, the major goal of this section is to present two new generalizations of ordinary and trigonometric polynomials, namely, *generalized n-adic polynomials* and *n-adic* S-*polynomials.*

### 15.5.1 Generalized *n*-adic polynomials: polynomials of polynomials

The definitions of ordinary and trigonometric polynomials can be extended recursively to address new types of special and generalized polynomials. The next characterization makes this precise.

**Definition 15.5.1 (Generalized *n*-adic Polynomials)** *For a nonnegative integer* $i \in \{1, \dots, l\}$, *let* $f_i$ *be polynomials (ordinary, trigonometric, or otherwise). For an integer* $n \geqslant 0$, *a generalized n-adic polynomial* $h_n$ *(in short, a generalized polynomial) can be characterized recursively as follows.*

(i) $n = 0 \Rightarrow h_0 = \prod_{i=1}^{l} f_i$,

(ii) $n > 0 \Rightarrow h_n = \sum_{\alpha=1}^{n} (a_\alpha \mathrm{U}^{k_\alpha}(\alpha h_{\alpha-1}) + b_\alpha \mathrm{V}^{k_\alpha}(\alpha h_{\alpha-1}))$,

*where U and V are trigonometric functions, the coefficients* $a_\alpha$ *and* $b_\alpha$ *are real constants, and* $k_\alpha$ *is a nonnegative integral exponent.*

Indeed, the coefficients $a_\alpha$ and $b_\alpha$ can be allowed to be complex. In order to clarify the matters, take U and V to be respectively the *cosine* and *sine* functions, and let $f_1(x) = 2$ and $f_2(x) = 3x^6$. Then $h_0 = \prod_{i=1}^{2} f_i = 6x^6$. Accordingly, for $a_1 = b_1 = 1$, $a_2 = b_2 = 2$, and $k_1 = k_2 = 9$, the generalized *monadic* and *dyadic* polynomials can be respectively given by

$$
\begin{aligned}
h_1(x) &= \cos^9(6x^6) + \sin^9(6x^6),\\
h_2(x) &= \sum_{\alpha=1}^{2} (a_\alpha \cos^{k_\alpha}(\alpha h_{\alpha-1}) + b_\alpha \sin^{k_\alpha}(\alpha h_{\alpha-1}))\\
&= (a_1 \cos^{k_1}(h_0) + b_1 \sin^{k_1}(h_0)) + (a_2 \cos^{k_2}(2h_1) + b_2 \sin^{k_2}(2h_1))\\
&= (\cos^9(6x^6) + \sin^9(6x^6))\\
&\quad + (2\cos^9(2(\cos^9(6x^6) + \sin^9(6x^6))) + 2\sin^9(2(\cos^9(6x^6) + \sin^9(6x^6)))).
\end{aligned}
$$

The following example will illustrate how to evaluate generalized real polynomials in InCLosure.

**Example 15.5.1 (InCLosure Evaluation of Generalized Real Polynomials)** Consider the generalized real polynomial

$$h(x) = 6\cos^9(3x^6) + 6\sin^9(3x^6).$$

To evaluate $h(x)$ at $x = 369$, we write the following InCLosure command

```
rEval "6*(cos(3*x^6))^9+6*(sin(3*x^6))^9" "x = 369"
```

which will result in 5.675 824 631 042 380 371 24.

### 15.5.2 Polynomials over S-semirings and *n*-adic S-polynomials

Polynomial algebras can be generalized beyond semiring structures. Here we provide a generalization of ordinary and *n*-adic polynomials to the case of *uncertain* structures (*set-valued* structures).

**Definition 15.5.2 (S-Polynomials)** *Let $\mathfrak{R} = \langle \mathcal{R}; +_{\mathcal{R}}, \times_{\mathcal{R}} \rangle$ be an S-semiring. An S-polynomial (set-valued polynomial, or a polynomial over $\mathfrak{R}$) is a polynomial all of whose coefficients belong to the universe set $\mathcal{R}$.*

**Definition 15.5.3 (Generalized *n*-adic S-Polynomials)** *We understand by a generalized n-adic S-polynomial (in short, a generalized S-polynomial) an S-polynomial that satisfies the recursion of definition 15.5.1.*

The preceding two definitions are so framed as to provide a rigorous algebraic foundation for interval and set-valued polynomials. For example, let $X$ be a variable symbol varying on the set of real closed intervals, and let $A$ and $B$ be interval constants. Then, the *interval polynomial*

$$f(X) = AX + BX^2,$$

is an S-polynomial, whereas the *generalized interval polynomial*

$$h(X) = A\cos^3(X^2) + B\cos^3(X^2),$$

is a generalized S-polynomial. Interval polynomials will be discussed, in some detail, in section 15.7.

## 15.6 Systems of generalized *n*-adic polynomial equations and inequalities

In practical applications, we often seek the simultaneous solutions of a system of multiple polynomials. A major goal of this chapter is to compute *guaranteed interval enclosures* of the solutions of a system of *generalized n-adic polynomials.* This will be discussed in detail in section 15.9. Towards this end, the first step is to characterize systems of generalized polynomial equations and inequalities.

**Definition 15.6.1 (Systems of Generalized Polynomial Equations)** *For $i \in \{1, \ldots, \alpha\}$, a system of a number $\alpha$ of generalized polynomial equations is defined to be*

$$\mathbf{F}_{(\alpha,=)} = \left\{ f_i(x) = 0 \middle| f_i \in \mathsf{P}_{\mathbb{R}} \right\}.$$

**Definition 15.6.2 (Systems of Generalized Polynomial Inequalities)** *Let $\prec$ be in $\{<, \leqslant\}$. For $i \in \{1, \dots, \alpha\}$, a system of a number $\alpha$ of generalized polynomial inequalities is defined to be*

$$\mathbf{F}_{(\alpha,\prec)} = \left\{f_i(x) \prec 0 \middle| f_i \in \mathsf{P}_{\mathbb{R}}\right\}.$$

Let $\mathbf{F}$ be in $\{\mathbf{F}_{(\alpha,=)}, \mathbf{F}_{(\alpha,\prec)}\}$ and $\mathfrak{R}$ be in $\{=, <, \leqslant\}$. The solution set of $\mathbf{F}$ is defined to be

$$\mathbf{F}^{-1} = \left\{x \in \mathbb{R} \middle| f_i(x) \mathfrak{R} 0\right\}.$$

## 15.7 Interval arithmetic and interval polynomials

In this section, we formalize the theories of interval algebra and interval polynomials and show how to evaluate interval functions (polynomials, generalized polynomials, or otherwise). The fundamental significance of an interval evaluation is that its result is a *verified* and *guaranteed* interval enclosure of the image a *whole family* of real-valued functions, including polynomial functions.

### 15.7.1 The algebra of real closed intervals

There are so many theories of interval algebras (see, e.g., [11, 12, 15–19, 35–41]). Here we characterize the classical interval theory as introduced in, e.g., [13, 16, 42, 43].

An algebra of real closed intervals can be characterized in the following definition (see [13, 17, 19]).

**Definition 15.7.1 (Theory of Real Interval Algebra)** *Take $\sigma = \{+, \times; -, ^{-1}; 0, 1\}$ to be a set of non-logical constants and let $\mathsf{R} = \langle \mathbb{R}; \sigma^{\mathsf{R}}\rangle$ be the totally $\leqslant$-ordered field of real numbers. The theory $\mathsf{Th}_{\mathcal{I}}$ of an interval algebra over the field $\mathsf{R}$ is the theory of a two-sorted structure $\mathfrak{I}_{\mathbb{R}} = \langle \mathcal{I}_{\mathbb{R}}; \mathbb{R}; \sigma^{\mathfrak{I}_{\mathbb{R}}}\rangle$ prescribed by the following set of axioms.*

(I1) $(\forall X \in \mathcal{I}_{\mathbb{R}})(X = \{x \in \mathbb{R} | (\exists \underline{x} \in \mathbb{R})(\exists \bar{x} \in \mathbb{R})(\underline{x} \leqslant_{\mathbb{R}} x \leqslant_{\mathbb{R}} \bar{x})\})$,

(I2) $(\forall X, Y \in \mathcal{I}_{\mathbb{R}})(\circ \in \{+, \times\} \Rightarrow X \circ_{\mathcal{I}_{\mathbb{R}}} Y = \{z \in \mathbb{R} | (\exists x \in X)(\exists y \in Y)$ $(z = x \circ_{\mathbb{R}} y)\})$,

(I3) $(\forall X \in \mathcal{I}_{\mathbb{R}})(\diamond \in \{-\} \vee \left(\diamond \in \{^{-1}\} \wedge 0_{\mathcal{I}_{\mathbb{R}}} \not\subseteq X\right) \Rightarrow \diamond_{\mathcal{I}_{\mathbb{R}}} X = \{z \in \mathbb{R} | (\exists x \in X)$ $(z = \diamond_{\mathbb{R}} x)\})$.

The sentence (I1) of the above definition is a definitional axiom characterizing what a real interval number (or an $\mathbb{R}$-interval) is. The sentences (I2) and (I3) are finite axiom schemata prescribing, respectively, two binary operations, namely *addition* ('+') and *multiplication* ('×'), and two unary operations, namely *negation* ('−') and *reciprocal* ('$^{-1}$'). In the sequel, the upper-case letters $X$, $Y$, and $Z$, or equivalently $[\underline{x}, \bar{x}]$, $[\underline{y}, \bar{y}]$, and $[\underline{z}, \bar{z}]$, will be used as variable symbols ranging over the set $\mathcal{I}_{\mathbb{R}}$ of interval numbers. A *point* (*singleton*, or *degenerate*) interval number $\{x\}$ will be denoted by $[x]$. Also, the letters $A$, $B$, and $C$, or equivalently $[\underline{a}, \bar{a}]$, $[\underline{b}, \bar{b}]$, and

$[\underline{c}, \bar{c}]$, will be used to designate constants of the set $\mathcal{I}_{\mathbb{R}}$. In particular, we will use $1_{\mathcal{I}}$ and $0_{\mathcal{I}}$ to designate, respectively, the singleton real interval numbers $\{1_{\mathbb{R}}\}$ and $\{0_{\mathbb{R}}\}$.

Equality of real interval numbers is defined by

$$[\underline{x}, \bar{x}] =_{\mathcal{I}} [\underline{y}, \bar{y}] \Leftrightarrow \underline{x} =_{\mathbb{R}} \underline{y} \wedge \bar{x} =_{\mathbb{R}} \bar{y}.$$

The next theorem is easily provable (see [17, 19, 44]).

**Theorem 15.7.1 (Interval Operations)** *For any two interval numbers* $[\underline{x}, \bar{x}]$ *and* $[\underline{y}, \bar{y}]$, *the binary and unary interval operations are formulated in terms of the intervals' endpoints as follows.*

(i) $[\underline{x}, \bar{x}] +_{\mathcal{I}} [\underline{y}, \bar{y}] = [\underline{x} +_{\mathbb{R}} \underline{y}, \bar{x} +_{\mathbb{R}} \bar{y}]$,

(ii) $[\underline{x}, \bar{x}] \times_{\mathcal{I}} [\underline{y}, \bar{y}] = \big[\min\{\underline{x} \times_{\mathbb{R}} \underline{y}, \underline{x} \times_{\mathbb{R}} \bar{y}, \bar{x} \times_{\mathbb{R}} \underline{y}, \bar{x} \times_{\mathbb{R}} \bar{y}\},$
$\max\{\underline{x} \times_{\mathbb{R}} \underline{y}, \underline{x} \times_{\mathbb{R}} \bar{y}, \bar{x} \times_{\mathbb{R}} \underline{y}, \bar{x} \times_{\mathbb{R}} \bar{y}\}\big]$,

(iii) $-_{\mathcal{I}}[\underline{x}, \bar{x}] = [-_{\mathbb{R}}\bar{x}, -_{\mathbb{R}}\underline{x}]$,

(iv) $0_{\mathcal{I}} \not\subseteq [\underline{x}, \bar{x}] \Rightarrow [\underline{x}, \bar{x}]^{-1_{\mathcal{I}}} = [\bar{x}^{-1_{\mathbb{R}}}, \underline{x}^{-1_{\mathbb{R}}}]$,

*where min and max are respectively the* $\leqslant_{\mathbb{R}}$*-minimal and* $\leqslant_{\mathbb{R}}$*-maximal.*

Interval subtraction and division can be characterized respectively as $X - Y = X + (-Y)$ and $X \div Y = X \times (Y^{-1})$. We can easily show, in accord with definition 15.2.4, that the structure $\langle \mathcal{I}_{\mathbb{R}}; +_{\mathcal{I}}, \times_{\mathcal{I}}; 0_{\mathcal{I}}, 1_{\mathcal{I}} \rangle$ of real interval numbers is a commutative S-semiring [19].

Throughout this chapter, we will make use of the following theorem and its corollary (see, [13, 17]).

**Theorem 15.7.2 (Inclusion Monotonicity in Classical Intervals)** *Let* $X_1$, $X_2$, $Y_1$, and $Y_2$ *be interval numbers such that* $X_1 \subseteq Y_1$ *and* $X_2 \subseteq Y_2$. *Then for any binary operation* $\circ \in \{+, \times\}$ *and any definable unary operation* $\diamond \in \{-, ^{-1}\}$, *we have*

(i) $X_1 \circ_{\mathcal{I}} X_2 \subseteq Y_1 \circ_{\mathcal{I}} Y_2$,

(ii) $\diamond_{\mathcal{I}} X_1 \subseteq \diamond_{\mathcal{I}} Y_1$.

**Corollary 15.7.1 (Membership Monotonicity for Classical Intervals)** *Let* $X$ *and* $Y$ *be real interval numbers with* $x \in X$ *and* $y \in Y$. *Then for any binary operation* $\circ \in \{+, \times\}$ *and any definable unary operation* $\diamond \in \{-, ^{-1}\}$, *we have*

(i) $x \circ_{\mathbb{R}} y \in X \circ_{c} Y$,

(ii) $\diamond_{\mathbb{R}} x \in \diamond_{c} X$.

### 15.7.2 Interval functions and interval polynomials: guaranteed interval enclosures of families of generalized real polynomials

In this section, we show how to evaluate interval functions (polynomials, generalized polynomials, or otherwise). The major significance of an interval evaluation is that its result is a *verified* and *guaranteed* interval enclosure of the image of a *whole family* of real-valued functions, including polynomial functions.

First, it is adequate to introduce some notational conventions. By a *finitary* real-valued function in real arguments (in short, a *real function* or $\mathbb{R}$-function), we understand a function $f_{\mathbb{R}} : \mathcal{D}_{\mathbb{R}} \subseteq \mathbb{R}^n \mapsto \mathbb{R}$, and by an *interval function* (or $\mathcal{I}$-function) we understand a function $f_{\mathcal{I}} : \mathcal{D}_{\mathcal{I}} \subseteq \mathcal{I}^n \mapsto \mathcal{I}$. If the type of function is clear from its arguments, we will usually drop the subscripts '$\mathbb{R}$' and '$\mathcal{I}$'. Thus, we may write $f(x_1, \ldots, x_n)$ and $f(X_1, \ldots, X_n)$ for, respectively, a real-valued function and an interval-valued function, which are both defined by the same rule. We will say that the interval function $f(X_1, \ldots, X_n)$ is an interval extension of the real function $f(x_1, \ldots, x_n)$. By virtue of the inclusion monotonicity theorem (theorem 15.7.2), $f(x_1, \ldots, x_n) \in f(X_1, \ldots, X_n)$, for all $x_i \in X_i$. That is, an interval extension is a *guaranteed interval enclosure* of the corresponding real function defined by the same rule (see, [19, 29]).

Recall definition 15.5.1 of a generalized $n$-adic real polynomial. Next, we define *generalized interval polynomials*.

**Definition 15.7.2 (Generalized *n*-adic Interval Polynomials)** *A generalized n-adic interval polynomial (in short, a generalized interval polynomial) is the interval extension of the corresponding generalized n-adic real polynomial defined by the same rule.*

In accord with definition 15.5.3, it can easily be shown that a generalized $n$-adic interval polynomial is a generalized $n$-adic S-polynomial.

In fact, more profoundly, an interval function is a guaranteed enclosure of a *whole family* of real-valued functions (polynomials, generalized polynomials, or otherwise). For example, the *generalized interval polynomial*

$$h(X) = A\cos^3(X^2) + B\cos^3(X^2),$$

is an interval extension of the *infinite* family

$$h(x) = a\cos^3(x^2) + b\cos^3(x^2),$$

of generalized real polynomials, with $a \in A$ and $b \in B$.

We close this subsection by illustrating how to evaluate a generalized interval polynomial using InCLosure. In other words, we will show how to compute a *reliable interval enclosure* of a family of generalized real polynomials.

**Example 15.7.1 (InCLosure Evaluation of Generalized Interval Polynomials)** Consider the generalized interval polynomial

$$h(X) = [2, 3]\cos(X^2) + [3, 4].$$

To evaluate $h(X)$ at $X = [3, 9]$, we write the following InCLosure command

```
iEval "[2,3]*cos(X^2)+[3,4]" "X=[3,9]"
```

which will result in the interval [0.0, 7.0]. This interval is a guaranteed interval enclosure of the image of the infinite real family

$$h(x) = a\cos(x^2) + b,$$

with $a \in [2, 3]$ and $b \in [3, 4]$.

### 15.7.3 The interval subdivision method: more refined enclosures of families of generalized real polynomials

The naive interval method for getting guaranteed enclosures of families of real functions (polynomials or otherwise) suffers from dependencies of the domain intervals. The resultant enclosures get larger as the width of the domain intervals increases. Recent investigations of the logical underpinnings of the interval dependency problem can be found in [17, 19]. Also a plausible definition was proposed in [25]. Many improvements to the naive interval evaluation have been proposed. Among these is the interval subdivision method (for more on the subdivision method and other methods see, e.g., [42, 43, 45–48]). The subdivision method can be described as follows. Let $f$ be a continuous real-valued function (polynomial or otherwise) on the interval $X$. The interval $X$ is subdivided into $m \geqslant 2$ subintervals $X_i$ of equal width, such that

$$X_i = [\underline{x} + (i - 1)w(X)/m,\ \underline{x} + (i)w(X)/m],$$

hence $X = \cup_{i=1}^{m} X_i$. The interval-valued function is then evaluated for each subinterval $X_i$, $f(X_i)$. The subdivision enclosure of the image is given by $\cup_{i=1}^{m} f(X_i)$. As a consequence of inclusion monotonicity of the interval-valued function $f$, the following result is easily provable

$$\mathrm{I}_f(X) \subseteq \bigcup_{i=1}^{m} f(X_i) \subseteq f(X).$$

The above result ensures that the resultant subdivision enclosure is sharper than the naive enclosure. Moreover, the larger the number of subintervals $m$, the better the subdivision enclosure.

Next we give an example in InCLosure illustrating the performance of the subdivision method.

**Example 15.7.2 (The Subdivision Method)** Consider the polynomial function $f(x) = (1 + x)(x + x^2)$ where $x \in X = [-0.1, 0.1]$. To compute an enclosure of the image of $f$ with the subdivision method we write the following InCLosure command

```
iEval "(1+X)*(X+X^2)" 100 "X=[-0.1,0.1]" 5
```

this will result in $[-0.081\,54, 0.121]$.

The parameter '100' in the above InCLosure command indicates the number of subintervals. The last parameter '5' is the precision of the result. InCLosure computes the subdivision enclosure of the image to an *arbitrary* number of subintervals and with an arbitrary precision.

## 15.8 Taylor models: sharper interval enclosures with infinite polynomials

This section is devoted to describing how to use infinite polynomials with Taylor models to compute finer interval enclosures (For further reading about Taylor models, see, e.g., [48–53]).

Let the function $f$ (polynomial or otherwise) be $(n+1)$-times continuously differentiable on the domain interval $X$. An $n$th order Taylor model of $f$, around a point $x_0 \in X$, is an enclosure remainder form representation of the function denoted $\mathrm{t}_f$ and defined as a pair

$$\mathrm{t}_f = \left(p_f,\, R_f\right),$$

where

$p_f(x) = \sum_{k=0}^{n} f^{(k)}(x_0)\frac{(x - x_0)^k}{k!}$ ($n$th degree Taylorpolynomial), and

$R_f$ is an enclosure interval of the Lagrange remainder $f^{(n+1)}(\xi)\frac{(x - x_0)^{n+1}}{(n+1)!}$,

$\xi$ between $x_0$ and $x$.

The enclosure interval $R_f$ can be computed using naive or subdivision interval methods. A Taylor model of the function $f$ is an enclosure of the function and we have

$$\mathrm{I}_f(X) \subseteq \mathrm{I}_{p_f}(X) + R_f.$$

The Taylor model enclosure of the image $\mathrm{I}_{p_f}(X) + R_f$ is sharper than the naive and subdivision enclosures when the width of the domain interval is small. Moreover, as the order of the model increases, the width of the remainder enclosure interval $R_f$ decreases and narrower enclosures are obtained.

Techniques based on Taylor models play an important role in many computational problems. Applications of Taylor models include range bounding, quadrature, solving implicit equations, solving differential equations, global optimization problems and many others (for more on the various applications of Taylor models, see, e.g., [49]).

Taylor models are implemented in Lisp as a part of version 4.0 of InCLosure [28]. Next we give some Taylor model examples in InCLosure.

**Example 15.8.1 (Taylor Model of a Trigonometric Polynomial)** Consider the trigonometric polynomial $f(x) = 3\sin(6x) + 6\cos(9x)$ where $x \in X = [-0.01, 0.01]$. To compute a 3rd order Taylor model of $f$ around 0 to precision 5, we write the following InCLosure command

```
TM "3*sin(6*x)+6*cos(9*x)" "x" "[-0.01,0.01]" 0 3 5
```

which will result in the pair $(6.0 + (18.0)x - (243.0)x^2 - (108.0)x^3;$ $[-1.0 \times 10^{-5}, 2.0 \times 10^{-5}])$.

**Example 15.8.2 (Taylor Model Enclosure of the Image)** Consider the polynomial function $f(x) = (1 + x)(x + x^2)$ where $x \in [-0.1, 0.1]$. To compute an enclosure of the image of $f$ with a 3rd order Taylor model to precision 5, we write the following InCLosure command

```
TMImg "(1+x)*(x+x^2)" "x" "[-0.1,0.1]" 3 5
```

This will result in [−0.081 03, 0.121].

With a Taylor model of order only 3, the above example gives a sharper enclosure interval than that of the subdivision method (see example 15.7.2) with 100 subintervals. Clearly, $w([-0.081\,03, 0.121]) < w([-0.081\,54, 0.121])$.

A Taylor model represents a real function with a pair of a symbolic polynomial as the first component and an interval as the second component. Definite integration of the function is then easily computed and is reduced to only integrating the polynomial instead. Taylor model integration is illustrated by the following example.

**Example 15.8.3 (Taylor Model Enclosure of Definite Integral)** Consider the trigonometric polynomial $f(x) = 3\sin(6x) + 6\cos(9x)$ where $x \in X = [-0.01, 0.01]$. To compute an enclosure of the definite integral of $f$ on X with a 3rd order Taylor model to precision 5, we write the following InCLosure command

```
TMInteg "3*sin(6*x)+6*cos(9*x)" -0.01 0.01 "x" 3 5
```

This will result in [0.119 83, 0.119 85].

## 15.9 Interval root finding: guaranteed interval enclosures of roots of real polynomials

Computing roots of real polynomials is of great importance in mathematics and scientific computing. There are many algorithms to solve polynomial equations or systems of polynomial equations. As examples we can mention *Newton–Raphson method*, *interval Newton–Raphson method*, and *interval branch-and-prune* techniques. Techniques of computation vary from computing approximations to the roots (e.g. *Newton–Raphson method*) to getting guaranteed enclosures of the roots (e.g. *interval Newton–Raphson method* and *interval branch-and-prune* techniques). Some techniques are also applied to verify the existence or uniqueness of solutions. These could include *Newton operator*, *Gauss–Seidel* method, *Miranda test*, *quadratic approximation*, *Borsuk test*, and many others.

Many applications require computing guaranteed enclosures of the roots with respect to a specific given precision. In what follows we discuss the interval Newton method and the interval branch-and-prune technique to get guaranteed enclosures of the roots of real polynomials. A new refined version of the interval branch-and-prune technique is then introduced that makes use of the interval Newton method to speed up convergence (for more on interval root finding methods see, e.g., [5, 25, 27, 47]).

### 15.9.1 Interval Newton–Raphson method for root enclosures

The interval Newton–Raphson method is first introduced by Moore [47] as an extension of the real Newton method for approximating the roots of real functions. Rather than computing just numerical approximations, the interval extension provides guaranteed enclosures of the roots. The interval Newton–Raphson method is figured in the following theorem (for more details on the interval Newton–Raphson method see, e.g., [27, 47]).

**Theorem 15.9.1 (The Interval Newton–Raphson Method)** *Let $f$ be a continuously differentiable real-valued function on the interval number $X$. If $X$ contains a root of $f$ and $0 \notin f^{(1)}(X)$ then*

$$X \bigcap \left( m(X) - \frac{f(m(X))}{f^{(1)}(X)} \right) \subset X,$$

*where $m(X)$ is the midpoint of $X$, is a narrower interval than $X$ that also contains the root of $f$.*

*Proof.* Let $x_0$ and $x^*$ be in $X$. From the mean-value theorem, we have

$$\frac{f(x_0) - f(x^*)}{x_0 - x^*} = f^{(1)}(c),$$

where $c$ is some point between $x_0$ and $x^*$. Now for $x^*$ to be a root of $f$, we must have $f(x^*)=0$ and then

$$x^* = x_0 - \frac{f(x_0)}{f^{(1)}(c)}.$$

Now, since $f^{(1)}(c) \in f^{(1)}(X)$ because $c \in X$, an interval enclosure of the root $x^*$ can be computed as follows

$$x^* \in x_0 - \frac{f(x_0)}{f^{(1)}(X)}.$$

Hence, being in the two intervals $X$ and $x_0 - \frac{f(x_0)}{f^{(1)}(X)}$, the root $x^*$ is also in their intersection

$$x^* \in X \bigcap \left( x_0 - \frac{f(x_0)}{f^{(1)}(X)} \right) \subset X.$$

The theorem then follows by choosing $x_0$ to be the midpoint of $X$. □

The initial point $x_0$, called the *point of expansion*, could be any point in $X$, however choosing $x_0$ to be the midpoint is most efficient to speed up the convergence to the root.

The interval Newton algorithm applies the technique prescribed by theorem 15.9.1 recursively until the desired enclosure interval is obtained. That is, start with the interval $X_1 = X \cap \left( x_0 - \frac{f(x_0)}{f^{(1)}(X)} \right)$, compute its midpoint, call $x_1$, then compute $X_2 = X_1 \cap \left( x_1 - \frac{f(x_1)}{f^{(1)}(X_1)} \right) \subset X_1$ to get a better enclosure interval of the root than $X_1$. If

$X_2$ is narrow enough, the algorithm stops and $X_2$ is the desired enclosure interval. One simple criterion to stop the algorithm is when the width of the enclosure interval is less than some given positive number $e$ called the *error* of the computation. For instance, to compute the root to two decimal places the error stop criterion could be $e = 10^{-2}$ which guarantees that the width of the resultant interval does not exceed $10^{-2}$.

The existence of the roots of $f$ in $X$ can be easily checked as follows.

**Theorem 15.9.2 (Interval Newton Existence)** *Let $f$ be a continuous real-valued function on the interval $X$. If $0 \in f(X)$ then $X$ contains at least one root of $f$.*

The condition $0 \notin f^{(1)}(X)$ in theorem 15.9.1 ensures that the function $f$ is monotonic yielding the following important result.

**Theorem 15.9.3 (Interval Newton Uniqueness)** *Let $f$ be a continuously differentiable real-valued function on the interval number $X$. If $X$ contains a root of $f$ and $0 \notin f^{(1)}(X)$ then the root is unique.*

The most important property of the interval Newton method is its fast convergence. Moore in [47] has proved that the method converges *quadratically*. This is made precise in the following theorem.

**Theorem 15.9.4 (Interval Newton Convergence)** *Let $f$ be a continuously differentiable real-valued function on the interval number $X$ and $X$ contains a root of $f$ with $0 \notin f^{(1)}(X)$. Let $X_n$ and $X_{n+1}$ be the enclosure intervals obtained from the nth and $(n+1)$th iteration of the interval Newton algorithm respectively. Then there exists a real constant $c$ such that*

$$w(X_{n+1}) \leqslant c(w(X_n))^2,$$

*where $w(X_n)$ and $w(X_{n+1})$ are the widths of the intervals $X_n$ and $X_{n+1}$ respectively.*

To illustrate the interval Newton algorithm, a simple example follows.

**Example 15.9.1 (The Interval Newton Technique)** Let $f(x) = x^2 - 2$, $x \in X = [1, 2]$. Then

$$\begin{aligned} f(X) &= f([1, 2]) \\ &= [-1, 2], \\ f^{(1)}(X) &= f^{(1)}([1, 2]) \\ &= [2, 4]. \end{aligned}$$

We then have $0 \in f(X) = [-1, 2]$ and $0 \notin f^{(1)}(X) = [2, 4]$ and, by theorems 15.9.2 and 15.9.3, the conditions of theorem 15.9.1 are satisfied. Assume we want to compute the root of $f$ to 3 decimal places. Let $x_0 = m(X) = 1.5$. Now computing $X_1$ yields

$$\begin{aligned} X_1 &= [1, 2] \cap \left(1.5 - \frac{(1.5)^2 - 2}{[2, 4]}\right) \\ &= [1, 2] \cap [1.375, 1.437\,5] \\ &= [1.375, 1.437\,5], \\ w(X_1) &= 0.062\,5. \end{aligned}$$

Similarly, computing $X_2$, with $x_1 = m(X_1) = 1.406\,25$, yields

$$\begin{aligned} X_2 &= [1.375, 1.437\,5] \cap \left(1.406\,25 - \frac{(1.406\,25)^2 - 2}{2[1.375, 1.437\,5]}\right) \\ &= [1.375, 1.437\,5] \cap [1.414\,06, 1.414\,42] \\ &= [1.414\,06, 1.414\,42], \\ w(X_2) &= 3.6 \times 10^{-4} < 10^{-3}. \end{aligned}$$

Hence the desired precision is obtained and the algorithm stops with the resultant enclosure interval $[1.414\,06, 1.414\,42]$.

### 15.9.2 Interval branch-and-prune method for root enclosures

Another simple technique for getting guaranteed enclosures of the roots of continuous real-valued functions is the interval branch-and-prune method. The interval branch-and-prune method can be described as follows (For more details on interval branch-and-prune techniques see, e.g., [5, 25, 47]).

**Definition 15.9.1 (Interval Branch-and-Prune Method)** *Let $f$ be a continuous real-valued function on the interval $X$. The interval branch-and-prune method to find the roots of* f *can be described as follows:*

**Branching Step.** *If $X$ contains a solution, subdivide $X$ into $n$ narrower subintervals in which the roots of $f$ may exist.*

**Pruning Step.** *For each subinterval, check the existence of a solution. If no solution exists, the subinterval is discarded.*

*The two steps are repeated recursively until a solution interval is obtained with the desired precision.*

The existence of a root of $f$ in $X$ can be easily verified by theorem 15.9.2. The Newton condition $0 \notin f^{(1)}(X)$ is not required by the interval branch-and-prune technique and the function $f$ may have more than one root in $X$. The interval branch-and-prune algorithm could stop when the first root is obtained or compute all roots of $f$ in $X$ to the desired precision.

The following simple example illustrates the interval branch-and-prune method.

**Example 15.9.2 (Interval Branch-and-Prune Method)** Let $f(x) = x^2 - 1$ be a real polynomial of degree 2 where $x \in X = [-1, 1]$. To find the roots of $f$ using the interval branch-and-prune technique we first check for the existence of roots in $X$. Since the evaluation of the interval-valued function $f$ at $X$ yields the interval $f(X) = [-1, 0]$ which contains 0, the interval $X$ is then guaranteed to contain a solution of $f(x) = 0$. To get better enclosures of the solutions, $X$ is subdivided into 2 subintervals of equal width $X_1 = [-1, 0]$ and $X_2 = [0, 1]$. The interval evaluations $f(X_1) = [-1, 0]$ and $f(X_2) = [-1, 0]$ ensure the existence of solutions in both $X_1$ and $X_2$. The subdivision is then repeated for $X_1$ and $X_2$ yielding $X_{11} = \left[-1, -\frac{1}{2}\right]$, $X_{12} = \left[-\frac{1}{2}, 0\right]$, $X_{21} = \left[0, \frac{1}{2}\right]$, and $X_{22} = \left[\frac{1}{2}, 1\right]$. Evaluations at the subintervals $X_{12} = \left[-\frac{1}{2}, 0\right]$ and $X_{21} = \left[0, \frac{1}{2}\right]$ yield intervals with no zeros and hence these two intervals are discarded. The branching process continues for $X_{11} = \left[-1, -\frac{1}{2}\right]$ and $X_{22} = \left[\frac{1}{2}, 1\right]$ until the required width (precision) is preserved. Figure 15.1 illustrates the interval branch-and-prune method to find the roots of $f$. The method is represented by a tree rooted at node containing $X$. The children (leafs) of each node represents subintervals of the interval at that node.

### 15.9.3 A more refined interval branch-and-prune method

The interval branch-and-prune technique can be more efficient if some subintervals contain only one root of the function. So instead of branching the subinterval, the

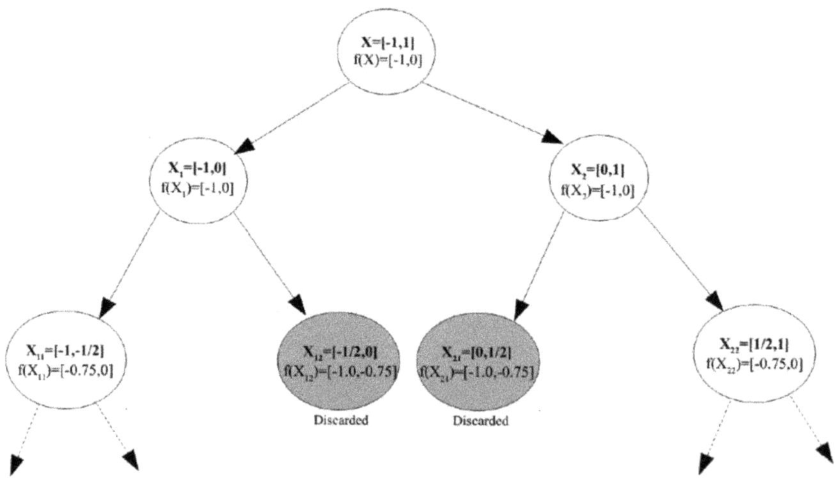


**Figure 15.1.** Interval branch-and-prune for $f(x) = x^2 - 1$, $x \in X = [-1, 1]$.

interval Newton method, with quadratic convergence, is applied to find the unique root of the function in the subinterval. In what follows we introduce a more refined interval branch-and-prune technique that makes use of the interval Newton method to speed up convergence. In order to illustrate the significance of our algorithm, we give many numerical examples, then we compute a finer interval enclosure for an example introduced by Eldon Hansen and G William Walster in [27]. We show that our interval enclosure is sharper than the optimal result given by Hansen and Walster.

**Algorithm 15.9.1 (Interval Branch-and-Prune with Newton Method)** *Let $f$ be a continuous real-valued function on the interval $X$. Let $e > 0$ be the error of the computation. Let L be the list of solutions initially empty. The interval branch-and-prune with Newton method algorithm with inputs $f$ and $X$, abbreviated* IBPN$(f, X)$, *is described by the following steps.*

**Step 1.** *Compute the interval-valued function $f(X)$.* If $0 \in f(X)$ *and* $w(X) < e$, *add $X$ to L and go to step 4. If $0 \notin f(X)$, go to step 4, otherwise go to step 2.*

**Step 2.** *Compute $f^{(1)}(X)$.*

**Step 2.1.** *If $0 \notin f^{(1)}(X)$, compute $X_1 = X \cap \left(m(X) - \frac{f(m(X))}{f^{(1)}(X)}\right)$ and update $X = X_1$ then go to step 2.2, otherwise go to step 3.*

**Step 2.2.** *If $w(X) < e$, add $X$ to L and go to step 4, otherwise go to step 3.*

**Step 3.** *Subdivide $X$ into two subintervals$X_1$ and $X_2$. Apply* IBPN$(f, X_1)$ *and* IBPN $(f, X_2)$.

**Step 4.** *Return L and exit. If L is empty report 'No Solution'.*

The above computational algorithm is implemented in Lisp as a part of version 4.0 of InCLosure [28]. Next we give some examples in InCLosure to illustrate the efficiency of algorithm 15.9.1 in solving systems of generalized real polynomials.

**Example 15.9.3 (IBPN Solution of Real Polynomials)** Consider the polynomial function $f(x) = x^2 + 4x + 3$ where $x \in [-3, 1]$. To solve $f(x) = 0$ with error $e = 10^{-5}$ we write the following InCLosure command

```
ISolve "x^2 + 4*x + 3 = 0" "x in [-3,1]" 5
```

This will result in the set of solutions $\{([-3.0, -2.999\,99])\ ([-1.000\,01, -0.999\,99])\}$.

In [27], Hansen and Walster introduced the following example. In the following we recompute the solution with our IBPN algorithm.

**Example 15.9.4 (IBPN Solution of Hansen Polynomial)** Consider the polynomial function $f(x) = 4567x^2 - 9134x + 4567$ where $x \in [0, 3]$. To solve $f(x) = 0$ with error $e = 10^{-4}$ we write the following InCLosure command

```
ISolve "4567*x^2-9134*x + 4567 = 0" "x in [0,3]" 4
```

This will result in {([0.999, 1.001])}.

The resultant interval [0.999, 1.001] is sharper than the optimal result [0.985 2, 1.015] given by Eldon Hansen and G William Walster in section 9.7 of [27]. Hansen and Walster computed the result with error $e = 0$, however with our error $e = 10^{-4} > 0$, $w([0.999, 1.001]) < w([0.985\,2, 1.015])$.

The following example illustrates the case of solving a system of two polynomials.

**Example 15.9.5 (IBPN Solution of a System of Polynomials)** Consider the polynomial functions $f(x) = x^2 + 4x + 3$ and $g(x) = x + 1$ where $x \in [-3, 1]$. To solve $f(x) = 0$ and $g(x) = 0$ with error $e = 10^{-5}$ we write the following InCLosure command

```
ISolve "x^2+4*x+3=0" "x+1=0" "x in [-3,1]" 5
```

This will result in {([−1.0, −1.0])}.

In what follows, we consider solving the generalized real polynomial $\sin(x^2)$ for various error values.

**Example 15.9.6 (IBPN Solution of Trigonometric Polynomials)** Consider the generalized polynomial function $f(x) = \sin(x^2)$ where $x \in [-2, 2]$. To solve $f(x) = 0$ with error $e = 10^{-1}$ we write the following InCLosure command

```
ISolve "sin(x^2) = 0" "x in [-2,2]" 1
```

**Table 15.1.** InCLosure result for solving a trigonometric polynomial.

| $e$ | `InCLosure` result of solving $\sin(x^2) = 0$ with $x \in [-2, 2]$ |
|---|---|
| $10^{-1}$ | {([−1.8, −1.7])([−0.1, 0.1])([1.7, 1.8])} |
| $10^{-2}$ | {([−1.78, −1.77])([−0.01, 0.01])([1.77, 1.78])} |
| $10^{-3}$ | {([−1.773, −1.772])([−0.001, 0.001])([1.772, 1.773])} |
| $10^{-4}$ | {([−1.772 5, −1.772 4])([$-1.0 \times 10^{-4}$, $1.0 \times 10^{-4}$])([1.772 4, 1.772 5])} |
| $10^{-5}$ | {([−1.772 46, −1.772 45])([$-1.0 \times 10^{-5}$, $1.0 \times 10^{-5}$])([1.772 45, 1.772 46])} |
| $10^{-6}$ | {([−1.772 454, −1.772 453])([$-1.0 \times 10^{-6}$, $1.0 \times 10^{-6}$])([1.772 453, 1.772 454])} |
| $10^{-7}$ | {([−1.772 4539, −1.772 4538])([$-1.0 \times 10^{-7}$, $1.0 \times 10^{-7}$])([1.772 4538, 1.772 4539])} |
| $10^{-8}$ | {([−1.772 453 86, −1.772 453 85])([$-1.0 \times 10^{-8}$, $1.0 \times 10^{-8}$])([1.772 453 85, 1.772 453 86])} |
| $10^{-9}$ | {([−1.772 453 851, −1.772 453 85])([$-1.0 \times 10^{-9}$, $1.0 \times 10^{-9}$])( [1.772 453 85, 1.772 453 851])} |
| $10^{-10}$ | {([−1.772 453 851, −1.772 453 8509])([$-1.0 \times 10^{-10}$, $1.0 \times 10^{-10}$])( [1.772 453 8509, 1.772 453 851])} |

which will result in the set of solutions $\{([-1.8, -1.7])([-0.1, 0.1])([1.7, 1.8])\}$. The exact real values of the roots of $f$ are 0, $-\sqrt{\pi}$, and $\sqrt{\pi}$. Table 15.1 summarizes InCLosure results of solving $f(x) = 0$ with various error values ranging from $e = 10^{-1}$ to $e = 10^{-10}$. It is clearly seen that, as the value of the error decreases, the width of the resultant enclosure interval decreases and hence converges to the exact real solution.

## 15.10 Concluding remarks

Having now reached the end of this chapter, it may not be amiss if we briefly consider our major results. As we detailed in the introduction and elsewhere, systems of polynomials are at the heart of a wide range of scientific fields. Because of their ever-increasing importance in many practical applications, polynomials are heavily and actively researched across various disciplines of mathematics, science and engineering. Over the past decades, research into the study of different types of polynomials has played an indispensable role and reached its most advanced and sophisticated forms in several areas, such as numerical analysis, algebraic geometry, tropical geometry, statistics, dynamical systems, global optimization, control theory, artificial intelligence, computer-aided design, structural dynamics, and quantum mechanics, to mention a few. In scientific and engineering applications, a recurring problem is to determine the roots of polynomials under uncertainty or imprecision. Interval mathematics is a subtle edifice of principles and methods for manipulating systems involving quantifiable uncertainties. Interval analysis has proven successful and reliable for computing guaranteed enclosures of roots of polynomials under parametric uncertainty. This 'self-validating' feature of interval mathematics makes it competitive and preferable to the ordinary approximation methods in many practical applications. By the pursuit of the quest of reliable knowledge amidst uncertainty, this chapter has been devoted to conducting a systematic investigation of *uncertain polynomials.*

In the first place, after formalizing some set-theoretical and algebraic notions of particular importance for our purpose, we laid out a rigorous algebraic foundation for ordinary polynomials and some of their generalizations, then we defined two new generalizations of ordinary and trigonometric polynomials, namely, *generalized n-adic polynomials* and *n-adic* S-*polynomials.* These two generalizations were so framed as to serve as a rigorous mathematical foundation for interval and set-valued polynomials. Secondly, we formalized the theories of interval algebra and interval polynomials and showed how to obtain verified interval enclosures of families of generalized real polynomials. Thirdly, we described how to use infinite polynomials with Taylor models to compute finer interval enclosures. Afterwards, we investigated some of the interval methods for root finding, with special attention is paid to the interval branch-and-prune method. Finally, we presented a more refined interval branch-and-prune algorithm. In order to illustrate the significance of our algorithm, we recomputed a finer interval enclosure for an example introduced by Eldon Hansen and G William Walster in [27] . We showed that our interval enclosure was sharper than the optimal result given by Hansen and Walster.

Some of the methods presented in this chapter are well known; some are believed to be new. We introduced the mathematical underpinnings, discussed the computational algorithms, and explored the impact of systems of uncertain polynomials in science and engineering. We also provided an implementation of the algorithms in InCLosure, along with a number of numerical examples, illustrating the efficiency and significance of the methods and algorithms presented in the body of this chapter. In summing up, it is hoped that this work will contribute to a better understanding of uncertain polynomials and their applications.

## Supplementary materials

The software and code files used in this chapter are freely accessible from CERN's archive. The algorithms discussed in the text are coded into Common Lisp as a part of the software package, InCLosure. To reproduce the results of the examples, latest version of InCLosure is accessible for free download from CERN's archive via https://doi.org/10.5281/zenodo.2702404. An InCLosure input file for each section of the chapter and its corresponding output containing, respectively, the code and results of the examples are also freely accessible as a supplementary material to this chapter, via https://doi.org/10.5281/zenodo.7030230.

## Acknowledgement

We would like to thank Prof. Snehashish Chakraverty, the editor of this book, for his patience and constructive comments, and for his outstanding help with this chapter. Also, it is a pleasure to acknowledge the courtesy and substantial help which we have received from the staff of IOP Publishing at all stages of writing this chapter.

## References

[1] Droste M and Kuich W 2009 Semirings and formal power series *Handbook of Weighted Automata (Monographs in Theoretical Computer Science: An EATCS Series)* ed M Droste, W Kuich and H Vogler (Berlin: Springer) pp 3–28

[2] Dumitrescu B 2017 *Positive Trigonometric Polynomials and Signal Processing Applications* 2nd edn *(Signals and Communication Technology)* (Berlin: Springer)

[3] Lang S 2002 *Algebra (Graduate Texts in Mathematics 211)* 3rd edn (New York: Springer)

[4] Lang S 2005 *Undergraduate Algebra (Undergraduate Texts in Mathematics)* (New York: Springer)

[5] Kubica B J 2019 *Interval Methods for Solving Nonlinear Constraint Satisfaction, Optimization and Similar Problems: From Inequalities Systems to Game Solutions (Studies in Computational Intelligence* vol 805*)* (Berlin: Springer)

[6] Salomaa A 1990 Formal languages and power series *Formal Models and Semantics (Handbook of Theoretical Computer Science)* ed J van Leeuwen (Amsterdam: Elsevier), pp 103–32

[7] Shestopalov Y 2020 Trigonometric and cylindrical polynomials and their applications in electromagnetics *Appl. Anal.* **99** 2807–22

[8] Sommese A J and Wampler C W 2005 *The Numerical Solution of Systems of Polynomials Arising in Engineering and Science* (Singapore: World Scientific)

[9] Allahviranloo T, Pedrycz W and Esfandiari A 2022 *Advances in Numerical Analysis Emphasizing Interval Data* (Boca Raton, FL: CRC Press)

[10] Beutner R, Ong L and Zaiser F 2022 Guaranteed bounds for posterior inference in universal probabilistic programming *Proc. 43rd ACM SIGPLAN Int. Conf. on Programming Language Design and Implementation, PLDI 22g* (San Diego, CA: ACM SIGPLAN, ACM)

[11] Dawood H 2011 *Theories of Interval Arithmetic: Mathematical Foundations and Applications* (Saarbrücken: LAP)

[12] Dawood H 2012 Interval mathematics: foundations, algebraic structures, and applications *Master's Thesis* Department of Mathematics, Faculty of Science, Cairo University, Giza

[13] Dawood H 2014 Interval mathematics as a potential weapon against uncertainty *Mathematics of Uncertainty Modeling in the Analysis of Engineering and Science Problems* ed S Chakraverty ch 1, pp 1–38

[14] Dawood H and Dawood Y 2016 Interval algerbras: a formalized treatment *Technical Report CU-Math-2016-06-IAFT* Department of Mathematics, Faculty of Science, Cairo University, June 2016

[15] Dawood H and Dawood Y 2016 The form of the uncertain: on the mathematical structures of uncertainty *Technical Report CU-Math-2016-09-FUMSU* Department of Mathematics, Faculty of Science, Cairo University, September 2016

[16] Dawood H 2019 On some algebraic and order-theoretic aspects of machine interval arithmetic *Online Math. J.* **1** 1–13

[17] Dawood H and Dawood Y 2019 A logical formalization of the notion of interval dependency: towards reliable intervalizations of quantifiable uncertainties *Online Math. J.* **1** 15–36

[18] Dawood H and Dawood Y 2019 Parametric intervals: more reliable or foundationally problematic? *Online Math. J.* **1** 37–54

[19] Dawood H and Dawood Y 2020 Universal intervals: towards a dependency-aware interval algebra *Mathematical Methods in Interdisciplinary Sciences* ed S Chakraverty (Hoboken, NJ: Wiley) ch 10, pp 167–214

[20] IEEE 1788 Committee 2018 IEEE standard for interval arithmetic (simplified) *IEEE Std 1 788.1-2017* pp 1–38

[21] Jiang C, Han X and Xie H 2021 *Nonlinear Interval Optimization for Uncertain Problems (Springer Tracts in Mechanical Engineering)* (Berlin: Springer)

[22] Kearfott R B 2021 Mathematically rigorous global optimization and fuzzy optimization *Black Box Optimization, Machine Learning, and No-Free Lunch Theorems* ed P M Pardalos, V Rasskazova and M N Vrahatis (Berlin: Springer) ch 10, pp 169–94

[23] Mahato N R, Rout S and Chakraverty S 2020 Affine-contractor approach to handle nonlinear dynamical problems in uncertain environment *Mathematical Methods in Interdisciplinary Sciences* ed S Chakraverty (Hoboken, NJ: Wiley) ch 11, pp 215–37

[24] Matanga Y, Sun Y and Wang Z 2022 Hybrid PSO-$\alpha$BB global optimisation for $C^2$ box-constrained multimodal NLPs *IEEE Access* **10** 805–18

[25] Shary S P and Moradi B 2021 Solving interval linear least squares problems by PPS-methods *Numer. Algorithms* **87** 41–75

[26] Zheng T, Zhang M, Li L, Wu Q and Zhou L 2020 Correlation coefficients of interval-valued pythagorean hesitant fuzzy sets and their applications *IEEE Access* **8** 9271–86

[27] Hansen E R and William Walster G 2003 *Global Optimization Using Interval Analysis 'Revised and Expanded'* 2nd edn (Boca Raton, FL: CRC Press)

[28] Dawood H 2022 *InCLosure (Interval enCLosure)–a language and environment for reliable scientific computing* Computer Software, Version 4.0, Department of Mathematics, Faculty of Science, Cairo University, Giza, Egypt, September 2022. InCLosure Support http://scholar.cu.edu.eg/henddawood/software/InCLosure
[29] Dawood H and Dawood Y 2017 Investigations into a formalized theory of interval differentiation *Technical Report CU-Math-2017-03-IFTID* Department of Mathematics, Faculty of Science, Cairo University
[30] Claudia Menini and Freddy Van Oystaeyen 2004 *Abstract Algebra: A Comprehensive Treatment* 1st edn (Boca Raton, FL: CRC Press)
[31] Dawood H and Megahed N 2019 A consistent and categorical axiomatization of differentiation arithmetic applicable to first and higher order derivatives *Punjab Univ. J. Math.* **51** 77–100
[32] van Hoorn Willy G and van Rootselaar B 1967 Fundamental notions in the theory of seminearrings *Compos. Math.* **18** 65–78
[33] Gunter P 1983 *Near-Rings: The Theory and its Applications (North-Holland Mathematics Studies* vol 23*)* (North-Holland: Amsterdam)
[34] James R 1992 *Nearrings: Geneses and Applications* (Oxford: Oxford University Press)
[35] Eldon R 1975 A generalized interval arithmetic *Interval Mathematics (Lecture Notes in Computer Science* vol 29*)* ed K L E Nickel (Berlin: Springer) pp 7–18
[36] Ulrich W 2013 *Computer Arithmetic and Validity: Theory, Implementation, and Applications (De Gruyter Studies in Mathematics* vol 33*)* 2nd edn (Berlin: Walter de Gruyter)
[37] Gardenyes E, Mielgo H and Trepat A 1985 Modal intervals: reason and ground semantics *Interval Mathematics (Lecture Notes in Computer Science* vol 212*)* ed K L E Nickel (Berlin: Springer) pp 27–35
[38] Markov S M 1995 On directed interval arithmetic and its applications *J. Univers. Comput. Sci.* **1** 514–26
[39] Kaucher E W 1980 Interval analysis in the extended interval space IR *Comput. Suppl.* **2** 33–49
[40] Shary S P 2002 A new technique in systems analysis under interval uncertainty and ambiguity *Reliab. Comput.* **8** 321–418
[41] Piegat A and Dobryakova L 2020 A decomposition approach to type 2 interval arithmetic *Int. J. Appl. Math. Comput. Sci.* **30** 185–201
[42] Alefeld G and Mayer G 2000 Interval analysis: theory and applications *J. Comput. Appl. Math.* **121** 421–64
[43] Moore R E 1979 *Methods and Applications of Interval Analysis (SIAM Studies in Applied Mathematics* no 2*)* (Philadelphia, PA: SIAM)
[44] Dawood H and Dawood Y 2017 On the mathematical foundations of algorithmic differentiation *Technical Report CU-Math-2017-01-MFAD* Department of Mathematics, Faculty of Science, Cairo University
[45] Alefeld G and Herzberger J 1983 *Introduction to Interval Computation* 1st edn (New York: Academic)
[46] Kulisch U W 2008 *Computer Arithmetic and Validity: Theory, Implementation, and Applications* (Berlin: Walter de Gruyter)
[47] Moore R E 1966 *Interval Analysis* (Englewood Cliffs, NJ: Prentice-Hall)
[48] Rokne J and Ratschek H 1984 *Computer Methods for the Range of Functions* (Chichester: Ellis Horwood)
[49] Makino K and Berz M 2003 Taylor models and other validated functional inclusion methods *J. Pure Appl. Math.* **4** 379–456

[50] Cornelius H and Lohner R 1984 Computing the range of values of real functions with accuracy higher than second order *Computing* **33** 331–47

[51] Makino K and Berz M 1996 Remainder differential algebras and their alocations *Computational Differentiation: Techniques, Applications, and Tools* ed M Berz, C Bischof, G Corliss and A Griewank (Philadelphia, PA: SIAM) pp 63–74

[52] Berz M 1997 From Taylor series to Taylor models *AIP Conference Proceedings* (Philadelphia, PA: AIP) pp 1–24

[53] Makino K 1998 Rigorous analysis of nonlinear motion in particle accelerators *PhD Thesis* (East Lansing, MI: Michigan State University)

www.ingramcontent.com/pod-product-compliance
Ingram Content Group UK Ltd.
Pitfield, Milton Keynes, MK11 3LW, UK
UKHW051906100926
8556IPUK00005B/782

9 780750 350655